Statistical Hypothesis Testing in Context

Fay and Brittain present statistical hypothesis testing and compatible confidence intervals, focusing on application and proper interpretation. The emphasis is on equipping applied statisticians with enough tools – and advice on choosing among them – to find reasonable methods for almost any problem and enough theory to tackle new problems by modifying existing methods. After covering the basic mathematical theory and scientific principles, tests and confidence intervals are developed for specific types of data. Essential methods for applications are covered, such as general procedures for creating tests (e.g., likelihood ratio, bootstrap, permutation, testing from models), adjustments for multiple testing, clustering, stratification, causality, censoring, missing data, group sequential tests, and noninferiority tests. New methods developed by the authors are included throughout, such as melded confidence intervals for comparing two samples and confidence intervals associated with Wilcoxon–Mann–Whitney tests and Kaplan–Meier estimates. Examples, exercises, and the R package asht support practical use.

MICHAEL P. FAY is a Mathematical Statistician at the National Institute of Allergy and Infectious Diseases, and previously worked at the National Cancer Institute. He has served as Associate Editor for *Biometrics* and is currently an Associate Editor for *Clinical Trials* and a Fellow of the American Statistical Association. He is a coauthor on over 100 papers in statistical and medical journals and has written and maintains over a dozen R packages on CRAN.

ERICA H. BRITTAIN is Deputy Branch Chief of Biostatistics Research at the National Institute of Allergy and Infectious Diseases and has well over three decades of experience as a statistician, with previous positions at the FDA, the National Heart, Lung, and Blood Institute, and a statistical consulting company. Her applied work at the NIH and her methodological publications in statistical journals focus on innovation in clinical trial design. She frequently serves on advisory panels for the FDA and the NIH and has served as Statistical Consultant for *Nature* journals and Associate Editor for *Controlled Clinical Trials*.

CAMBRIDGE SERIES IN STATISTICAL AND PROBABILISTIC MATHEMATICS

This series of high-quality upper-division textbooks and expository monographs covers all aspects of stochastic applicable mathematics. The topics range from pure and applied statistics to probability theory, operations research, optimization, and mathematical programming. The books contain clear presentations of new developments in the field and also of the state of the art in classical methods. While emphasizing rigorous treatment of theoretical methods, the books also contain applications and discussions of new techniques made possible by advances in computational practice.

A complete list of books in the series can be found at www.cambridge.org/statistics.
Recent titles include the following:

Statistical Hypothesis Testing in Context

Reproducibility, Inference, and Science

Michael P. Fay
National Institute of Allergy and Infectious Diseases

Erica H. Brittain
National Institute of Allergy and Infectious Diseases

CAMBRIDGE
UNIVERSITY PRESS

CAMBRIDGE
UNIVERSITY PRESS

University Printing House, Cambridge CB2 8BS, United Kingdom

One Liberty Plaza, 20th Floor, New York, NY 10006, USA

477 Williamstown Road, Port Melbourne, VIC 3207, Australia

314–321, 3rd Floor, Plot 3, Splendor Forum, Jasola District Centre,
New Delhi – 110025, India

103 Penang Road, #05–06/07, Visioncrest Commercial, Singapore 238467

Cambridge University Press is part of the University of Cambridge.

It furthers the University's mission by disseminating knowledge in the pursuit of
education, learning, and research at the highest international levels of excellence.

www.cambridge.org
Information on this title: www.cambridge.org/9781108423564
DOI: 10.1017/9781108528825

First published 2022

Printed in the United Kingdom by TJ Books Ltd, Padstow Cornwall

A catalogue record for this publication is available from the British Library.

Library of Congress Cataloging-in-Publication Data
Names: P. Fay, Michael, author. | Brittain, Erica H., author.
Title: Statistical hypothesis testing in context : reproducibility, inference, and science /
Michael P. Fay and Erica H. Brittain.
Description: Cambridge, United Kingdom ; New York, NY, USA :
Cambridge University Press, 2022. | Series: Cambridge series in statistical and
probabilistic mathematics | Includes bibliographical references and index.
Identifiers: LCCN 2021044789 | ISBN 9781108423564 (hardback) |
ISBN 9781108437677 (paperback) | ISBN 9781108528825 (epub)
Subjects: LCSH: Statistical hypothesis testing. | BISAC: MATHEMATICS /
Probability & Statistics / General
Classification: LCC QA277 .P43 2022 | DDC 519.5/6–dc23/eng/20211117
LC record available at https://lccn.loc.gov/2021044789

ISBN 978-1-108-42356-4 Hardback

Contents

Preface

This book is about statistical hypothesis testing and related statistics such as p-values and confidence intervals (CIs), which together are known as the *frequentist* (or *classical*) school of statistics. The focus is both scientific and practical, so there is substantial discussion of study design and causality with a broad scope to attack many types of applications. The frequentist school of statistics is founded on bounding the probability of errors from false claims and its theory is based on imagining replicating studies. Currently, there is a crisis in reproducibility in that scientists and others are becoming more aware that many claims made from scientific studies do not turn out to be reproducible.

A problem is that statistical hypothesis tests and p-values, the basic tools of frequentist statistics, are often misunderstood and misused, and this misuse has contributed to the reproducibility crisis. The American Statistical Association has joined the conversation in a strong way by publishing a statement on p-values with accompanying commentaries (Wasserstein and Lazar, 2016), as well as publishing a special issue of *The American Statistician* on "Moving to a World Beyond '$p < 0.05$'" (Wasserstein et al., 2019). While the statistical community can agree that p-values are often misunderstood and can unify in correcting some common mistakes, there is less unity on the proper response. Many voices in those commentaries and the special issue propose that the answer to this reproducibility crisis and the misuse of p-values is to use different methods of statistics to make scientific inferences. Others state that the issue is not inherent in the p-value statistic itself, and most of its misuse problems would accompany many if not all of the alternative statistics proposed (see, e.g., Benjamini, 2016). We are in the latter camp. We believe if the problem is reproducibility[1] then the answer is not to replace p-values, but to better educate users about their correct use and limitations. After all, the frequentist school is founded on the idea of keeping the rate of false claims low, and hence will, *if properly applied*, lead to better reproducibility in science. This book is largely about the proper use of p-values and confidence intervals in scientific research.

The level of mathematical difficulty in the book is mixed. There are some very heavy mathematics that are appropriate for a graduate-level course in statistics (so we assume some familiarity with basic probability and statistics at the level of, for example, Casella and Berger (2002)), yet there are many practical and simple examples that are used to convey the

---

[1] The term reproducibility has multiple meanings in the science literature (Plesser, 2018), our use of the term is nontechnical. In the book, we will use precise technical language about certain reproducibility properties. For example, we emphasize the validity of the confidence interval, where a valid 95% confidence interval procedure means that if we repeat the procedure multiple times, we expect that at least 95% of the time we will cover the true parameter.

ideas without using mathematical notation. Our professional experience is not from teaching statistics in a classroom, but from extensive experience in applying statistics in the medical environment. We have over 60 years of combined experience, primarily working for the US National Institutes of Health (NIH). Many of the applied problems we have encountered have motivated us to develop valid and appropriate statistical methods to address those problems, and thankfully the environment at the NIH encouraged that work and allowed us the time to write this book. We envision the use of this book as a reference book for applied statisticians or for graduate students of statistics.

The book is structured with the applied researcher in mind. The chapters are defined so that researchers handling a specific type of data may skip to the appropriate chapter, as long as they are already aware of the material in Chapters 2 and 3. Each chapter begins with an overview, so that readers may get a flavor of what is in the chapter and decide what sections will be helpful to them at that time. Specifically, before mathematically theoretical sections, we discuss why we thought it was useful to go into that level of abstraction. This gives readers insight into whether it makes more sense for them to delve into the slower reading of mathematical notation, or perhaps skip that section. Finally, each chapter ends with a summary section that outlines the methods introduced therein and provides context and recommendations for use, and this is written at a relatively nontechnical level. Because the applied statistician must be able to handle all kinds of data inference problems, the focus of the book is for a generalist who needs to know some about many areas of statistics. A solid theoretical foundation is useful for the applied statistician as well, because by knowing the theory well, new tests and confidence intervals may be derived using slight adjustments to meet new situational needs. There will be few mathematical theorems, and when provided we will only give the conditions of their applicability, and no proofs. For the applied researcher, many chapters will serve as introductions or reviews of statistical areas. For example, Chapter 10 has only a few pages on the bootstrap, Chapter 14 covers testing from models, and Chapter 15 covers causal inference. These chapters are there because the applied researcher needs to be aware of them, but it is expected that to learn these topics in a deep way, the reader will need to supplement them with further study with the references given.

In terms of the use of this book for a graduate course in statistical hypothesis testing, we contrast it to one of the best books on the topic, Lehmann and Romano (2005). When using statistical hypothesis tests in a scientific problem, there are both scientific issues and mathematical issues. Loosely speaking, the mathematical issues relate to determining the appropriate statistical test and its properties given the assumed probability models represented by the null and alternative hypotheses. The scientific issues relate to the appropriateness of those probability models for answering the scientific question, and the presentation of the results such that the scientific community will properly interpret those results. Lehmann and Romano (2005) is an excellent book that focuses on the mathematical issues of statistical hypothesis testing. Their book discusses important properties of tests once one has already defined the null and alternative hypotheses (for example, validity, consistency, uniformly most powerful test, asymptotic relative efficiency). Lehmann and Romano (2005) has extensive discussions on how one determines the optimal test given the hypotheses assumed, including different ways that optimality may be defined (e.g., uniformly most powerful unbiased [UMPU] tests). In this book, we focus on some of those mathematical issues (validity), but much less on others (UMPU tests). We will spend more time on expanding the scope of applicability. For example, our book covers censoring and causality which are not covered in Lehmann and Romano (2005). In other words, we discuss more scientific issues – how to design a study to answer a scientific question, and how to find a set of probability models (encompassing the null and alternative hypotheses) that can plausibly be applied to the study.

We have tried to include all of the essential ideas needed for applied frequentist statistical hypothesis testing. But, of course, the selection of what is essential is subjective, so we will try to be clear about our biases upfront. We are biased toward ensuring valid inferences, so some may think we overly emphasize exact methods and de-emphasize asymptotics; however, we have included asymptotics where we think it is helpful. We are biased toward experiments over observational studies, so many of the causal discussions will be heavily focused on experiments. Part of this is our professional experience that is shaped heavily by randomized clinical trials. Our bias is toward calculating two one-sided hypotheses instead of one two-sided hypothesis, because, for example, if we find two treatments differ, we usually want to know which one is better. Our bias in applications is toward biomedical ones, rather than other scientific, economic, or business applications. Finally, we hope the reader will forgive our bias in presenting methods that we have worked on ourselves, but we believe that many of those methods have wide applicability. Despite these biases, we believe a quick perusal of the Table of Contents would convince most statisticians that we have included most of the important ideas needed for the applied statistician wanting to make frequentist inferences, and we even have included Chapter 21 at the end that compares frequentist hypothesis testing to some Bayesian approaches.

Although every effort has been made to be accurate, please inform us if you find a mistake (mfay@niaid.nih.gov). We will keep a list of corrections at https://github.com/michaelpfay/StatHypTest.

We have both had the good fortune to work in the Biostatistics Research Branch of the National Institute of Allergy and Infectious Diseases since the early 2000s, where the environment is intellectually vibrant, as well as warm and cooperative. In addition to engaging collaboration with doctors and other scientists at the Institute, we have time and space to work on statistical methodology; allowing us to write this book. We particularly want to thank our Branch Chief Dean Follmann who supported this endeavor, and who has always been a great collaborator and source of wisdom. Michael Proschan also stands out; he has a deep knowledge of mathematical statistics, probability, and clinical trials and has been wonderful to work with. We are grateful for those two and all our Biostatistics Research Branch colleagues from whom we have learned so much and have shared many enjoyable lunches.

We thank several anonymous reviewers who provided comments on early drafts of the book. We thank Dianna Gillooly for encouragement throughout the several years it took to finish this book, and a great title.

Erica Brittain would also like to thank all her colleagues from throughout her career for making her a better statistician, but most of all, for enhancing her love of statistics. She feels especially blessed by the many amazing mentors she had during her education and in the early years of her career, particularly Tom Fleming, Ed Davis, and Janet Wittes. And, finally, she is beyond grateful to her family: her son Tim, her daughter Rebecca, and her husband John Hyde (who doubles as her favorite and wisest statistical consultant) for making everything worthwhile.

Michael Fay wants to thank Rocky Feuer, Barry Graubard, Janet Wittes, and Larry Freedman for great mentorship early in his career, and Joanna Shih for being a wonderful collaborator. He thanks Sally Hunsberger for her dual role as a practical advisor on all issues related to clinical trials, and as a life partner. On the latter, the list of thanks needed is too long; the short version is that his life is much fuller and more fun with her around. Finally, he wishes that his and Sally's children, Paul and Sara, will find their work life as interesting and fulfilling as he has through biostatistics.

1

Introduction

1.1 A Cautionary Tale

On March 20, 2020, an online version of Gautret et al. (2020) about the treatment of COVID-19 with hydroxychloroquine was posted. This was the beginning of the pandemic, and we (the authors) were just beginning the start of over a year of working from home. The world was anxious for a treatment for this deadly disease, and the article (to some) looked like a promising treatment. The first results table of the paper (Table 2), showed that by Day 3 post inclusion, 10/20 (50%) of patients treated with hydroxychloroquine had negative nasopharyngeal tests for the SARS-CoV-2 virus, compared to only 1/16 (6.3%) of control patients. A chi-squared test reveals a significant difference (with a two-sided p-value[1] of $p = 0.005$), suggesting the results are much more different between arms than would be expected to occur if there was no true treatment effect. Some interpreted that result and similar ones from that paper as implying that it is strong evidence that hydroxychloroquine helps clear the body of the virus that causes COVID-19; however, we and many others could see from reading Gautret et al. (2020) that there were many potential problems with the presented results.

We begin with this cautionary tale, because we do not want the users of this book to use it to compute p-values such as those in Gautret et al. (2020). It would be very easy for a user to think: "I have a two-sample study" (e.g., a new treatment and a standard of care) "and my responses are binary" (e.g., presence of virus three days after treatment) "so I will just look into Chapter 7 to find the appropriate test, and to find how to calculate my p-value." Unfortunately, the user could use the book that way, but if they do that, they are not assured that they have done an appropriate statistical and scientific analysis. There is much more to a statistical analysis than computing a p-value.

Before we get into the concerns about Gautret et al. (2020), we want to emphasize that we are not implying any intention to mislead from the authors. Remember, at this time the world was desperate for a treatment of this new and deadly disease. The authors may have thought that they were doing good science by looking at the data many ways until a pattern with a very reasonable explanation became very clear to them. Scientists and other researchers are all susceptible to deceiving themselves that their explanations of data patterns are supported

---

[1] We define the p-value formally in Chapter 2. Briefly, the p-value, p, is a statistic used for statistical hypothesis testing with possible values from 0 to 1, with lower values indicating more evidence to reject the null hypothesis, and 0.05 being the traditional boundary (in many disciplines) for deciding between the null hypothesis ($p > 0.05$) or the alternative ($p \leq 0.05$).

by the data, especially when the need is so great. That is why the methods and principles presented in this book are important, as they will help avoid such self-deception.

We review some of the weaknesses in Gautret et al. (2020), many of which are outlined in Rosendaal (2020). First, the study was a nonrandomized clinical trial and there may have been systematic ways in which the two groups differed. In this study, some of the control group were people that declined the treatment, and others in the control group were from a different treatment center. So there may have been critical differences between the two groups besides their treatment that caused the apparent treatment effect. In general, researchers should be very careful when comparing groups of individuals, in those scenarios where individuals (or their doctors) choose for themselves which group they are in. We may even get a difference effect between the two groups that is repeatable if the repeat study allows the groups to be chosen in the same way, but that does not mean that the repeatable difference can be interpreted such that the treatment caused the different outcomes (see Section 3.2 and Chapter 15).

A second problem is that not everyone that started out in the two arms of the study is analyzed in Table 2 of Gautret et al. (2020). Six people in the treatment arm were excluded from those analyses, and importantly, some of those that were excluded had worsening outcomes suggesting that the virus was not controlled in their bodies (one died, and three were transferred to the intensive care unit of the hospital). In other words, the reason that those individuals were missing from the analysis is very likely related to the outcome we are trying to measure. Such missing data can lead to very misleading results (see Section 3.5.2 and Chapter 17).

A third problem is that the prespecified primary endpoint of the study is listed as SARS-CoV-2 virus presence at Day 1, 4, 7, and 14; however, the endpoints listed in Table 2 are Days 3, 4, 5, 6. Similar results to the Day 3 results appear for Day 4 (12/20 vs 4/16, $p = 0.04$), Day 5 (13/20 vs 3/16, $p = 0.006$), and Day 6 (14/20 vs 2/16, $p = 0.001$). It is concerning that Day 4 is the only prespecified day that was ultimately presented, and that is the one with the largest p-value of the presented results. Prespecifying a primary endpoint is a way of avoiding problems of multiple testing due to looking at the data many ways (see Chapter 13).

A fourth problem is that the statistical test used in the table was Pearson's chi-squared test, but the statistical methods states that "statistical differences were evaluated by Pearson's chi-squared or Fisher's exact tests as categorical variables, as appropriate." In our view, Fisher's exact test is more appropriate than Pearson's chi-squared test, because Fisher's exact test is valid even for the small sample size, but Pearson's chi-squared test is an approximation (see Chapter 7). Further, for every case in Table 2, Pearson's chi-squared has a lower p-value than the p-value from Fisher's exact test, so that has the appearance of choosing the test with the lowest p-value (see Example 13.5, page 238).

We will not go through all the concerns with the publication, but all the flaws are important because this Gautret et al. (2020) paper is a critical part of the history of hydroxychloroquine treatment for COVID-19. Gould and Norris (2021) review that history, including the Food and Drug Administration emergency use authorization (EUA) (March 28, 2020) and the EUA retraction (June 15, 2020). By February 2021, published meta-analyses of several randomized clinical trials estimated that hydroxychloroquine treatment for COVID-19 actually leads to a lower rate of negative PCR tests at Day 7 than standard care (rate ratio 0.86 [95% confidence interval: 0.68, 1.09], Analysis 1.4), and more importantly it shows that hydroxychloroquine treatment "does not reduce deaths from COVID-19, and probably does not reduce the number of people needing mechanical ventilation" (Singh et al., 2021, p. 3). This is a cautionary tale because hydroxychloroquine treatment for COVID-19 turned out not to be helpful for the treatment of COVID-19, and resources were used to find out that information that may have been used more efficiently on other potential treatments.

This is just one example of a study that could have been improved by following the principles and methods discussed in this book. But each study may have its own set of challenges that make scientific inferences difficult. For example, how do we estimate the effect of vaccines? We want to get enough scientific information about the vaccine efficacy to know whether it is a viable public health response, without wasting too much time collecting more information than is needed (see, e.g., Section 7.8 and Chapter 18). Before going into those details, we start with the more basic question: what is science?

1.2 Science, Reproducibility, and Statistical Inferences

Science is about developing theories that both explain reality and reproducibly predict outcomes from studies. If a theory does not predict something reliably, and it is not possible to disprove it with data, then it is not a scientific theory. This book is about the most common school of statistical thought in making scientific claims from data, the frequentist school of statistics, whose basic building block is the statistical hypothesis test.[2] This type of hypothesis testing uses data to decide between two sets of probability models (or two hypotheses): a null hypothesis and an alternative hypothesis. The fundamental idea of these methods is that if we repeatedly apply a test using a specific method then we will rarely make the mistake of deciding in favor of the alternative hypothesis when in fact the null hypothesis is true. We set up the null hypothesis as something we are trying to disprove by our study, so that the interesting results from a study occur when we decide in favor of the alternative hypothesis. Typically, such interesting results are called *statistically significant effects* (see Note N1). Restating, the core fundamental idea of frequentist statistical hypothesis testing is about keeping our error rate of concluding interesting results low *for any specific study*.

In the last decade there has been renewed awareness that a substantial proportion of scientific discoveries published in peer reviewed scientific journals have later turned out to not be reproducible. For example, The Open Science Collaboration (2015) conducted replications of 100 experiments in psychological science published in 3 important psychology journals. Of the 97 studies that originally found a significant effect (a p-value with $p \leq 0.05$), only 35 found a significant effect when the same study design was replicated. Given this perceived crisis in reproducibility, one might think that there would be a clamoring for more books and research about statistical hypothesis testing, a method whose fundamental core is about reproducibility. In fact, it seems like the opposite is happening, and there appears to be an increase in voices calling for replacements of frequentist statistical hypothesis testing.

---

[2] One can test statistical hypotheses using Bayesian statistics, the other main school, but this book is not about that. We compare the two schools' approaches to statistical hypothesis testing in Chapter 21.

The thinking seems to be: if most scientific studies use these hypothesis tests and define significance as p-values less than 0.05, and we have a reproducibility crisis, then we need to find a new way of doing science that does not use $p \leq 0.05$ to define "significant" effects. We disagree, but not totally.

First, we question whether the reproducibility crisis is truly a full-blown crisis, as it may partly be a problem of incorrect interpretation of published scientific discoveries. For example, if of all hypotheses tested: (1) there is a high proportion where the null hypothesis is true; (2) of the small proportion of null hypotheses that are false the average power to detect that falsity is low; and (3) if only significant effects are published, then it is likely that most published research findings will be false (Problem P2). So part of the reproducibility "crisis" may be an unrealistic expectation that almost all published scientific discoveries will be true. Another part of the explanation of the low reproducibility rate is that many researchers may not be properly using p-values. In this book, we emphasize the tools of frequentist statistics that address the latter problem. While the causes of lack of reproducibility are varied and complex, the proper use of these tools should enhance reproducibility. We still find it acceptable to use a threshold such as $p \leq 0.05$ to define significant effects, but we find it helpful to use other frequentist tools as well. For example, we recommend reporting the results of your test with not just the binary decision, significant at the 0.05 level or not, but with the p-value, which allows the reader to determine if the null hypothesis would have been rejected at other significance levels besides 0.05. It is important, whenever feasible, to accompany the p-value with an estimate and a confidence interval. Whenever possible, you should prespecify your primary hypothesis test before beginning your study, but when that has not been done, you should adjust for all the ways you selected which tests to do, and how many of those tests you performed. If you are interested in results which allow an interpretation of causality (e.g., this drug caused that effect), then you need to pay attention to how you design your study.

Why are the frequentist tools used so much, and how can they be used better? Consider the case of the Wilcoxon–Mann–Whitney test. This is a popular test, and this book presents it in a different way to increase its proper applicability to science. Its popularity comes from its simplicity and lack of strong assumptions. Suppose you want to compare two groups of individuals, where one group is given one drug and the other group is given a placebo, and the response is measured on a numeric scale. The Wilcoxon–Mann–Whitney test can be applied by only assuming independence of responses and a null hypothesis probability model. The null model does not need to assume a specific distributional form for the responses (like a normal distribution or a Poisson distribution), it only needs to assume that the drug does not affect the response, so that the distribution of the responses for individuals given the drug is the same as the distribution for those given the placebo. The p-value is calculated by ranking all responses and permuting the treatment labels to find the distribution of the difference in the means of the ranks in the two groups, assuming no treatment difference. The permutation is conceptually simple, and the computer can handle the details. Thus, the Wilcoxon–Mann–Whitney test can be easily applied to many situations with minimal assumptions. Our book is different, as illustrated in how we take the Wilcoxon–Mann–Whitney test inferences to the next level. We present methods for estimating a parameter from the test, discuss the causal interpretation of that parameter, explain how to calculate the confidence interval on that parameter, and discuss the additional assumptions needed to ensure validity of the confidence interval. Further, because the confidence intervals are new (Fay and Malinovsky, 2018), we provide an R package **asht** that has a function to do all of these calculations. That function has exact versions suitable for small sample sizes, and approximations suitable for larger sample sizes.

This book focuses on applying frequentist statistical methods, which requires both knowledge of the mathematics and of scientific ideas. Mathematical ideas relate to how to define probability models and find valid p-values and confidence intervals given the probability models. Scientific ideas relate to deciding on the appropriate probability model and properly interpreting the result. The interpretation will include determining what kind of causal statements we can make about the result. Loosely speaking, applying frequentist statistical methods requires asking what are we trying to learn about, how can we design a study from which we can learn that, and how do we analyze our data and present the results to properly and fairly convey what we have learned from the study. Proper application requires many tools to meet each specific situation, hence a book is appropriate.

1.3 Using the Book

Here is an outline of the rest of the book. We give the mathematical theory of statistical hypothesis tests in Chapter 2 and go over some of the scientific theory in Chapter 3. Many of the subsequent chapters are based on specific types of data, with increasingly complicated data coming later (Chapters 4, 5, 6, 7, 9, 11, 12, and 16). Chapter 8 addresses robustness in testing, and is presented just before the chapter on two-sample tests (Chapter 9), so that in that chapter we can discuss, for example, how the two-sample t-test does not necessarily require normally distributed data to be approximately valid. Chapter 10 gives a brief introduction to general methods for calculating p-values or confidence intervals (e.g., asympototic likelihood-based methods, bootstrap, permutation tests). Chapter 11 covers k-sample tests including multiple comparison adjustments for subsequent tests made after the k-sample test, while Chapter 13 covers more general multiple comparison adjustments (e.g., Holm's adjustment or false discovery rate). Chapter 14 provides a brief overview of some very useful standard models (e.g., generalized linear models) and how testing may be done using those models. Chapter 15 gives a brief introduction to causality. Chapter 16 discusses censored data, which is common with time-to-event responses. Chapter 17 covers missing data. Chapter 18 addresses group sequential testing. Chapter 19 covers situations where what we want to show is not the typical alternative hypothesis. For example, even though we want to show that the data fit a model well, the goodness-of-fit test defines the null hypothesis as a correctly specified model, so in rejecting the null hypothesis we can only show lack-of-fit by that test. Also, if we want to show two parameters are equal, we must reformulate the problem so that the alternative is that the parameters are close (to within a prespecified margin). We also study the related noninferiority testing, when we want to show one treatment is not inferior to another within a prespecified margin. Chapter 20 deals with power and sample size calculations. Chapter 21 reviews some Bayesian approaches to hypothesis testing.

At the end of most chapters we have a summary of the main points, a section with extra references, a section on R packages, and Notes and Problems. The extra references are not meant to be comprehensive, nor to point to the first one that developed an idea, but to be a useful place to find needed extra details. In some cases (e.g., Chapter 15 on causality), the references given will be for entire books about the subject of the chapter. The lists of R packages are not comprehensive either, especially in the latter part of the book (e.g., Chapter 14 on testing from models), where it would be difficult to list all the relevant packages. In fact, there is an extreme bias in the R package lists, since one of us (Fay) maintains several packages (e.g., asht, bpcp, exact2x2) to calculate some of the methods emphasized in this book. For packages that we do not maintain, we try to recommend only packages that have existed for a while and with a reputation for quality. The Notes section provides extra details, and the Problems may be used in teaching a course, but sometimes also contain valuable extra information as well.

To find a particular topic, first turn to the Concept Index. The **bold** entry for a topic is often the most relevant, such as its definition. If you know an important reference for a topic, then you can look at the end of the Bibliography entry for that reference and it will list the page numbers where that reference was cited. Notation is defined as it is introduced. Because there are so many concepts, most letters (Greek or Roman) will be used to represent multiple different types of values; however, as a general rule, notation for any letter does not change within a chapter. If you have trouble finding the definition of any notation, first look in the previous paragraph, then go to the Notation Index (see page 420).

1.4 Notes

N1 **Statistically significant:** The summary article of the American Statistical Association's special issue on "Moving to a World Beyond '$p < 0.05$'" states that "it is time to stop using the term 'statistically significant' entirely." (Wasserstein et al., 2019, Section 2). The main reason is that "using bright-line rules for justifying scientific claims or conclusions can lead to erroneous beliefs and poor decision making" (see Problem P1). That summary recommends using a p-value and not dichotomizing it into significant or nonsignificant. We agree that it is generally better not to dichotomize p-values, but we do not recommend never doing that.

1.5 Problems

P1 Suppose two studies explored the same effect, and one study found a significant effect and the other did not, where significant effect was defined for both as rejecting the null hypothesis that the risk ratio equals 1 at the 5% significance level.

   (a) Do these studies contradict each other?
   (b) What if we additionally told you that the estimates and 95% confidence intervals on the risk ratios were 1.20 (95% CI: 1.09, 1.33) for one study and 1.20 (95% CI: 0.97, 1.48) for the other. Now do the studies contradict each other? Explain. (This is a real example from the medical literature, see Greenland (2017, p. 640) for references and other examples.)

P2 Suppose we have a scientific program where we test many hypotheses (null versus alternative). Let p_A be the proportion of hypotheses where the alternative is true.

For hypotheses where the alternative hypothesis is true, let $1 - \beta$ be the probability that we reject the null hypothesis. (We can interpret $1 - \beta$ as the average power from a random selection of studies with different powers, among studies where the alternative is true.) For hypotheses where the null hypothesis is true, let α be the probability that we mistakenly reject the null hypothesis. In this ideal scenario, if we test many hypotheses and only publish the significant ones (those that reject the null hypothesis), then show that most of those published significance claims will be false if $p_A(1 - \beta) < \alpha(1 - p_A)$ (Ioannidis, 2005).

2

Theory of Tests, *p*-Values, and Confidence Intervals

2.1 Overview

In this chapter we outline the theory of statistical hypothesis tests. This is important for defining terminology. As we introduce terms, we first show in Section 2.2.1 how they apply to a simple single sample binomial example. In Section 2.2.2 we explore some of the generality of the theory by outlining three different ways of formulating the null and alternative hypotheses for a second data example: a two-sample study with numeric responses. This shows that there is not one "correct" way of formulating a problem. These three formulations can get abstract, for example, the hypotheses can be represented by sets of infinite dimensional parameters. Yet even in this representation, the statistical hypothesis problem can still be solved. We present this level of abstraction this early in the book to show the breadth of application of the statistical hypothesis test theory.

In Section 2.3 we show how hypothesis tests are intimately related to confidence intervals. This relationship is important, because the scientific information in finding a significant effect (i.e., rejecting the null hypothesis) is enhanced when supplemented with important information such as an effect estimate and its confidence interval. In Section 2.4 we define some important properties of hypothesis tests.

Readers who are less theoretically inclined can skip to the end-of-chapter summary (Section 2.5). This summary will provide the essential terminology and information needed for the rest of the book.

2.2 Components of a Hypothesis Test

2.2.1 Main Definitions with a Simple Example

A hypothesis test has three basic parts: (1) the set of possible values of data if the study is repeated, (2) a set of probability models that is partitioned into two sets of models, called the *null hypothesis set* and *alternative hypotheses set*, and (3) the decision rule which uses the observed data to decide whether or not you reject the null hypothesis (i.e., whether or not you reject the statement that the true data generating probability model is in the null hypothesis set). The decision rule is a function of the data and the α-level. We describe each part separately.

Suppose we observe a vector of data, say $\mathbf{x}$. For a statistical hypothesis test, we envision that the data comes from a data generating process, or probability model. If we reran the data generating process, then we would get a different vector of data. Let $\mathcal{X}$ be the set of all possible data vectors associated with this process. The set $\mathcal{X}$ is called the *sample space* or *support*.

Example 2.1 *Suppose we observe x = 6 successes out of n = 10 tries. Suppose the data generating process is binomial with n = 10 and parameter θ with $0 < \theta < 1$. Then here* $\mathbf{x} = 6$ *and* $\mathcal{X} = \{0, 1, \ldots, 10\}$.

Let the probability model associated with the data generating process be denoted P_θ, where θ represents the parameter vector, which may be infinite dimensional (see Example 2.4). We do not know the value of θ but assume that it is one member in a set of possible values, Θ, and therefore, P_θ is one member of a set of probability models, $\mathcal{P} = \{P_\theta : \theta \in \Theta\}$ (see Note N1 for set notation explanation). We partition the probability models into two disjoint sets, the null hypothesis models, $\mathcal{P}_0 = \{P_\theta : \theta \in \Theta_0\}$ and the alternative hypothesis models, $\mathcal{P}_1 = \{P_\theta : \theta \in \Theta_1\}$. The null ($H_0$) and alternative ($H_1$) hypothesis statements are

$$H_0 : \theta \in \Theta_0$$
$$H_1 : \theta \in \Theta_1,$$

where θ is the true parameter.

Example 2.1 *(continued) For the binomial example, P_θ is a binomial distribution with parameters n and θ. Let $\Theta = \{\theta : \theta \in [0,1]\}$. We might want to know if θ is bigger than 0.5. In this case we partition Θ as $\Theta_0 = \{\theta : \theta \in [0,0.5]\}$ and $\Theta_1 = \{\theta : \theta \in (0.5,1]\}$. We can write the hypotheses in this case as $H_0 : \theta \leq 0.5$ and $H_1 : \theta > 0.5$.*

The final piece of the hypothesis test is the decision rule, which we write as δ. The *decision rule* is a function of $\mathbf{X}$ (a possible element of $\mathcal{X}$) and the *significance level*. The significance level is set by the researcher, and it is usually denoted α, and hence is sometimes called the α-level.[1] The result of the decision rule is the probability of rejecting the null hypothesis. For example, if we observe $\mathbf{x}$, the probability of rejecting the null hypothesis using a significance level of α is $\delta(\mathbf{x}, \alpha)$. Typically, we only use decision rules that only give two values, 0 (do not reject the null) or 1 (reject the null). Other decision rules are called *randomized decision rules* and are not discussed in this chapter and are rarely used in practice (see Note N2). A *valid decision rule* is defined such that, the probability of rejecting the null hypothesis for any one of the null hypothesis models is less than or equal to α. Mathematically, a valid (nonrandomized) decision rule has (for any $\alpha \in (0,1)$)

$$\sup_{\theta \in \Theta_0} \Pr[\delta(\mathbf{X}, \alpha) = 1 | \theta] \leq \alpha. \tag{2.1}$$

The left-hand side of the above equation is called the *size* of the hypothesis test.

When the alternative represents parameters only in one direction, we call this a one-sided hypothesis test. Example 2.1 is a one-sided hypothesis test, because the alternatives are all $\theta > 0.5$. We can formulate one-sided hypotheses in the other direction: $H_0 : \theta \geq 0.5$ and

---

[1] Lehmann and Romano (2005) use the term significance level, but Greenland (2019) prefers "α-level" because Fisher and other British authors sometimes use "significance level" to mean p-value. We never use "significance level" to mean p-value, and we use "significance level" and "α-level" interchangeably. We do not always use α-level because it can sometimes lead to awkward phrases like "for the one-sided test we use $\alpha/2$ for the α-level."

$H_1 : \theta < 0.5$. For the binomial problem, a two-sided formulation is $H_0 : \theta = 0.5$ and $H_1 : \theta \neq 0.5$. For setting significance levels, in medical publications the tradition has been to use $\alpha = 0.05$ for two-sided tests. Often the two-sided test rejects at the $\alpha = 0.05$ level if either of the one-sided tests rejects at the $\alpha = 0.025$ level. Thus, we would argue that a significance level of 0.025 for a one-sided test is appropriate when an $\alpha = 0.05$ for two-sided tests is appropriate. Of course, not all situations demand using two-sided $\alpha = 0.05$ (see Note N3).

Aside from tradition, the α-level is somewhat arbitrary. Because of this, a useful approach is to give the smallest α-level for which we would reject the null hypothesis for the observed data at that level and all larger levels. This is known as the *p-value*. In mathematical notation, for nonrandomized decision rules the *p*-value is

$$p(\mathbf{x}) = \inf\{\alpha : \delta(\mathbf{x}, \alpha^*) = 1 \text{ for all } \alpha^* \geq \alpha\}. \tag{2.2}$$

So that for a valid hypothesis test, we have

$$\sup_{\theta \in \Theta_0} \Pr[p(\mathbf{X}) \leq \alpha | \theta] \leq \alpha. \tag{2.3}$$

We call a *p*-value function, $p(\cdot)$, that satisfies Equation 2.3 a valid *p*-value.[2]

Except with rare exceptions, applied decision rules are monotonic in α, rejecting more often as α increases (see Note N4). For monotonic decision rules, we can think of the *p*-value function as providing an ordering of the possible data from the most extreme (least likely under the null) to the least extreme (most likely under the null), where least and most likely are defined using a $\theta \in \Theta_0$ that maximizes the left-hand side of expression 2.3. Then the *p*-value can be interpreted as the probability of observing equal or more extreme data under the null hypothesis (see Note N5).

We always want valid decision rules but sometimes we settle for approximately valid decision rules. For example, we might base a decision rule on asymptotics, so that as the sample size gets larger and larger the decision rule gets closer and closer to valid. These are asymptotically valid decision rules. These are discussed in Section 2.4.

Sometimes when using nonasymptotic decision rules or tests we use the term *exact* to mean *valid*. In some statistical literature, *exact* means that the inequality in the definition of validity (expression 2.1) is an equality; however, in this book, we use the term *exact* as synonymous with *valid*. This latter usage is more common in the biostatistics literature (for example, Fisher's exact test is valid, but the size is generally not equal to the α-level).

Now consider decision rules related to our example.

Example 2.1 *(continued) For the binomial testing $H_0 : \theta \leq 0.5$ versus $H_1 : \theta > 0.5$, we use the binomial distribution to define δ. First, note that larger values of x suggest larger values of θ, so we want to reject when $x \geq c$, where c is a constant that depends on n and α. We want to choose c to give a valid test and have as much power as possible, so we want the smallest c that gives a size less than or equal to α. To calculate the size, in this case we only need to calculate the probabilities under the model on the boundary between the null and alternative hypotheses, the binomial with $\theta = 0.5$. When $n = 10$ and $c = 8$ or $c = 9$ we get sizes (i.e., probabilities of rejection at the boundary) of $\Pr[X \geq 8 | \theta = 0.5] = 0.0547$*

---

[2] The term *p-value* can refer to either $p(\mathbf{x})$ (the value of the *p*-value function evaluated at the data $\mathbf{x}$), or $p(\cdot)$ (the *p*-value function itself); the meaning is inferred from the context.

and $\Pr[X \geq 9 | \theta = 0.5] = 0.0107$. *So when* $\alpha = 0.05$, *a valid test uses the decision rule,* $\delta(X, 0.05) = 1$ *if* $X \geq 9$ *and* 0 *otherwise.*

2.2.2 Hypothesis Test Components for a Two-Sample Example

Suppose we have two samples with five independent responses in each sample. Let the data be $\mathbf{x} = [\mathbf{y}, \mathbf{z}]$, where $\mathbf{y}$ is a vector of responses, and $\mathbf{z}$ is the vector of associated group indicators. For example,

$$\mathbf{y} = \{3.1, 5.2, 6.6, 5.2, 1.3, 6.8, 9.2, 11.1, 9.5, 2.2\},$$
$$\mathbf{z} = \{2, 1, 2, 1, 1, 1, 2, 2, 2, 1\}. \tag{2.4}$$

Suppose we want to know if the responses from Group 1 (i.e., those with $z_i = 1$) are in general larger or smaller than the responses from Group 2. There are many ways to formulate the null and alternative hypotheses; we present three formulations as Examples 2.2–2.4. Here we do not examine all the features that determine the process that created the data (for that see Chapter 3). The data generating and analysis context may favor one formulation over another, (for that see Chapter 9), although there are situations where many different formulations would be acceptable. The purpose of presenting three hypothesis test formulations here is to get a flavor for the different ways the components of the hypothesis test may be formulated, and to emphasize that there is a choice for each data set.

The first example formulation is the *t*-test. We give a standard set of assumptions associated with this test, although other formulations of the *t*-test are possible (see Chapter 9).

Example 2.2 *In this t-test formulation, we assume that the group indicators are fixed, and the responses are samples from a normal distribution. Each normal response can be any real number, so we write* $Y_i \in \Re$ *for the ith response, or* $\mathbf{Y} \in \Re^{10}$ *for all 10 responses. The sample space is* $\mathcal{X} = \{\mathbf{Y}, \mathbf{z} : \mathbf{Y} \in \Re^{10}\}$.

If $z_i = 1$, *then we assume* Y_i *is normal with mean* μ_1 *and variance* σ_1^2, *or* $Y_i \sim N(\mu_1, \sigma_1^2)$, *while if* $z_i = 2$, *then* $Y_i \sim N(\mu_2, \sigma_2^2)$. *Here* P_θ *represents the probability model for the* 10 *responses of the* $\mathbf{Y}$ *vector. The parameter* θ *is four-dimensional,* $\theta = [\mu_1, \sigma_1, \mu_2, \sigma_2]$. *The possible values of* θ *are*

$$\Theta = \{\mu_i, \sigma_i : \mu_i \in (-\infty, \infty), \sigma_i \in (0, \infty), i = 1, 2\}.$$

One formulation of the null and alternative hypotheses is $H_0 : \mu_1 = \mu_2$ *and* $H_1 : \mu_1 \neq \mu_2$. *This is shorthand for expressing the two sets of possible parameters,*

$$\Theta_0 = \{\mu_i, \sigma_i : \mu_1 = \mu_2, \mu_i \in (-\infty, \infty), \sigma_i \in (0, \infty), i = 1, 2\}$$
$$\Theta_1 = \{\mu_i, \sigma_i : \mu_1 \neq \mu_2, \mu_i \in (-\infty, \infty), \sigma_i \in (0, \infty), i = 1, 2\}.$$

We can calculate p-values by using Welch's t-test, which takes the difference in means of the groups, divides that difference by an estimate of the standard error of that difference, and compares the resulting ratio to a t-distribution. We show in Chapter 9 that Welch's t-test is an approximately valid way to calculate p-values and confidence intervals when we allow $\sigma_1 \neq \sigma_2$ *as in this formulation.*

A second way to formulate statistical hypotheses for the data of expression 2.4 is as a randomization test. This switches the randomness from the responses to the group indicators.

Example 2.3   *Suppose the data generating process keeps the responses the same every time new data is generated, but the group indicator vector is randomly shuffled. For example, suppose we do an experiment where we randomize 10 individuals to two interventions ($z = 1$ or $z = 2$). We imagine redoing the experiment such that the group indicators were different each time, but the interventions had the same effects, so that the ith response would be the same no matter which of the two interventions was applied. Then the sample space is $\mathcal{X} = \{\mathbf{y}, \mathbf{Z} : \mathbf{Z} \text{ is a vector with five ones and five twos}\}$. There are 10 choose 5, or 252 elements in $\mathcal{X}$.*

For this example, P_θ represents the probability model for the $\mathbf{Z}$ vectors, vectors of length 10 with 5 zeros and 5 ones. So θ could be a vector of length 252 giving the probability for each of the 252 elements in $\mathcal{X}$. Typically, Θ_0 is a set with one value, a vector of length 252 with each element of the vector equal to $1/252$. The set Θ_1 would be all other possible probability vectors, in other words,

$$\Theta_1 = \left\{\theta = [\theta_1, \ldots, \theta_{252}] : 0 \le \theta_i \le 1 \text{ for all } i, \sum \theta_i = 1, \text{ and } \theta \neq \left[\frac{1}{252}, \ldots, \frac{1}{252}\right]\right\}.$$

In this case, typically we do not refer to θ in describing the hypothesis test (since it is not a parameter that is useful for describing the data), but only use it in the probability calculations under the null hypothesis.

In this example, we have that under the null hypothesis each of the 252 ways of picking the vector $\mathbf{Z}$ are equally likely. To choose a decision rule, we first choose a test statistic, T. The test statistic is a function of the possible values of the data, $\mathbf{X}$, that returns a scalar value. So T allows us to order the sample space. Let us assume that larger values of $T(\mathbf{X})$ indicate more extreme values under the null hypothesis. For example, we may use T which is the mean of responses from one group ($z_i = 2$) minus the mean of responses from the other group ($z_i = 1$). Note that because θ is such a large dimension, we cannot look at the null and alternative hypotheses and automatically see how to define more extreme values. There are many choices for defining that extremeness, and choosing T implicitly defines a direction for measuring extremeness under the null hypothesis. Given T we want the smallest value c such that under the null hypothesis

$$\Pr[T(\mathbf{X}) \ge c] \le \alpha.$$

The calculations that find c depend on $\mathbf{x}$ and can be computationally difficult. Tests using this kind of decision rule are called permutation tests (see Chapter 10 for more details).

Finally, we consider a proportional odds formulation. This allows us to estimate a parameter for an effect (the proportional odds parameter) yet does not require nearly as many assumptions as the t-test (Example 2.2). In fact, the number of parameters is infinite, yet we can still use a permutation test to calculate the p-value.

Example 2.4   *Again, suppose we have observed data vector as listed in expression 2.4. Now, let the ith elements of $\mathbf{y}$ and $\mathbf{z}$ be y_i and z_i. Assume that if $z_i = 1$, then Y_i, the random variable associated with y_i, is distributed with distribution F. We write this as $Y_i \sim F$, so $\Pr[Y_i \le y] = F(y)$ for any y. Also, if $z_i = 2$, then $Y_i \sim G$. We assume F and G are unspecified distributions with support $[0, \infty)$. In other words, because somehow we know that the response variable in this setting cannot have a negative value, the possible values of*

Y_i are in $[0,\infty)$. Thus, $\mathcal{X} = \{\mathbf{Y}, \mathbf{z} : \mathbf{Y} \in \left(\Re^+\right)^{10}\}$. There are an uncountably infinite number of elements in $\mathcal{X}$.

The probability model, P_θ, is represented by the two distributions F and G with support on $[0,\infty)$. Consider a proportional odds model where the odds of one group are proportional to the odds of the other group. Letting F and G be cumulative distribution functions (cdfs), assume that for each $y \in (0,\infty)$,

$$\frac{G(y)}{1 - G(y)} = \beta \frac{F(y)}{1 - F(y)},$$

where $\beta \in (0,\infty)$. This implies that

$$G(y) = \frac{\beta F(y)}{1 - F(y) + \beta F(y)}.$$

So the parameters, θ, can be represented by the scalar β which is in $(0,\infty)$, and the (infinite dimensional) cdf, F. We can write $\Theta = \{\beta, F\}$. A typical partition is $\Theta_0 = \{\beta, F : \beta = 1\}$ and $\Theta_1 = \{\beta, F : \beta \neq 1\}$, with hypotheses usually written as $H_0 : \beta = 1$ and $H_1 : \beta \neq 1$.

Under the null hypothesis that $\beta = 1$, then $F = G$, and we can use a test statistic, T, similarly to what was done in Example 2.3. Under the set of null hypothesis probability models, it is not possible to calculate $\Pr[T(\mathbf{X}) \geq c]$ for a constant c, without making some assumptions about F. But we can calculate the conditional probability given the ordered vector of responses. When we use the conditional probability to calculate the p-value, this is a permutation test (just as was done in Example 2.3). For the proportional odds model, we usually rank all of the 10 responses and use the difference in means of the ranks between the groups as the test statistic, T. This gives us the Wilcoxon-Mann-Whitney test (see Chapter 9).

So, although it is a bit complicated to show, we can calculate p-values using the permutation test methods, and this creates valid p-values for all sample sizes (see Note N6).

So the same permutation method for deriving decision rules can be applied to the assumptions of Example 2.3 as well as Example 2.4.

When we are being formal, we write the hypothesis test as the set $\{\delta, A\}$, where $A = \{\mathcal{X}, \mathcal{P}_0, \mathcal{P}_1\}$ is the set of assumptions defining the sample space, $\mathcal{X}$, the null hypothesis model or models, $\mathcal{P}_0 = \{P_\theta : \theta \in \Theta_0\}$, and the alternative hypothesis models, $\mathcal{P}_1 = \{P_\theta : \theta \in \Theta_1\}$. This formality is often not necessary, and the hypothesis test is sometimes defined by the δ with the assumptions A implied. The formality is useful when talking about how the same decision rule may be used for different sets of assumptions. For example, a permutation test with the same test statistic, $T(\mathbf{x})$, is valid for the assumptions of either Example 2.3 or Example 2.4 (see also Chapter 8).

2.3 Confidence Sets: Inverting a Series of Hypothesis Tests

It is possible to create a confidence set by inverting a series of hypothesis tests. For one-dimensional parameters, in many cases, the set is an interval and is known as a *confidence interval*. We start with confidence sets for the general case, where θ may be a vector of parameters. Then most of the rest of this section focuses on the case when θ is a scalar.

2.3.1 Inverting Point Null Hypotheses

First, consider the general case where θ may be a vector. Let

$$H_0 : \theta = \theta_0$$
$$H_1 : \theta \neq \theta_0$$

be a series of point null hypotheses and their alternatives indexed by θ_0. Suppose you have a series of valid (nonrandomized) decision rules associated with each set of hypotheses, $\delta(X, \alpha; \theta_0)$. Let the p-value associated with $H_0 : \theta = \theta_0$ be $p(X, \theta_0)$ then $\delta(X, \alpha; \theta_0) = I(p(X, \theta_0) \leq \alpha)$. For each X let the set of θ_0 values for which we fail to reject the null hypotheses be

$$C_S(X, 1 - \alpha) = \{\theta_0 : \delta(X, \alpha; \theta_0) = 0\}$$
$$= \{\theta_0 : p(X, \theta_0) > \alpha\}. \tag{2.5}$$

Then $C_S(X, 1 - \alpha)$ is a $100(1 - \alpha)\%$ *confidence set* (also known as a confidence region) and

$$\Pr[\theta \in C_S(X, 1 - \alpha)|\theta] \geq 1 - \alpha \text{ for all } \theta, \tag{2.6}$$

so that $C_S(X, 1 - \alpha)$ is valid.

When θ is a scalar, the alternative hypothesis ($H_1 : \theta \neq \theta_0$) associated with the point null hypothesis ($H_0 : \theta = \theta_0$) is called the *two-sided hypothesis*, and the confidence set given by Equation 2.5 will often be an interval and is called the *confidence interval*. But the confidence set is not guaranteed to be an interval for scalar θ, since there may be gaps. In that case, a confidence interval can be defined by filling in the gaps. In other words, the $100(1 - \alpha)\%$ confidence interval is

$$C(X, 1 - \alpha) = (\theta_L, \theta_U), \tag{2.7}$$

where $\theta_L = \min[C_S(X, 1 - \alpha)]$ if it exists and $-\infty$ (or the minimum possible value of θ) otherwise, and similarly $\theta_U = \max[C_S(X, 1 - \alpha)]$ if it exists or ∞ (or the maximum possible value of θ) otherwise. Since $C_S(X, 1 - \alpha)$ is valid by Equation 2.6, the associated confidence interval, $C(X, 1 - \alpha)$, will be valid as well. For a given p-value function for a series of two-sided hypotheses, $p(X, \theta_0)$, define the *matching confidence set* as $C_S(X, 1 - \alpha)$ of Equation 2.5, and the *matching confidence interval* as $C(X, 1 - \alpha)$ of Equation 2.7. If the matching confidence set is an interval, i.e., $C_S(X, 1 - \alpha) = C(X, 1 - \alpha)$, then we say that $p(X, \theta_0)$ and $C(X, 1 - \alpha)$ are *compatible*, meaning that for any $\mathbf{x}$, the rejection of $H_0 : \theta = \theta_0$ when $p(\mathbf{x}, \theta_0) \leq \alpha$ occurs if and only if $C(\mathbf{x}, 1 - \alpha)$ excludes θ_0.

For some p-value functions the matching confidence intervals are not compatible. In other words, sometimes there is a gap in the matching confidence sets. To show that this noncompatibility problem can occur in practice, we consider an example with Fisher's exact test, which can be used to test for differences between two groups when the responses are binary.

Example 2.5   *For the two-sided test there are two versions of Fisher's exact test: the Fisher–Irwin test and the central Fisher's exact test. The noninterval confidence set can occur with the Fisher–Irwin version (the default version in* fisher.test *in R, see Note N7). Suppose we observe proportions of 7/262 from one group and 30/494 from another. We are interested in comparing the population proportions, say θ_1 and θ_2. Let $\beta = \frac{\theta_2(1-\theta_1)}{\theta_1(1-\theta_2)}$ represent* *an odds ratio. The value $\beta = 1$ if $\theta_1 = \theta_2$ and $\beta > 1$ if $\theta_2 > \theta_1$. If we test $H_0 : \beta = 0.99$ using the Fisher–Irwin test we fail to reject at the 0.05 level ($p = 0.05005$), when we test $H_0 : \beta = 1.00$ we do reject ($p = 0.04996$), but when we test $H_0 : \beta = 1.01$ we fail to reject again ($p = 0.05006$). So the 95% confidence set for the odds ratio has a hole in it: $C_S(x, 0.95) = \{\beta : \beta \in (0.177, 0.993) \text{ or } \beta \in (1.006, 1.014)\}$.*

The details of the example are not important at this juncture (see Chapter 7). The critical point is that for a commonly used test, inverting a series of two-sided hypothesis test can give a confidence set that is not an interval. However, noninterval confidence sets rarely occur, even when using a procedure (e.g., the Fisher–Irwin test) with this possibility. If the confidence set is not an interval, we create one by filling in the gap. For Example 2.5 this gives a 95% confidence interval of $(0.177, 1.014)$.

2.3.2 Central Intervals: Inverting Two One-Sided Hypotheses

An alternative strategy to create a two-sided confidence interval than inverting a series of point null hypotheses (such as $H_0 : \theta = \theta_0$) is to combine the confidence intervals from two series of one-sided hypothesis tests. Typically, we combine two $100(1 - \alpha/2)\%$ one-sided confidence intervals, to get one $100(1 - \alpha)\%$ two-sided confidence interval, which we call a *central* confidence interval. This strategy has the advantage that we can automatically infer either one-sided hypothesis test by using either the lower or the upper limits of the interval. Start with the alternative-is-greater series of one-sided hypotheses,

$$H_0 : \theta \leq \theta_B$$
$$H_1 : \theta > \theta_B,$$

indexed by θ_B, the parameter value on the boundary between the null and alternative hypotheses. Suppose the associated (nonrandomized) decision rule associated with the valid hypothesis test is $\delta_{AG}(\mathbf{X}, \alpha/2; \theta_B)$, where AG represents the alternative-is-greater hypotheses. Using the validity property of the hypothesis test, we can show that for any $\theta \in \Theta_0$, the probability of failing to reject the null hypothesis is at least $1 - \alpha/2$ (see Problem P1). We can use this property to define a confidence set

$$C_L(\mathbf{x}, 1 - \alpha/2) = \{\theta_B : \delta_{AG}(\mathbf{x}, \alpha; \theta_B) = 0\}. \tag{2.8}$$

The set $C_L(X, 1-\alpha/2)$ will almost always be an interval (see Problem P2). If it is not an interval we can write an interval that contains it as $[\theta_L(X, 1 - \alpha/2), \infty)$, where $\theta_L(X, 1 - \alpha/2) = \min C_L(\mathbf{x}, 1-\alpha/2)$ (or $-\infty$ if no minimum exists) and if appropriate, we may replace ∞ with θ_{max}, the maximum possible value of θ. Similarly, we can create an analogous confidence set starting with the alternative-is-less series of one-sided hypothesis tests (i.e., with $H_0 : \theta \geq \theta_B$ and $H_1 : \theta < \theta_B$), and its decision rule δ_{AL}. The associated confidence set is

$$C_U(\mathbf{x}, 1 - \alpha/2) = \{\theta_B : \delta_{AL}(\mathbf{x}, \alpha/2; \theta_B) = 0\}, \tag{2.9}$$

and let the smallest interval that contains it be $(-\infty, \theta_U(\mathbf{x}, 1 - \alpha/2)]$. Then a *central* $100(1 - \alpha)\%$ confidence interval is the two-sided interval that is the intersection of both one-sided intervals, giving

$$C_c(\mathbf{x}, 1 - \alpha) = [\theta_L(x, 1 - \alpha/2), \theta_U(x, 1 - \alpha/2)]. \tag{2.10}$$

A central confidence interval created this way is valid, since

$$\Pr\left[\theta \in C_c(\mathbf{X}, 1 - \alpha) | \theta\right] \geq 1 - \alpha.$$

Now we return to the Fisher's exact test example.

Example 2.5 *(continued) Now consider the confidence interval created by inverting the central Fisher's exact test, the test created from two one-sided Fisher's exact tests (unlike the two-sided test, for each pair of the one-sided hypotheses, there is only one version of a one-sided Fisher's exact test). Again, using the data, 7/262 from one group and 30/494 from the other, we get a p-value of $p = 0.0518$ and a 95% central confidence interval on the odds ratio of* $(0.155, 1.006)$.

2.3.3 Compatibility

On page 14 we defined compatibility between p-value functions and confidence intervals, where the p-value was based on a series of point null hypotheses. Here we define compatibility more formally so that it can apply to other types of tests and p-values.

Consider first a single hypothesis test using the formal definition given on page 13). Let the hypothesis test be $\{\delta, A\}$, where $A = \{\mathcal{X}, \mathcal{P}_0, \mathcal{P}_1\}$ is the set of assumptions defining the sample space, $\mathcal{X}$, the null hypothesis model or models, $\mathcal{P}_0 = \{P_\theta : \theta \in \Theta_0\}$, and the alternative hypothesis models, $\mathcal{P}_1 = \{P_\theta : \theta \in \Theta_1\}$. Define the (nonrandomized) decision rule in terms of a p-value, so that $\delta(\mathbf{x}, \alpha; \Theta_0) = I(p(\mathbf{x}, \Theta_0) \leq \alpha)$. Let $C_S(\mathbf{x}, 1 - \alpha)$ be a confidence set procedure.

Definition 2.1   Compatibility of test and confidence set: *A hypothesis test $\{\delta, A\}$ is compatible with a confidence set procedure $C_S(\mathbf{x}, 1-\alpha)$ when, for any $\mathbf{x}$, $\delta(\mathbf{x}, \alpha; \Theta_0) = 1$ (or equivalently $p(\mathbf{x}, \Theta_0) \leq \alpha$) if and only if $\theta \notin \Theta_0$ for all $\theta \in C_S(\mathbf{x}, 1 - \alpha)$. The definition is the same if the confidence set is a confidence interval.*

Now consider a series of hypothesis tests. Let t be the index for each specific hypothesis test and let the associated null and alternative parameter spaces be $\Theta_0(t)$ and $\Theta_1(t)$. For example, a series of alternative-is-greater hypotheses could be represented by parameter spaces $\Theta_0(t) = \{\theta : \theta \leq t\}$ and $\Theta_1(t) = \{\theta : \theta > t\}$. Let the associated p-value function for the series be $p(\mathbf{x}, \Theta_0(t))$. Then analogously to Definition 2.1, the p-value function is *compatible* with a confidence set procedure $C_S(\mathbf{x}, 1-\alpha)$ when for any t and $\mathbf{x}$, $p(\mathbf{x}, \Theta_0(t)) \leq \alpha$ if and only if $\theta \notin \Theta_0(t)$ for all $\theta \in C_S(\mathbf{x}, 1 - \alpha)$.

A closely related property of a p-value function is *coherence*. A p-value function $p(\mathbf{x}, \Theta_0(t))$ is coherent if for every $\mathbf{x} \in \mathcal{X}$, then $p(\mathbf{x}, \Theta_0(s)) \leq p(\mathbf{x}, \Theta_0(t))$ if $\Theta_0(s) \subseteq \Theta_0(t)$ (see Röhmel (2005) or Fay and Hunsberger (2021)).

2.3.4 Three-Decision Rules

For many situations, we want to know not just whether $\theta = \theta_B$, but if it is not equal, we want to know if $\theta > \theta_B$ or $\theta < \theta_B$. For example, suppose θ represents the difference in means of the response from a randomized trial of a new treatment compared to the standard treatment, and $\theta_B = 0$ represents equal means. At the design stage, we do not know if standard is better than new or vice versa, so it will be difficult to choose one of the one-sided hypothesis tests in advance. In this case we consider the *three-decision rule*, where the three decisions are: (1) fail to reject $H_0 : \theta = \theta_B$; (2) reject $H_0 : \theta = \theta_B$ and conclude $\theta > \theta_B$; or (3) reject $H_0 : \theta = \theta_B$ and conclude $\theta < \theta_B$.

The three-decision rule can be tested by using a central confidence interval together with its associated two-sided p-value, we can then combine the two-sided hypothesis with both one-sided hypotheses to create a three-decision rule. It works like this. Let p_c be the p-value created from doubling the minimum of the one-sided p-values,

$$p_c(\mathbf{x}, \theta_B) = \min\{1, 2p_{AG}(\mathbf{x}, \theta_B), 2p_{AL}(\mathbf{x}, \theta_B)\}, \tag{2.11}$$

where $p_{AG}(\mathbf{x}, \theta_B)$ is the p-value associated with the one-sided null hypothesis $H_0 : \theta \leq \theta_B$ (AG represents the alternative-is-greater hypotheses). If $p_c \leq \alpha$, we reject $H_0 : \theta = \theta_B$ at level α. Alternatively, we say we reject $H_0 : \theta = \theta_B$ with a two-sided p-value equal p_c, which implies that we would reject for any $p_c \leq \alpha$. Then if the $100(1-\alpha)\%$ central confidence interval, say (θ_L, θ_U), is compatible with p_c, this means the following: either $\theta_B < \theta_L$ in which case we reject $H_0 : \theta \leq \theta_B$ with p-value $p_c/2 = p_{AG}$, or $\theta_U < \theta_B$ in which case we reject $H_0 : \theta \geq \theta_B$ with p-value $p_c/2 = p_{AL}$.

2.4 Properties of Hypothesis Tests

It is useful to present the Type I and II errors in Table 2.1.

A Type I error is when we reject the null hypothesis when the null is true, while a Type II error is when we fail to reject the null hypothesis when the alternative is true.

Consider first the Type I error. Ideally, we design our decision rules to be valid, meaning the rate of Type I errors under any probability model in the null hypothesis is less than or equal to the α-level, α. Rejecting the null hypothesis is known as finding a *significant effect* (regardless of the truth, which is not known). It can be difficult to develop a decision rule that is valid for all sample sizes. In some cases, we can get approximate validity. One important property for these kinds of decision rules is that you have asymptotic validity, so that the approximation gets better as you collect more data. We give more details about asymptotics in Section 3.7 and Chapter 10.

Now consider the Type II error, failing to reject the null hypothesis when it is false. When we do not reject the null, we often use the phrase "fail to reject the null" (or equivalently "fail

		Truth	
		Null	*Alternative*
Decision	*Reject Null*	Type I Error	
	Fail to Reject Null		Type II Error

Table 2.1 *Tabular description of Type I and Type II errors*

to find a significant effect") to emphasize that we cannot make a strong statement regarding either hypothesis. We can fail to reject the null because either (1) the null hypothesis is true, or (2) even though the null hypothesis is false, the data do not appear extreme under the null perhaps due to randomness or not enough data. A common mistake is to interpret failure to reject the null as showing that the null is true, when it often means that there are insufficient data to say anything practically meaningful about the hypotheses.

Let $\beta(\theta)$ be the Type II error rate when the probability parameter vector is θ. We want $\beta(\theta)$ to be small for some reasonably likely values of $\theta \in \Theta_1$. Typically, instead of talking about Type II error rates, we talk about *power*, the probability of rejecting the null given θ. The implications for minimizing Type II error rates and maximizing power are the same, since the power under θ is equal to $1 - \beta(\theta)$. For applied work, we typically require that the power be large, about 80% or 90%, under a reasonably likely alternative, or an alternative that is based on minimum clinical significance. If the power is small, then it is likely that at the end of the study we will fail to reject the null hypothesis, which may not give us much new scientific information.

Most reasonable hypothesis tests are *consistent* for some $\theta \in \Theta_1$, meaning that the power, $1 - \beta(\theta)$ increases to 1 as the sample size approaches infinity. For these hypotheses, we can determine the smallest sample size, say n, such that the power is at least the target power (say 90%) for a specific θ. This is an important part of designing a study (see Chapter 20).

Ideally, for any set of assumptions, A, we want to pick the decision rule that is most powerful among all possible decision rules for any $\theta \in \Theta_1$. The resulting test, $\{\delta, A\}$, is called the *uniformly most powerful* (UMP) test. For many sets of assumptions, there does not exist a UMP test. In this case, when choosing between two decision rules (assuming both are valid), one would choose the one that has larger power under the alternative model of interest for some fixed sample size. Alternatively, one could fix $\theta \in \Theta_1$ and compare the minimum sample size needed to achieve a specific power, $1 - \beta(\theta)$, between the two decision rules. For example, if N_1 and N_2 represent those sample sizes for decision rules δ_1 and δ_2, then the relative efficiency of δ_2 with respect to δ_1 is defined as the ratio, $r = N_1/N_2$. In other words, δ_2 will require r times the sample size of δ_1 to get the same power (given θ).

Another way to choose between decision rules is to choose the one that has larger relative efficiency asymptotically. Since most tests are consistent for all $\theta \in \Theta_1$, the way to measure the asymptotic relative efficiency (ARE) is to consider a sequence of parameters, θ_n, that approach the boundary between the null and alternative as n goes to infinity. In most cases, the ARE is useful because it does not depend on the power or on which θ in the alternative is chosen.

Two properties of decision rules that are useful for applied work are the very closely related properties of invariance and equivariance. Invariance under a set of transformations is when the decision rule does not change after applying the transformation to the data. For example, one might want a decision rule to give the same result if you make a monotonic transformation of the responses, such as taking the log of the response. Equivariance is when the decision rule changes appropriately if you transform the data and the hypotheses in the same way. For example, consider a binomial response, where x represents the number of successes with parameters n and θ. Let $\delta(x, \alpha; \theta_0)$ be a decision rule testing the null $H_0 : \theta = \theta_0$. Suppose that instead you measure failures, $n - x$. Then the decision rule would be equivariant to switching the successes and failures if $\delta(x, \alpha; \theta_0) = \delta(n - x, \alpha; 1 - \theta_0)$ for all

x and θ_0. Another type of equivariance is palindromic equivariance, defined as equivariant to systematically switching the order of the responses (and parameters) so that the largest response (or parameter) becomes the smallest, the next largest becomes the next smallest, and so on. In much of the literature, no distinction is made between invariance and equivariance, and both are known as invariance. For example, "invariant to monotonic transformations" usually means "equivariant to monotonic transformations," and this is usually clear from the context. In the rest of this book, we usually use the term invariance to include equivariance.

Robustness is an important idea for applied work. A test is robust if the results do not change drastically for minor violations of the assumptions. For example, if we perform a one-sample t-test on data that is only approximately normally distributed, the results will often be reasonable. For some types of approximately normal data, however, the t-test may have poor power. For example, if the population is roughly normal except for a few extreme outliers, the t-test will have very poor power for testing the mean (see Chapter 5). Another way to improve the robustness is to change your hypotheses and base your tests on parameters that are not highly dependent on a small percentage of outliers. For example, you might test inferences about the median rather than the mean. These robustness ideas will be emphasized throughout this book.

Other properties, not emphasized much in this book are mentioned in Section 2.6.

2.5 Summary

A hypothesis test has several components:

- A set of possible data values, known as the sample space or support;
- A set of probability models that is partitioned into two models (the null and alternative hypotheses); and
- A decision rule, which determines whether the null hypothesis is rejected. It is a function of the data and the α-level.

An α-level, α, is set by the researcher. If the probability of the decision rule rejecting the null hypothesis is always less than or equal to α, given that the null is true, then the rule is said to be valid. Another useful quantity is the p-value which is the smallest α such that we would reject the null hypothesis for a given data set at level α (and additionally we would reject for all $\alpha^* > \alpha$). We desire valid decision rules, but sometimes settle for approximately valid rules.

We can create central confidence intervals by inverting two one-sided hypothesis tests, and these intervals are good for making directional inferences (i.e., making three-decision rules).

The Type I error rate is the probability of rejecting the null hypothesis when it is true. The Type II error rate is the probability of failing to reject the null when the alternative is true. Failing to reject the null hypothesis often occurs when there are insufficient data to say anything meaningful about the hypothesis, so it is important not to declare that the null is true. Power is the probability of rejecting the null hypothesis, usually under an alternative hypothesis model. For applied work, we typically require that power be large, such as 80 or 90%, whenever a reasonably likely alternative is true; otherwise, the research could provide little new information.

The following properties are desirable when selecting among possible decision rules: high power, invariance (i.e., the inferences are unchanged with certain transformations to the data), and robustness (i.e., the test performs well even with some violations of assumptions).

2.6 Extensions and Bibliographic Notes

Lehmann and Romano (2005) give a comprehensive theoretical treatment of statistical hypothesis tests. Their text has an emphasis on optimal tests (e.g., uniformly most powerful tests). For example, it covers the Neyman–Pearson lemma, that defines the most powerful test for comparing simple null and alternative hypotheses (a simple hypothesis is one that has only one element, e.g., the simple null has Θ_0 that is one parameter not a set of parameters). Further, their text defines an unbiased test: a test where the power to reject is at least as large as the size for all $\theta \in \Theta_1$. By restricting the set of tests to unbiased tests, one can sometimes find a test that is uniformly most powerful among those unbiased tests; this is called the *UMP unbiased test*. Similarly, we can consider an invariance property, and find the UMP tests among those with that invariance property.

2.7 Notes

N1 **Set notation:** A set is a collection of objects. For example, the even numbers greater than 0 and less than 10 are a set, $\{2, 4, 6, 8\}$. We can write the same set in the following way: $\{x : x \text{ is even and } x > 0 \text{ and } x < 10\}$. This notation is read as, the collection of all xs such that (the ":" means such that) x meets all three conditions mentioned. This notation can be useful if you are defining a set in relation to some function. For example, $\{x : f(x) > 5\}$, means all xs such that $f(x) > 5$. If $\mathcal{X}$ is a set, the notation $x \in \mathcal{X}$ means x is in that set.

N2 **Randomized decision rules:** An important theoretical use of randomized decision rules is for getting hypothesis tests with sizes (*size* is defined below Equation 2.1) exactly equal to α when the data are discrete. Consider Example 2.1, a binomial with $n = 10$. If we set

$$\delta(x, 0.05) = \begin{cases} 1 & \text{if } x \geq 9 \\ \frac{0.05 - \Pr[X \geq 9|\theta = .5]}{\Pr[X = 8|\theta = .5]} = 0.893 & \text{if } x = 8 \\ 0 & \text{if } x \leq 7, \end{cases}$$

then the size will be exactly equal to 0.05. The problem is that if $x = 8$, then sometimes we reject the null hypothesis but sometimes we do not, and whether we reject or not does not depend on the data. So two analysts could do the proper analysis on the same data and come up with different answers. This makes many applied statisticians uncomfortable, especially since the randomness is obviously coming from the random number generator not the data. Sometimes applied statisticians will use methods that use random number generators to implement a method with many samples (for example the use of bootstrap sampling, or Monte Carlo sampling), but in those cases the randomness is added to aid in calculations and the probability of getting very different answers by two different implementations can be kept small by using a large number of samples in the calculation.

N3 **Setting α-level:** Ebola is a very serious disease, and as of the beginning of 2015 there had not been a vaccine that had been tested for efficacy in humans. There was a very large outbreak of Ebola in West Africa that started in 2014, and a study was quickly designed to test some of the best vaccines in development. During the protocol development, some questioned about whether to use an α-level of 0.05 for a two-sided test. If the vaccine showed promise, but the significance was between 0.05 and 0.10, it might make sense to stop the study and start supplying the vaccine to the population if the outbreak was still continuing at that point. In this case, the Type I error (calling a useless vaccine effective) might not be as big a problem as a Type II error (failing to detect a significant effect on an effective vaccine).

N4 **Monotonic decision rules:** Formally, a monotonic decision rule has $\delta(\mathbf{x}, \alpha_1) \leq \delta(\mathbf{x}, \alpha_2)$ for all $\mathbf{x}$ and $\alpha_1 < \alpha_2$. For some simple examples that show nonmonotonic decision rules are possible (but not necessarily practical) see (Lehmann and Romano, 2005, Problems 3.17, 3.58). One reason nonmonotonic decision rules have been proposed is to increase power for unbiased (see Section 2.6) tests, but this can lead to some strange behaviour (Perlman and Wu, 1999).

N5 **Surprisal:** For cases when the p-value can be interpreted as the probability of observing equal or more extreme data under the null, a transformation has been proposed: using the surprisal, $s = -\log_2(p)$. The surprisal gives a way to envision small probabilities. An $s = 6$ is the probability associated with seeing six heads in a row from fair coin tosses. A p-value of $p = 0.05$ corresponds to $s = 4.3$ which is close to seeing four heads in a row (see Greenland, 2017).

N6 **Two-sample permutation test:** Here is how to show that a two-sample permutation test is valid under the assumptions of Example 2.4. First consider conditioning on the order statistics of $\mathbf{y}$: the values of $\mathbf{y}$ ordered from smallest to largest. This is simply a way of conditioning on the 10 values of $\mathbf{y}$, but not on the original order of the indexes associated with them. This conditioning makes the calculation of the p-values tractable. Let $\mathbf{y}_o$ be the vector of order statistics, and $\mathbf{Y}_o$ the associated vector of random variables. For clarity, we write the constant c as $c(\mathbf{y}_o)$ to emphasize that it depends on $\mathbf{y}_o$. Then unconditionally we have,

$$\int \Pr[T(\mathbf{X}) \geq c(\mathbf{y}_o)|\mathbf{Y}_o = \mathbf{y}_o]\Pr[\mathbf{Y}_o = \mathbf{y}_o],$$

where the integration is over all the possible values of $\mathbf{y}_o$ (and $T(\mathbf{x})$ is a function of the data like the difference in means or difference in mean ranks). We do not need to worry about the details of the integration because we solve for $c(\mathbf{y}_o)$ using the methods of Example 2.3. To see this, note that when y_o is fixed, then the only possible values of $\mathbf{y}$ are the permutations of the indices for the y_i. And it turns out that permuting the indices of the y_i and keeping z_i fixed, gives the same possibilities as keeping the y_i fixed and permuting the z_i. But because there are 5 ones and 5 twos in z_i, we can group the 10! different permutations of the indices of z_i into 252 groups of 5! 5! permutations, where within each group all 5! 5! permutations give the same $\mathbf{z}$. Thus, we can solve for $c(y_o)$ so that $\Pr[T(\mathbf{X}) \geq c(\mathbf{y}_o)|\mathbf{Y}_o = \mathbf{y}_o] \leq \alpha$ for all $\mathbf{y}_o$. So since each conditional term is $\leq \alpha$, when we integrate over the set of possible values of $\mathbf{y}_o$, this integral is also $\leq \alpha$. Mathematically,

$$\Pr[T(\mathbf{X}) \geq c(\mathbf{y}_o)] \leq \int \alpha \Pr[\mathbf{Y}_o = \mathbf{y}_o] = \alpha.$$

N7 **Fisher–Irwin test and central Fisher exact tests in R:** The default in fisher.test
in R (at least for version 4.0.4 or earlier), is to use the Fisher–Irwin two-sided
p-value, but to invert the central Fisher's exact test to get the confidence intervals (hence,
those confidence intervals are neither matching nor compatible with the Fisher–Irwin
p-values). To get matching confidence intervals use the **exact2x2** R package. The
Fisher–Irwin test cannot get compatible confidence intervals but can get matching
confidence intervals. See Fay (2010a) or Fay and Hunsberger (2021) for discussion of
the Fisher's exact tests and their matching confidence intervals. For the general issue of
matching confidence intervals see Fay (2010b). For another example, see Problem P2.

2.8 Problems

P1 Show that for a valid decision rule, for any $\theta \in \Theta_0$, the probability of failing to reject
the null hypothesis at level α is at least $1 - \alpha$. Hint: start with the definition of a valid
decision rule, Equation 2.1. Multiply the expression by -1 and add 1 to both sides, then
use $\Pr[\delta = 0] = 1 - \Pr[\delta = 1]$ to get

$$\min_{\theta \in \Theta_0} \Pr[\delta(\mathbf{X}, \alpha) = 0 | \theta] \geq 1 - \alpha.$$

P2 **Matching confidence interval not compatible:** For an example of a confidence set
derived from inverting a one-sided hypothesis test that does not give an interval,
consider the one-sided exact unconditional test for the two-sample binomial problem
using the score method (see Section 10.7). Using the **exact2x2** R package explore
this. For example, testing 130/248 against 76/170, and test different null hypotheses,
$H_0 : \beta \geq \beta_0$, for $\beta_0 = 0.02, 0.024$, or $0.0.26$. Are the p-values monotonic? If they are
not what does this mean in terms of confidence sets possibly being confidence intervals?
(Fay and Hunsberger, 2021).

3

From Scientific Theory to Statistical Hypothesis Test

3.1 The Statistician's Role

In science we create theories about how the world works and test those theories using scientific studies. The applied statistician's role in this process can be broken down into three important steps:

(1) Help design a study to learn something about the scientific theory.
(2) Help decide on a reasonable probability model for data that comes from the study.
(3) Use statistics (functions of the data) to make inferences about the probability model. Interpret the inferences about the probability model in terms of the scientific theory.

Although other types of statistical inferences are used in science, for this book, the third step will be statistical hypothesis tests and closely related statistics such as parameter estimates and confidence intervals (see Chapter 2).

The goal of this chapter is to introduce some important issues and highlight potential pitfalls if those issues are ignored. Some of the topics will be explored more in depth later in the book.

Before describing the three steps of the applied statistician's role, we first discuss the important point of causality versus association. This distinction is key for being clear about what we want to learn and what we can learn from scientific studies.

3.2 Causality versus Association

3.2.1 Observational versus Experimental Studies

In scientific discourse, we are usually interested in explaining our observations by a theory about what causes them. The problem is that causality is a concept that is outside the realm of probability theory. The theory of statistical hypothesis tests by itself (see Chapter 2) is not sufficient for determining causality. In general, if we show some relationship between two variables through statistical inference such as a hypothesis test (for example, the two variables are correlated, or the group membership variable indicates different means for the response variable), then we say that the two variables are *associated*. If Y is associated with X, this does not say anything about presence or absence of a causal relationship between the two variables.[1]

---

[1] Here Y and X represent any two different variables. Notation is defined differently between chapters, see page 420.

To clarify the point that causality cannot be represented in a standard probability model, consider a hypothetical experiment testing a new experimental vaccine for anthrax in monkeys. First, randomize the monkeys to get either the new vaccine or a placebo vaccine and then give the appropriate vaccine. Wait 28 days and give a second shot using the same type of vaccine as the original shot. Wait one day and measure a variable, X. Then one week later challenge the monkeys by forcing them to inhale anthrax spores. After two more weeks, see which monkeys survive the anthrax challenge. Let Y be a random binary indicator for survival. Suppose we assume a probability model for Y and X, and we test whether monkeys that survived had higher values of X than monkeys that did not survive. The details of the probability model are not important for this discussion. If we find, through a statistical hypothesis test, that the X values for monkeys who survived tended to be higher than values of those that died, this implies that there is an association between X and Y, but the test result by itself does not imply that X caused the monkeys to survive the challenge.

To see that the association does not imply causation, consider two hypothetical values for X. If X measures an anti-anthrax-toxin antibody level then it may be biologically plausible to suppose that high values of X acted in causing the monkeys to survive, because the antibodies help the immune system fight off the deleterious effects of the anthrax toxin. On the other hand, if X measures the size of the inflammation at the sight of the second vaccination shot, then the association of higher levels of X with survival does not seem biologically plausible to explain the causal mechanism of that survival. It is more likely that the vaccination may set off a series of effects in the monkeys, some of which help to protect them from anthrax challenge (like antibody level) and others that are incidental byproducts (like the inflammation size) that are not an important part of the protective mechanism but are associated with a strong active immune system that is protective. The point is that the same probability model could describe a variable that is an integral part of the protective mechanism or describe a variable that is an incidental byproduct that has no part in the protective mechanism. We cannot show causality through the probability model, it must be shown through information that is outside of the probability model.

The gold standard for demonstrating causality is the randomized experiment. In the above experiment, let Z be a random variable that denotes whether a monkey is randomized to the experimental vaccine or a placebo vaccine. The value, Z, for a monkey is determined by a pseudorandom number generator that is independent of any information about the monkey before the study is started. Thus, if we find a significant association between survival, Y, and vaccination type, Z, from the statistical hypothesis test, we can infer that the experimental vaccination did somehow have a causal effect on the survival. In order to infer the causation, we need the information that randomization was used and the fact that the intervention (i.e., which vaccine) depended on the randomization. That added information is not part of a probability model (in Chapter 2 we did not use randomization or intervention information).

An important distinction between the variables X (variable measured after the experimental vaccine or placebo is given) and Z (random group assignment of a monkey) is that for X we observe its value only, while for Z we intervene based on its value (for example, whether a monkey is in the experimental vaccine group or placebo vaccine group is actively determined by the researcher through the randomization). Pearl (2009b) makes the distinction explicit by writing $do(Z = z) = do(z)$ to denote that the value of Z is forced to be z. In Section 3.2.2 and Chapter 15, we use a different notation, potential outcome notation, to make rigorous causal inferences. For now, note that variables related to interventions are special kinds of variables.

We will classify studies into two types, *observational* studies and *experimental* studies or experiments. For the former we just watch things as they are, while for the latter, we intervene and do something on the objects we are observing.

Experiments are much better than observational studies at determining causation. We show this by example using a high-profile scientific controversy. Prentice et al. (2005) describe results from the Women's Health Initiative (WHI), a set of studies that included both an observational study and several randomized clinical trials (experimental studies). Previous observational studies had repeatedly shown an association between lower cardiovascular disease risk in women and use of hormone replacement therapy (HRT). The WHI was designed to test several hypotheses simultaneously; here we focus on one primary hypothesis of whether HRT (specifically, estrogen plus progestin) affects risk of coronary heart disease. Prior to the start of the WHI, many experts questioned the need for these expensive WHI trials, noting the consistently observed cardiovascular benefit of HRT (Wittes et al., 2007). Thus, it was a stunning result when a randomized clinical trial of the WHI showed that when you randomly assign women to HRT or placebo, the women who got HRT had an *increased* risk of coronary heart disease compared to those that got the placebo. Fortunately, the WHI additionally had an observational study. When that observational study was analyzed separately, the women who got HRT had significantly *less* coronary heart disease, agreeing with the previous observational studies. Because the effect is statistically significant, the decreased heart disease for women that use HRT from the WHI observational study was very unlikely to be due to chance. Besides, it repeated an effect that had been seen in other observational studies. This effect represents a significant association that is repeatable, but this association does not necessarily mean that the HRT directly led to reduced risk.

Why do the observational and the experimental studies of the WHI not agree? One might suspect that there are variables in the observational study that are positively associated with taking hormones but negatively associated with coronary heart disease; in other words, those who are likely to take hormones are at lower risk for heart disease at the time they make the decision to take hormones than those who do not choose to take hormones. However, even if you adjust for the effect of many covariates on the hazard rate (such as age, body mass index, smoking status), the WHI observational study still appears to show that hormones protect against coronary heart disease, while the experimental study shows that hormones cause that disease. A potential explanation for resolving this conflict, at least in part, is seen when examining the time from starting hormones. For the first five years after starting HRT, women on HRT tended to have higher rates of heart disease relative to those who did not, while for women who had been taking hormones for over five years, HRT appeared to reduce the risk. This effect is reasonably consistent with the data regardless of whether the effect was measured in the observational study or the experimental study. The problem is that the number of women in each stratum (of time since starting HRT) was different in the observational and experimental studies, so that the overall experimental result was primarily based on early use, and the overall observational result was primarily based on long term use. We reproduce data excerpted from table 5 of Prentice et al. (2005) in Table 3.1 to show this (the number at risk can be inferred from the cases observed and the hazard ratios). Notice that more cases of coronary heart disease occur in the experimental study for women who

Time from hormone initiation	Randomized clinical trial		Observational study	
	Number of cases in hormone group	Hazard ratio	Number of cases in hormone group	Hazard ratio
<2 years	80	1.68	5	1.12
2–5 years	80	1.25	27	1.05
>5 years	28	0.66	126	0.83

Table 3.1 *Hazard Ratios for coronary heart disease for women on hormones (estrogen plus progestin) compared to those not on hormones. The hazard ratios come from a Cox proportional hazards model that controls for age, race, body mass index, education, smoking, age at menopause, and a measure of physical functioning*

recently started HRT, while more cases occur in the observational study for women who started therapy over five years ago. The results in this table shed some light on what might have led to the striking difference between the randomized result and the observational result. It is important to recognize, however, that without the randomized results as a guide to the "truth," the observational data might never have been analyzed as they were in this table.

The example illustrates how hard it is to measure effects that act over a long period of time (see Note N1 for some of the subtleties). But the example highlights a more important point, showing how an experiment provided evidence of risk during the first five years of HRT, which was not apparent after many observational studies. This failure to detect the harm in HRT (at least in the first five years of use) was not corrected by the scientific community until the experiment was done. In fact, women at risk for cardiovascular disease were often advised to consider HRT based on the misleading observational study conclusions before the results of the experiment were publicized and the advice sharply shifted. There are better analytic methods designed to measure causal estimands and highlight the assumptions used, in an attempt to avoid misleading interpretations. However, sophisticated analyses do not necessarily correct misleading interpretations. For example, Peikes et al. (2008) show that even when using propensity score matching (see Section 3.5) on an observational study when there existed both observational and experimental studies on a specific topic (employment and education programs), the observational study results did not necessarily match the experimental results. These methods will be discussed in Section 3.5 and Chapter 15.

3.2.2 Causal Estimands

A causal estimand is a parameter that measures a causal effect. A parameter from a probability model may or may not be measuring a causal effect. To tell the difference, it will be helpful to introduce the potential outcome notation which is useful for being precise about causal relationships. It is easiest to think of it in terms of an experiment, where an intervention is performed. Suppose we are interested in a response, Y, measured after we give an intervention, Z. Let $Y_i(z)$ be the response the ith individual would have if they had gotten the intervention, z. Here are some examples.

Example 3.1 *Suppose we have two interventions, an anthrax vaccine ($z = 1$) and a placebo vaccine ($z = 0$) applied to monkeys. After intervention, we wait 28 days and we*

challenge the monkeys with inhalation anthrax and see which monkeys survive the challenge.
If the ith monkey got the anthrax vaccine, then we observe the response, $Y_i(1)$, which equals
0 if the monkey survived and 1 if they died. The ith monkey also has a potential outcome,
$Y_i(0)$, that is not observed, i.e., the outcome that monkey would have gotten if they had gotten
the placebo.

The intervention does not need to be a categorical variable.

Example 3.2 *Suppose in the anthrax vaccine study, z represents the dose of the experimen-*
tal vaccine. Then $Y_i(z)$ represents whether the ith monkey survives or not given that their
dose is z. We only observe one dose for the ith monkey, say z_i. So we observe $Y_i(z_i)$ and
$Y_i(z)$ for $z \neq z_i$ represents an unobserved potential outcome for that monkey.

Let $\mathcal{Z}$ be the possible values of z, then the potential outcome set for the ith individual
is $\{Y_i(z), z \in \mathcal{Z}\}$. For example, for a binary z, the pair, $\{Y_i(0), Y_i(1)\}$, is the set potential
outcomes for the ith individual. It is helpful to think of the potential outcome set as a set of
baseline variables for the ith individual that is not observed before the intervention, then z_i
is performed and part of the potential outcome set is observed after the intervention.

A common causal estimand when z only has two possible values is the average causal
difference effect (or just average causal effect), the expected value of $Y_i(1) - Y_i(0)$. In other
words, we want to know on average how much a typical individual's response changed
if they had gotten the intervention $z = 1$ instead of $z = 0$. For continuous interventions
like dose, we can estimate a similar parameter. For example, the average causal effect of
increasing the dose from 10 to 20 units could be written as the expected value of $Y_i(20) -$
$Y_i(10)$ when z represents dose. For ease of explanation, in this subsection we consider only
the case when z is binary and we are interested in the average causal effect,

$$\Delta = \mathrm{E}\left\{Y_i(1) - Y_i(0)\right\} = \mathrm{E}\left\{Y_i(1)\right\} - \mathrm{E}\left\{Y_i(0)\right\}. \tag{3.1}$$

The important difference between experimental studies and observational studies is that
for experimental studies we can manipulate z. If we make sure that the manipulation of z
is independent of the potential outcome pair (e.g., $\{Y_i(0), Y_i(1)\}$), then individuals that have
$z = 1$ have the same probability model for their potential outcome pair, as individuals that
have $z = 0$. Then we can estimate the average causal effect using the difference in the mean
responses in both groups, i.e.,

$$\hat{\Delta} = \left(\frac{1}{n_1} \sum_{i:z_i=1} y_i(1)\right) - \left(\frac{1}{n_0} \sum_{i:z_i=0} y_i(0)\right), \tag{3.2}$$

where n_1 and n_0 are the number of individuals that had $z_i = 1$ or $z_1 = 0$, respectively. So
although for each individual, we only observe one of the potential outcomes ($Y_i(1)$ or $Y_i(0)$),
we can estimate Δ (the expected difference within an individual) by taking the average
response of those with $z_i = 1$ minus the average of those with $z_i = 0$.

For observational studies, we do not manipulate z_i, and there can be problems in using the
simple difference in means, $\hat{\Delta}$, to estimate Δ. For example, suppose z_i is an indicator that
tells whether the ith individual eats at least a certain amount of beta-carotene in their diet
on average at baseline. Let y_i be whether that individual has been diagnosed with a cancer
in the five years since baseline. Beta-carotene is found in many fruits and vegetables, and

observational studies have shown that people that consume more beta-carotene tend to have lower cancer rates. Using the difference in cancer rates by $\hat{\Delta}$ from an observational study may show that there is an association between higher beta-carotene consumption and lower cancer risk. This may not be a good estimate of the causal estimand, Δ, because people that consume more beta-carotene tend to consume more fruits and vegetables, so the lower cancer rate could be from eating more fruits and vegetables (or consequently less meat or fat) and not from the added beta-carotene itself. In other words, although we want to estimate the expected difference of the potential outcomes within an individual, by just observing the beta-carotene levels, there are many differences between people with $z_i = 1$ and those with $z_i = 0$ besides the level of beta-carotene in their diet.

In the beta-carotene example, a better estimate of Δ would be from an experiment where the researcher randomly chose individuals to add beta-carotene supplements to their diet. Then $\hat{\Delta}$ would estimate Δ well. In fact, when randomized studies were done like this, there was no apparent benefit from beta-carotene supplements (Druesne-Pecollo et al., 2010).

3.3 Designing Any Study

3.3.1 Guiding Principles

We perform scientific studies to learn from the data that we collect from those studies. For learning to happen, the data must inform the scientific theory. In order to be a scientific theory, the theory or model must be falsifiable. In other words, if you have a theory to explain some phenomenon, then in order for it to be a scientific theory it must be disprovable by collecting data. The classic example of a nonscientific theory is explaining human behavior by Freudian psychology (see Popper, 1963, chapter 1). Each new patient's behavior is explained in light of the theory, and there is no behavior that a patient (or even hundreds of patients) could exhibit that could disprove the theory. Then the theory does not predict anything, and it is not science. Beware of theories that explain everything and predict nothing.

The overall guiding principle in designing an experiment is to think through the process that creates the data and try to be aware of factors that may affect that process or the measurement of the responses. To avoid subtle pitfalls that may affect the results, we discuss a number of guiding principles and recommendations that arise repeatedly in study design. Each study is different, and properly designing a study involves knowing the subject matter; however, many of the principles discussed are applicable to many types of studies. For example, randomization and blinding can remove muddying effects of confounding variables without the researcher even needing to be aware of those variables. Because of this, the statistician can be a valuable member of a collaborative team even though the statistician may lack specialized subject matter knowledge. Sometimes the new eyes and naïve questions of the statistician can lead to substantially better study designs and analyses, and hence better science.

3.3.2 Have a Clear Focus About Study Questions

Studies can be roughly categorized into exploratory studies and confirmatory studies. Exploratory studies are used to generate hypotheses to explain phenomena, while confirmatory studies test specific hypotheses.

For a confirmatory study, it is usually important to prespecify the primary endpoint and the analysis that will be used on that endpoint. If this is not done, then there may be the appearance (or the fact!) that the primary endpoint was chosen post hoc because it confirmed the theory in the most striking way. If there are many ways to measure an important study outcome, then picking the most striking one is a strategy that will lead to more false positives. The problem is that if you test many statistical hypotheses, the probability that at least one will be significant even when all null hypotheses are true can be quite large (see Chapter 13). A more subtle problem occurs if you pick only one hypothesis to test, but pick it after looking at the data, as you may be likely to pick the one with the biggest apparent effect. So although only one test was formally done, if you informally select the effect that looks most significant, then this may be the essentially the same as testing all the effects and keeping only the most significant.

Example 3.3   *Consider a randomized study comparing an experimental malaria vaccine to a control vaccine in a village in Africa. The malaria parasite* plasmodium falciparum *first infects humans in the liver cells (the asymptomatic liver stage), then infects the red blood cells (blood stage) which can cause fevers and other problems including death. Malaria is treatable and it is possible to become reinfected after treatment and after the treatment drugs have been eliminated from the body. Suppose the vaccine is intended to boost antibodies to the malaria parasite in the blood stage. To keep this simpler, suppose all the subjects are enrolled and vaccinated at around the same time and all are followed for the same amount of time. Subjects have periodic scheduled exams and are encouraged to additionally report to the clinic if they feel sick. There are many possible responses to measure effectiveness of the malaria vaccine. Here are eight possible endpoints: (1) number of times a subject has detectable parasite in their blood during only scheduled visits; or (2) during any visit, scheduled or unscheduled; (3) number of times the parasite is detected in the blood **and** when fever is present for scheduled visits only; or (4) for any visit; (5–8) the time to the first event defined any of the four previous ways. In reality, all subjects are not followed for the same amount of time, and there are many more choices for the primary endpoint. For example, if we measure the rate of events per time at risk, then we must decide if subjects under treatment for malaria are at risk during treatment, and if not, how long after treatment are they not at risk. In designing such a study, it is important to prespecify the primary endpoint, since there are so many choices.*

The choice of a primary endpoint can be the most critical decision in the study design. Sometimes it will be a compromise about what is considered the perfect outcome measure versus a practical measure. For example, researchers might consider a composite endpoint of mortality and progression to various disease states to be an ideal metric to test the treatment effect. But, if the disease progression is difficult to assess and also vulnerable to patients failing to come to study visits, mortality alone might be the better choice for the primary outcome.

For an exploratory study, it is helpful to outline how many hypotheses are being tested. This will help in the interpretation of the results. If you are testing for a difference between two groups and find a significant effect with a p-value of $p = 0.04$, then the interpretation of that p-value depends very much on how many effects were tested. If only one effect was tested, then the chance of making a Type I error would be low, but there is a very different interpretation if 20 or 2 000 effects were tested. Of course, in a very exploratory setting it may not be feasible to be this formal, but any conclusions from such a study require strong caveats, such that any reported p-values are only descriptive.

If there is one prespecified primary endpoint and test, then your sample size should be large enough to show interesting effects. If you run your study but find that the primary effect is not significantly different from chance, then the study may not be giving you much information. The probability of finding a significant effect is known as the power (see Section 2.4), and studies should be designed such that the sample size is large enough to have a large power under a reasonable probability model in the alternative hypothesis set (see Chapter 20).

3.3.3 Reliability and Validity

Reliability refers to the repeatability or reproduciblility of your study. In science it is important that your results are repeatable. A problem is that you can reproduce the results of a study with the same mistaken interpretation. Recall the observational hormone replacement therapy studies of Section 3.2.1. Many studies appeared to show that HRT reduced risk of coronary heart disease, but this repeatable effect was misleading since it missed the early risk of HRT. So together with reliability, we need validity in a study.[2]

Validity refers to how well the conclusions from your study relate to the scientific theory you are trying to elucidate. Validity can be partitioned into internal validity and external validity. Internal validity means that the conclusions from your study are correct. Much of the rest of this chapter explores ways to avoid internal validity mistakes, in other words, ways in which choices in study design lead to conclusions from the study that may be misleading and not represent reality. External validity refers to how generalizable your conclusions from your study are to other situations.

Example 3.4   *The Physicians' Health Study was an experimental study to determine if low doses of aspirin can reduce the risk of deaths from heart disease (it also included a study of beta-carotene and cancer) (Steering Committee for PHS, 1989). The study randomized to either aspirin or aspirin-placebo groups, male physicians between the ages of 40 and 84 who were willing to participate in the study and who had shown in a run-in phase that they had good adherence to taking daily pills. The study showed a highly significant reduction in risk of heart attacks among those assigned to aspirin groups compared to those assigned to the placebo groups. The analysis using group assignments, randomization, blinding, placebo arm, and a run-in phase helped ensure the internal validity of the study so that the statistically significant differences were most likely caused by the aspirin. The external validity of the study refers to how generalizable the results are to other people unlike the study participants (for example, women).*

---

[2] In this section, we use *validity* as a general concept, while in Chapter 2 we used *valid* more precisely and formally to define a statistical property of p-values, confidence intervals, and decision rules. We use both words (which are closely related) because both concepts are important and have solidified terminology. If we need to distinguish between the two meanings, we will call the former *scientific validity* and the latter *statistical validity*, but except in this section we usually mean statistically valid when we use the term valid.

Since the external validity is extrapolating the scientific theories onto objects that were not measured in the study, external validity is not addressed by the statistical hypothesis tests of this book. Therefore, although external validity is important for science, we spend very little time discussing it in this book. So, for example, none of the methods of this book will help in determining if we can apply the results of the Physicians' Health Study to women. But the methods used in the study to ensure internal validity are explored further in this chapter.

Internal and external validity are important at different stages of the scientific process. In the design stage, we think about how focused our study will be. What individuals or objects will we study? What specific interventions will we measure or values will we observe? The design determines much of the external validity of the study. However, once the study has been designed, our control of external validity ends and the focus shifts to interval validity. At this stage, it is helpful to ignore the external validity. For example, there is no need to think about how men and women differ when making inferences from the Physicians' Health Study, because for internal validity that does not matter. At the implementation and analysis phases, internal validity is at the forefront. Careful implementation and appropriate assumptions and analyses ensure proper inferences from the study at hand. Only after the analysis has been completed, do we again return to speculating about the external validity of the results.

3.3.4 *Are You Measuring What You Want?*

In both observational and experimental studies, you may not be measuring what you think you are measuring. The placebo effect and the Hawthorne effect are two classic examples of when an experimenter is not measuring the intended effect.

Example 3.5 *Placebo effect: This refers to the effect of a treatment that is due to the expectations of the individuals getting the treatment. This effect can be measured experimentally by giving individuals an inert drug or a sham treatment. Typically, the placebo effect refers to positive effects, although there can be negative placebo effects (sometimes called nocebo effects).*

Example 3.6 *Hawthorne effect: This effect results from individuals in a study knowing that they are being observed. For example, in a diet study, the fact that the individuals know they will be weighed at the end of the study may affect their eating habits and hence the final study weight.*

In medical treatment studies, we want to show that the treatment effects are better than the placebo and Hawthorne effects combined. That is partly why treatments are compared to placebos or controls (to account for the placebo effect) and studies are blinded and randomized (so both the treatment and placebo arms are susceptible to the Hawthorne effect).

Many times the intervention that the researcher wants to measure is not easily implemented or not ethical, so a closely related intervention may be studied instead. So, the researcher may not be evaluating what they think they are.

Example 3.7 *A scientific theory is that eating a low-fat diet reduces the risk of breast cancer in women. The clinical trial component of the Women's Health Initiative included an intervention for dietary modification; women were randomized to either an intense behavior* *modification group that included group and individual sessions designed to help reduce fat intake $(z = 1)$ or a control group that only received diet-related educational materials $(z = 0)$. One outcome measure was time to first coronary heart disease event after the start of the intervention, Y. If the ith woman was randomized to the behavior modification group, then her observed time to event is $Y_i(1)$, while her unobserved potential outcome if she had gotten randomized to the control group is $Y_i(0)$. Although our scientific interest is in a low-fat diet, the intervention is not a low-fat diet but the option to partake in a more intense effort to get the participant to eat a low-fat diet. This points to a common issue in experimental studies with humans, often you cannot intervene exactly with the effect that is of scientific interest. You may want to force people to eat a low-fat diet, but that is not ethical. Nonetheless, the study is asking a clear question that has relevance to public health strategy.*

For measuring observations by survey, you can get very different answers depending on how and when you ask the question. This is well known to political analysts who often use survey wording to their advantage. But it is also important in measuring difficult things like diet. Asking people to keep track of what they eat daily will give systematically different answers than if you ask them to remember all that they ate during the past week.

Measuring length of time for events can be particularly tricky.

Example 3.8 *Suppose we want to know if mammography is helpful for increasing survival time of women with breast cancer. It may seem that measuring survival is straightforward, it is time from diagnosis to death. But what if mammograms lead to earlier diagnosis, but do not prolong life compared to a woman who did not get a mammogram? Then time from diagnosis to death will be longer; however, it is not longer because the death has been delayed (which is what we wanted), but because the diagnosis has come earlier (which does not help unless earlier diagnosis leads to earlier treatment and prolonged life). For a discussion of this and other issues on screening, see Grimes and Schulz (2002).*

3.3.5 Regression to the Mean

Regression to the mean is simplest to envision in terms of repeated observations from individuals in a sample. Suppose responses from each individual in a sample come from some steady state with variability around that steady state. Then the individual with the highest response will likely have a subsequent response that is lower. This is because the highest response was probably due to a combination of a high steady state and a random high, and the subsequent response would likely not have its random component as high as the previous one. Similarly, lowest responses will tend to be followed by slightly higher ones. The extreme responses tend to "regress to the mean" over time.

Example 3.9 *Kahneman (2011, chapter 17) gives a clear example of regression to the mean. He was trying to convince flight instructors that rewards worked better than punishments for motivating students. An instructor disagreed, responding that generally when a student did poorly on an exercise and they were reprimanded, they did better the next time, while if a student did well on an exercise and was praised, they did worse the next time. The flight instructor's counter examples may not be related to his feedback but may be solely due to regression to the mean.*

One-sample pre-post studies with population selected by their pretest values are highly vulnerable to regression to the mean.

Example 3.10 *Blood pressure is variable over time, and there is measurement error as well. Thus, if we select patients based on having high systolic blood pressure (say greater than 130), and we measured them one week later, the mean systolic blood pressure would likely have dropped with no intervention. Had there been an intervention in all of these patients, we might have wrongly concluded that the intervention led to this decline.*

The original example (and one of the first uses of the term *regression*) comes from Galton (1886), who noted that children of particularly tall parents tended to be shorter than their parents.

3.4 Designing an Experimental Study

3.4.1 Randomization

As discussed in Section 3.2.2, it is desirable to have the intervention assignment be independent of all potential outcomes. Since we almost never know all the factors that are related to the potential outcomes, the simplest way to achieve that goal is to use randomization. That is, we choose z_i as a random process independent of all data related to subject i. We can use computer-based pseudorandom number generators for this task.

The beauty of randomization is that we do not need to worry about any known or unknown relationships between the intervention assignment and the response. Here are some examples of those relationships that might not be obvious during the design stage.

Example 3.11 *Suppose the experiment requires performing one of two different interventions on each mouse from a set of 10 genetically identical mice all raised in the same cage. If the researcher applies the first intervention to the first five mice chosen by haphazardly reaching into the cage and picking one out at a time. The next five mice are chosen in the same manner and the second intervention is applied to each of them. Although all the mice appear the same (genetically identical, raised in the same cage, fed the same food, etc.), it could be that in the process of picking each mouse the slower mice have a higher probability of being selected earlier, so that on average the first intervention has slower mice. If the slowness is related to the health of the mice in a way that is also related to the outcome, then the apparent effect of the intervention could be entirely due to the differences apparent through the slowness.*

Example 3.12 *Suppose the experiment is for treatment of a disease in humans that come to a clinic, and half of the patients will get treatment A and half will get treatment B. A nonrandom but seemingly reasonable procedure is to switch treatments for each patient, so that the first patient gets treatment A, the second gets treatment B, the third gets treatment A, and so on. There is a problem if the researcher in charge of deciding who is the next patient knows which treatment is coming next and has a bias toward treatment A. The researcher may (consciously or unconsciously) look at the next patient and think, "This patient is very sick, I better make sure they get treatment A." This systematic treatment assignment may* *result in the group that gets treatment A being systematically sicker than those assigned to treatment B, and any differences in response could be due to the selection rather than the treatments.*

Randomization can be very simple (i.e., assign to either of two treatments with a flip of a coin) or it can be more complex in order to ensure some type of balance. For example, one approach is block randomization, where for each sequential set of four or six subjects that enter the study we randomly assign half to each treatment. With block randomization, we can ensure that the final sample size in both treatments is equal. Stratified randomization is like blocked randomization, except the blocks are defined by baseline covariates that may be important predictors of the response. Other more complicated randomization schemes are possible that either try to randomize more subjects to the better treatment (e.g., response-adaptive randomization) or balance very many baseline covariates (minimization). The more complicated the randomization, the more care must be made in the interpretation and analysis (Begg, 1990; Proschan et al., 2011).

3.4.2 Masking or Blinding

It is important that the intervention is the only thing that is substantially different between intervention groups. For example, you would not want to run all the assays for the primary endpoint in a study for one intervention group on one day, and then run all the assays for the other intervention group on a different day. Then you will not be able to tell if any significant effects you may see between groups are due to differences caused by the intervention or caused by differences in how the assays were done.

Even if the two intervention groups are handled and measured similarly after intervention, just grouping them together can affect the results.

Example 3.13 (Edgington, 1987)   *Consider a mouse experiment with two intervention groups. If after the intervention the mice are segregated into cages by intervention, this can inadvertently affect the results. For example, suppose that one mouse had an infectious disease. If all the mice from the same intervention group are housed together, then the group with the infected mouse may all become infected. Then there could be big differences in responses for the groups, but it may be due entirely to the infectious disease not the intervention. Troublingly, this problem could occur even if the disease is undetected. Similar problems could occur if one cage is closer to the heater or the air ventilator.*

Blinding or masking is when people involved in a study are unaware of which individual has gotten which intervention. For clinical trials (studies on people), a *single-blind* study is when only the subjects of the experiment are blinded to their intervention and a *double-blind* study is when both the subjects and the people carrying out the intervention (those with contact with the subjects) are blinded. Let us return to Example 3.12, where every other subject got treatment A. If that study design was changed to a randomized treatment allocation, the interpretation of the results would still be questionable if the researcher picking the next individual to enter the trial knew which intervention the next individual is scheduled to get, and if the researcher has a bias toward one of the treatments. Blinding avoids even the appearance of those kinds of biases. Although it may be much harder to accomplish, it is often useful to mask (or blind) researchers to which intervention each individual in an experiment received, from the time of intervention until the time that the final response is measured. If blinding is not done, some unintended effects may occur. For example, a researcher may be worried that one of the interventions causes nausea, and just the facial expression on the researcher as he asks a subject if she has any nausea may bias the results. The subject may think "Well, the doctor looks concerned. Maybe this does cause nausea. And now that I think of it, my stomach does not feel so well."

There is a common misconception that as long as the primary endpoint is objectively determined that blinding is not necessary. It is true that a subjective endpoint and lack of blinding is an especially bad combination. However, even if the endpoint were mortality, knowledge of the treatment assignment might affect other drugs provided to the patient leading to a potentially biased result of whether the test treatment affects mortality. In another example, consider a trial with one year of follow-up. If the researcher is worried about elevated levels of white blood cell counts for treatment A and consequently orders more blood tests only for those on treatment A, this more frequent testing could change subsequent care in those patients. The differences in outcomes between treatments could be due to more frequent blood tests, rather than difference in the treatment effects themselves.

3.4.3 Intent-to-Treat Principal

In Section 3.3.4, we discussed making sure you are measuring what you want. For experimental studies, there is an important special case related to a problem of estimating effects when not all individuals complete the study. Even for interventions that seem straightforward, the actual intervention may be more subtle.

Example 3.14 *Consider a trial where individuals are randomized to being offered daily treatment with a new drug or with a placebo for a chronic disease. Suppose the new drug has no effect except an unpleasant side effect that causes 40% of the subjects randomized to the new drug group to immediately stop taking the drug, and all of the subjects randomized to the placebo group continue taking their placebo pills. It could be that the 40% who discontinued treatment are more sensitive to pain. If the main response is pain level at the end of the study, and if the researcher only analyzes patients still taking their interventions at the end, it may appear that the new drug is having a beneficial effect. In fact, the entire effect could be due to the new drug effectively removing the subjects most sensitive to pain.*

In the previous example, the intervention is not the new drug or the placebo, but rather *to be offered* the new drug or the placebo. Although this effect may not be the one that is most scientifically interesting, it is the one that is easiest to measure the causal effect in a randomized trial. Thus, statisticians often use the *intention-to-treat* principle, and compare all subjects randomized to either group to measure this effect. The randomization is what ensures the unbiased estimation of the causal estimand. If you only compare individuals that complete the intervention, there may be systematic differences between those that complete the intervention and those that do not. This can bias the results. That is why it is critical to measure the response, when possible, even on subjects who are no longer taking the study drug. We explore this and related ideas in Section 15.4.4.

Note that even when you follow the intent-to-treat principle, randomize, blind, and treat all subjects the same after randomization, the effect you are measuring may still not be what you are interested in.

Example 3.15   *Malaria is a treatable disease caused by a parasite that is spread by mosquito bite, and a symptom of malaria is fever but not everyone infected with a malaria parasite has a fever. Consider a double-blind trial where subjects are randomized to either an experimental malaria vaccine or a control vaccine (such as a Hepatitis vaccine). All subjects are followed for the same period of time, and the endpoint is how many times a subject has a malaria parasite detected. You detect the malaria parasite if a blood sample shows the parasite, and you take blood samples at scheduled times throughout the trial plus additionally when subjects come into the clinic at unscheduled times with fever or other malaria symptoms. Suppose at the end of the trial, the control vaccine group has more malaria detected on average per individual. Because of randomization, we can infer that the malaria vaccine causes less malaria parasite detection. There are at least two plausible interpretations. First, the malaria vaccine is working and preventing some malaria. Alternatively, the two vaccines act the same with respect to preventing malaria, but the control vaccine is causing fevers (sometimes vaccines do cause fever reactions). Under the second interpretation, subjects in the control group are getting tested more often, and even though there is an equal chance of having parasites in the blood at any time, the control vaccine group has malaria parasites detected more often. This illustrates a vulnerability in the way that the primary endpoint was formulated.*

Although the intent-to-treat causal estimand is the easier one to study, sometimes it is important to estimate the effect of treatment when taken as directed. This is much more difficult to estimate, and we must use techniques developed for analyzing observational studies in order to account for the biases previously mentioned (see Hernán and Hernández-Díaz, 2012 and Section 15.4.3).

3.5 Designing an Observational Study

3.5.1 Types of Observational Studies

Observational studies can be purely for estimating population parameters. For example, the goal could be to find the population rate of a certain disease at a certain time frame. There is an entire theory of survey methods when estimating population parameters, where often it can be more efficient to over-sample from a minority in the population than to simply take a simple random sample from the whole population. This book has very little about survey methods.

We can also classify observational studies according to how individuals are followed over time. In prospective studies, we collect data at a baseline time and watch changes in individuals over time. In retrospective studies, we collect data about individual's past experience. In cross-sectional studies, we collect data at a single point in time.

Much of this chapter is about studies intended to determine causes of outcome; these studies require careful design. Since causal inference is about things from the past determining the future, the natural type of design for causal inference is a prospective study. In this case, we can see what is different about people at baseline and see if those differences are related to outcomes that will occur in the future. Even prospective studies do not automatically have causal interpretation (see Section 3.5.3). Further, it is sometimes helpful to use a retrospective study to estimate causal estimands (see Section 3.5.4).

3.5.2 Selection Bias

For an observational study, since the researcher is not intervening, great care must be made in deciding what individuals are included in the study. For example, when comparing two groups it is important to be aware of how the groups are assigned and whether this is related to the response of interest.

Example 3.16 *Consider the problem of determining whether private or public schools provide a better education in the United States. We cannot just compare the test results of students from the same areas in the same grade who go to the two different types of schools. Since public schools are free to all and private schools are not, the students who go to private schools may come from more affluent or more religious families than those who go to public school. The differences in test scores may be more related to the differences in the home life than in differences in the schools.*

Ellenberg (2014) describes a different sort of selection bias example. In fighter planes, there is a trade-off in how much armor to add, the more armor the safer, but also the less maneuverable and fuel efficient. So, one strategy is to armor only the part of the plane which is most effective for safety. During WW II, US military officers had data on planes from the war: the engine had the fewest bullets per square foot (1.1), the fuselage had 1.7, the fuel system had 1.6, and other parts had 1.8. The officers thought they should put armor on the parts that got hit the most (such as the fuselage) but wanted advice from an expert statistician. Their expert, Abraham Wald, (yes, the namesake of the Wald statistic) had the opposite conclusion. He figured that damage would be uniform over the plane. He also recognized that they were only seeing planes that returned from battle, and that the reason that planes were coming back with few hits on engines is that planes with hit engines were less likely to come back. So where the officers had seen the problem as which part of the plane is most vulnerable to bullets, Wald recognized that the planes that were "selected" (i.e., returned from battle) were different than the planes that were not selected, and that this selection process was the key to interpreting the data.

Ellenberg (2014) describes another selection bias example. In this artificial scenario, diabetes and high blood pressure are not associated in the full population of a town. But what if all patients with diabetes or high blood pressure in the town landed in the hospital, and nobody without these conditions was in the hospital? And, then what if we did a study of patients in the hospital? We would find that some fraction of patients with diabetes have high blood pressure, but that *all* nondiabetics have high blood pressure. Thus, our study would erroneously lead us to conclude that diabetes protects against high blood pressure. Obviously, this is an extreme and contrived example, but it illustrates how a spurious association might be observed purely due to the selection process.

3.5.3 *Matching on Covariates*

Suppose we are trying to study the causal effect of something, say a treatment, on a response. Let $z_i = 1$ if the ith individual got the treatment and $z_i = 0$ if not. One observational study design would be to collect pairs of individuals with one individual in the pair getting the treatment and the other not. The ideal situation is if the treatment allocation in each pair is determined by randomization, so that the z_i will be independent of all information (i.e., all variables, measured or unmeasured) prior to treatment. For an observational trial this is not possible, but we could have approximate independence. Let us introduce notation to be precise. Let Z_i be the random variable for treatment for the ith individual. Let $\mathbf{x}_i$ and $\mathbf{u}_i$ be the set of measured and (possibly unknown) unmeasured baseline variables for the ith individual. Let $\mathbf{y}_i(\cdot) = [y_i(0), y_i(1)]$ be the set of potential outcomes for the ith individual (see Section 3.2.2). The ideal pair of individuals, i and j, have

$$\Pr[Z_i = 1|\mathbf{y}_i(\cdot), \mathbf{x}_i, \mathbf{u}_i] = \Pr[Z_j = 1|\mathbf{y}_j(\cdot), \mathbf{x}_j, \mathbf{u}_j],$$

and the probabilities are not 0 or 1. In this ideal situation, because we have only one of the pair that was treated ($z_i + z_j = 1$), we can analyze the observational study as if we picked the member of the pair to get the treatment by the flip of a fair coin. Then after the study is done and we observe one of the potential outcomes from each individual, we can analyze the responses as if the data came from a randomized experiment.

The problem for observational studies is that we never observe $\mathbf{y}_i(\cdot)$ and $\mathbf{u}_i$ so we must make assumptions. One simplifying assumption is that given the measured covariates, $\mathbf{x}_i$, the treatment assignment, Z_i, is independent of $\mathbf{y}_i(\cdot)$ and $\mathbf{u}_i$. Under that assumption, an ideal (i, j) pair has

$$\Pr[Z_i = 1|\mathbf{x}_i] = \Pr[Z_j = 1|\mathbf{x}_j]. \tag{3.3}$$

If that independence assumption and Equation 3.3 are both true and $z_i + z_j = 1$ then we can still analyze the data as if it came from a trial with paired one-to-one randomization. One way to design a study like this is to only include pairs of subjects with $\mathbf{x}_i = \mathbf{x}_j$ and $z_i + z_j = 1$. Of course, the results of the study may be criticized for making an incorrect assumption, especially if the experts in the field know of an unmeasured covariate that is strongly associated with both treatment and response.

When there are many observed covariates (i.e., the dimension of $\mathbf{x}_i$ is large), then finding pairs with $\mathbf{x}_i = \mathbf{x}_j$ may be difficult. In this case we can match pairs that have similar values of $\lambda(\mathbf{x}_i) \equiv \Pr[Z_i = 1|\mathbf{x}_i]$. The scalar value, $\lambda(\mathbf{x}_i)$, is known as the *propensity score*. Typically, the propensity score is estimated by logistic regression. We can compare individuals with different values of z_i within each quintile of the estimated propensity scores (as long as there are enough individuals with each treatment value within each quintile) and get an overall causal effect by a weighted average of those five strata of scores (see, e.g., Rubin, 1997).

3.5.4 *Other Observational Study Designs and Analysis Issues*

Besides matching on covariates there are many other ways of designing an observational study if the interest is in causal effects. Section 15.6 gives a more detailed account. Here we list some ideas to give a feeling for the breadth of design choices.

In Section 3.5.3 we mentioned comparing weighted averages of differences within strata of individuals matched on covariates. Another strategy is to weigh each individual inversely by the propensity scores. In this case, responses from individuals that got the treatment despite being very unlikely to get the treatment (according to their propensity score) have a large weight, while responses from individuals that got the treatment and were likely to get it (based on their propensity score) have a relatively smaller weight. Similarly, the propensity scores can be used to weigh individuals that did not get the treatment. Then we can estimate the causal expected difference (e.g., Equation 3.1) in responses using *inverse probability weighting*. A simple version of this is

$$\hat{\Delta}_{ipw} = \frac{1}{n} \sum_{i=1}^{n} \frac{z_i y_i}{\hat{\lambda}(\mathbf{x}_i)} - \frac{1}{n} \sum_{i=1}^{n} \frac{(1-z_i)y_i}{1 - \hat{\lambda}(\mathbf{x}_i)} \tag{3.4}$$

(see, e.g., Lunceford and Davidian, 2004, p. 2941). Other inverse probability weighted estimators are given in Section 15.6.

When you are interested in the effect of a certain exposure on a rare disease, a useful design is a case-control or case-referent study. In this type of design, we may sample equally from those with the disease and those without to see if the two groups have similar levels of the exposure of interest. This may seem like it introduces selection bias (see Section 3.5.2), since we are over-sampling from those with the disease compared to their proportion in the population. Further, it appears we are estimating different probabilities than the causal effects of interest. The design will estimate the probability of exposure given disease or not (i.e., $\Pr[E|D]$ or $\Pr[E|\bar{D}]$) when the causal effects we are interested in are the probability of disease given exposure or not (i.e., $\Pr[D|E]$ or $\Pr[D|\bar{E}]$). It turns out that when we use the odds ratio as the causal estimand, then the proper causal odds ratio is estimable from this design (see Example 7.2 on page 107 and Section 15.6).

3.5.5 *Be Aware of How You Measure Responses*

As discussed earlier, in experimental studies, you should treat the groups similarly, when studying a difference between groups (Section 3.4.2). Similarly, for an observational study, you should measure variables from the groups similarly.

Example 3.17 *Baggerly et al. (2004) were unable to reproduce an exciting, published result that was able to distinguish sera from healthy women and women with ovarian cancer using protein patterns. They found that the ways of processing the protein pattern data was done differently in the normal and cancer groups. So the original "significant" effect was an effect of differences in processing, not the important difference between women with cancer and those without.*

For measuring effects, it is important to realize that sample size can be very strongly correlated with the variance of a measure. For example, if you look at rates of brain cancer by county within a state, it may be that the highest and lowest rates are from those counties with the smallest population. If just one person in a small county gets brain cancer (a rare type of cancer) this may give a very high rate, but if no one in that county got brain cancer then the rate would be 0. Especially when looking for extremes this is an important issue.

3.6 A Probability Model for the Study

3.6.1 Frequentist Inference

Statistical hypothesis testing is the backbone of the frequentist school of statistical inference. In the frequentist school, we ask: if we repeat our study, what proportion of the time will we reject the null hypothesis? For valid methods, if the null hypothesis assumptions are true, then the proportion of rejections will be bounded by the level α. Additionally, if the assumptions hold, then a valid 95% confidence interval will cover the true parameter at least 95% of the time if we repeat our study. Using a frequentist approach, background information may be used from any source to design a study and choose the test and hypotheses to be tested, but ideally once the study starts no further information outside the study is used to adjust the estimates. The goal of the frequentist analysis is not to get the best estimate of the study effect given all prior and outside information, but it is to set up a fair system so that the resulting estimate of the study effect uses only data from the study itself. The frequentist analysis wants each study to stand on its own as much as is reasonable. In contrast, a Bayesian analysis has different goals, it wants to combine the information from the study with the outside prior information to make coherent inferences (see Chapter 21). Another difference is that for the frequentist, the process that produces the data is important information, while for many Bayesians that process is not necessarily vital. Consider the following classic example.

Example 3.18 *Suppose we observe* 6 *successes out of* 20 *independent runs of an experiment. Can we reject the hypothesis that the probability of a typical run of the same type is less than* 0.5? *The frequentist must know how the data were collected to perform this test. If the experiment was designed to perform 20 runs regardless of the responses, then the distribution is binomial and the one-sided p-value is* $p = 0.058$, *while if the experiment was designed to keep performing runs until* 6 *successes, then the distribution is negative binomial and the one-sided p-value is* 0.032 *(see Chapter 18 for a much more general discussion of stopping rules). For many Bayesians (see Carlin and Louis, 2009, Section 1.4), but not all (see Gelman et al., 2013, Section 8.1), both experiments would be analyzed the same way (see also Example 21.4 in Chapter 21).*

As the example shows, for the frequentist, the stopping rule (when to stop sampling new data) is an important part of the study design. For medical experiments, often it is important to stop the study early if there is a large apparent effect. Chapter 18 discusses how to plan and carry out multiple testing over time allowing for early stopping and yet still get proper frequentist inference.

3.6.2 Independence and Dependence

The decision about the probability model for a scientific study is substantially impacted by the dependence between observations. Loosely speaking, two observations are independent when knowing the value of one does not tell you any information about the other. To get an intuition about this idea, let us look at some examples of dependence.

Example 3.19 *Suppose we are testing an experimental vaccine against an infectious disease in two villages. Suppose we vaccinate 100 people in each village and wait for a* *year to see how many of the vaccinated have gotten the disease by the end of the year. The complication is that because the disease is infectious, the probability that any one person in a village gets the disease is related to whether or not others in the village have gotten the disease. Because of this problem of dependence, sometimes in these types of trials, researchers will randomize villages instead of people (Halloran et al., 2010).*

Example 3.20   *Consider a randomized experiment comparing 2 treatment arms of 20 individuals each and the response is some cytokine in the blood. It is expensive to run the test on the blood, so the researcher pooled all the blood for each treatment arm together. Then the researcher tested each pool three times because the assay was not perfect. Suppose the researcher assumed that within each arm the three replicates are independent, and under this assumption found a "statistically significant" difference between the mean amount of cytokine in the two treatment arms with $n = 3$ in each arm. It is a mistake to infer that this statistically significant difference means that there is a statistically significant difference between the means in the cytokine in two arms of individuals, because the difference between the arms could be caused by a single individual who has a relatively large amount of the cytokine in his blood. Although there may be 40 independent individual blood samples, once the blood samples within each arm are pooled together, the independence between those individual samples within an arm is lost. Here the test is on replications of independent measurements of pools of blood, not on independent samples of blood from different individuals.*

Dependence can ruin an experiment in the sense that it can induce effects that look like the effect of interest (see Example 3.11 on page 33). Put another way, dependence can increase the probability of a false positive result, and the Type I error rate can be much higher than the nominal level. However, it is important to realize that as long as randomization is performed at the individual level (as opposed to at the community level) and patients are handled and evaluated independently of their random assignment, then the incidental dependence (e.g., from infectious diseases) does not always increase the Type I error rate. Proschan and Follmann (2008) show that in this situation, since the randomization is independent of all variables (including future infection clustering that can cause dependence), a permutation test that treats the individuals as independent bounds the Type I error rate under the null assumption that all interventions have identical effects. Further, even in this situation the two-sample permutation test on the difference in means is asymptotically equivalent to the pooled variance *t*-test (see Proschan and Follmann, 2008, p. 800, result 2). (In Chapter 10 we discuss permutation tests and in Chapter 9 compare permutation tests and *t*-tests.) Once again, randomization is very helpful for eliminating unforeseen problems that could have caused interpretational problems from an experiment.

3.6.3 Choosing the Probability Model

There are many concerns when choosing a probability model. In most cases in our work, the primary concern is that the assumptions of the hypothesis test are plausible and that given those assumptions the *p*-value is, at least, approximately valid (see Equation 2.3). The problem for the statistician is that although simpler convenience assumptions imply that a simple *p*-value is valid, those convenience assumptions may be less plausibly true in the application than more complicated assumptions which require more difficult testing methods.

Example 3.21   *Consider a randomized experiment comparing two treatments for malaria, where the response is binary, whether or not the patient had a positive blood smear for the malaria parasite on Day 28 after treatment. Suppose that 20 people were randomized to each treatment group, but 3 people in one treatment group did not show up for the Day 28 blood smear. A simple assumption is that the reasons those subjects did not show up on Day 28 has nothing to do with whether the subjects would have had a positive blood smear. A more plausible assumption is that if the subjects had very many parasites in their blood at Day 28, then they would be much more likely to show up on Day 28, since they would also be likely to be feeling sick. But the latter assumption is more difficult to formulate into a probability model from which we can calculate a valid p-value. (See Chapter 17 for handling missing data.)*

Part of the goal of this book is to arm researchers and statisticians with many tools in order to allow their probability models to get as complicated as needed in order for the associated assumptions to be more plausible to real-world applications. This book will provide ways of calculating valid or approximately valid p-values for many complicated probability models.

The desired plausibility of the assumptions does not mean that we need to think the null hypothesis is likely. Recall the assumptions of the hypothesis test posit two sets of models that are associated with the null and alternative hypotheses. So to say that the assumptions of the hypothesis test are plausible, means that one of the probability models (either in the null or alternative set) appears to apply to the data-generating process reasonably well. It is not necessary to believe that the set of null hypothesis models are equally as plausible as the set of alternative models, since the null set is often just a theoretical hypothetical representing the "no effect" model or the "suppose it is all noise" model. In other words, in some situations "no effect" is not that plausible, and that is okay.

For most applications the result of the hypothesis test inference will be three statistics: a p-value for the test; an estimate of the main parameter of interest; and a confidence interval on the parameter of interest. Sometimes only the p-value is needed, such as certain permutation tests (see Example 2.3 and Section 10.9), but in this section we focus on the typical case where the three statistics are used. Ideally, we want the p-value and the confidence interval to be compatible (see Section 2.3). For example, if the p-value for testing the null that the proportions of two groups' binary responses are the same is less than α, then we want a $100(1 - \alpha)\%$ confidence interval on the difference in proportions to exclude 0.

A problem is that we will often need to make more assumptions in order to calculate the confidence interval than to just calculate the p-value. In other words, we often need to add structure to the alternative hypothesis in order to create interpretable parameters and confidence intervals, and that added structure requires added assumptions. For example, to calculate a p-value for a Wilcoxon–Mann–Whitney test (a two-sample test based on ranks) we only need to assume that the distributions of the two groups are the same, but in order to give an estimate and confidence interval to accompany that p-value we need extra assumptions (see Chapter 9).

When we are interested in presenting estimates and confidence intervals with p-values, we have a choice about which parameter to estimate. For example, with the Wilcoxon–Mann–Whitney test, we could use the proportional odds parameter, a location shift parameter, or the Mann–Whitney parameter (see Chapter 9), each of which requires different sets of assumptions. One criterion for choosing the parameter is which model more accurately reflects the reality of the data-generating process, and less restrictive assumptions can of course be more plausibly applied. The accuracy criterion is often opposed to the simplicity criterion (simpler models are better). Generally, more complex models can better reflect reality, but may be harder to explain. Two guiding questions for balancing these two criteria in choosing a probability model are: who is my audience and is this model the best one to explain the science to that audience? In other words, if I have two competing models, does the added complexity of the more complex model help the audience understand the reality of what we should learn from the study better?

Sometimes the complexity is vital to the proper interpretation of the study.

Example 3.22 *Recall the presentation of the observational portion of the Women's Health Initiative in Table 3.1. If we did not break down the rates by time since starting hormone replacement therapy, then the one might infer from the study results that there is a benefit of hormones on coronary heart disease regardless of how long a patient has been on it.*

Sometimes the complexity of a model comes with added assumptions that are more restrictive. Other times the added complexity is not important for the main message.

Example 3.23 *Coulibaly et al. (2009) describe an experiment where subjects with a parasitic infection (M. perstans microfilaremia) were randomized to doxycyline or no treatment. After 12 months, the proportion of subjects that cleared the parasite from their blood was $\hat{p}_1 = \frac{67}{69} = 0.971$ for the doxycycline group and $\hat{p}_0 = \frac{10}{63} = 0.159$ for the no treatment group. The results could be expressed as rate ratios $\left(\frac{\hat{p}_1}{\hat{p}_0}\right)$ or odds ratios $\left(\frac{\hat{p}_1(1-\hat{p}_0)}{\hat{p}_0(1-\hat{p}_1)}\right)$. Using the former, the doxycycline group appears about 6 times better, while using the latter doxycycline group appears about 178 times better. The odds ratio has the advantage that the results have similar interpretation if we instead switch to measuring proportions of people with parasites detected and (to keep the same direction of effects) invert the ratios: for the rate ratios the doxycycline now appears about 29 times better, while for the odds ratios the doxycycline continues to appear about 178 times better. Despite this theoretical advantage, the odds ratio is not as easy to understand, so the rate ratios were used in the paper. The important point is that the doxycycline had a huge beneficial effect!*

Once the probability model for the data-generating mechanism is posited, then we can explore using one of several appropriate statistical methods (see Section 3.7). At this point, if the properties of the statistical method are not well known (or even if they are just not known to you), it is often useful to run some computer simulations to get a feel for how the methods will work. For example, suppose you believe there will be some correlation among the responses, and you are considering a method that is known to be asymptotically valid when there is correlation, but you are not sure if your sample size will be large enough to give approximately valid inferences. Then you can simulate data with different correlations and sample sizes under the null hypothesis and see what proportion of the time the considered method rejects the null hypothesis.

3.7 Statistics Methods Overview

Most of this book is about statistical methods. Those methods are used to calculate *p*-values and confidence intervals once you have a probability model and have delineated the null and alternative hypotheses. There are many ways to perform these calculations, and some are easier for different probability models than others. Here we present a broad overview of the methods, a more detailed overview is presented in Chapter 10. The presentation here is to give a flavor (not an exhaustive list) of the many techniques that are used for doing these calculations.

The simplest formulation has the probability model completely described by one scalar parameter. In this case, the probability calculations may be straightforward (see Example 2.1 and Chapter 4). One level more complicated is when the probability model is completely described by two scalars, with one parameter being the parameter of interest (i.e., the differentiation between the null and alternative can be described by partitioning possible values of this parameter) and the other being a nuisance parameter. The two-sample binary response model (see Chapter 7) or the two-sample t-test for normal data (see Chapter 9) are examples of this type of model. To calculate p-values in this situation, there are several tricks. First, we could condition on a statistic that will eliminate the nuisance parameter, so that a resulting conditional distribution on another statistic only depends on the parameter of interest, and that distribution can be used to calculate the p-value (for example, for Fisher's exact test). Second, we could base our inferences on a *pivot*, a statistic that does not depend on the nuisance parameter (for example, in the two-sample t-test assuming a shift in normal distributions). Third, we can sometimes calculate the p-value by going back to the definition of the p-value (Equation 2.2) and calculating the maximum of equal or more extreme values of a test statistic over all values of the nuisance parameter (for example, unconditional tests for two-by-two tables, see Chapter 7). The above three tricks give valid p-values regardless of sample size, but often we obtain approximately valid p-values using asymptotic methods.

There are several asymptotic methods based on the likelihood, which is defined by the probability model. The *likelihood* is a function of the parameters and the observed data, although many of the uses of the likelihood fix the observed data and treat the likelihood as a function of the parameters only. For discrete data the likelihood gives the probability of observing the data from the probability model. For continuous data, there are an infinite number of possible values and the probability for any one value is 0. To calculate meaningful probabilities for continuous data, we integrate the likelihood over possible regions in the sample space in order to give the probability that the response is in that region. Because of the likelihood's relationship to probabilities, one intuitive class of estimators of parameters, are the parameter values that maximize the likelihood for the observed data. These *maximum likelihood estimators* play an important role in much statistical theory.

The traditional asymptotic likelihood methods are based on independent observations, so that the likelihood is a product of the likelihood piece for each observation. Then the log likelihood is a sum, and we can use a central limit theorem to show that the certain functions of the maximum likelihood estimators (under the restrictions of the null or alternative hypotheses) are asymptotically normal. Three important tests to show asymptotic normality are the Wald tests, the score tests, and the likelihood ratio tests (see Section 10.7). These asymptotic methods provide tests for parameters of interest when there is a finite set of nuisance parameters. If the number of nuisance parameters increases with the sample size, then other methods need to be used such as conditioning on statistics so that the resulting conditional likelihood does not depend on the nuisance parameters (for example, in conditional logistic regression, see Chapter 14).

If the nuisance parameter is infinite dimensional, we call this a *semiparametric model*, since it has a parametric part (the parameters of interest) and a nonparametric part (the infinite dimensional nuisance parameter). For the semiparametric model conditioning can also work (for example, permutation tests). Another strategy for handling this model is to use a *profile likelihood*. The profile likelihood is a function of the parameter of interest and the data, where we replace the unknown nuisance parameter with an estimator that maximizes the full likelihood. It turns out that the likelihood ratio test can be asymptotically normal even for semiparametric models (Murphy and van der Vaart, 2000). A closely related method is the partial likelihood of Cox (1975), that is used in the proportional hazard model (see Chapter 16).

There are cases where we do not need to fully specify the likelihood and can still get asymptotically valid p-values. For example, if we only want to assume a model for the mean and a model for the variance, we can create a quasi-likelihood (for independent observations) or generalized estimating equations (for clustered observations). Even if we use a working model of the likelihood that is not correct (i.e., it is misspecified), then we can still get inferences that are asymptotically valid (see Section 10.8).

Another important technique for getting asymptotic inferences in complicated situations is the bootstrap. In the nonparametric version, we sample from the data set with replacement and recalculate a statistic of interest for each of those "bootstrap" samples. Parametric bootstrap inferences are also possible. Details are in Section 10.10.

There are many more techniques for getting inferences. The bulk of the book is a collection of these techniques applied to specific situations. Chapter 10 gives some theory for many of the important methods.

3.8 Summary

The scientific method is often concerned with explaining causality, but this concept is outside the realm of probability theory. That is, statistical hypothesis tests do not help in determining causality, as they only demonstrate association. Causality must be established through other information, such as study design. We classify studies into two types: observational studies or experiments, and experiments are much better than observational studies at determining causation.

We discuss the instructive example of the Womens Health Initiative, where the results of its randomized study showed a harmful effect of hormone replacement therapy in postmenopausal women on coronary heart disease, which was in complete opposition to conclusions drawn from many prior observational studies. (In fact, before the study was done, some questioned its appropriateness, given that the evidence of benefit from hormone replacement therapy on coronary heart disease from the observational data was so compelling.) We discuss a possible explanation that has been proposed for the contradictory conclusions by considering the varying effect of therapy on coronary heart disease as a function of how long the woman has been treated. We point out that this nuanced understanding almost certainly would not have been realized without the results of the randomized study that provided the true causal results.

A causal estimand is a parameter that measures a causal effect. Causal estimands are often expressed using potential outcomes. For example, each individual has two potential outcomes, the one if they had gotten the intervention and the one if they had gotten the control. Unfortunately, it is rare to be able to observe both potential outcomes in an individual. Nevertheless, in experiments, where we manipulate who gets what treatment, we can estimate the average causal effect across individuals with the difference between the group means. In observational studies this simple estimator may not work well. For example, perhaps sicker patients are more likely to be treated with a particular regimen, in which case a simple comparison of the group means would not lead to a good estimate of the average causal effect. Thus, if we want to make conclusions about causal effects of treatments from observational studies, we recommend using strategies that take observed covariates into account, such as propensity score matching, or inverse probability weighting; however, even these relatively rigorous methods require assumptions that are not verifiable.

We discuss considerations for good study design, whether experimental or observational. For example, it is critical to have a clear focus on the study question, and, if possible, to prespecify a primary endpoint and analysis to avoid the appearance of choosing the endpoint and analysis post hoc that provides the most interesting results. Excessive missing data on the primary endpoint can defeat the purpose of randomization, as there may be differential drivers for missing data by group, in which case we would no longer have two systematically balanced groups. Thus, in a randomized experiment, instead of eliminating patients who do not comply with their assigned treatment, we typically follow the intent-to-treat principle where all randomized patients are included in the primary analysis.

Broadly, there are several ways to design studies to estimate causation. Here we give a short list of study designs, generally from the designs that give the most confidence in the results to those that give the least.

(1) Randomized, blinded, experiments.
(2) Randomized, nonblinded experiments.
(3) Experiments that try to balance factors thought to be related to outcome without randomizing, or experiments with haphazard assignment to intervention groups, or with assignment seemingly independent of factors that influence the response (e.g., every other individual is assigned to one group).
(4) Thoughtful matched observational studies, or unmatched observational studies with careful analysis based on causal theory (e.g., propensity score matching with statements of the assumptions needed for causal interpretation).
(5) Observational study without examining treatment assignment mechanisms.

This order is not absolute. Even a randomized, blinded experiment can be so poorly implemented that it gives less reliable inferences than a well-done observational study.

3.9 Extensions and Bibliographic Notes

Pearl (2009b) is an important book on causality. It is quite theoretical but has some very readable extended discussions. For those just starting out, the Epilogue gives a nice introduction to the ideas, especially the fact that causality cannot be explained from within a standard probability model.

Mayo (1996) gives a modern view of the philosophy of science, defending the fact that scientists continue to use *p*-values and frequentist statistics more than Bayesian statistics, even though Bayesian statistics, with its updating of prior beliefs, seems like an ideal way to update knowledge with new data. For a discussion of the Bayes/frequentist controversy from the Bayesian perspective see Basu (1980) or Lindley and Phillips (1976).

Hand (1994) and discussants address "mapping from the client's domain to a statistical question."

For more on the issue of postmenopausal hormone therapy and coronary heart disease, see Hernán et al. (2008) who analyze the Nurses' Health study, an important observational study, to try and answer this question. The results generally agree with the Women's Health Initiative study (Prentice et al., 2005). Some of the causal examples come from work done on anthrax vaccines in animals (Fay et al., 2012). For studies of beta-catotene supplements and risk of cancers see Hennekens et al. (1996) and Druesne-Pecollo et al. (2010). For discussion of placebo effects see Finniss et al. (2010).

Rosenbaum (2002, 2010) are excellent books on observational studies with a focus on matching, propensity scores, sensitivity, and causal interpretation. Rubin (2006) collects many of the important papers by Rubin and colleagues about matching for estimating causal effects from observational studies (including the joint work with Rosenbaum introducing propensity scores, and some accessible reviews). Korn and Graubard (1999) is an excellent book on survey methods, especially appropriate for those with some prior non-survey-based statistical knowledge. See De Veaux and Hand (2005) for a discussion of Example 3.17 and other data error examples.

3.10 Notes

N1 **Frailty:** When the response is time to the first occurrence of an event (like coronary heart disease), then those that have the event are removed from the risk set, and typically the population hazard rate cannot be interpreted as individual hazard rates (hazard rates are instantaneous risks, see Chapter 16). To take a simple example, suppose that each person in the population has their own frailty parameter, with frail people having higher values and stronger people having lower values. The hazard rate at any time may be estimated as the expected proportion of the people at risk who will have the event. Suppose the population hazard rate over time increases until one year then declines after that. A naïve interpretation is that if you survive past one year then your hazard of having an event goes down. But that is interpreting the population hazard rate as an individual hazard rate. What is more likely happening is that the hazard for each person at risk continues to increase (after all, heart disease is a disease of older people), but the frail people, who are more likely to have the event early, are consequently more likely to be removed from the risk set early. This means that the population hazard goes down over time since only the stronger people have not had an event yet. For a discussion of this see chapter 6 of Aalen et al. (2008). Of course, in contrast, declining hazard rates could reflect that the risk is truly declining. This is best determined by constructing a Kaplan–Meier curve for each group. This is straightforward to compute from a randomized study. Some observational studies might have sufficient data to estimate the Kaplan–Meier curves; imagine a cohort study where each individual is followed from birth.

3.11 Problems

P1 Sometimes the statistician needs to take a request and reframe it into something that can be answered. Suppose you got the following email (this is similar to an actual request, but the details have been changed). "We have a study of 550 children. Using an expensive assay we have measured 20 of them, and all 20 have genetic trait G. To determine if all the children in the study have trait G, how many do we have to measure? We only want to look at a subpopulation, and how big should that subpopulation be?" The short answer might be, you need to run the assay on all 550 to be 100% sure that all 550 are have trait G. But that is clearly not the answer the researcher was looking for. It is often helpful to come up with a reframing of the question into a new question that you know how to answer. Reformulate that email into a question that can be answered statistically that gets at the scientific essence of the question. It does not have to perfectly match the original question, because this reformulating is just a way to get the conversation going with the researcher.

4

One-Sample Studies with Binary Responses

4.1 Overview

In this chapter we deal with measuring binary observations from a single sample. For Sections 4.2 and 4.3 we handle the case where we sample n independent observations from the same Bernoulli distribution. For example, we might measure a disease from a simple random sample of people from a village and be interested in the proportion of people in the village that have that disease. This is known as the *prevalence* of the disease. If the assay for measuring the disease is perfect (it always correctly differentiates people with the disease from those without), an appropriate analysis is the sample proportion with an exact 95% confidence interval on the population proportion using the Clopper–Pearson method (Section 4.2). We call this an *exact central confidence interval*, because the probability of an error on each side of the 95% confidence interval is bounded by 2.5%. Another binary example is when we have pairs, and one member of the pair is randomized to treatment arm A, and the other member to treatment arm B. If there are no ties, then each pair can be represented as a binary response: treatment A is preferred ($x = 1$) or treatment B is preferred ($x = 0$). We can assume independence if the preference for any individual is unlikely to be affected by the preference of another individual. In this case, we may be interested in testing the null hypothesis that both treatments are equally preferred. The exact central method of Section 4.2 is appropriate for that situation as well. For the independent binomial situation, the exact central method will meet most users' needs.

Section 4.3 explores several other methods for independent Bernoullis, with Section 4.3.1 giving reasons why each of the other methods might be useful. Section 4.4 gives several methods for calculating sample size for the independent binomial case, depending on the study purpose.

Section 4.5 summarizes the independent binary response cases, and Section 4.6 deals with some common non-independent binary response cases. Some examples are: surveys using sampling that is not a simple random sample; cases when the responses may be misclassified; and settings where we only observe binary responses pooled across groups. As an example of the last scenario, sometimes blood samples are pooled when testing for a rare disease, and we would like to make inferences about the prevalence using the results from the pooled samples.

4.2 Key Method: Exact Central Method (Clopper–Pearson)

We already have shown the derivation of the exact one-sided test for the binary response (Chapter 2, Example 2.1). That test is a key one for binary responses, and here we give the expressions for the associated p-values and confidence intervals.

Suppose we observe n independent binary responses, and the probability that each one has a positive response is θ. Let x be the number of positive responses out of n, and X the associated random variable. Then $X \sim Binomial(n, \theta)$. Typically, we either want to test a two-sided hypothesis,

$$H_0 : \theta = \theta_B$$
$$H_1 : \theta \neq \theta_B,$$

or one of these two one-sided hypotheses,

Alternative is less	Alternative is greater
$H_0 : \theta \geq \theta_B$	$H_0 : \theta \leq \theta_B$
$H_1 : \theta < \theta_B$	$H_1 : \theta > \theta_B,$

where θ_B is the parameter value on the boundary between the null and alternative hypotheses.

First, consider the one-sided hypotheses. The exact p-value associated with the alternative-is-less hypotheses is

$$p_{AL}(x, \theta_B) = \Pr[X \leq x | \theta_B], \tag{4.1}$$

and the one associated with the alternative-is-greater hypotheses is

$$p_{AG}(x, \theta_B) = \Pr[X \geq x | \theta_B]. \tag{4.2}$$

If we invert the hypotheses, we can get confidence limits. The $100(1 - \alpha)\%$ one-sided upper confidence limit is the value θ_B such that we just barely reject for the alternative-is-less hypotheses,

$$\theta_U(x, 1 - \alpha) = \{\theta_B : \Pr[X \leq x | \theta_B] = \alpha\},$$

unless $x = n$, in which case we let $\theta_U(x, 1 - \alpha) = 1$. Similarly, the $100(1 - \alpha)\%$ one-sided lower confidence limit is the value θ_B such that we just barely reject for the alternative-is-greater hypotheses,

$$\theta_L(x, 1 - \alpha) = \{\theta_B : \Pr[X \geq x | \theta_B] = \alpha\},$$

or $\theta_L(x, 1 - \alpha) = 0$ when $x = 0$. The confidence limits can be calculated by using the relationship between the binomial and beta distributions (see Note N1). The confidence limits have the properties,

$$\Pr[\theta \leq \theta_U(X, 1 - \alpha) | \theta] \geq 1 - \alpha$$

and

$$\Pr[\theta_L(X, 1 - \alpha) \leq \theta | \theta] \geq 1 - \alpha.$$

Typically, we combine the one-sided lower and upper confidence limits to get a two-sided confidence interval. It is straightforward (see Note N2) to show that a $100(1 - \alpha)$ two-sided confidence interval is $\{\theta_L(X, 1 - \alpha/2), \theta_U(X, 1 - \alpha/2)\}$ and that

$$\Pr[\theta_L(X, 1 - \alpha/2) \leq \theta \leq \theta_U(X, 1 - \alpha/2)] \geq 1 - \alpha. \tag{4.3}$$

This confidence interval is an exact central confidence interval and is called the *Clopper–Pearson* interval. The central *p*-value matches (and is compatible with) the Clopper–Pearson interval and is twice the minimum of the two one-sided *p*-values. Mathematically,

$$p_C(\mathbf{x}, \theta_B) = \min\left(1, 2p_{AG}(\mathbf{x}, \theta_B), 2p_{AL}(\mathbf{x}, \theta_B)\right), \tag{4.4}$$

and we reject $H_0 : \theta = \theta_B$ at level α if $p_C(\mathbf{x}, \theta_B) \leq \alpha$.

With an exact central confidence interval, whenever $\theta_L(X, 1 - \alpha/2) > \theta_B$ then we can reject the null $\theta \leq \theta_B$ at significance level $\alpha/2$, and similarly when $\theta_U(X, 1 - \alpha/2) < \theta_B$ we can reject $\theta \geq \theta_B$ at level $\alpha/2$. This is true because the confidence interval is just an inversion of the two one-sided hypothesis tests. Thus, as a shorthand, we call the exact one-sided tests and the exact central confidence intervals, the *exact central method*. This method is very useful and has some other nice properties. The confidence intervals are *nested*, meaning that the exact central $100(1-\alpha_1)\%$ confidence interval is a subset of the $100(1-\alpha_2)\%$ confidence interval whenever $(1 - \alpha_1) < (1 - \alpha_2)$. Additionally, the intervals behave reasonably when we add an additional data point. Let $\theta_U(x, 1 - \alpha; n) = \theta_U(x; n)$ be the exact central upper limit. Then we have $\theta_U(x; n + 1) \leq \theta_U(x; n) \leq \theta_U(x + 1; n + 1)$ for any $(1 - \alpha)$ level. In other words, adding a positive response cannot decrease θ_U and adding a negative response cannot increase it. The lower limit has an analogous property. Thus, for all these reasons, we recommend it as the default for most situations.

We give a brief overview of when other types of tests and confidence intervals are useful in Section 4.3.

4.3 Other Methods for Independent Binary Responses

4.3.1 Reasons for Using Other Methods

Now we outline some reasons why one would use other tests and confidence intervals for independent binary responses besides the exact one-sided tests and the Clopper–Pearson interval described in the previous section. We will refer to both these tests and their matching confidence intervals as exact central methods. Alternate tests address some shortcomings of the exact central methods, but for the applied researcher, those weaknesses may not be important. Further, addressing the shortcomings of the exact central methods introduces other disadvantages. Here are the details.

First, the exact central method can be very conservative. Specifically, the coverage (probability that the confidence interval covers the true value of θ) of the exact $100(1 - \alpha)\%$ central interval is larger than $(1 - \alpha)$ for nearly all values of θ to ensure that it is not less than $(1 - \alpha)$ for any value of θ. Conservatism is caused by the discreteness of the data. In addition, we have guaranteed that the error on each side of the exact central confidence interval is less than $\alpha/2$, which we think is important for clear interpretation when drawing any conclusion specific to a lower or upper limit. On the other hand, if we do not require the latter property then we can reduce the width of the confidence interval and still ensure at least $(1 - \alpha)$ coverage overall. This idea is discussed in Section 4.3.2. The problem is that when we make these adjustments, then we can have problems like nonnestedness and confidence limits that behave unreasonably when adding a new point (e.g., we add a negative response but the upper limit increases: $\theta_U(x; n + 1) > \theta_U(x; n)$).

To address the conservatism of the exact central interval, we may relax the exactness property and not require that the $100(1 - \alpha)\%$ confidence intervals have coverage of at least $(1 - \alpha)$ for all θ. We may pick intervals that "on average" over different values of θ have coverage of about $1 - \alpha$. This adjustment can be made by using the mid p-value for the hypothesis tests with the matching confidence intervals called *mid-p confidence intervals* and is discussed in Section 4.3.3.

We may also relax the exactness property and use confidence intervals and tests using approximations based on large sample theory, or asymptotics. These intervals are discussed in Section 4.3.4. There are several interesting tests based on normal approximations. These tests are approximations, are very easy to calculate, and tend to work well if the sample size is large. Since this situation is very simple, it is an easy example to demonstrate asymptotics, and hence the alternative tests are very useful for teaching large sample theory. An added benefit of using some of these approximations is that, like the mid-p confidence intervals, they may give coverage which is *on average* closer to the nominal level than the exact central ones. Further, since the intervals are simpler and have closed form expressions, they can be easily used to create simple sample size calculations (see Chapter 20). Finally, the asymptotic methods are faster to calculate. However, current computer speed generally makes this advantage irrelevant, so we typically recommend exact methods over asymptotic methods. The computational speed advantage of the asymptotic methods only comes into play when many intervals need to be calculated.

4.3.2 Two-Sided Noncentral Exact Methods

In Figure 4.1, we graphically explore the error from the 95% exact central confidence intervals for $n = 30$. Let that 95% interval when $X = x$ be $\{\theta_L(x), \theta_U(x)\}$. In the bottom panel, we see that the probability that the lower confidence limit is greater than θ is less than 2.5% for all θ. The value $\theta_L(30) \approx 0.884$, and for any $\theta > \theta_L(30)$ there is no possibility that the lower limit will be larger than θ so the error goes to 0. Also, because of discreteness, the error function is not a smooth function of θ. For the upper error, the error is 0 for $\theta < \theta_U(0) \approx 0.116$. When we add the two errors together, only in the middle parts does the total error (where the nominal is 0.05) get much above 0.025. So there is room for decreasing the width of the confidence intervals while still guaranteeing that the coverage is at least the nominal level, for those who are comfortable with allowing greater than 0.025 for either lower or upper error.

One method for decreasing the width of the two-sided interval is Blaker's exact confidence interval. The Blaker intervals have the nice property that for any confidence level, the Blaker intervals are completely within the exact central intervals. The errors for Blaker's intervals are given in Figure 4.2. The total error appears much closer to 0.05 for the Blaker than for the Clopper–Pearson intervals. But look at the actual intervals in Figure 4.3. We see that the Blaker intervals, although narrower, are not too much different from the exact central intervals (the average decrease in width is about 5.5%).

The error plots also illustrate that the Blaker procedure is exploiting the low error on the upper limit for low values of θ and vice versa. So, for values of $\theta < 0.10$, the Blaker method

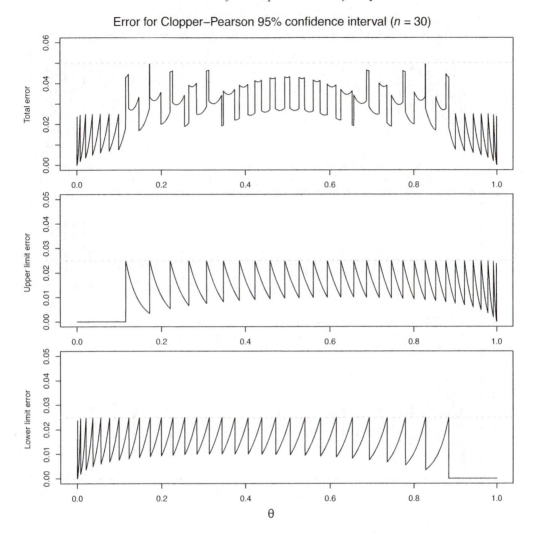

Figure 4.1 Probability of lower error ($\theta_L > \theta$), upper error ($\theta_U < \theta$), and total error (lower plus upper error) for the 95% exact central confidence interval for binomial response with $n = 30$.

is allowing essentially double error on the lower limit. While this is very appropriate if truly the only goal is total error under 0.05, if there is interest in one-sided inferences, then the exact central method is better. To see this, recall the result from page 51. For a $100(1 - \alpha)$ central interval, say (θ_L^c, θ_U^c), if $\theta_L^c > \theta_B$, then we can reject the null $\theta \leq \theta_B$ at significance level $\alpha/2$. In contrast, for a $100(1 - \alpha)$ noncentral interval, $(\theta_L^{nc}, \theta_U^{nc})$, if $\theta_L^{nc} > \theta_B$, then we can only reject the null $\theta \leq \theta_B$ at significance level α. Thus, if our interest is in one-sided inferences, then to put the intervals on equal footing (i.e., both have the one-sided errors bounded at α), we should compare the $100(1 - 2\alpha)$ central interval to the $100(1 - \alpha)$ noncentral interval. We expect that generally the $100(1 - 2\alpha)$ exact central interval will be narrower than the $100(1 - \alpha)$ (noncentral) Blaker interval.

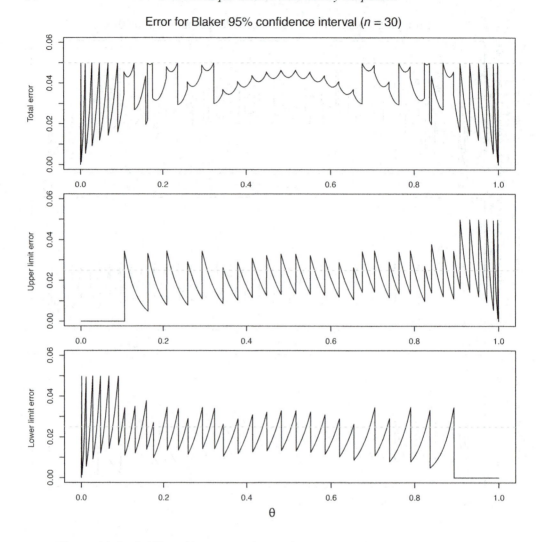

Figure 4.2 Probability of lower error ($\theta_L > \theta$), upper error ($\theta_U < \theta$), and total error (lower plus upper error) for the 95% Blaker's exact confidence interval for binomial response with $n = 30$.

We explain how to create a general two-sided decision rule based on an ordering of the data. Suppose we are interested in a two-sided hypothesis: $H_0 : \theta = \theta_0$ and $H_1 : \theta \neq \theta_0$. Suppose we define some function of the data, $T(\mathbf{X})$, such that larger values indicate more evidence against the null hypothesis (i.e., more "extreme" values). Then we can define the two-sided p-value using that function as

$$p(\mathbf{x}, \theta_0) = \Pr[T(\mathbf{X}) \geq T(\mathbf{x}) | \theta = \theta_0]. \qquad (4.5)$$

The decision rule is then $\delta(\mathbf{x}, \alpha) = I[p(\mathbf{x}) \leq \alpha]$, where $I[A]$ is an indicator function that is 1 when A is true and 0 otherwise.

We consider the binomial case with $n = 10$, $x = 3$, and $\theta_0 = 0.21$ to explore several ways of defining T. We plot the probability mass function in Figure 4.4. We consider two

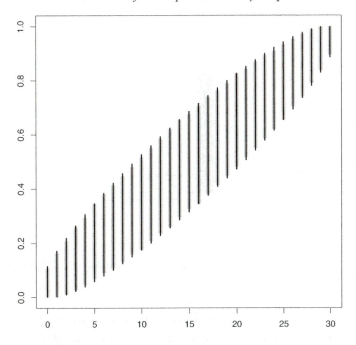

Figure 4.3 Blaker's 95% exact confidence intervals (thick gray) and 95% exact central intervals (thin black) for binomial response (x) with $n = 30$.

ordering functions, both based on measuring how likely observing **x** or "more extreme." First, define extreme by the probability mass function evaluated at θ_0, which we write as $\Pr[X = x|\theta_0] = f(x; \theta_0)$. In other words, let $T(x) = 1 - f(x; \theta_0)$. Then the p-value, the probability of observing X that is equal to or more extreme than x under the null, is

$$p_M(x) = \Pr[f(X) \le f(x)|\theta_0].$$

For the case depicted in Figure 4.4, $p_M(3) = \Pr[X \ge 3] + \Pr[X = 0] = 0.353 + 0.095 = 0.447$. This method was proposed by Sterne (1954) and is the default in the R function binom.test, and we call it the *minimum likelihood method*.

Blaker (2000) defined extreme in a different way. Let $F(x) = \Pr[X \le x|\theta_0]$ and $\bar{F}(x) = \Pr[X \ge x|\theta_0]$ define the lower tail and upper tail of the distribution from the observed value. Blaker's function, $T_B(x) = \min\{F(x), \bar{F}(x)\}$. The p-value is given by substituting T_B for T in Equation 4.5. Mathematically, it can be expressed as

$$p_B(x, \theta_0) = \min\{F(x) + \bar{F}(x^*), F(x^{**}) + \bar{F}(x)\},$$

where

$$x^* = \text{ is the smallest } u \text{ such that } \bar{F}(u) \le F(x)$$

and

$$x^{**} = \text{ is the largest } u \text{ such that } F(u) \le \bar{F}(x).$$

In words, Blaker's p-value is the sum of the smaller of the two tails plus a tail from the other side that is the largest tail value that is not larger than the smaller tail. For the situation

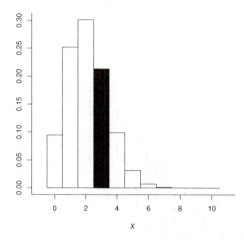

Figure 4.4 The probability mass function for a binomial with $n = 10$ and $\theta = 0.21$. The black rectangle is $x = 3$.

depicted in Figure 4.4, we have $0.353 = \bar{F}(3) < F(3) = 0.861$. Further, $F(0) = 0.095$, $F(1) = 0.346$, and $F(2) = 0.647$. So $p_B(3) = \bar{F}(3) + F(1) = 0.699$.

Either of the hypothesis tests associated with the p-values, $p_M(x,\theta_0)$ or $p_B(x,\theta_0)$, can be inverted to create a $100(1 - \alpha)\%$ confidence set,

$$C(\mathbf{X}, 1 - \alpha) = \{\theta : p(\mathbf{X},\theta_0) > \alpha\}.$$

These confidence sets are nested in the sense that $C(x,q_1) \subset C(x,q_2)$ for $q_1 < q_2$ and for all x (see Blaker, 2000, Lemma 1).

A problem with both the minimum likelihood method and the Blaker method is that the confidence sets created by inverting the p-value may not be a confidence interval. For example, when $x = 10$ and $n = 108$ the 95% confidence set from inverting p_M is approximately $\{t : t \in (0.0494, 0.1617) \text{ or } t \in (0.1636, 0.1653)\}$. A simple fix is to "fill in the hole" to create the matching confidence interval, so that for the example the 95% confidence interval would be $(0.0494, 0.1653)$. Unfortunately, this fix can lead to incompatible inferences between tests that use a p-value procedure and its matching confidence interval procedure. For example, if we test $\theta_0 = 0.163$ then $p_M(10, 0.163) = 0.04993$ but the 95% confidence interval contains 0.163. Despite this problem, a nice property of the Blaker confidence interval is that it is always a subset of the exact central interval (see Blaker, 2000, Corollary 1).

Here is another example where the confidence limits for both the minimum likelihood method and the Blaker method act unexpectedly (Vos and Hudson, 2008). Suppose we observe 7 successes out of 63 tries, then the 95% confidence interval that matches Blaker's exact method is $(0.0503, 0.2104)$. That data shows $1/9 = 0.111$ of tries are successes, but if we observe more data with the same proportion, 8 out of 72 tries, Blaker's 95% confidence interval is $(0.0495, 0.2047)$. Notice that the lower limit decreases when we get more data. This decrease happens for the minimum likelihood method also. This is non-intuitive, and an undesirable property. When we collect more data showing the same proportion, we might expect the lower limit to increase and the upper to decrease; in this case, we only see the latter. The same problem is evident if we look at the p-values from testing $H_0 : \theta = 0.05$ using Blaker's exact method. At the standard $\alpha = 0.05$ level, we reject for $7/63$ ($p_B = 0.037$), but fail to reject for $8/72$ ($p_B = 0.052$).

In Section 4.6 we discuss other two-sided noncentral exact methods proposed to minimize the confidence interval length.

4.3.3 Mid-p Methods

In Figure 4.1 we see that the lower and upper errors from the 95% exact central confidence interval are bounded by 2.5%, and the total error is very often much less than 5%, so the 95% confidence interval has coverage that is strictly larger than 95% for almost all θ. Suppose that we wanted a confidence interval that *on average* has about 95% coverage, where for some values of θ it has slightly less than 95% coverage and for other values of θ it has slightly more than 95% coverage. One simple adjustment is to a *mid-p* method to account for the fact that the responses are discrete. Whereas the typical exact p-value measures the probability of equal or more extreme under the null, the *mid p-value* measures one half the probability of equality plus the probability of more extreme. For example, the mid-p version of $p_{AG}(x)$ is

$$p_{AGmid}(x, \theta_B) = \Pr[X > x | \theta_B] + \frac{1}{2}\Pr[X = x | \theta_B]$$

and similarly, the mid-p version of $p_{AL}(x)$ is

$$p_{ALmid}(x, \theta_B) = \Pr[X < x | \theta_B] + \frac{1}{2}\Pr[X = x | \theta_B],$$

so that $p_{ALmid}(x, \theta_B) = 1 - p_{AGmid}(x, \theta_B)$. The mid p-value analogous to the central p-value is $p_{cmid}(x, \theta_B) = \min\{1, 2p_{ALmid}(x, \theta_B), 2p_{AGmid}(x, \theta_B)\}$. The $100(1-\alpha)\%$ mid-p confidence interval is $C_{cmid}(x, 1-\alpha) = \{\theta_B : p_{cmid}(x, \theta_B) > \alpha\}$. In this case, $C_{cmid}(x, 1-\alpha) = (\theta_{Lmid}, \theta_{Umid})$, where $p_{AGmid}(x, \theta_{Lmid}) = \alpha/2$ and $p_{ALmid}(x, \theta_{Umid}) = \alpha/2$. For $x = 2$ and $n = 10$ this 95% mid-p confidence interval is $(0.035, 0.520)$.

We see in Figure 4.5 that the lower limit error has the same jagged pattern as the exact central method, but the errors cross back and forth over the target of 0.025, instead of staying always below the targeted nominal error.

If we randomly select the θ value, then the expected coverage of the $100(1-\alpha)\%$ mid-p confidence interval may be about $1 - \alpha$; however, for frequentists, θ is fixed, so depending on the value of θ, the coverage may be slightly less or more than $1 - \alpha$. Thus, using mid-p methods when we are not interested in a specific θ may be acceptable in some situations. There can be a problem when the primary interest is testing a specific value of θ (as opposed to estimation of θ), and the researcher can reliably match the planned sample size. The following illustrates that problem.

Prior to conducting a study, if we fix n, θ_B (the value of θ to be tested), and α (the significance level), we can determine whether there is a difference in the threshold for significance between the exact central and the mid-p method. For example, if $n = 30$ and you wish to do a one-sided test of $\theta = 0.15$, you must have no events to have a significant (i.e., $p < 0.025$) result for both procedures. However, if $n = 31$, you can have up to one event for the mid-p, but still only 0 events for the exact central (note: in all cases, the decision rules for the two procedures are the same, or differ by only a single event). Thus, if the sample

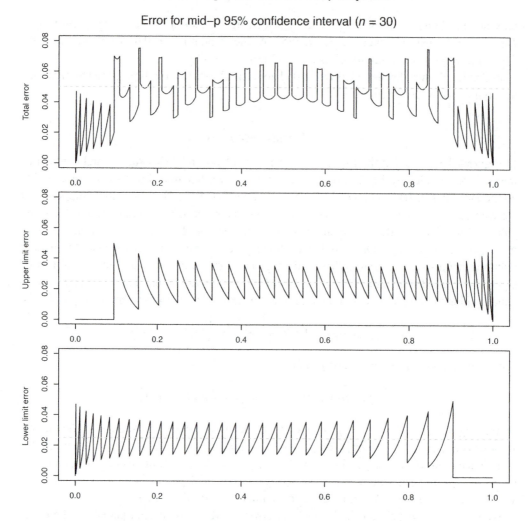

Figure 4.5 Probability of lower error ($\theta_L > \theta$), upper error ($\theta_U < \theta$), and total error (lower plus upper error) for the 95% exact mid-p confidence interval for binomial response with $n = 30$.

size were fixed in advance at 30 by some outside constraint, there would be no motivation to plan to use a mid-p procedure; in contrast if the sample size were fixed at 31, there would be motivation. In this constrained sample size setting, planning to use mid-p only when it makes a difference, will lead to an inflated Type I error rate, on average.

Now we consider the more common scenario where the investigator chooses the planned sample size. Figure 4.6 illustrates the Type I error rate and power of the one-sided mid-p procedure as a function of sample size for the test described above, where the assumed θ in the power calculation is 0.02. This one-sided test designed to demonstrate that $\theta < 0.15$, and the nominal Type I error is 0.025. As shown in the figure, a sample size of 31 appears to be the obvious choice because it has much more power than $n = 30$ and even more power than sample sizes of 32 through 35. However, the figure shows that the Type I error

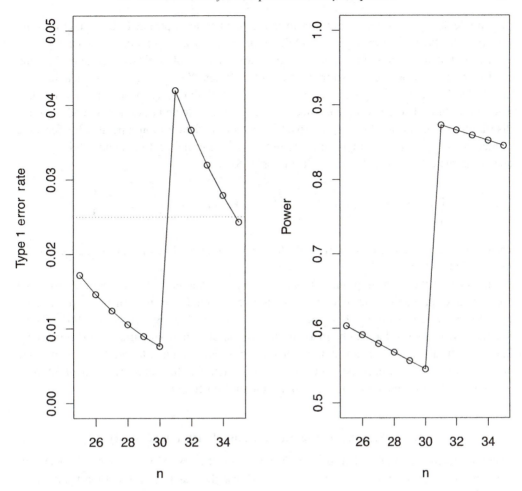

Figure 4.6 Type I error rate and power for mid-p procedure testing $H_0 : \theta \geq 0.15$ versus $H_1 : \theta < 0.15$, as a function of sample size; the power calculations assume $\theta = 0.02$.

rate for $n = 31$ is also at a local high (about 0.042, as opposed to the nominal 0.025). In fact, these plots illustrate, not surprisingly, that there appears to be a strong correspondence between Type I error rate and power. That is, selection of the sample size that corresponds with a local high in power is likely to be a case where the mid-p method is associated with an inflated Type I error, at least for modest sample sizes. Since there would be a natural inclination to pick such a sample size, we doubt that use of mid-p would truly lead to correct coverage on average.

4.3.4 Asymptotic Methods

With modern computers, the role of asymptotic methods is now less important than in the past, as the exact central method or its mid-p variant can serve most needs. Nonetheless, in this section, we talk about asymptotic methods: approximations that become better as the sample size gets large. In addition to being of historic importance, these methods are important for the insight they give about distributions of means with large samples.

The central limit theorem is an important idea for asymptotics (see Section 10.5) and the binary case is one of the simplest examples. When $X_1, \ldots, X_n$ are independent and distributed Bernoulli with parameter θ, then $E(X_i) = \theta$ and $Var(X_i) = \theta(1 - \theta)$ for all i. The central limit theorem says that since the variance is finite and does not depend on n, the distribution of the mean of the X_i approaches the normal distribution. Specifically, for large n, then $\bar{X} = \frac{1}{n}\sum_{i=1}^{n} X_i$ is approximately normal with mean θ and variance $\frac{1}{n}\theta(1 - \theta)$. So an approximate $100(1 - \alpha)\%$ confidence interval for θ is

$$\left(\hat{\theta} - \Phi^{-1}(1 - \alpha/2)\sqrt{\frac{\hat{\theta}(1 - \hat{\theta})}{n}}, \hat{\theta} + \Phi^{-1}(1 - \alpha/2)\sqrt{\frac{\hat{\theta}(1 - \hat{\theta})}{n}} \right),$$

where $\Phi^{-1}(q)$ is the qth quantile of the standard normal distribution. For example, $\Phi^{-1}(0.975) = 1.96$.

This normal theory asymptotic interval is an approximation for large sample sizes, but the coverage (probability of covering θ) of the 95% confidence interval may not be good even for large n: it is only 0.928 when $n = 98$ and $p = 0.2$ (Brown et al., 2001). There are many adjustments that can be made to this asymptotic approximation (see Brown et al., 2001 and Problem P3), but for applications, there is rarely a need to use these asymptotic approximations; if you need guaranteed coverage, then use the exact central method, or if you want 95% coverage on average, then use the mid-p version.

4.4 Designing a One-Sample Binary Response Study

There are many situations when one might want to test or estimate a proportion. Here we tell how to design a study for three situations. All use the exact central method for the final analysis.

4.4.1 Set the Power to Reject a Specific Proportion

Consider a situation where each binary measurement is a preference for one of two choices. Let the response for the ith individual be $x_i = 1$ if they prefer choice A and $x_i = 0$ if they prefer choice B. Let $\Pr[X_i = 1] = \theta$. Then we might want to test the simple two-sided null hypothesis that there is no difference in preference,

$$H_0 : \theta = 0.5$$
$$H_1 : \theta \neq 0.5.$$

In the simplest design, we sample n individuals then use the exact central test to see if we can show that $\theta \neq 0.5$ with level α. Because we are using the central test, this is similar to showing either $\theta < 0.5$ or $\theta > 0.5$ at the level $\alpha/2 = 0.025$. At the design stage we pick n so that at a particular point in the alternative space, for example $\theta = 0.7$, the power to reject the null hypothesis is larger than some value, typically 80% or 90%.

In Figure 4.7 we give the exact power to reject the null described by $\theta_B = 0.5$ when the true θ is 0.70. For each n, the power is the probability of rejecting the null hypothesis when

$\theta = 0.70$. When using the exact central method, we get nearly identical results whether we use the one-sided 0.025 level or the two-sided 0.05 level exact central test (see Note N3). Notice the jagged nature of the power curve. This is caused by the discrete nature of the problem. So the power at $n = 49$ is about 81% while the power when adding another observation decreases to about 78%. For planning a study with a targeted 80% power, you might want to use an $n = 54$, since all the higher n values have power greater than 80% (see Figure 4.7); however, it should be recognized that this sort of strategy is problematic if using the mid-p test (see Section 4.3.3). In some settings, achieving a precise planned sample size is not completely under the control of the investigator; for example, the outcome for one observation in the study might become missing over the course of a long follow-up.

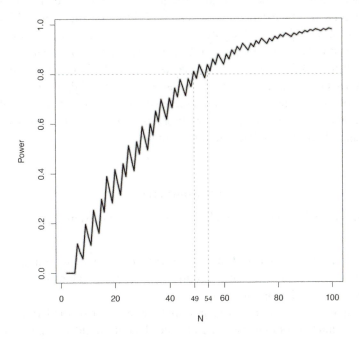

Figure 4.7 Power when $\theta = 0.7$ to reject the null (when $\theta_B = 0.5$) at the one-sided 0.025 level (thick gray line) or two-sided 0.05 (thin black line) using the exact central method. The two lines are so close that they are on top of each other. Dotted lines highlight the minimum N with power greater than 80% ($N = 49$) and the minimum N such that all greater than or equal to N have power greater than 80% ($N = 54$).

4.4.2 For a Given Sample Size, How Likely Are We to See an Event?

For a Phase I clinical trial, when a new drug is first given to humans, researchers typically give the new drug to a very small number of people, then if there are no apparent immediate serious adverse events, more and more people are tested on the drug. When designing such a study with n total individuals included by the end, it is useful to know how likely it is to see a serious adverse event out of those n individuals if the true proportion in the population was θ. This is useful for Phase II trials as well, that also study adverse events. This probability is $1 - \text{Pr}[\text{Observe 0 events}] = 1 - (1 - \theta)^n$. To find the minimum n such that the power to observe any serious adverse event is at least $1 - \beta$, let n be the smallest integer greater than or equal to $\log(\beta)/\log(1 - \theta)$. For example, in order to have at least a 90% power to see at least one event, when the probability of that event is 0.15, we need at least $n = 15$.

4.4.3 How Precisely Can We Estimate the Population Proportion?

Sometimes there is no clear hypothesis about the binary parameter θ, we just want to gather enough data to estimate it well. In this case one strategy for determining the sample size is to pick the sample size that will give an expected confidence interval length less than some predetermined value. For example, if one wanted the expected length of the 95% exact central confidence interval to be less than 0.10, then one needs to sample $n = 401$ or larger. This calculation is done by calculating the length of the confidence interval for all possible values of x out of n, and then taking the weighted average of those lengths with weights for each x proportional to the binomial probability mass function at that x value under the assumption that $\theta = 0.5$, since $\theta = 0.5$ appears to give the largest expected lengths (see Problem N4). This is one case where the asymptotic formula might be a reasonable tool for determination of sample size, and it is extremely simple to compute (even in the middle of a meeting without access to a computer, see Problem P1). In addition, one can very simply compute a graph of power as a function of n, or of θ.

4.5 Summary

We discussed the following two-sided procedures for one-sample inference of binary responses:

- Exact central method (Clopper–Pearson): Its p-value is twice the minimum of the two one-sided intervals. Its $100(1 - \alpha)$% confidence interval guarantees that the one-sided error of both lower and upper limits is $\leq \alpha/2$.
- Two noncentral exact methods: The minimum likelihood method (the default testing method in R function binom.test, whereas the default confidence interval is based on the exact central method) and Blaker's method. These two methods control the total error at α but not the lower and upper errors at $\alpha/2$. Thus, although the $100(1 - \alpha)$% Blaker confidence intervals are narrower than the $100(1 - \alpha)$% Clopper–Pearson intervals, if one-sided inferences are important, we should compare the $100(1 - \alpha)$% Blaker confidence intervals to the $100(1 - 2\alpha)$% Clopper–Pearson intervals (which tend to be narrower). Further, both noncentral exact methods have matching confidence sets that are not intervals, and hence their p-values are not compatible with their matching confidence intervals. In addition, both procedures can change in unsatisfying ways with the addition of a new observation.
- Mid-p method: While usual exact p-values represent the probability of observing equal or more extreme responses, the mid p-values represent the probability of observing more extreme plus one half the probability of equal responses. While the resulting mid-p confidence interval procedure does not guarantee coverage, *on average* over multiple parameter values, it can have closer to correct coverage than the exact central method. Briefly, exact confidence intervals have coverage greater than the nominal value for

almost all parameter values to ensure that the coverage is at least the nominal value for all parameter values, while mid-p confidence intervals have coverage that is either a little more or a little less than the nominal coverage depending on the parameter value.

- Asymptotic methods: These give often simple formulas and are approximations that work well for large sample sizes.

The exact central method, the mid-p method, and the asymptotic method all have natural one-sided analogues.

Authors' Take: In our era of modern computing, we see little role for asymptotic methods in this setting unless sample sizes are very large or very many tests are being done. We also believe the minimum likelihood method and Blaker procedure have too many disadvantages to be preferred in practice.

Therefore, we believe the exact central method and the mid-p procedure are the two best options. The advantage of the exact central method is that it has guaranteed protection of both upper and lower error rates. The advantage of the mid-p procedure is higher power and narrower intervals. While the error rates are not protected, researchers may argue that the mid-p approach is reasonable because *on average* it provides the correct Type I error rate.

Our personal inclination is to employ a valid test, so that when we reject the null, we have complete confidence in the inference, and thus we prefer the exact central method (i.e., Clopper–Pearson), in general.

A more neutral stance might be to use the exact central method only when precise control of the Type I error rate is deemed critical, such as in a trial to garner licensure of products by the Food and Drug Administration. Then, in other research settings, one might opt for the mid-p procedure. One practical consideration is that the Clopper–Pearson is a widely accepted procedure, and the mid-p is not currently in much use; but this may change in the future.

One thing to keep in mind, in any given study, the difference in the outcome of the decision rule of the two procedures (exact or mid-p) will only occur for a single borderline outcome, and in some settings, the decisions are identical. Furthermore, as sample sizes increase, these two methods will converge.

4.6 Extensions and Bibliographic Notes

Crow (1956) modified the confidence sets of Sterne (1954) to always produce intervals, creating a class of exact confidence intervals with the shortest length (i.e., the sum of the lengths of the $n + 1$ confidence intervals are minimized). Casella (1986), expanding on the work of Blyth and Still (1983), finds the class of exact shortest length confidence intervals with monotonically increasing (in x) lower and upper endpoints and that are equivariant to switching the definition of successes and failures. Schilling and Doi (2014) introduce the Length/Coverage Optimal confidence intervals that are members of the exact shortest length class of Casella (1986) that maximize coverage. A problem with all of these intervals is that they do not have the nestedness property (see Blaker, 2000, p. 789). In addition, none of these methods are central, i.e., ensure that error for both lower and upper limits are bounded by $\alpha/2$.

Most of this chapter has assumed that the samples of binary responses are independent. Sometimes they are not. A common example is in survey data, where the probability of observing some individuals may be higher than others. For example, when estimating prevalence of a disease one might plan to have a higher probability of sampling from some subgroup of the population that may be expected to have higher levels of disease than the general population. For inferences about the total population prevalence then the sampling scheme should be taken into account. Another survey technique is to sample from groups of individuals. For example, you might randomly sample households, then sample from within a household. The individuals within a household may be highly correlated with respect to the binary response (especially if the response is the presence of an infectious disease). So adjustments must be made for this correlation. For discussion of these and other issues with surveys, see Korn and Graubard (1999), a nice comprehensive book on statistical methods for surveys.

Another problem is that the binary response of interest may inherently be measured with error. For example, if you measure disease with an assay, the assay may not be able to correctly identify those with the disease all of the time. A simple model of the assay response is to assume that the sensitivity (probability that the assay will call a subject diseased, given the subject does have the disease) is a constant (say β_{Se}), and to assume that the specificity (probability that the assay will call a subject free from disease given the subject does not have the disease) is another constant (say β_{Sp}). Then if one observes a proportion who tested by the assay to have the disease, say $\hat{\theta} = x/n$, then an adjusted estimate of prevalence is

$$\hat{\theta}_{adj} = \frac{\hat{\theta} + \beta_{Sp} - 1}{\beta_{Se} + \beta_{Sp} - 1}. \tag{4.6}$$

(see Problem P2). Reiczigel et al. (2010) discuss how to give exact confidence intervals for θ based on $\hat{\theta}_{adj}$ when β_{Sp} and β_{Se} are known, and Lang and Reiczigel (2014) gives approximate confidence intervals when they are estimated.

Another interesting example not covered in this chapter is estimating prevalence after pooling samples. For example, Anderson et al. (2011) estimated the prevalence of a particular parasite (Leishmania major) infection in a species of sand fly in some villages in Mali. To estimate the prevalence, it is easier to pool many sand flies together and test for any parasites in the pool. Then one can estimate the prevalence after adjusting for the pooling according to the methods of Hepworth (1996, 2005).

4.7 R Functions and Packages

The **binom.test** function in the main **stats** package gives the Clopper–Pearson confidence intervals, but the test it uses is the minimum likelihood one of Sterne (1954), which has several disadvantages as noted in Section 4.3.2. In order to get the recommended Clopper–Pearson interval and the p-value that has matching inferences in the two-sided case (i.e., the exact central method), use the **binom.exact** function in the **exactci** R package. The **tsmethod** argument to **binom.exact** additionally allows testing and matching confidence intervals for the minimum likelihood method and Blaker's exact method. The **midp** option for **exactci** allows testing and confidence intervals using the mid-p method. The **powerBinom** calculates sample size and power based on the methods of Section 4.4.

The epi.prev function in the epiR R package gives exact confidence intervals for binary responses measured with error as outlined by Reiczigel et al. (2010). The binGroup R package has the methods of Hepworth (1996, 2005) for pooled data.

4.8 Notes

N1 **Relationship of Binomial and Beta:** One can repeatedly use integration-by-parts to show the following relationship between the cumulative distribution functions for the binomial, $F_{bin}(x,n,p)$, and the beta, $F_\beta(x,a,b)$,

$$F_{bin}(x,n,p) = F_\beta(1-p,n-x,x+1).$$

N2 Start with the $100(1-\alpha/2)\%$ one-sided confidence limits, $\theta_L(X) \equiv \theta_L(X,1-\alpha/2)$ and $\theta_U(X) \equiv \theta_U(X,1-\alpha/2)$. The error bounds are:

$$\Pr[\theta > \theta_U(X)|\theta] \le \frac{\alpha}{2}$$

and

$$\Pr[\theta_L(X) > \theta|\theta] \le \frac{\alpha}{2}.$$

Typically, we want to show both confidence limits at the same time. Since $\theta_L(X) < \theta_U(X)$ then $\Pr[\theta_U(X) < \theta$ and $\theta < \theta_L(X)] = 0$, and

$$\Pr[\theta_L(X) > \theta \text{ or } \theta > \theta_U(X)] = \Pr[\theta_L(X) > \theta] + \Pr[\theta > \theta_U(X)]$$
$$\le \frac{\alpha}{2} + \frac{\alpha}{2} = \alpha.$$

Then multiplying the expression by -1 and adding 1 we get Equation 4.3.

N3 **Directional inferences:** Consider the power plot of Figure 4.7 for testing the null with the boundary parameter value $\theta_B = 0.5$. Although the power looks the same for the one-sided 0.025 level and the two-sided 0.05 level exact central test, it can be different. The two-sided test can count rejections in the "wrong" direction. For example, when $n = 6$ we reject at the two-sided 0.05 level if $x = 0$ or $x = 6$. For the power calculation when $\theta = 0.7$, then we usually do not want to count the rejection at $x = 0$, because it implies that $\theta < \theta_B$ when under the alternative of interest $0.7 = \theta > \theta_B = 0.5$. So the power of the two-sided test is $\Pr[x = 0|\theta = 0.7] = 0.0007$ larger than the one-sided test. Because we generally do not want to count those rejections in the wrong direction, it is preferable to estimate the power from the one-sided $\alpha/2$ level test, even if you are using a two-sided α-level test.

N4 **Maximization of average confidence interval length:** It is hard to show rigorously that the average exact central confidence interval length is maximized when $\theta = 0.5$. However, we can show that for large samples, using the normal theory approximation the 95% confidence interval length is approximately $2\Phi^{-1}(0.975)\sqrt{\theta(1-\theta)/n}$, where $\Phi^{-1}(0.975) = 1.96$ is the 0.975th quantile of the standard normal distribution, which is maximized at $\theta = 0.5$. Further, we can program the average confidence interval length and try some values to see that it is maximized at 0.5.

4.9 Problems

P1 Show that the sample size that gives an average 95% confidence interval length of q is about $(2/q)^2$ when $\theta = 0.5$. So if $q = 0.10$ then $n \approx 400$. Hint: use the normal approximation with $\Phi^{-1}(0.975) \approx 2$.

P2 Motivate Equation 4.6. Hint: write out the probability of observing a positive test result in terms of the sensitivity and specificity and the prevalence in the population, then solve for the prevalence.

P3 **Wilson interval:** The Wilson confidence interval for a binomial uses a score test, which is a very general method (see Section 10.7). For the binomial case with $X \sim Binomial(n, \theta)$, consider testing $H_0 : \theta = \theta_0$. Then under the null hypothesis using asymptotic normality, we have $\hat{\theta} = x/n$ is approximately normal with mean θ_0 and variance $V(\theta_0) \equiv \theta_0(1 - \theta_0)/n$. The Wilson $100(1 - \alpha)\%$ confidence interval is the set of θ_0 values that would fail to reject $H_0 : \theta = \theta_0$ at level α, or the set that solve

$$\frac{\left|\hat{\theta} - \theta_0\right|}{\sqrt{V(\theta_0)}} < \Phi^{-1}(1 - \alpha/2),$$

Write this confidence interval in closed form. Answer: found in Agresti and Caffo (2000).

P4 Reproduce the power calculations of Figure 4.7. Show that the difference in power between the one-sided 0.025 level and two-sided 0.05 level exact central tests gets very small as N gets large.

5

One-Sample Studies with Ordinal or Numeric Responses

5.1 Overview

In this chapter we consider one sample tests on ordinal (i.e., ordered categorical) or numeric data. Both of these types of data have a natural ordering. An example of ordinal data are answers to a survey such as: *Extremely Dislike, Moderately Dislike, No Preference, Moderately Favorable,* and *Extremely Favorable.* For ordinal data, we can assign an order to different responses, but we will not use information about the difference between responses. For example, we will not assume the difference between *Extremely Dislike* and *Moderately Dislike* is the same size as the difference between *Moderately Dislike* and *No Preference.* If the categories have numeric scores and the differences between scores have inherent meaning we call the data *numeric.* We also use numeric to refer to discrete, continuous, or nearly continuous data represented by numbers.

In this chapter we are primarily concerned with tests on measures of central tendency, such as the median and mean. See Chapter 19 for testing if data from one sample are distributed as a member of a certain class of distributions (like a normal distribution). For one-sample tests that assume symmetry, see Chapter 6, since differences in paired responses are a straightforward motivation for the symmetry assumption. For symmetric distributions on numerical data the mean equals the median, so the discussion on choosing between the two under symmetry is given in Chapter 6.

When we cannot assume symmetry, when do we use the median and when the mean? First, consider the advantages of the median. For ordered categorical data the mean is not well defined, so we use the median. If you have numeric data, but only care about the ranking of the responses, then the median is appropriate since it can be calculated only using the ranks. Because of this, the median is invariant to monotonic transformations. For numeric data where you care about the actual values not just the ranks, the choice is less straightforward. The median is preferred when you do not want very extreme values to have a large effect on your parameter.

Example 5.1 *Suppose we are measuring income for a population. We do not typically use mean for summarizing income because the distribution tends to be very skewed, with a small percentage of the population often making a very large amount of money. If a billionaire moves into a middle-class neighborhood, the mean income of the neighborhood will change drastically, but the median income will hardly change at all. The median provides a more stable summary of the income distribution that does not change a lot based on one or two extremely large values.*

Sometimes in this situation with a small percentage of very large responses, we use the geometric mean. This is the same as taking the mean of the (natural) log transformed data and then taking the antilog (e.g., exponential) of that mean. For example, geometric means are often used with antibody measurements in the blood.

The median is also useful if it is hard to measure extreme values, since we do not need to know the extreme values to estimate the median.

Example 5.2   *Suppose we follow n individuals from time of diagnosis with cancer until death. We may be interested in the median survival time. We cannot calculate the sample mean time from diagnosis until death unless we observe the death time on everyone; however, we can calculate the sample median if we observe deaths in at least half of the sample. For example, if we follow all subjects for at least two years after diagnosis, and if at least half of the death times occur before two years, then we can calculate the sample median time from diagnosis until death, since that value is the same regardless of how we define the values larger than 2. (See Chapter 16 for handling the case when some individuals drop out of the study before the median death time.)*

Example 5.3   *Suppose we measure the viral load in a cohort of patients with HIV. Some patients may have values below the limit of detection, which might be 40 copies/ml in one laboratory, for example. As long as the median value is above this threshold, we can calculate the sample median, but we cannot calculate the sample mean without arbitrary imputation for values below the limit.*

The mean is useful because of its simplicity. Sometimes you want your summary to account for the very large values.

Example 5.4   *Suppose you are interested in how much you spend each month on electricity. If every year you spend much more in August because you run your air conditioning much more during that month, then the value for August is an outlier. Although a median monthly expenditure might be closer to what you actually spend in most months, that is not the important statistic for planning your yearly expenditures. For that, the mean monthly expenditures easily translate to yearly expenses.*

A problem is that the mean can be affected by a very small percentage of the population or sample being very extreme, and so we need to make some assumptions about the distribution of the responses if we are to make inferences about the mean. This is discussed in Section 5.3.1. Despite this problem, the test derived assuming normally distributed responses, the t-test, has good properties when the sample size is large or the distributions are approximately normal (Section 5.3.2).

An important class of distributions are the Poisson distribution and over-dispersed versions of it. These distributions have the variance depend on the mean, and there are special methods to handle this situation. In Section 5.3.3 we talk about the types of data that may lead to distributions of this class. We only briefly discuss how to handle this type of data and delay the major discussion of it until Chapter 10.

Finally, we discuss measures of dispersion such as the variance (Section 5.4.1) and the coefficient of variation (Section 5.4.2). For these methods we make the convenience assumption of normality on the responses for inferences. Convenience assumptions are made to make inferences tractable, but they are not really expected to be true. (When convenience assumptions are made, there is usually room for improved statistical methods in the future.) We mention more robust measures of dispersion in Section 5.7.

5.2 Exact Test on the Median or Other Quantiles

We motivate an exact test on the median that requires no assumptions on the distribution of the observations, except that the observations are orderable (i.e., ordinal or numeric) so that the median is defined. Suppose the data are ordinal with k possible categorical responses that are ordered. Let X_i^* be a $k \times 1$ vector, with the jth element being an indicator function giving 1 if the response was in the jth category and 0 otherwise. So that we can treat numeric and ordinal variables similarly, we reexpress X_i^* by assigning integers to the categories. Let

$$X_i = \begin{cases} 1 & \text{if } X_i^* = [1,0,0,\dots,0] \\ 2 & \text{if } X_i^* = [0,1,0,\dots,0] \\ \vdots & \vdots \\ k & \text{if } X_i^* = [0,0,\dots,0,1]. \end{cases} \tag{5.1}$$

For example, *Extremely Dislike* is 1, *Moderately Dislike* is 2, etc. Then we let $X_1,\dots,X_n$ be random variables for either type of orderable (ordinal or numeric) responses. Suppose the n responses are independent orderable responses, and $\mathbf{x} = [x_1,\dots,x_n]$ are the observed responses. Let the distribution associated with X_i be F. Let θ denote the median of F. The *median* is the value θ such that $F(\theta) = \Pr[X \le \theta] \ge 0.5$ and $\bar{F}(\theta) = \Pr[X \ge \theta] \ge 0.5$. Strictly speaking we should say "a median," since the median may not be unique. We avoid this technicality by defining the median as the middle of the interval of possible medians. For example, if any value between 2 and 3 is a median, we define "the" median as 2.5 (or if a median is between *Moderately Dislike* and *No Preference*, we define it as a new category: *Moderately Dislike/No Preference*). Here are two equivalent ways to write the one-sided hypotheses to show the alternative that the median is greater than θ_B,

Median is greater	Distribution is less
$H_0 : \theta \le \theta_B$	$H_0 : F(\theta_B) \ge 0.5$
$H_1 : \theta > \theta_B$	$H_1 : F(\theta_B) < 0.5,$

$$\tag{5.2}$$

where θ_B is some predefined value for the median on the boundary between the hypotheses.

For testing these one-sided hypotheses, we use the number of observations less than or equal to θ_B, $y(\theta_B)$, where $y(\theta_B) = n\hat{F}(\theta_B) = \sum_{i=1}^n I(x_i \le \theta_B)$, with $I(A) = 1$ if A is true and 0 otherwise. The boundary null hypothesis model, the one that is closest to the alternative, has $\theta = \theta_B$ and $F(\theta_B) = 0.5$, and $Y(\theta_B) \sim Binomial(n, F(\theta_B))$, where $Y(\theta_B)$ is the random variable associated with $y(\theta_B)$. We can test this null using the methods of Chapter 4, where the hypotheses are similar to the right-hand expression of (5.2) which test the alternative-is-less hypotheses on $F(\theta_B)$. So that the p-value for testing the alternative that the median is greater than θ_B is $p_{MG}(\mathbf{x}, \theta_B) = \Pr[Y(\theta_B) \le y(\theta_B); 0.5]$ (this is just a one-sided p-value for a binomial random variable see Equation 4.1).

We test the other one-sided hypotheses analogously. The hypotheses are

Median is less	$\bar{F}$ is less
$H_0 : \theta \geq \theta_B$	$H_0 : \bar{F}(\theta_B) \geq 0.5$
$H_1 : \theta < \theta_B$	$H_1 : \bar{F}(\theta_B) < 0.5.$

$$(5.3)$$

Specifically, we use $y^*(\theta_B) = n\hat{\bar{F}}(\theta_B) = \sum_{i=1}^{n} I(x_i \geq \theta_B)$, so that under the boundary null hypothesis $Y^*(\theta_B) \sim Binomial(n, \bar{F}(\theta_B))$ with $\bar{F}(\theta_B) = 0.5$ and $p_{ML}(\mathbf{x}, \theta_B) = \Pr[Y^*(\theta_B) \leq y^*(\theta_B); 0.5]$.

Now consider the two-sided hypotheses in both forms,

Median is not equal	In terms of F and $\bar{F}$
$H_0 : \theta = \theta_B$	$H_0 : F(\theta_B) \geq 0.5$ and $\bar{F}(\theta_B) \geq 0.5$
$H_1 : \theta \neq \theta_B$	$H_1 : F(\theta_B) < 0.5$ or $\bar{F}(\theta_B) < 0.5.$

$$(5.4)$$

We see that the right-hand expression is not so simple. Nevertheless, we can use the two one-sided p-values in the usual way (see, for example, Equation 4.4 where there p_{AL}, p_{AG}, and p_c refer to the p-value functions for the binomial tests) to get the central p-value, $p_c(\mathbf{x}, \theta_B) = \min \{1, 2p_{MG}(\mathbf{x}, \theta_B), 2p_{ML}(\mathbf{x}, \theta_B)\}$.

To calculate the exact central confidence interval, we just invert the central p-value. So the 95% confidence interval is the set of θ_B such that $p_c(\mathbf{x}, \theta_B) > 0.05$. We can rewrite it in a more intuitive form (see Note N1).

The median is the qth quantile of the distribution, where $q = 0.50$. For arbitrary $0 < q < 1$, a qth *quantile* is defined as any value θ such that $\Pr[X < \theta] \equiv F(\theta-) \leq q \leq F(\theta)$ (Problem P1). For continuous data quantiles are unique, but just as with the median, for discrete data quantiles are not necessarily unique. We can calculate similar tests and confidence intervals on other quantiles (Problem P2).

If you can assume that the data are symmetric, then more powerful tests on the median are possible, see Section 6.3.

5.3 Tests on the Mean

5.3.1 Why There Is No Nonparametric Test of the Mean

Let X be an random variable from the distribution, F, and let the mean of F be $\mu \equiv \mu(F)$. Consider the one-sided hypotheses,

Mean is greater
$H_0 : \quad \mu \leq \mu_B$
$H_1 : \quad \mu > \mu_B.$

$$(5.5)$$

Testing this simply stated hypothesis is not practically possible without making some assumptions about F. In other words, if you have a valid nonparametric test of the hypotheses (5.5) at the α level, then the power under any alternative cannot be substantially more than α. The issue is that there can be distributions under the null hypothesis that look like distributions under the alternative except for having a tiny probability of having a very small value which pulls the mean into the null space (see Note N2). In fact, even if we restrict the distributions F to the class of distributions with finite variance, the t-test has a size of 1 for every sample size, n (see Note N3).

If we add the assumption of finite variance for F, then we can show that for any fixed F with $\mu = \mu_B$ the size of the t-test approaches the nominal size asymptotically (see Lehmann and Romano, 2005, p. 445). However, this does not guarantee an asymptotically nominal size if we do not fix F as n changes. For example, we can show that for each n and $\epsilon > 0$ there exists an F_n with mean μ_B and finite variance, such that the size of the t-test is greater than $1 - \epsilon$ (see Lehmann and Romano, 2005, Theorem 11.4.3). Even with the finite variance, we still have the same issue as described above and in Note N2: a tiny probability with a very small value can cause problems.

If we assume normality of the distribution, then we do not have any of these problems. In fact, the one-sample t-test is exact and provides the uniformly most powerful unbiased test (see Lehmann and Romano, 2005, p. 87). If we relax the assumptions on normality and allow distributions that are skewed or have fatter tails than the normal, then the t-test may still be approximately valid. So we can often still recommend the t-test even if the data are statistically significantly different from the normal distribution. We discuss these issues in Section 5.3.2.

5.3.2 Key Test: t-test

Consider the hypothesis of expression (5.5) with the added assumption that F is normally distributed. Then the uniformly most powerful unbiased test for any alternative is the *one-sample t-test*.

One-sample t-test: Suppose the data are $x_1, \ldots, x_n$. Let $\bar{x}$ be the sample mean, and $s^2 = \frac{1}{n-1} \sum_{i=1}^{n} (x_i - \bar{x})^2$ be the (unbiased) sample variance. Then the alternative-is-greater p-value is

$$p_{AG}(\mathbf{x}, \mu_B) = 1 - F_{t,n-1}\left(\frac{\bar{x} - \mu_B}{\sqrt{\frac{s^2}{n}}} \right),$$

where $F_{t,n-1}$ is the cumulative distribution of a t-distribution with degrees of freedom equal to $n - 1$. To test the hypotheses with the alternative that $\mu < \mu_B$, the one-sided p-value is $p_{AL}(\mathbf{x}, \mu_B) = 1 - p_{AG}(\mathbf{x}, \mu_B)$ and the two-sided p-value testing with the alternative that $\mu \neq \mu_B$ is $p_{TS}(\mathbf{x}, \mu_B) = 2 \min(p_{AL}, p_{AG})$. The $100(1 - \alpha)\%$ confidence interval for μ is

$$\bar{x} \pm F_{t,n-1}^{-1}(1 - \alpha/2)\sqrt{\frac{s^2}{n}}, \tag{5.6}$$

where $F_{t,n-1}^{-1}(q)$ is the qth quantile of the t-distribution with $n - 1$ degrees of freedom. Note N4 outlines the derivation.

When the data are not quite normal, the test may still be approximately correct for large samples. In fact, the t-test will have asymptotically nominal size for testing the mean of a distribution F, as long as the variance of F is bounded away from 0 and $E(X^4)$ is

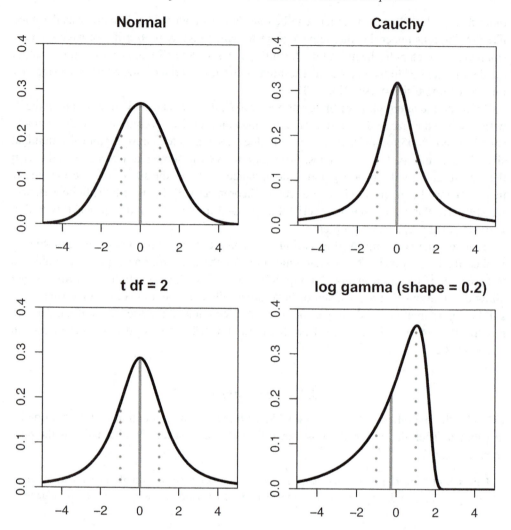

Figure 5.1 Four distributions used in the simulation. All distributions are scaled so that the 25th percentile and 75th percentile are at −1 and 1 respectively (vertical gray dotted lines). The means are the vertical solid gray lines (the means are 0 except for −0.266 for the log gamma).

bounded away from ∞ (see Note N5). To see how the t-test does for small or moderate sample sizes ($n = 20$, 50, and 200), we run some simulations. We consider four distributions: (1) a normal distribution; (2) a t-distribution with 1 degree of freedom, also known as the Cauchy distribution; (3) a t-distribution with two degrees of freedom; and (4) a log-gamma distribution with shape parameter 0.2. In Figure 5.1 we plot the distributions. Since the t-test is invariant to linear transformations (Problem P3), the simulation results will be identical for shifted and scaled results. For example, the normal simulation results hold for all normal distributions. We standardize the distributions as noted in the figure legend.

The simulation had 10^6 replications. In Figure 5.2 we see that the simulated size matches the nominal size for the normal as it theoretically should. For the Cauchy distribution, the

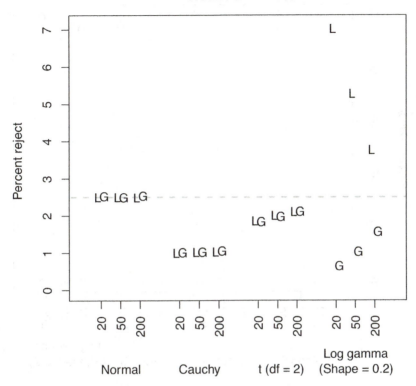

Figure 5.2 Simulated Type I error rate for testing the alternative that the mean is less (L) than $\mu_B = 0$ or greater (G) than $\mu_B = 0$. Numbers on the x-axis are sample sizes.

Type I error rate is less than the nominal, and because the variance is infinite, the error rates do not get better with larger sample size. In contrast, t with two degrees of freedom, which does have finite variance, has the Type I error rate getting closer to the nominal as the sample size increases. For the log gamma which is skewed towards the lower values, the Type I error rate for showing the mean is less than μ_B is inflated, while the test in the other direction is conservative. In both cases, the error rates approach the nominal as the sample size increases.

In Figure 5.3 we plot the power. The normal distribution is the most powerful as seen by its comparison to the two t-distributions (the comparison to the log gamma is not fair because the Type I error rate is inflated, and the mean is different from 0). The infinite variance of the Cauchy leads to quite diminished powers.

In summary, these simulations show that if the distribution is not normal, but still has a normal-looking shape and is symmetric, then the t-tests appear to bound the Type I error rate and can give reasonable powers. On the other hand, if the distribution is skewed, then the sample size should be large before using the t-test, since it may be anti-conservative for small samples. With large skewness, sometimes the median test is more relevant anyway. If your sample size is small and you have no knowledge about the distribution, then you may want to consider an alternate test.

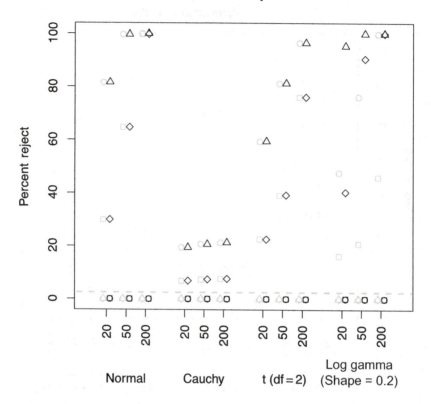

Figure 5.3 Simulated power for testing the alternative that the mean is less than (gray) or greater than (black) different values: -1 (circle), -0.5 (square), 0.5 (diamond), or 1 (triangle).

There are methods that are designed to be more nonparametric, such as the nonparametric bootstrap. We discuss this more in Section 10.10. We mention here that although several of the bootstrap methods do not do substantially better than the t-test for the skewed data simulation, the bootstrap-t interval does perform substantially better (see Note N6).

5.3.3 Expectation of a Single Rate or Count

Consider measuring a response that can be written in terms of a rate, $r = x/t$, where r is the rate, x is the count, and t is the units on which we are measuring. For example, x could be the amount of antibody per t units of blood. Or x could be the number of times one person got malaria when followed for t units of time. Under some basic assumptions, X, the random variable associated with x, is distributed Poisson. The Poisson distribution has a sample space of $\{0, 1, \ldots\}$, and has its mean equal to the variance.

Here are those basic assumptions. First, suppose we can assume homogeneity of the process that produces X. For example, suppose we got a 1 ml sample of blood from a 5 ml blood draw from a person. If we can assume that the count from that 1 ml will be expected to be (on average) the same as the count from another 1 ml aliquot from the same 5 ml draw, then the process is homogenous. Similarly, if we expect that the number of cases of malaria for a subject during the first 30 days of will be on average the same as the number from the next 30 days, then that process is homogeneous (that assumption may be less likely to be true in places with seasonal malaria). The second assumption (loosely speaking) is that we can measure the counts when the units approach infinitesimal amounts, and at those amounts the counts will be either 0 or 1. In the antibody example, this means that when we take smaller and smaller aliquot of blood from the 5 ml sample, the counts of antibodies in the infinitesimal amount of blood with be either 0 or 1. So the antibodies do not clump together. The assumptions will not typically be met exactly in practice but are often met approximately. We give a more mathematically precise description in Note N7. In general, for evenly mixed analytes that do not clump, or evenly distributed events in time where the probability of an event does not depend on the time since the last event and events happen one at a time, then the Poisson distribution is a good approximation for the distribution of the counts.

The exact central p-value and confidence interval for a Poisson distributed random variable are calculated using the same technique as for the binomial (see Equations 4.1 and 4.2 and the text following). Let the mean of the Poisson variate be $\theta = t\lambda$, and to show that $\theta < \theta_B$, the p-value is $p_{AL}(\mathbf{x}, \theta_B) = \Pr[X \le x | \theta_B]$. The one-sided $100(1 - \alpha/2)$ upper confidence limit is the value θ_U such that $p_{AL}(\mathbf{x}, \theta_U) = \alpha/2$. The probability distributions of the Poisson and the gamma distributions are related so that the confidence limits can be written as quantiles of the gamma distribution (see Note N8). If we want a confidence interval on the rate, we simply divide by the units, t. For example, θ_U/t is the upper limit for λ.

Example 5.5 *The global program to eliminate lymphatic filariasis (which may cause elephantiasis), is a mass treatment program where the total population of specific African villages are treated for filariasis. The treatment is typically not harmful even if the subject does not have lymphatic filariasis, but if the subject is infected with Loa loa parasites, the treatment can cause a bad immune reaction as the treatment kills those parasites. In one program, given in an area where Loa loa was endemic, 2 people out of 17,877 treated became comatose for several days after treatment. Using a Poisson assumption, we calculate a 95% confidence interval for the rate of this complication in this Loa loa endemic area: we estimate the rate as 11.2 per 100,000 (95% CI: 1.35 to 40.4 per 100,000).*

A more natural model might be the binomial model, which gives nearly the same confidence interval (in Example 5.5 the binomial confidence intervals are exactly the same to the first 2 decimals of the rate per 100,000). For small rates and large sample sizes, we can show the two methods are asymptotically equivalent (see Note N9). Because there is the close relationship to the binomial, it is not surprising that there are some of the same issues. For example, there are exact noncentral two-sided Poisson tests that can decrease the width of the $100(1 - \alpha)\%$ confidence interval at the cost of not guaranteeing the bounding of the error on each side at $\alpha/2$ (see Section 4.3.2 for discussion in the binomial case, and Casella (1989), Blaker (2000), and Fay (2010b) for the application to the Poisson case). Further, the mid p-values and mid-p confidence intervals can be calculated analogously to the binomial case (see Section 4.3.3). We do not discuss the details since they are very similar to the binomial case, but we do give the R functions in Section 5.8 to calculate the exact (central or noncentral) confidence intervals and optional mid-p adjustments.

5.3.4 Multiple Rates or Counts

Suppose we have a sample of n independent Poisson counts, $X_1, \ldots, X_n$, with the units for the ith count equal to t_i. Suppose $\mathrm{E}(X_i) = t_i \lambda$. Then $n\bar{X} = \sum_{i=1}^{n} X_i \sim Poisson\left(\lambda \sum_{i=1}^{n} t_i\right)$. So an exact central $100(1 - \alpha)$% confidence interval for λ is

$$\left(\frac{F^{-1}(\alpha/2; n\bar{X}, 1)}{\sum t_i}, \frac{F^{-1}(1 - \alpha/2; n\bar{X} + 1, 1)}{\sum t_i} \right),$$

where the summations are from 1 to n and $F^{-1}(q; a, b)$ is the qth quantile of a gamma random variable with mean ab and variance ab^2. One application of this is for indirectly standardized rates (Fleiss et al., 2003). Direct standardization is more complicated (see Section 12.3.6).

Often in practice we cannot assume that the λ are equal for all n. An alternative assumption is that the rates for each individual are random effects. In other words, the rate for each individual in the population varies around a true overall mean. This is a more involved type of analysis, but and the description of the R functions to handle this are given in Section 5.8.

Here are the details on the random effects model. Let $\mathrm{E}(X_i | W_i) = t_i W_i \lambda$, where W_i are independent random variables with mean 1 and variance a. In this case, without conditioning on W_i, we have $\mathrm{E}(X_i) = t_i \lambda$ and

$$\mathrm{Var}(X_i) = t_i \lambda + a t_i^2 \lambda^2. \tag{5.7}$$

Because the rates change for each individual, this induces overdispersion, which is variance that is more than you would expect from the Poisson model. The overdispersion as given in Equation 5.7 is called *quadratic* overdispersion. An alternative model is *linear* overdispersion, where $\mathrm{E}(X_i | W_i^*) = t_i W_i^* \lambda$ and W_i^* has mean 1 and variance $a^*/(t_i \lambda)$. This results in unconditional mean the same as in the quadratic overdispersion model, $\mathrm{E}(X_i) = t_i \lambda$; however, the unconditional variance is

$$\mathrm{Var}(X_i) = t_i \lambda + \left(\frac{a^*}{t_i \lambda} \right) t_i^2 \lambda^2$$

$$= (1 + a^*) t_i \lambda. \tag{5.8}$$

In the linear overdispersion model, the Poisson variance is multiplied by the overdispersion factor $\sigma^2 = (1 + a^*)$. Note that if $t_1 = t_2 = \cdots = t_n = t$ then the linear and the quadratic overdispersion models are just reparameterizations of the same model, with $a = a^*/(t\lambda)$.

A simple estimate of λ is the mean of the observed rates, $r_i = x_i/t_i$, say $\hat{\lambda}_0 = \frac{1}{n} \sum r_i$. The problem with that estimate is that each subject is weighed equally in the estimator, but some subjects have more variability in their contribution. A typical way to account for this variability is to use a weighted least squares estimator, with the weights for each individual inversely proportional to its variance. This does not apply straightforwardly since the variance is a function of the mean. A way to handle this situation is to use a quasi-likelihood (Chapter 10). Note that in general, the quasi-likelihood method does slightly better than the t-test (see Note N10). Perhaps the take-home message is that even in this case the t-test does adequately.

5.4 Testing Other Parameters from One-Sample Numeric Data

Throughout this section let $X_1, \ldots, X_n$ be random variables representing independent observations from a distribution F with mean μ and variance σ^2. Let $\bar{X}$ be the random variable for the associated mean and S^2 be the random variable for the unbiased variance estimator,

$$S^2 = \frac{1}{n-1} \sum_{i=1}^{n} \left(X_i - \bar{X}\right)^2.$$

The variance is defined as $\mathrm{E}\left\{(X_i - \mu)^2\right\} = \mathrm{E}(X_i^2) - \mu^2$. For making inferences about the variance there are similar problems as with the mean: a small portion of the population with very large values can make the variance very large, so we need to make assumptions about the distribution. In fact, the situation is worse for the variance than the mean, since the variance is the expectation of the **squared** distance from the mean. Since tests on the variance are more rare, we only consider the simple case where we can assume normality. When we cannot (even approximately) assume normality, we typically look for a transformation of the data that gives approximately normally distributed responses, then test the variance of that transformation, and then use an inverse transformation to return to the original scale.

Sometimes we are concerned only with how large the variance is with respect to the mean. In this case, we use the ratio σ/μ, known as the *coefficient of variation*, to measure the spread with respect to the mean. We use the standard deviation, σ, in the coefficient of variation instead of the variance, σ^2, because σ/μ is invariant to scale changes in the responses. For making inferences about the coefficient of variation, we again make the convenience assumption of normality. See Section 5.7 for other measures of dispersion.

5.4.1 Variance or Standard Deviation Assuming Normality

If we assume that F is normal, then

$$\frac{(n-1)S^2}{\sigma^2} \sim \chi^2_{n-1}. \tag{5.9}$$

where χ^2_{n-1} is a chi-squared distribution with $n - 1$ degrees of freedom. Let $q_{n-1}(a)$ be the ath quantile of that chi-squared distribution. Then a $100(1 - \alpha)\%$ confidence interval for σ^2 is

$$\left(\frac{(n-1)s^2}{q_{n-1}(1 - \alpha/2)}, \frac{(n-1)s^2}{q_{n-1}(\alpha/2)} \right). \tag{5.10}$$

These are the most accurate unbiased confidence intervals for σ^2 (see Lehmann and Romano, 2005, p. 165). The one-sided p-values to test the alternative $H_1 : \sigma^2 < \sigma_0^2$ is $p_{AL}(\mathbf{x}, \sigma_0^2) = F_{\chi^2_{n-1}}\left((n-1)s^2/\sigma_0^2\right)$, where $F_{\chi^2_{n-1}}$ is the cumulative distribution of a chi-squared distribution with $n - 1$ degrees of freedom. The other p-values are defined similarly.

Confidence intervals on the standard deviation, σ, are calculated by simply taking the square root of the confidence intervals on the variance.

5.4.2 Coefficient of Variation or Signal-to-Noise Ratio

Suppose that we are interested in the precision of an assay that measures some value that is positive. We want the variability of the assay to be small with respect to the mean. In other words, we want a small coefficient of variation, $\tau = \sigma/\mu$. The inverse of the coefficient of variation, $\theta = \mu/\sigma$, is called the *signal-to-noise ratio*.

Usually, we want to show that the coefficient of variation is small or equivalently the signal-to-noise ratio is large, written in hypothesis form as:

Coefficient of variation	Signal-to-noise ratio	
$H_0 : \tau \geq \tau_B$	$H_0 : \theta \leq \theta_B$	(5.11)
$H_1 : \tau < \tau_B$	$H_0 : \theta > \theta_B.$	

Usually, we are interested in distributions that measure nonnegative values, and the log-normal distribution is often a reasonable distribution for this type of data. If $X_1, \ldots, X_n$ are log normal, then $Y_1 = \log(X_1), \ldots, Y_n = \log(X_n)$ are normally distributed with mean ξ and variance v. Then

$$\mathrm{E}(X) = \mu = \exp\left(\xi + \frac{v}{2}\right)$$

$$\mathrm{Var}(X) = \mu^2 \{\exp(v) - 1\}$$

so that

$$\tau = \{\exp(v) - 1\}^{1/2}, \tag{5.12}$$

which is a function of v alone. So if all the data are positive, then we can get confidence intervals for v using the methods of Section 5.4.1 (use Equation (5.10) but replace S^2 with the sample variance of the Y_i values). Then we use Equation (5.12) to transform the confidence intervals to confidence intervals for τ.

Suppose some of the data have negative values, but we can still assume that $\mu > 0$.

Example 5.6 *Consider an assay that is expected to give nonnegative values, but due to measurement error or biological variation, some values may be negative. For example, consider a growth inhibition assay for malaria. We want to measure how much certain antibodies reduce malaria parasite growth in blood in vitro. Consider a test sample of antibodies targeted to attach to the parasite. To measure the growth inhibition, we see if parasites grown in blood with that test sample of antibodies slow parasite growth more than parasites grown in a control sample of blood. Let $X_i = 100\left(1 - \frac{M_{i,test}}{M_{i,control}}\right)$, where $M_{i,test}$ and $M_{i,control}$ are the measures of parasite growth in the ith test sample and the ith control sample. We want the assay to be able to detect growth inhibition ($\mathrm{E}(X_i) = \mu > 0$) when it exists. Biologically, we expect that the targeted antibodies will slow parasite growth more than the control antibodies, so $\mathrm{E}(M_{i,test}) < \mathrm{E}(M_{i,control})$, where here $\mathrm{E}(M)$ is a theoretical quantity denoting the expected value over an infinite number of repeats of the assay. But even though we may be able to assume $\mathrm{E}(M_{i,test}) \leq \mathrm{E}(M_{i,control})$, when we actually run the assay we can have $M_{i,test} > M_{i,control}$ for some i, so that $X_i < 0$.*

Assume that the X_i are independent normally distributed responses with $\mu > 0$. Let $T_n = \sqrt{n}\bar{X}/S$. Under normality of the X_i, T_n has a noncentral t-distribution with $n - 1$

degrees of freedom and noncentrality parameter $\sqrt{n}\theta = \sqrt{n}/\tau$. To test the hypotheses in expression (5.11), we reject when T_n is large. Let t_n be the observed value of T_n, then the one-sided p-value is

$$p_{AG}(\mathbf{x}, \theta_B) = p_{AL}(\mathbf{x}, \tau_B) = \Pr[T_n \geq t_n | \tau = \tau_B].$$

So a one-sided $100(1-\alpha)\%$ upper confidence limit is the smallest τ_B, such that $p_{AL}(\mathbf{x}, \tau_B) \geq \alpha$ which is the value τ_B where $p_{AL}(\mathbf{x}, \tau_B) = \alpha$. This can be found computationally by numerical search (for example using the uniroot function in R).

5.5 Designing a One-Sample Study

For designing a one-sample study for the median or the mean, see Chapter 20. For designing a study to bound the variance or the coefficient of variation, often the studies will be more complicated since there may be more than one source of variability. For example, if you want to validate a biological assay, you want the within-run variation to be low, but you also want the between-run variation to be low. See Fay et al. (2018b) for a specific type of validation study where the interest is in showing that the variance or coefficient of variation is bounded.

5.6 Summary

This chapter presents tests for one-sample data where the data are ordinal or numeric. When the data are either ordinal or ranked, only a median is applicable. Often we use a mean when the data are numeric. However even when data are numeric, the median may be preferable when data are highly skewed, have outliers, or are not measured well at the extremes. We present a simple exact method for testing the population median against some null value θ_B. A one-sided test that the median is greater (or less) than θ_B simply involves the binomial distribution where $p = 0.5$. So, for example, for the one-sided test, where the alternative is the median is greater than θ_B, we count the number of values in our sample less than or equal to θ_B. Under the null, this count is distributed Binomial $(n, 0.5)$. Thus, if $n = 20$, and there are three observations less than or equal to θ_B, the one-sided p-value is 0.001, and we conclude the median is greater than θ_B. The two-sided test is analogous, and the 95% exact central confidence interval is the set of θ_B with p-values greater than 0.05.

There is no nonparametric test of the mean. Fortunately, as we illustrate by simulation, as long as distributions are normal-looking in shape and not very skewed, the t-test appears to have good properties. On the other hand, if the distribution is skewed and the sample size is relatively small, it would be better to either transform the data before using the t-test, or perform a test of the median instead, which may be a more appropriate measure anyway given the skewing. We also describe the use of the Poisson distribution for testing the mean of a rate (i.e., a count divided by the number of units measured) and producing the corresponding confidence intervals. We extend this to the setting where we have a sample of n independent Poisson counts, and conveniently the sum of independent Poisson variables is still a Poisson variable. We also show R code relevant to the case where it is not reasonable to assume that each individual has the same Poisson rate, known as random effects. Finally, we present simple tests for other parameters. First, we provide a one-sample test of the variance; if data are not normal, we recommend transforming the data to achieve approximate normality. Second, we illustrate a test for the coefficient of variation.

5.7 Extensions and Bibliographic Notes

Koopmans et al. (1964) derive the confidence intervals for the coefficient of variance on the normal and log-normal distributions.

There are other measures of the spread of a distribution that do not require assumptions about the distribution. A common measure used is the *interquartile range* defined as $F^{-1}(0.75) - F^{-1}(0.25)$, where F^{-1} is the inverse cumulative distribution function. Another measure is the median absolute deviation from the median, which is the median of the distribution of $|X - median(X)|$. For some asymptotic results about that estimator see Serfling and Mazumder (2009). Some other measures used for income inequality are described in Dagum (2006).

5.8 R Functions and Packages

The t.test function that comes with base R performs both the one-sample and the two-sample *t*-test. The medianTest function in the asht R package, does the exact *p*-values and confidence intervals discussed in Section 5.2.

The poisson.exact function in the exactci R package can calculate the exact central Poisson confidence intervals and other exact intervals. It provides *p*-values and confidence intervals with matching inferences, while the poisson.test function in base R, does not have matching inferences (Fay, 2010b).

To make inferences about the mean for the quasi-Poisson model with a vector of data, x, of length n, with a vector of units t, use the following code (available in base R):

```
g<-glm(x~1+offset(log(t)),family="quasipoisson")
lambdahat<- exp(coef(g))
## Profile method for confidence interval
ci.95.lambda<-exp(confint(g,level=0.95))
## Wald method for confidence interval: it matches p-value
ci.95.Wald<- exp( coef(g) + c(-1,1)*summary(g)$coef[2]*qt(0.975,n-1) )
```

See Chapter 10 for a description of the generalized linear model, what the offset means, and the difference between the two confidence interval methods.

The var1Test and cvTest functions in the asht R package make inferences about the variance and the coefficient of variation, respectively, assuming normality of the responses.

5.9 Notes

N1 **Exact central confidence interval on median:** We give the exact one-sided lower and upper confidence limits for the median. Let the $k \leq n$ unique values of x_i be $t_1 < t_2 < \cdots < t_k$. We can write the lower limit of the $100(1 - \alpha)\%$ central interval (or equivalently the lower limit of the one-sided $100(1 - \alpha/2)\%$ interval) as

$$L_\theta(1 - \alpha/2) = \begin{cases} -\infty & \text{if } 0 \leq \alpha/2 < P_0 \\ t_1 & \text{if } P_0 \leq \alpha/2 < P_1 \\ \vdots & \vdots \\ t_j & \text{if } P_{j-1} \leq \alpha/2 < P_j \\ \vdots & \vdots \\ t_k & \text{if } P_{k-1} \leq \alpha/2 < 1, \end{cases} \qquad (5.13)$$

where $P_j = \Pr[Y(t_j) \leq y(t_j)]$ and $Y(t_j) \sim Binomial(n, 0.5)]$ and we let $t_0 \equiv -\infty$. Analogously, we can write the upper limit of the $100(1 - \alpha)\%$ central interval (the upper limit of the one-sided $100(1 - \alpha/2)\%$ interval). Let $\bar{P}_j = \Pr[Y^*(t_j) \geq y^*(t_j)]$ and $Y^*(t_j) \sim Binomial(n, 0.5)]$, with $t_{k+1} = \infty$ so $y^*(t_{k+1}) = 0$. Then,

$$U_\theta(1 - \alpha/2) = \begin{cases} t_1 & \text{if } \bar{P}_2 < \alpha/2 \leq \bar{P}_1 = 1 \\ t_2 & \text{if } \bar{P}_3 < \alpha/2 \leq \bar{P}_2 \\ \vdots & \vdots \\ t_j & \text{if } \bar{P}_{j+1} < \alpha/2 \leq \bar{P}_j \\ \vdots & \vdots \\ t_k & \text{if } \bar{P}_{k+1} < \alpha/2 \leq \bar{P}_k \\ \infty & \text{if } \bar{P}_{k+1} < \alpha/2, \end{cases} \qquad (5.14)$$

Fay et al. (2015, web Appendix G) show that expressions (5.13) and (5.14) are the inversions of the central median test procedure.

N2 **Why we cannot do nonparametric tests on the mean:** Here we show that hypotheses (5.5) cannot be tested without making some assumptions about F. We show this by assuming there is a valid decision rule, say $\delta(\mathbf{x}, \alpha)$, that has power substantially greater than α for some alternative, say F_1, then showing that this cannot be true. By assumption

$$\Pr_{F_1}[\delta(\mathbf{X}, \alpha) = 1] = 1 - \beta > \alpha,$$

where $1 - \beta$ is substantially greater than α. Let $F_1^{(\epsilon)}$ be a series of mixture distributions indexed by ϵ, that each have a mean μ_B. Specifically, suppose we sample data from F_1 (having mean μ_1) with probability $(1 - \epsilon)$ and sample data from G_ϵ with probability ϵ, with the mean of G_ϵ being $\epsilon^{-1}(\mu_B - (1 - \epsilon)\mu_1)$. Then $F_1^{(\epsilon)}$ is a distribution in the null hypothesis set, so for any ϵ,

$$\Pr_{F_1^{(\epsilon)}}[\delta(\mathbf{X}, \alpha) = 1] \leq \alpha,$$

by the validity assumption, but because as $\lim_{\epsilon \to 0} F_1^{(\epsilon)} = F_1$,

$$\lim_{\epsilon \to 0} \Pr_{F_1^{(\epsilon)}}[\delta(\mathbf{X}, \alpha) = 1] = 1 - \beta \leq \alpha,$$

which contradicts the assumption. Lehmann and Romano (2005, Theorem 11.4.3) discusses a similar idea with respect to the t-test.

N3 **Pointwise asymptotic validity:** If we add the assumption of finite variance for F, then we can show that for any fixed F with $\mu = \mu_B$ the size of the t-test approaches the nominal size asymptotically (see Lehmann and Romano, 2005, p. 445). A test that meets this condition is called *pointwise asymptotically valid*. However, this does not guarantee an asymptotically nominal size if we do not fix F as n changes. In other words, for each n and $\epsilon > 0$ there exists an F_n with mean μ_B and finite variance, such that the size of the t-test is greater than $1 - \epsilon$ (see Lehmann and Romano, 2005, Theorem 11.4.3).

N4 **Student's t-test derivation outline:** Many textbooks show the derivation of the t-test (see, e.g., Dudewicz and Mishra, 1988, Section 9.5). Briefly, using the notation of Section 5.4 and assuming normality, the derivation uses the following facts: (1) $Z = \sqrt{n}(\bar{X} - \mu)/\sigma \sim N(0,1)$, (2) $U = (n-1)S^2/\sigma^2 \sim \chi^2_{n-1}$; (3) Z and U are independent; and (4) $T = Z/\sqrt{U/(n-1)}$ is distributed t with $n-1$ degrees of freedom.

N5 **Uniformly asymptotically valid:** We call a test that guarantees asymptotically nominal size, a *uniformly asymptotically valid* (UAV) test. If we assume that $\text{Var}(X) \geq \epsilon > 0$, for some ϵ, and $0 < \text{E}(X^4) \leq B$ for some finite constant B, then the t-test is UAV. The proof of this is similar but simpler than Appendix C of Fay and Proschan (2010).

N6 **Bootstrap methods on mean:** We repeat the simulation of Section 5.3.2 on the log-gamma distribution with nominal error rates of 2.5%, but now evaluate the nonparametric bootstrap instead of the t-test. The nonparametric bootstrap is a flexible, robust method that is asymptotically valid in many situations (see Chapter 10). This simulation took much longer than the t-test simulation; we used 10^6 simulated data sets with 1,999 bootstrap replications on each. We first use the simple percentile confidence intervals and invert them to test the hypotheses. For the case with $n = 20$ we get similar simulated Type I error rates as seen for the t-test (for alternative: $\mu < \mu_B$ is 7.7% and for alternative: $\mu > \mu_B$ is 1.7%). If we use the bias-corrected and accelerated (BCa) percentile interval that adjusts for skewness we do slightly better (for alternative: $\mu < \mu_B$ is 7.0% and for alternative: $\mu > \mu_B$ is 2.1%). The best in terms of coverage is the bootstrap-t interval. For this we only use 10^4 simulated data sets (for alternative: $\mu < \mu_B$ is 4.0% and for alternative: $\mu > \mu_B$ is 1.7%).

N7 **Poisson process:** Consider counting events over time. Let X_t be the random variable denoting the number of events in $(0,t]$. Suppose X_t is defined for any $t \in (0,T]$. Suppose that $\text{E}(X_t) = t\lambda$. Then the stochastic process $\{X_t : t \in (0,T]\}$ is a Poisson process if the following are true:

1. For any two time intervals, $(a,b]$ and $(c,d]$ with $b \leq c$, then $X_b - X_a$ and $X_d - X_c$ are the counts in the two intervals, and those counts are independent.
2. The mean of $X_b - X_a$ only depends on the difference, $b - a$, not on the values a or b themselves.
3. $\lim_{b \to a} \frac{\Pr[X_b - X_a > 0]}{b - a} = \lambda$.
4. $\lim_{b \to a} \frac{\Pr[X_b - X_a > 1]}{b - a} = 0$.

The Poisson process has $\text{E}(X_b - X_a) = (b-a)\lambda$ and $\Pr[X_b - X_a = 0] = exp(-(b-a)\lambda)$. For a proof of this see Karlin and Taylor (1975, p. 22–26).

N8 **Poisson and gamma relationship:** Note that if $X \sim Poisson(\theta)$ and $G \sim Gamma(x+1,1)$ (so that $\text{E}(G) = \text{Var}(G) = x+1$), then $\Pr[X \leq x] = \Pr[G \geq \theta]$.

N9 **Binomial and Poisson relationship:** If the binomial parameters are n and λ/n, for fixed $\lambda > 0$ then as n goes to infinity, the distribution becomes the Poisson distribution with mean λ (see, e.g., Lehmann, 1999, p. 12).

N10 **Quasi-Poisson versus t-test:** We ran a simulation, where the responses, $X_1, \ldots, X_n$ were independent with $X_i | W_i, t_i \sim Poisson(t_i W_i \lambda)$, with $t_i = 1 + 2 * U_i$ and $U \sim U(0, 1)$, and $W_i \sim Gamma(1/a, scale = a)$. We did the simulation under 12 scenarios: all combinations of $n = 20, 50$, $\lambda = 3, 10$, and $a = 0, 2, 0.8, 2$. We compared the coverage of the 95% confidence intervals for λ using confidence intervals derived from either the t-test or the quasi-Poisson method using Wald statistics. We ran 10,000 replications. The t-test intervals had simulated coverage from 89.3% to 94.7% with mean 92.7%. The quasi-Poisson method (using linear overdispersion) had slightly better coverage from 91.6% to 94.5% with mean 93.2%. The latter method had larger simulated coverage in 7 out of 12 scenarios.

5.10 Problems

P1 Show that $F(t-) \le q \le F(t)$ is equivalent to $F(t) \ge q$ and $\bar{F}(t) \ge 1 - q$.

P2 Let θ be the qth quantile. To test $H_1 : \theta > \theta_B$ show that we can create an exact test, and the one-sided p-value is $p_{QG}(\mathbf{x}, \theta_B) = \Pr[Y(\theta_B) \le y(\theta_B)]$, where here $Y(\theta_B) \sim Binomial(n, q)$, and $y(\theta_B)$ is defined in Section 5.2. Similarly to test $H_1 : \theta < \theta_B$ the one-sided p-value is $p_{QL}(\mathbf{x}, \theta_B) = \Pr[Y^*(\theta_B) \le y^*(\theta_B)]$, where here $Y^*(\theta_B) \sim Binomial(n, 1 - q)$.

P3 Show that the one-sample t-test is invariant to linear transformations. In other words, if you transform each value to $x_i^* = a + bx_i$ and the parameters similarly to $\mu^* = a + b\mu$, then the t-test decision rule gives the same result.

6

Paired Data

6.1 Overview

In this chapter we consider pairs of responses on independent units. Within each unit we want to compare one type of response (say X) to the other (Y). In many applications we want to determine if X is generally larger or smaller than Y.

Example 6.1 *One very common paired data setting involves a pre-post comparison, such as: vaccinate n individuals and measure antibody levels at baseline (X) and after 210 days (Y). We want to know if the values at 210 days are generally larger than at baseline.*

Another common setting involves pairs, where one member of the pair is randomized to one treatment and the other member gets the other treatment. Another example is when a researcher creates pairing by matching individuals on a set of key covariates such as age and weight. Sometimes the units are inherently paired such as right eye and left eye, or identical twins.

Example 6.2 *A researcher matches n pairs of mice on important characteristics, within each pair randomly assigns one mouse to diet A and the other to diet B, then after a month weighs the mice. We want to know if the weights of mice on diet A (X) are more than the weights of mice on diet B (Y).*

In these two examples, we can summarize the responses from each pair by using a difference: $D = Y - X$. Then we can treat the n differences as a one-sample problem, since showing $Y > X$ is the same as showing $D > 0$. Similarly, if the responses are positive, we can use the n ratios. Since the log of the ratios is the same as the difference in log transformed responses, we can work with those differences instead. We can use the one-sample methods of Chapter 5 for these problems; however, the special nature of these paired differences motivates two new tests, the sign test (Section 6.2) and the Wilcoxon signed-rank test (Section 6.3).

We first motivate the sign test by comparing it to the median of the differences. An exact test that the median of the differences is equal to 0 (see Section 5.2) does not make any additional assumptions about the distribution of the differences, so it can be used for continuous or discrete differences. An issue with the median test is that often we are not really interested in the median. For example, in Section 6.2 we discuss a situation where the median of the differences is 0, but there are more values with a positive difference than with a negative difference (e.g., 30% of pairs had a positive difference, 60% had no difference, and 10% had a negative difference). In this case, a test that the median is 0 may not be appropriate, and you may instead want to show that there are more positive differences than negative differences (or vice versa). That latter situation calls for the sign test (which equals the median test when there are no ties at 0). When the responses for each member of the pair are binary, then the sign test on the differences is also known as the exact *McNemar's test.*

For paired data, it is often reasonable to assume that the paired differences are symmetric about 0 under the null hypothesis (see Example 6.1 or Example 6.2). One useful test under this assumption is the Wilcoxon signed-rank test which is discussed in Section 6.3. The symmetry assumption also allows us to motivate different tests. For example, since a symmetric distribution has the mean equal to the median, we can use tests on the mean, like the one-sample t-test on the paired differences. We discuss that test and compare it to the Wilcoxon signed-rank test in Section 6.4.

In some paired data settings, we might be interested not in whether Y is larger or smaller than X, but how closely the responses X and Y track each other. This situation is examined in Section 6.5. For example, if we measure two different cytokines in the blood, we may be interested in how highly correlated those two cytokines are to each other. A related but different problem is determining if two test instruments agree (Section 6.6). For example, say there are two assays designed to measure the same thing, but neither is the gold standard. In this case even if both assays were perfectly linearly related or perfectly correlated, they do not perfectly agree (i.e., yield identical answers) unless the line relating them is the identity line. So we need different parameters for measuring agreement than for measuring correlation.

Finally, consider a different type of pairing, where within the two responses in a pair there is no clear designation of which response is X and which is Y. For example, in a twin study, each set of twins is a pair, but there is no inherent reason to define the responses of one twin as X and the other as Y. In that case, we must deal with the pairing because it induces within-pair correlation, but the statistical issues are closer to those of stratification and will be discussed in Chapter 12.

6.2 Sign Test

Suppose $d_1, \ldots, d_n$ represent the differences in responses within n independent pairs. If the responses are discrete so that there are a lot of 0 responses (no difference within a pair) then the exact test on the median may not the very powerful.

Example 6.3 *Consider a paired experiment, where one member of each pair is assigned to treatment A and the other to treatment B. The responses for the pair are ordered, (-1) the individual on treatment A responded better, (0) both individuals in the pair responded the same, and (1) the individual on treatment B responded better. Suppose 100 pairs are tested and the response counts for the 3 responses are $[10, 60, 30]$. If we tested that the median is significantly different from 0 (the middle response) using the methods of Section 5.2, then $p_c = 1$ and the associated 95% confidence interval is $[0, 0]$. So although there are three times more pairs that favor treatment B than treatment A, the test of the median is not detecting this. The reason is clear, the data are indicating a median of 0, but maybe that is not the most important statistic in this case.*

Let w_ℓ, w_e, and w_g represent the number of responses out of n less than, equal to, or greater than 0, with W_ℓ, W_e, and W_g the associated random variables. Let $\mathrm{E}(W_\ell) = n\theta_\ell$,

$E(W_e) = n\theta_e$, and $E(W_g) = n\theta_g$. Instead of the median, a more useful parameter may be $\Delta = \theta_g - \theta_\ell$. To test whether there are more positive than negative responses, we test

$$H_0 : \Delta \leq 0$$

$$H_1 : \Delta > 0.$$

In order to test this, we use w_g after conditioning on nonzero responses. Let $m = w_g + w_\ell$. Define new parameters, $\mu = (1/n)E(W_g + W_\ell) = \theta_g + \theta_\ell$, $\theta_g^* = \theta_g/\mu$, $\theta_\ell^* = \theta_\ell/\mu$, and $\Delta^* = \theta_g^* - \theta_\ell^*$. Then

$$W_g | (W_g + W_\ell = m) \sim Binomial\left(m, \theta_g^*\right).$$

Since the conditional distribution of W_g is binomial, we can use the methods for binomial from Chapter 4. Further, since $\Delta > 0$ is equivalent to $\Delta^* > 0$, getting inferences about Δ^* are useful. We can get two-sided p-values and confidence intervals about θ_g^* after conditioning on $W_g + W_\ell = m$ using the exact central methods for a binomial response with parameters m and θ_g^* (Section 4.2). The usual null hypothesis is that $\theta_g^* = 0.5$. This is called the *sign test*. Then we can transform those intervals into ones about Δ^* using $\Delta^* = \theta_g^* - \theta_\ell^* = 2\theta_g^* - 1$, because $\theta_g^* + \theta_\ell^* = 1$. Obtaining confidence intervals for Δ is not as straightforward (see Note N1).

Example 6.3 *(continued) When $w_\ell = 10, w_e = 60$, and $w_g = 30$, the two-sided p-value from the sign test is $p = 0.002$, which is based on the Binomial(40,0.5) distribution. The associated 95% confidence interval for Δ^* is [0.176, 0.746], derived from transforming the confidence interval for θ_g^*, [0.588, 0.873], and the confidence interval for θ_g^* is just the exact central confidence interval computed with 30 events out of 40.*

Example 6.4 Exact McNemar's test: *Now consider the special case when X_i and Y_i are both binary. Then the data may be summarized by the following 2×2 table:*

	$Y = 0$	$Y = 1$
$X = 0$	a	b
$X = 1$	c	d

For example, there are b pairs with $X_i = 0$ and $Y_i = 1$. Then $W_g = b$, $W_\ell = c$, and $W_e = a + d$. If we assume that for each pair, $\Pr[X_i = 1]$ may be different but the odds ratio is the same, then that odds ratio is $\beta = \theta_g^*/(1 - \theta_g^*)$ (Problem P1). For example, if $b = 9$ and $c = 2$, then $\hat{\theta}_g^* = 9/11$ and $\hat{\beta} = 9/2 = 4.5$. The exact central McNemar p-value is $p = 0.065$ with the 95% confidence interval on θ_g^* equal to $(0.482, 0.977)$ and the 95% confidence interval on β equal to $(0.931, 42.8)$.

6.3 Wilcoxon Signed-Rank Test

6.3.1 Testing

The sign test does not use information about how much larger or smaller than 0 are the differences in the paired responses. Usually, you can get a more powerful test by using that information. In this section we introduce the Wilcoxon signed-rank test that only uses the ranks of the absolute value of the differences.

Let the random variables $D_1, \ldots, D_n$ be the paired differences, and let their distribution be F. The null hypothesis is that F is symmetric about 0, since under the null hypothesis each difference, D_i, could equally likely be equal to $-D_i$ by switching the labels of the pairs (e.g., in Example 6.3 switching the treatment labels). We can write the two-sided hypotheses as:

$$H_0 : F \text{ is symmetric about } 0$$
$$H_1 : F \text{ is not symmetric about } 0. \tag{6.1}$$

The representation of one-sided hypotheses requires more assumptions and will be discussed in Section 6.3.2.

Let R_i be the rank of $|D_i|$ among $|D_1|, \ldots, |D_n|$. If there are ties among the $|D_1|, \ldots, |D_n|$, break the ties arbitrarily and then average the ranks among each tied value. The ranks using this adjustment for ties are called the *midranks*. Although Wilcoxon recommended discarding the values with $D_i = 0$ before ranking and it is still common practice to do that, following Pratt (1959), we keep them in because this will lead to more sensible matching confidence intervals. Let S_i be the sign of D_i (when $D_i = 0$ is allowed, $S_i = 0$). Let $T = \sum_{i=1}^{n} S_i R_i$ be a test statistic representing the sum of the positive (i.e., when $S_i = 1$) ranks minus the sum of the negative (when $S_i = -1$) ranks.

Under the null hypothesis and conditioning on the values $|D_1|, \ldots, |D_n|$, the distribution of T can be represented by a discrete distribution. Given $|D_i| > 0$, then under the null hypothesis S_i is equally likely to be -1 as $+1$, while for $|D_i| = 0$ then $S_i = 0$ always. We express the possible values of T as

$$T(\delta) = \sum_{i=1}^{n} (-1)^{\delta_i} S_i R_i,$$

where $\delta = [\delta_1, \ldots, \delta_n]$ and each $\delta_i = 0$ or 1. There are 2^n possible values of δ and each is equally likely under the null hypothesis. Let $T_1, \ldots T_{2^n}$ be the associated set of values of T. If we let T_0 be the observed value of T, then the one-sided p-values are $p_L = \Pr[T \leq T_0]$ and $p_G = \Pr[T \geq T_0]$, where the probability is over the conditional distribution of T under the null hypothesis. Specifically,

$$p_L = \frac{\sum_{j=1}^{2^n} I(T_j \leq T_0)}{2^n}$$

and

$$p_G = \frac{\sum_{j=1}^{2^n} I(T_j \geq T_0)}{2^n}. \tag{6.2}$$

A two-sided p-value is $p_c = \min(1, 2p_L, 2p_G)$. Simplifications of the calculation are possible when there are values with $|D_i| = 0$ (see Note N2). Asymptotic approximations are sometimes used (see Problem P2 and Note N3).

Notice that we have not given the null and alternative hypotheses in terms of a finite dimensional parameter. The reason we have not is that the discussion of the parameter associated with this test is more involved and leads to a point that will come up again. We discuss this in Section 6.3.2.

6.3.2 Confidence Intervals and Estimates

Often there is a clear motivation of the symmetry assumption under the null hypothesis (see Example 6.1 or Example 6.2). If we are calculating differences in pairs and there is no difference in the distribution of responses in each member of the pair, that implies that there is no difference between the distribution of the positive paired differences and the distribution of the negative paired differences. Unfortunately, if the null hypothesis is rejected, there is usually less motivation to assume symmetry under the alternative. Nevertheless, we will often make simplifying assumptions in order to present effect parameters. In general, it will often be necessary to make more assumptions in order to make confidence intervals about a parameter than are needed to test the null hypothesis. In the Wilcoxon signed-rank case, the null hypothesis assumption of symmetry will make calculation of the p-values (calculated under the null hypothesis) tractable, but this symmetry assumption may be harder to motivate under the alternative.

Here is a model that allows symmetry under the alternative. Let $[X_i, Y_i]$ be the random vector for the ith paired response. Assume $X_i = \mu_i + \epsilon_{xi}$ and $Y_i = \mu_i + \beta + \epsilon_{yi}$. Assume that ϵ_{xi} and ϵ_{yi} are independent and both come from the same continuous distribution, F_ϵ. The ith difference is

$$D_i = Y_i - X_i = \beta + (\epsilon_{yi} - \epsilon_{xi}).$$

Under this model, the null hypothesis is that $\beta = 0$ and the two-sided alternative is that $\beta \neq 0$. Let F_0 be the distribution of $\epsilon_{yi} - \epsilon_{xi}$ for every i. Then the distribution of D_i under the alternative is $F_\beta(y) = F_0(y - \beta)$. In other words, under the alternative hypothesis we still assume a symmetric distribution, but now it is symmetric about β. Now we can create a series of alternative hypothesis models indexed by β_0, with the symmetry of F_β about β implied:

$$H_0 : \beta = \beta_0$$
$$H_1 : \beta \neq \beta_0.$$

We can test these hypotheses using $D_i^* = D_i - \beta_0$, and using the same algorithm to calculate p-values on D_i^* as we did on D_i. Let $\mathbf{d}$ be the vector of observed differences and $p_c(\mathbf{d}, \beta_0)$ be the two-sided p-value testing $H_0 : \beta = \beta_0$. Then by inverting the series of hypothesis tests, a $100(1 - \alpha)\%$ confidence set for β is the set of all β_0 that we fail to reject at the two-sided α level:

$$C(\mathbf{d}, 1 - \alpha) = \{\beta_0 : p_c(\mathbf{d}, \beta_0) > \alpha\}.$$

Because this method depends on a location shift assumption (i.e., $F_\beta(y) = F_0(y - \beta)$), we call this type of confidence interval a *shift* confidence interval.

It is important that we have included the 0s in the ranking. Although it is barely apparent in Figure 6.1, if we exclude 0s, then the one-sided p-values in this example are not monotonic for β_0 at about 0 and 9. The 95% confidence interval is between the places where the one-sided p-values cross 0.025, and monotonicity ensures that each one-sided p-value only crosses that value once.

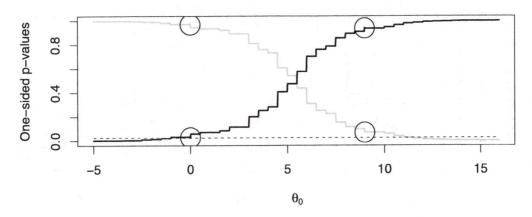

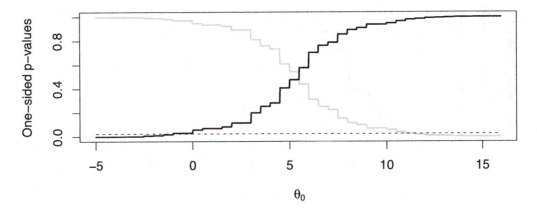

Figure 6.1 Wilcoxon signed-rank test one-sided *p*-values. The data are the differ-ences: 0, 0, −2, −3, −5, 6, 9, 11, 12, 15, 16 (Pratt, 1959, equation 5). Top panel, using the original Wilcoxon method where the 0s are discarded before ranking, and the bottom panel the 0s are ranked. We circle areas where the top method does not give monotonic one-sided *p*-values. The 95% confidence interval for both methods is (−1, 11.5).

The Hodges–Lehmann estimator associated with the Wilcoxon signed-rank test is the value of β_0 where there is no preference between $\beta > \beta_0$ and $\beta < \beta_0$. This is the value β where $p_L(\mathbf{d}, \beta)$ and $p_G(\mathbf{d}, \beta)$ cross, and for the data in Figure 6.1 this is 5.5 (see expanded scale in Figure 6.2). Note that this is not the median of the differences, which is 6 in this case. This Hodges–Lehmann estimator has a simple form

$$\hat{\beta} = \text{median}\left\{ \frac{D_i + D_j}{2}, i \leq j, j = 1, \ldots, n \right\}.$$

Wilcoxon method: do not rank 0s

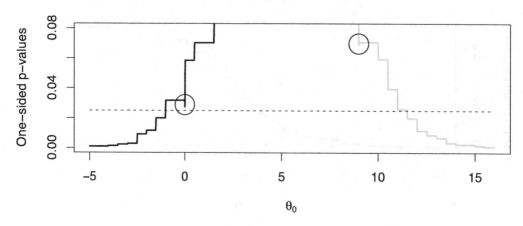

Pratt method: rank 0s

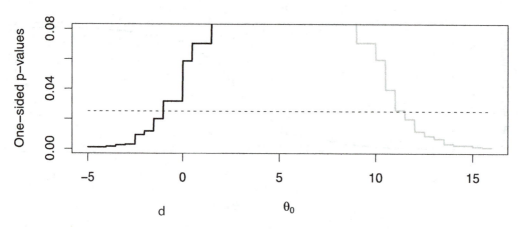

Figure 6.2 Same data and methods as Figure 6.1 but with a different scale. Wilcoxon signed-rank test one-sided p-values. The data are the differences: 0, 0, −2, −3, −5, 6, 9, 11, 12, 15, 16 (Pratt, 1959, Equation 5). The top panel shows the original Wilcoxon method where the 0s are discarded before ranking, and the bottom panel shows the Pratt method where the 0s are ranked. Only the original Wilcoxon method does not give monotonic one-sided p-values.

6.4 Choosing a Paired-Difference Test under Symmetry

6.4.1 Efficiency Considerations

Once we assume symmetry, then the mean equals the median, and we can consider using the one-sample t-test (see Section 5.3.2) to test paired-difference data. So which test should we use: a t-test, a sign test, or a Wilcoxon signed-rank test? To address this question, we introduce the term *efficiency*.

Start with the general hypotheses of Expression 6.1, so that the distribution of the paired differences under the alternative can be any distribution that is not symmetric about 0. Consider the one-sided tests that upon rejection imply that D_i is generally larger than 0. For example, the t-test rejection would imply $E(D_i) > 0$, the sign test rejection would imply $\Delta > 0$, and the Wilcoxon signed-rank rejection would imply $\beta > 0$ (i.e., would use the one-sided p-value p_G of Equation 6.2). Let F_1 be one specific distribution under the alternative. Let $N_W \equiv N_W(F_1, \alpha, 1 - \gamma)$ be the smallest sample size such that when $D_i \sim F_1$, the power to reject the null hypothesis at the α level is at least $1 - \gamma$ with the Wilcoxon signed-rank test. Similarly define $N_t \equiv N_t(F_1, \alpha, 1 - \gamma)$ and $N_S \equiv N_S(F_1, \alpha, 1 - \gamma)$ for the t- and sign tests. So if $N_W > N_t$ this would imply that the t-test is more efficient at showing significant differences at F_1 (with significant level α and power $1 - \gamma$). If we define efficiency as proportional to $1/N$, then $N_W > N_t$ could be written as $(1/N_W) < (1/N_t)$, emphasizing the better efficiency of the t-test. Define the *relative efficiency* of test A with respect to test B, as

$$e_{A,B} = \frac{1/N_A(F_1, \alpha, 1 - \gamma)}{1/N_B(F_1, \alpha, 1 - \gamma)} = \frac{N_B(F_1, \alpha, 1 - \gamma)}{N_A(F_1, \alpha, 1 - \gamma)}.$$

For example, if $e_{A,B} = N_B/N_A = 2$ the sample size needed to ensure a power of $1 - \gamma$ with test A is half as large as the sample size needed to get the same power with test B.

For testing location shifts for symmetric distributions where the shift is a function of n and approaches 0 as the sample size increases (specifically, the shift is $\frac{\beta}{\sqrt{n}}$ for fixed β), then we can compare the tests asymptotically. By measuring the efficiency this way, the results do not depend on the value of β, significance level, or on the power, but it still may depend on the alternative distribution, F_1. Denote this *asymptotic relative efficiency* (ARE) of test A with respect to test B as $e_{A,B}^{\infty}$. For example, the ARE of the Wilcoxon signed-rank test with respect to the t-test is $e_{W,t}^{\infty} = 0.955$ when F_1 is normal, 1.097 when F_1 is logistic, 1.5 when F_1 is a double exponential, and ∞ when F_1 is a Cauchy distribution, where we only need to specify the class of distributions for F_1 (see e.g., Hollander et al., 2014, p. 113). It can also be shown that for any continuous symmetric distribution, $e_{W,t}^{\infty} \geq 0.864$.

6.4.2 Application: Choice of Parameterization

In this section, we will show that considerations other than power are also important when selecting the test and the appropriate metric for relating X and Y. To explore some ideas related to paired-difference data concretely, we consider some data from a malaria vaccine study (Sagara et al., 2009). The analysis presented here is only a small part of the study, and not even the most important part. See Sagara et al. (2009) for details. Twelve adults who lived in a malaria endemic village in Mali were vaccinated with an experimental vaccine at Day 0 and Day 28. Typically, antibody levels associated with vaccines rise after each of the two vaccine doses, then decline over time. We see that in this study. This vaccine study is atypical because the disease is so prevalent in the study setting that most of the adults in the village had previously been exposed to the malaria parasite. One question is: does the level of antibody in the blood against the particular proteins in the vaccine (called AMA1-C1 proteins) continue to be larger at Day 210 than at Day 0? The data are plotted in Figures 6.3 and 6.4 and listed in Note N4.

6 *Paired Data*

(a)   **(b)**

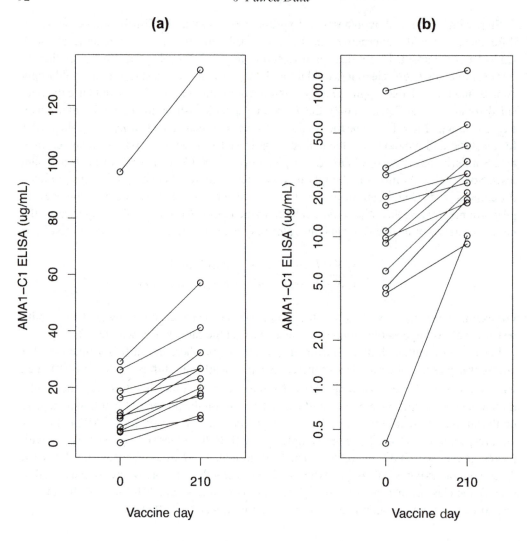

Figure 6.3 Anti-AMA1-C1 antibody levels in the blood as measured by ELISA at baseline (vaccine Day 0) and 210 days later on 12 Malian adults. Lines connect measurements from the same individual. Panel (a) is plotted on the arithmetic scale and panel (b) is plotted on the log scale.

First, notice that the data look very different if we plot on an arithmetic scale (Figure 6.3a and Figure 6.4a) or on a log scale (Figure 6.3b and Figure 6.4b). There is one subject who has very large values at both Day 0 and Day 210, and these values look very much like outliers on the arithmetic scale but not on the log scale. Conversely, the subject with the nearly undetectable value at Day 0 appears an outlier on the log scale but not on the arithmetic scale. Depending on the type of data, there may be more concern about one or the other kind of outliers and that can motivate the log transformation or not. Generally, for antibody data there is more concern about extremely large outliers than extremely small ones, and that type of data is more often presented on the log scale (Figure 6.3b and Sagara et al., 2009, figure 1).

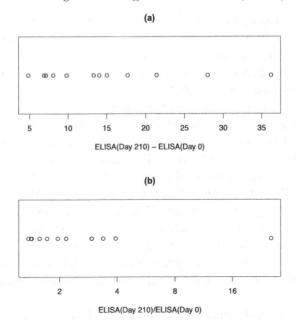

Figure 6.4 Paired differences in Anti-AMA1-C1 antibody levels in the blood as measured by ELISA. Panel (a) is the difference on an arithmetic scale and panel (b) is the ratio on a log scale.

The two types of plots emphasize different types of changes over time. Let the baseline and ending values for the ith individual be x_i and y_i. If biologically the absolute change in antibody level ($d_i = y_i - x_i$) is important, then the arithmetic scale would be appropriate. If the proportion of the baseline ($r_i = y_i/x_i$) is important, then the log scale is appropriate, since $\log(r_i) = \log(y_i) - \log(x_i)$ and the difference in log values is highlighted when plotted on the log scale. So either scale may be appropriate depending on the application.

Conveniently, both effects can be written as differences, either the arithmetic differences or the log transformed differences. That is why we have emphasized paired differences throughout the Chapter. Another operationalization of the effect is percent change from baseline:

$$p_i = 100 \left(\frac{y_i}{x_i} - 1 \right).$$

For this effect, we recommend doing inference on the log transformed differences (which equals the log of the ratio of the untransformed values). The corresponding estimates and confidence limits, in turn, can be transformed back to the p_i scale for presentation. If you do not work on the log scale, there is an asymmetry between percentage increase and decrease: for example, 100 to 150 is a 50% increase but 150 to 100 is a 33% decrease; this asymmetry does not exist on the log scale. If the percentage change is plotted, it should be on the log(ratio) scale. In other words, a 50% decrease should be the same distance from no change as a 100% increase.

Suppose the absolute changes in antibody are the most biologically meaningful. If we ignored the pairing and performed a two-sample t-test on the arithmetic scale (see Chapter 9) we would not find a significant difference in means between Day 0 and Day 210 (two-sided p-value is $p = 0.23$ using Welch's t-test). Figure 6.3a highlights why we can get an improvement in power by working with the differences. The Day 0 and Day 210 responses are very highly correlated (Pearson's correlation is 0.984). The paired t-test will give better power in this case (two-sided $p = 0.00016$).

If the percentage change was more meaningful biologically, then we would test for a difference on the log scale. The mean of the (natural) log response is 2.302 for Day 0 and 3.256 for Day 210, giving a significant difference in means of 0.955 (two-sided p-value by paired t-test, $p = 0.0017$) with 95% confidence interval of $(0.444, 1.465)$. By exponentiating, we can translate those values into geometric means (9.99 at Day 0 and 25.96 at Day 210) and a ratio (ratio = 2.60, 95% CI: 1.56, 4.33). We can also present the results in terms of percentage change: 160% increase (95% CI: +56.0%, +333%).

An alternative analysis would use the Wilcoxon signed-rank test on the log transformed responses. We find the median difference is significantly (two-sided $p = 0.00049$) different from 0, with the Hodges–Lehmann estimate of the median difference in the log responses of 0.802 (95% CI: 0.496, 1.296). We could say the fold-change is $exp(0.802) = 2.23$ with 95% CI: 1.64, 3.65.

The Wilcoxon signed-rank results on the ratio are similar to the t-test results. Some might prefer the Wilcoxon signed-rank results, since that analysis is not overly influenced by the one very high ratio (see Figure 6.4). However, the t-test results are within the realm of acceptable statistical practice, and may be preferred if geometric means are being presented with the analysis.

6.5 Correlation and Related Types of Association

6.5.1 Parameters and Estimates

As before, let the random variables for the ith paired response be $[X_i, Y_i]$, and $[X, Y]$ be the paired responses from an arbitrary individual. In this section we are not interested in showing that Y is larger or smaller than X, but we want to show that the responses are associated in the sense that larger values of X tend to be paired with larger values of Y. Loosely speaking we want to know if X and Y are correlated. Before talking of tests, we examine several methods for measuring this type of association between X and Y.

Let F_{XY} be the bivariate distribution for $[X, Y]$. *Pearson's correlation* (the conventional correlation index) is defined as

$$\rho_p = \text{Corr}(X, Y) = \frac{\text{Cov}(X, Y)}{\sqrt{\text{Var}(X)\text{Var}(Y)}} = \frac{\text{E}\left[(X - \text{E}(X))(Y - \text{E}(Y))\right]}{\sqrt{\text{E}\left[(X - \text{E}(X))^2\right]\text{E}\left[(Y - \text{E}(Y))^2\right]}}.$$

The last expression motivates the sample estimate of Pearson's correlation, where we replace expectations of (X, Y) with means (or summations since the factor of $1/n$ cancels out) over all of the data:

$$r_p = r_p(\mathbf{x}, \mathbf{y}) = \frac{\sum_{i=1}^{n}(x_i - \bar{x})(y_i - \bar{y})}{\sqrt{\left[\sum_{i=1}^{n}(x_i - \bar{x})^2\right]\left[\sum_{j=1}^{n}(y_j - \bar{y})^2\right]}}, \quad (6.3)$$

where $\bar{x}$ and $\bar{y}$ are the sample means.

Pearson's correlation is invariant to linear transformations (see Problem P3), but it is not invariant to monotonic transformations. So for constants a and b, $\text{Corr}(aX + b, aY + b) = \text{Corr}(X, Y)$, but, for example, $\text{Corr}(X, Y) \neq \text{Corr}(\log(X), \log(Y))$. In contrast, Spearman's correlation is invariant to monotonic transformations, where the sample Spearman's correlation, say r_s, is defined by replacing x_i and y_i in Equation 6.3 with the ranks (or midranks if there are ties).

To motivate the population value of Spearman's correlation for large n, we consider a different expression of r_s. First, note that the ranks of a sample are linearly related to the empirical distribution (see Problem P5). Similarly, the midrank for x_i is $n\hat{F}^*(x_i) + \frac{1}{2}$, where $\hat{F}^*$ is the *normalized empirical distribution* (see e.g., Brunner and Munzel, 2000) given by

$$\hat{F}_{\mathbf{x}}^*(x) = \frac{1}{n} \sum_{i=1}^{n} \left\{ I(x_i < x) + \frac{1}{2}I(x_i = x) \right\}. \tag{6.4}$$

Define $\hat{F}_{\mathbf{x}}^*(\mathbf{x}) = [\hat{F}_{\mathbf{x}}^*(x_1), \dots, \hat{F}_{\mathbf{x}}^*(x_n)]$ and similarly define $\hat{F}_{\mathbf{y}}^*(\mathbf{y})$. Then Spearman's (sample) correlation is

$$r_s = r_p \left(\hat{F}_{\mathbf{x}}^*(\mathbf{x}), \hat{F}_{\mathbf{y}}^*(\mathbf{y}) \right). \tag{6.5}$$

Define $\rho_s = \text{Corr}(F_X^*(X), F_Y^*(Y))$, where $F_X^*(x) = 0.5F_X(x) + 0.5F_X(x-)$ and $F_Y^*(y) = 0.5F_Y(y) + 0.5F_Y(y-)$ are the normalized marginal distributions of F_{XY}, F_X and F_Y are the usual marginal distributions of F_{XY}, and $F_X(x-) = \lim_{\epsilon \to 0} F_X(x - \epsilon) = \Pr[X < x]$. When there are no ties then $F_X(x) = F_X^*(x)$, $F_Y(y) = F_Y^*(y)$ and $\rho_s = \text{Corr}(F_X(X), F_Y(Y))$, and in this case it has been shown that $\lim_{n \to \infty} \text{E}(R_s) = \rho_s$, where R_s is the random variable associated with r_s (see, e.g., Gibbons, 1971, p. 240).

Another association invariant to monotonic transformations is *Kendall's tau*. To motivate this, we rewrite Pearson's sample correlation in a different form (see Problem P4) as

$$r_p = \frac{\sum_{i=1}^{n} \sum_{j=1}^{n} (x_i - x_j)(y_i - y_j)}{\sqrt{\left[\sum_{i=1}^{n} \sum_{j=1}^{n} (x_i - x_j)^2 \right] \left[\sum_{h=1}^{n} \sum_{k=1}^{n} (y_h - y_k)^2 \right]}}. \tag{6.6}$$

Replace $(x_i - x_j)$ with $\text{sign}(x_i - x_j)$ and $(y_i - y_j)$ with $\text{sign}(y_i - y_j)$ in Equation 6.6 to give, r_k, the sample Kendall's tau (see Gibbons, 1971, p. 214). Let R_k be the random variable associated with r_k. Let the population Kendall's tau be

$$\tau = \Pr[(X_j - X_i)(Y_j - Y_i) > 0] - \Pr[(X_j - X_i)(Y_j - Y_i) < 0], \tag{6.7}$$

which can be shown is $\tau = \text{E}(R_k)$. The right-hand side of Equation 6.7 represents the probability of a concordant pair minus the probability of a discordant pair.

We plot Kendall's tau and ρ_s for the normal distribution case in Figure 6.5. Notice that ρ_s matches the Pearson correlation, ρ_p, much more closely than τ. Because Kendall's tau is a much different parameter in this case, it is used less often than Spearman's correlation. Because $\rho_p \approx \rho_s$ when the data are bivariate normal, we may think of ρ_s as the correlation that would be observed if we could monotonically transform the responses X and Y so that both were approximately normal. Although we are not sure that such a transformation exists, this idea provides some motivation to why ρ_s can be interpreted similarly to ρ_p.

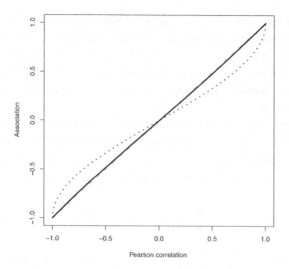

Figure 6.5 Kendall's tau (dotted black) and ρ_s (solid black) for the bivariate normal case. The value ρ_s is calculated by simulation (10^5 replications for each of the ρ values $-1, -0.99, \ldots, 0.99, 1$ and then applying a nonparametric smoother, R function supsmu). The dashed gray line is the identity. Kendall's tau is $(2/\pi)\arcsin(\rho)$ (Gibbons, 1971, p. 208).

6.5.2 Testing

A nonparametric test related to the associations of the previous section (ρ_p, ρ_s, or τ) is a test of independence. We write our test about F_{XY} as

$$H_0 : \quad \text{X and Y are independent}$$
$$H_1 : \quad \text{X and Y are not independent}$$

(see Note N5 for definition of independence in relation to F_{XY}). We can create a permutation test with respect to any of the three association statistics (r_p, r_s, or r_k). Let $T_0 = T(\mathbf{x}, \mathbf{y})$ be one of the three association statistics, and let $T_1, \ldots, T_{n!}$ be the those statistics calculated after permuting the order of the $\mathbf{y}$ in all $n!$ ways, so that the first X observation in the data set is now paired with a Y observation from the data set selected at random, and so on. Then the one-sided p-values are

$$p_L = \frac{\sum_{i=1}^{n!} I(T_i \leq T_0)}{n!}$$

$$p_G = \frac{\sum_{i=1}^{n!} I(T_i \geq T_0)}{n!}$$

and a two-sided p-value is $p_c = \min(1, 2p_L, 2p_G)$. Alternatively, you can define a two-sided p-value using the square of one of the three association statistics for T and using p_G with that statistic as a two-sided p-value.

6.5.3 Confidence Intervals

These permutation tests do not require any assumptions about F_{XY} in order to retain the Type I error rate under the null hypothesis; however, if you reject the null hypothesis, this says little with clarity about the association parameter related to the test statistic. For example, if you use $T_0 = r_p$ and reject because r_p is unusually large in the permutation distribution (i.e., p_G is small), this does not necessarily imply that $\rho_p > 0$ (although it is very likely for many distributions). To make precise inferences about ρ_p we need to make more assumptions about the distribution F_{XY}. Also, if the true ρ_p is large, this does not necessarily mean that we have large power to reject the null hypothesis using the permutation test with r_p (although in many situations it will give large power).

If we want to create confidence intervals about ρ_p we often make the additional assumption that F_{XY} is bivariate normal. Let $\rho = \rho_p$, and we test $H_0 : \rho = \rho_0$ versus $H_1 : \rho \neq \rho_0$ for a series of values of ρ_0. Even with the normality assumption, the distribution of r_p is complicated (see Johnson et al., 1995, chapter 32). A good and common approximation is to use the variance stabilizing transformation,

$$z_p = \tanh^{-1}(r_p) = \frac{1}{2} \log \left\{ \frac{1 + r_p}{1 - r_p} \right\}.$$

Then assume that the random variable associated with z_p, Z_p, is approximately normal with mean $\tanh^{-1}(\rho_p)$ and variance $1/(n-3)$, so that the $100(1-\alpha)\%$ confidence interval for ρ_p is approximated by

$$\left[\tanh \left\{ z_p - \frac{\Phi^{-1}(1 - \alpha/2)}{\sqrt{n-3}} \right\}, \tanh \left\{ z_p + \frac{\Phi^{-1}(1 - \alpha/2)}{\sqrt{n-3}} \right\} \right]. \tag{6.8}$$

There are other large sample approximations that are also used in practice. For example, it can be shown that when $\rho = 0$ and still under the normality assumption,

$$\frac{r_p \sqrt{n-2}}{\sqrt{1 - r_p^2}} \sim t_{n-2}, \tag{6.9}$$

where t_{n-2} represents a t distribution with $n-2$ degrees of freedom. More details when $\rho \neq 0$, as well as other approximations, are discussed in (Johnson et al., 1995, chapter 32). Hu et al. (2020) developed an asymptotic confidence interval based on the empirical likelihood that simulations show has coverage better than Equation 6.8 when the data are not multivariate normal (and similar when the data are multivariate normal).

Ideally, for the Spearman's correlation, we would make the assumption that there exists a monotonic transformation for each response that causes the bivariate distribution to be normal. Then we could use Equation 6.8 on r_s instead of r_p. It turns out that this strategy works reasonably well (Caruso and Cliff, 1997). Another strategy is to replace the ranks with order statistics from the standard normal distribution, then take the Pearson's correlation of that and use approximations like Equation 6.8 for the confidence interval (see, e.g., Gibbons, 1971, section 12.5).

For confidence intervals on Kendall's tau, Hollander et al. (2014, p. 415) recommend a modified jackknife-like method. The motivation is given in Samara and Randles (1988).

6.6 Agreement Coefficients

6.6.1 Parameters and Estimators

Suppose X and Y represent two different ways of measuring the same thing. For example, they could represent two different assays for measuring antibody level in a blood sample. If neither of the two assays were the gold standard, then we may be interested in how well the two assays agree. Let F_{XY} represent the bivariate distribution. Then we want an agreement coefficient to have large values when much of the probability distribution lies close to the line of equality. This is not the same as having correlation values close to 1 (Problem P6). Most agreement coefficients are scaled similarly to correlations: 1 represents perfect agreement, 0 represents chance agreement, and -1 represents the worst agreement. The agreement coefficients we will discuss all take the form,

$$\frac{\text{(chance cost of disagreement)} - \text{(expected cost of disagreement)}}{\text{chance cost of disagreement}},$$

with different definitions of disagreement cost and of chance. Some common agreement coefficients that take this form are Cohen's kappa and Scott's pi for categorical responses, and concordance correlation coefficients for continuous responses.

Let $c(X, Y)$ be a function that represents the cost of disagreement, where $c(X, Y) = 0$ when $X = Y$. Then a general form for an agreement coefficient is

$$A_F(c) = 1 - \frac{E_{F_{XY}}\{c(X, Y)\}}{E_{F_X} E_{F_Y}\{c(X, Y)\}}, \tag{6.10}$$

where F_X and F_Y are the marginal distributions of F_{XY}. In the denominator, the X and Y are independent, and represent chance agreement. Agreement coefficients that follow Equation 6.10 are called the *fixed marginal agreement coefficients* (FMAC). The less common *random marginal agreement coefficients* (RMAC) are

$$A_R(c) = 1 - \frac{E_{F_{XY}}\{c(X, Y)\}}{E_{F_{Z_1}} E_{F_{Z_2}}\{c(Z_1, Z_2)\}}, \tag{6.11}$$

where Z_1 and Z_2 represent independent random variables representing chance measurements, with $Z_1 \sim F_Z$ and $Z_2 \sim F_Z$, where $F_Z = \frac{1}{2}F_X + \frac{1}{2}F_Y$. The RMACs introduces an added bit of randomness to the chance variables, representing randomly choosing X or Y. The advantage of RMACs over FMACs is that for RMACs greater differences in F_X and F_Y while keeping $E_{F_{XY}}\{c(X, Y)\}$ constant cannot induce greater agreement, while for FMACs this can induce greater agreement.

There is an alternate interpretation of a transformation of the RMAC. Let $A_R(c)$ be an RMAC, then $(1 - A_R(c))/2$ represents the proportion of $\text{Var}(Z_1)$ due to disagreement. The value $\text{Var}(Z_1)$ may be partitioned into $\text{Var}(0.5X + 0.5Y)$ plus another term representing disagreement ($1/4$ times the expected squared difference between X and Y). Thus, when $A_R(c)$ is close to 1, almost all of $\text{Var}(Z_1)$ is due the variance of the average of X and Y and almost none is due to disagreement, while if $A_R(c)$ is close to -1 then the converse is true.

As an application of agreement coefficients, consider two assays designed to measure whether an individual has a disease or not. If $X_i = Y_i = 1$, then both assays diagnose the ith individual with the disease, while $X_i = Y_i = 0$ denote not diagnosed with the disease. Let $\pi_{ab} = \Pr[X = a, Y = b]$, where the probability is measured over a population of interest.

Then a simple naïve agreement measure in a population is $\Pr[X \text{ and } Y \text{ agree perfectly}] = \pi_{11} + \pi_{00}$. The problem with this agreement is that if the disease is very rare, the agreement could look good for two useless assays. Suppose the disease occurs in only 1 per 100 individuals, and for each individual you did not measure any value and just let X_i be a random Bernoulli response with parameter $1/100$ and let Y_i come independently from the same distribution. In this case, the probability of perfect agreement would be $0.01^2 + (0.99)^2 = 0.9802$. Compare this to the FMAC or RMAC (they are the same in this case) with $c(X, Y) = 0$ if $X = Y$ and 1 otherwise. In this case the chance cost of disagreement for calling every individual disease free is $2 * 0.01 * 0.99 = 0.0198$, which is the same as the expected cost of disagreement for the measures used. Thus, the agreement coefficient is $A(c) = 0$ denoting that the assays do not agree better than you would expect from chance. The FMAC in this situation is called *Cohen's kappa*, while the RMAC is called *Scott's pi* (or intraclass kappa).

Cohen's kappa and Scott's pi can be used to measure agreement for categorical responses with $k > 2$ categories. Suppose the possible responses are represented by $k \times 1$ vectors: $e_1, \ldots, e_k$, where e_j has a 1 in the jth row and 0's elsewhere. If there is no ordering of the categories, then the previous cost function would apply. If there is some inherent ordering in the categories such that we can treat $e_1 < e_2 < \cdots < e_k$, then possible cost functions could be $c(e_i, e_j) = (i - j)^2$ or $c(e_i, e_j) = |i - j|$. Agreement coefficients for categorical data are often written using positive weights for agreement (see Problem P7).

Now consider two assays that measure a numeric value. Let $c(X, Y) = (X - Y)^2$. Let μ_x and μ_y be the marginal means, and σ_x^2 and σ_y^2 be the marginal variances with ρ equal to the correlation. Then the FMAC and RMAC are given by

$$A_F(c) = \frac{2\rho\sigma_x\sigma_y}{\sigma_x^2 + \sigma_y^2 + (\mu_x - \mu_y)^2} = \frac{2\rho}{\gamma + (1/\gamma) + \delta^2}$$

and

$$A_R(c) = \frac{2\rho\sigma_x\sigma_y - 0.5(\mu_x - \mu_y)^2}{\sigma_x^2 + \sigma_y^2 + 0.5(\mu_x - \mu_y)^2} == \frac{2\rho - 0.5\delta^2}{\gamma + (1/\gamma) + 0.5\delta^2},$$

where $\gamma = \sigma_x/\sigma_y$ and $\delta = (\mu_x - \mu_y)/\sqrt{\sigma_x\sigma_y}$. The FMAC is the concordance correlation coefficient of Lin (1989), while the RMAC was introduced by Fay (2005). Notice that when $\mu_x = \mu_y$ and $\sigma_x = \sigma_y$ then $A_F(c) = A_R(c) = \rho$. This means that if you have two assays that you can standardize to have the same mean and variance, the agreement coefficients are just the (Pearson's) correlation. The agreement coefficients are needed when you do not know if the two assays have the same mean and variance.

To estimate the agreement coefficients, we replace expectations with means in Equations 6.10 and 6.11. So, for example, the FMAC is estimated with

$$\hat{A}_F(c) = 1 - \frac{n^{-1}\sum_{i=1}^n c(x_i, y_i)}{n^{-2}\sum_{i=1}^n \sum_{j=1}^n c(x_i, y_j)}$$

and the RMAC with

$$\hat{A}_R(c) = 1 - \frac{n^{-1}\sum_{i=1}^n c(x_i, y_i)}{(2n)^{-2}\sum_{i=1}^{2n} \sum_{j=1}^{2n} c(z_i, z_j)},$$

where $\mathbf{z} = [z_1, \ldots, z_n, z_{n+1}, \ldots, z_{2n}] = [\mathbf{x}, \mathbf{y}]$.

6.6.2 Testing and Confidence Intervals

To make inferences about an agreement coefficient, one strategy is to use bias-corrected and accelerated (BC_a) bootstrap confidence intervals. The details of this general method are discussed in Chapter 10. Alternatively, we can use the $\tanh^{-1}$ transformation, and use the delta method (see Chapter 10) to get asymptotically normal confidence intervals (see Fay, 2005, for details).

6.7 Summary

Paired data can be reduced to a one-sample problem. For example, this can be done by subtracting the baseline value from the postintervention value to get a single value, d_i, per subject. Inference about the mean or median of the corresponding distribution can be done using a sign test, a Wilcoxon signed-rank test, or a one-sample t-test.

The sign test simply compares the number of positive d_i values to the number of negative d_i values and is analyzed with respect to the $Binomial(m,0.5)$, where m is the sum of positive and negative d_i values.

The Wilcoxon signed-rank test considers the ranks of the $|d_i|$ values. So that even if the number of positive and negative d_i values were the same, if the positive d_i values were much greater than 0, and the negative d_i values were all close to 0, the signed-rank procedure could detect that the distribution is not symmetric around 0. Exact calculations use a permutation approach, and there are asymptotic approximations. A test can also be conducted for a specific difference (or shift), β, instead of just 0. This can be inverted to find the confidence interval for β, and the Hodges–Lehmann estimator provides an estimate for the shift.

We present relative efficiency for the paired t-test versus the Wilcoxon signed-rank under various distributions. For example, if the differences truly follow a normal distribution, then asymptotically, the relative efficiency is 0.955; that is, one would need a sample size that is about 5% larger if using the Wilcoxon signed-rank relative to a paired t-test, to yield the same power. Of course, in practice one will not know the true distribution, but consideration of a range of possible distributions should be useful.

As an illustration of the practical considerations for the methods presented in this chapter, we analyze a small set of paired data on antibody levels at Day 0 and Day 210 in a variety of ways. We analyze the raw differences, but also differences on the log scale, and discuss the considerations for making this choice, which usually will depend on the more natural and meaningful biological interpretation. We also analyze these data with the t-test and the Wilcoxon signed-rank test, in terms of both hypothesis testing and construction of confidence intervals.

We then discuss correlation, where we want to assess not whether Y is larger or smaller than X, but rather whether they track together. We introduce the usual correlation index, Pearson's, which is invariant to linear transformations to the data, but can yield a different result for monotonic transformations.

Spearman's correlation can be computed using the formula for Pearson's, but where the data for X and Y are replaced by their respective ranks. Spearman's correlation has the advantage over Pearson's correlation in that it is invariant to monotonic transformations (such as taking logs or square roots). Another nonparametric correlation measure is Kendall's tau which only uses information about the sign of the differences. When data follow a bivariate normal distribution, Spearman's correlation tends to match Pearson's correlation much more closely than Kendall's tau. Spearman's correlation is more commonly used than Kendall's tau.

We show that independence of the pairs can be evaluated with respect to any of three correlation indices with a permutation approach. The most common available methods for confidence interval calculation require an assumption of bivariate normality for F_{XY} although the recent work of Hu et al. (2020) provides an asymptotic method that does not assume the bivariate normality.

We then discuss agreement between X and Y; agreement means the values do not only track together (such as height and weight) but that they measure the same value (say each individual's weight is measured on two different scales, and we want to know if the scales agree.) We show a general form for the agreement coefficient. There are two general approaches to this, the fixed marginal agreement coefficient (FMAC) or the random marginal agreement coefficient (RMAC), where the RMAC has some better properties. When X and Y are binary (such as disease present or absent), the FMAC is Cohen's kappa, and the RMAC is Scott's pi (or intraclass kappa). When X and Y are numeric, the FMAC with a squared difference cost function is Lin's concordance correlation coefficient, and the associated RMAC was introduced by Fay.

6.8 Extensions and Bibliographic Notes

For discussion and motivation of the asymptotic relative efficiency comparing the t-test, the sign test and the Wilcoxon signed-rank test see Lehmann (1999) (pp. 173–176).

Another measure of association that is sometimes used is Goodman and Kruskal's gamma. This is Kendall's tau divided by the sum of the probability of concordance plus the probability of discordance (see Agresti, 2013).

With agreement coefficients, it is often very useful to perform a Bland–Altman plot (Bland and Altman, 1999). This method plots the average of the two measurements versus the difference, with 95% limits of agreement on the difference. This plot highlights where the methods agree or disagree. For other measures of agreement for continuous responses see Barnhart et al. (2007).

6.9 R Functions and Packages

The asht R package has functions to do sign tests (signTest) and exact implementations of the Wilcoxon signed-rank test (wsrTest). The signTest includes options for getting compatible confidence intervals for Δ (see Note N1), and for binary responses (i.e., McNemar's test) see the mcnemarExactcDP or the exact2x2 (with paried=TRUE) functions in the exact2x2 R package.

The cor.test function in base R uses the confidence interval method of Equation 6.8, but calculates the p-value using Equation 6.9. A simple way to get confidence intervals for the Spearman correlation is to use cor.test(rank(x),rank(y),conf.int=TRUE), which uses Equation 6.8 on the ranks. To calculate confidence intervals on Kendall's tau using the method recommended in Hollander et al. (2014) use the kendall.ci function in the NSM3 R package.

6.10 Notes

N1 Parameters for sign test: Note, $\Delta = \mu \Delta^*$. It is easy to get a confidence interval for μ. Letting $M = W_g + W_\ell$, we have

$$M \sim Binomial(n, \mu),$$

and we can use exact binomial intervals for μ. What is not straightforward is combining the confidence intervals for Δ^* and μ. The issue is that when Δ^* is positive, then Δ is an increasing function of μ, but when Δ^* is negative, then Δ is a decreasing function of μ. Fay and Lumbard (2021) used a slight generalization of the melding method (see Section 10.11) to get purportedly exact confidence intervals for Δ that are compatible with the sign test (or McNemar's test when responses are binary).

N2 Wilcoxon signed-rank p-values with zero differences: If only m of the n values $|D_i|$ are greater than 0, then you only need to calculate the 2^m different values for δ, switching δ_i only when $|D_i| > 0$. Let $T_1^*, \ldots, T_{2^m}^*$ be those values for T, and you can show that

$$p_L = \frac{\sum_{j=1}^{2^m} I(T_j^* \le T_0)}{2^m}.$$

N3 Wilcoxon signed-rank asymptotic method: The Wilcoxon signed-rank test often uses a different statistic, $T^* = \sum_{i=1}^n I(S_i = 1) R_i$. Using the theory of permutation tests (Chapter 10) we can show that permuting this statistic will give equivalent p-values. Here are the expressions of the mean and variance. Let m be the number of $|D_i| > 0$, then

$$E(T^*) = \frac{m(m+1)}{4},$$

and let $t_1, \ldots, t_g$ be the number of tied ranks for the g groups of ranks that are tied. Then

$$Var(T) = \frac{1}{24} \left\{ m(m+1)(2m+1) - \frac{1}{2} \sum_{j=1}^g t_j(t_j - 1)(t_j + 1) \right\}.$$

Asymptotic normality can be used for inferences (see Hollander et al., 2014, p. 42)

N4 ELISA data: Here are the ELISA data as (Day 0 value, Day 210 value) for the 12 subjects:

$$\begin{array}{cccc}
(10.96, 32.39) & (96.47, 132.68) & (18.72, 26.8) & (9.08, 26.78) \\
(9.85, 16.98) & (0.40, 10.21) & (5.87, 19.92) & (29.19, 57.18) \\
(26.21, 41.25) & (16.3, 23.17) & (4.15, 8.99) & (4.54, 17.88).
\end{array}$$

N5 Independence: Suppose (X, Y) come from a bivariate distribution F_{XY}, with probability density function (for continuous data) or probability mass function (for discrete data) given by $f_{XY}(x, y)$. Let the marginal pdfs (or pmfs) be $f_X(x)$ and $f_Y(y)$. Then X and Y are independent if and only if $f_{XY}(x, y) = f_X(x) f_Y(y)$. Alternatively, X and Y are independent if and only if $\Pr[X \in A, Y \in B] = \Pr[X \in A]\Pr[Y \in B]$ for all Borel sets A and B (see, e.g., Casella and Berger, 2002).

6.11 Problems

P1 **Odds ratio for McNemar's test:** Assume that we have n pairs of binary observations with $X_i \sim Binomial(1, \theta_{x_i})$ and independently $Y_i \sim Binomial(1, \theta_{y_i})$. Assume that every pair has the same odds ratio, so that $\beta = \theta_{y_i}(1 - \theta_{x_i}) / \{\theta_{x_i}(1 - \theta_{y_i})\}$ for $i = 1, \ldots, n$. Show that for any i,

$$Pr[Y_i = 1 | X_i + Y_i = 1] = \frac{\beta}{1 + \beta},$$

so that the conditional likelihood of $B | B + C = m$ (where B and C are the random variables associated with b and c of Example 6.4) is binomial with parameters m and $\frac{\beta}{1+\beta}$. Show that the maximum likelihood estimator of β from that conditional likelihood is $\hat{\beta} = b/c$.

P2 Using the notation on page 87, show that under the null hypothesis $E(T) = 0$ and $V = \text{Var}(T) = \sum_{i=1}^{n}(S_i R_i)^2$.

Then using a version of the central limit theorem (the Lindeberg–Feller theorem), as long as we can assume

$$\frac{\max_{1 \leq i \leq n}(S_i R_i)^2}{\text{Var}(T)} \to 0$$

as $n \to \infty$, then $T/\sqrt{V}$ is asymptotically standard normal under the null hypothesis (O'Brien and Fleming, 1987).

P3 Show that $\text{Corr}(aX + b, aY + b) = \text{Corr}(X, Y)$ for constants a and b. Hint: $\text{Var}(aX + b) = a^2 \text{Var}(X)$.

P4 Show that Equation 6.3 equals Equation 6.6. Hint: show that $\sum_{i=1}^{n}(x_i - \bar{x})^2 = \frac{1}{2n} \sum_{i=1}^{n} \sum_{j=1}^{n}(x_i - x_j)^2$.

P5 The *empirical distribution* of $x_1, \ldots, x_n$ is $\hat{F}(x) = \frac{1}{n} \sum_{i=1}^{n} I(x_i \leq x)$ for any x. Show that if there are no ties, the rank of x_i is $n\hat{F}(x_i)$.

P6 Give an example of a bivariate distribution with correlation nearly equal to 1, but with poor agreement.

P7 Let $w_{ij} = 1 - \frac{c(e_i, e_j)}{\max_{g,h} c(e_g, e_h)}$, and show that Cohen's kappa can be written as $(\Pi_o - \Pi_e)/(1 - \Pi_e)$ where $\Pi_o = \sum_{i=1}^{k} \sum_{j=1}^{k} w_{ij} \pi_{ij}$ and $\Pi_e = \sum_{i=1}^{k} \sum_{j=1}^{k} w_{ij} \pi_{i.} \pi_{.j}$, where $\pi_{i.} = \sum_j \pi_{ij}$ and $\pi_{.j} = \sum_i \pi_{ij}$. The RMAC replaces Π_e with

$$\sum_i \sum_j w_{ij}(0.5\pi_{i.} + 0.5\pi_{.i})(0.5\pi_{.j} + 0.5\pi_{j.})$$

(see Fay, 2005).

7

Two-Sample Studies with Binary Responses

7.1 Overview

This chapter deals with two-sample problems where the responses are binary and are all independent within each sample. One design associated with this type of data is an experimental design where n_1 individuals are assigned to Group 1 and n_2 to Group 2, and we collect binary responses on all $n_1 + n_2$ individuals.

Example 7.1 (Coulibaly et al., 2009) *The parasite Mansonella perstans infects people in parts of Africa. The parasite has a symbiotic relationship with the bacteria, Wolbachia, that helps the M. perstans live. Researchers suspected that if infected people were given the antibiotic, doxycycline, it would kill the bacteria which would, in turn, help eliminate the parasite. Individuals with M. perstans were randomized to receive either treatment (multiple doses of doxycycline) or control (no treatment). After 12 months, 67 of 69 subjects who received treatment were "cured" meaning they had no detectable parasites in their blood, while 10 of 63 subjects who received no treatment were cured. In the actual experiment there were missing data issues, but we ignore that for this chapter (Chapter 17 deals with missing data). We assume that $n_1 = 63$ and $n_2 = 69$ are fixed sample sizes and write the responses as $x_1 = 10$ and $x_2 = 67$.*

In Example 7.1 we treat the data as if the total number of cured in each group is binomial with fixed sample size. For Group a, $X_a \sim Binomial(n_a, \theta_a)$, for $a = 1, 2$. We call this probability model a *two-sample binomial model*. In addition to randomized trials, some observational studies fit that model as well, for example, comparing prevalence of chronic disease between males and females in a cross-sectional study. Most of this chapter will discuss analysis of the two-sample binomial model; however, the last section (Section 7.8) considers different designs, focusing on the one where not only n_1 and n_2 are fixed but the total number of events is fixed as well (this type of design is often used in vaccine trials).

Usually, one of three effect parameters are used in the two-sample binomial study: the difference in proportions ($\beta_d = \theta_2 - \theta_1$), the ratio of proportions ($\beta_r = \theta_2/\theta_1$), or the odds ratio of the proportions ($\beta_{or} = \theta_2(1 - \theta_1)/\{\theta_1(1 - \theta_2)\}$). We discuss that choice in Section 7.2.

Besides the choice of effect parameter, there are also many ways of creating tests and confidence intervals (CIs) for the two-sample binomial problem. There are several issues that make this problem not straightforward. The first issue is one that occurs with nearly all discrete tests on *composite hypotheses* (tests where the null hypothesis space has more than one probability model). In order to ensure that the test is valid, the Type I error rate must be less than the nominal for almost all values of the parameters in the null (see, for example, Figure 4.1). This means that valid tests (i.e., exact tests) for this problem will be conservative, and hence exact confidence intervals will be conservative as well. Because of this conservativeness, many authors suggest noncentral exact tests and their associated confidence intervals in order to get Type I error rates for two-sided tests and coverage for two-sided confidence intervals closer to the nominal value. We do not recommend this when directional inferences are desired (e.g., in Example 7.1 showing that the treatment is better than control rather than just showing the treatment is not the same as control). Recall that with a central $100(1 - \alpha)\%$ confidence interval, we can reject the one-sided test at the $\alpha/2$ level whenever the null parameter values are excluded from the central interval on the appropriate side. Alternatively, we can reduce conservativeness by not requiring the test and associated confidence interval be exact. In other words, unlike the valid $100(1 - \alpha)\%$ confidence intervals that cover the true parameter with probability at least $100(1 - \alpha)\%$, we could create nominally $100(1 - \alpha)\%$ confidence intervals that cover the true parameter "on average" about $100(1 - \alpha)\%$ of the time so that for some parameters the coverage is slightly more than $100(1 - \alpha)\%$ and for other parameters it is slightly less. These types of procedures can be done using mid-p methods, or asymptotically normal methods.

A second issue is that there are two parameters that describe the probability space, but we only really care about the effect parameter of interest. So we can reparametrize the problem into a β parameter (the parameter of interest, such as β_d or β_r) and a nuisance parameter. For example, if we care about β_d, then θ_1 could be a nuisance parameter, since if we know β_d and θ_1, we know the entire parameter space, $[\theta_1, \theta_2]$. Two main ways to create a valid test are to ensure that the Type I error rate is less than nominal by searching over the entire null hypothesis space (this is called an *unconditional* exact test, see Section 7.4), or to condition on the total number of events to eliminate the nuisance parameter (a *conditional* exact test, Section 7.3). Fisher's exact test, one of the first tests developed for this setting, is a conditional exact test and it is much quicker to calculate than the unconditional tests, but the more computationally difficult unconditional exact tests tend to have greater power. Further, the conditional exact tests only simplify the calculations if the parameter of interest is the odds ratio, otherwise (for β_d and β_r) we need to use the melding technique (which is a technique for combining one-sample confidence intervals for two-sample inference, see Section 10.11) to create confidence intervals that are compatible with Fisher's exact test (see Section 2.3.3 about compatible confidence intervals).

Another major issue with the two-sample binomial problem is that it is not straightforward to create compatible p-values and confidence intervals. The p-values from the central Fisher's exact test and its associated confidence interval are compatible; however, there is no compatible confidence interval to the commonly used Fisher–Irwin version of the two-sided Fisher's exact test (see Example 2.5 and Section 7.3.2). Further, many of the common versions of the unconditional exact test (including those that yield the largest power) do not have compatible confidence intervals.

After discussing effect parameters, we start with the central Fisher's exact test in Section 7.3.1. Its p-values and confidence intervals are compatible, valid, give directional inferences, and are easy to calculate. A more powerful method is a modifed Boschloo test, which is in the class of unconditional exact tests. The resulting p-values from this modified Boschloo test and its associated confidence intervals are also compatible, valid, and give directional inferences, but they are more difficult to calculate (see Section 7.4.2).

The nonvalid (i.e., nonexact) methods are in Sections 7.5 and 7.6. Section 7.5 gives the mid-p methods, which are nonasymptotic methods (e.g., calculated without using any approximations) that give Type I error rates that are closer to the nominal value "on average" over the null parameter space, but do not bound the Type I error rate for all values of the parameter in the null parameter space (see Section 4.3.3 for mid-p methods in the one-sample case). Section 7.6 gives some asymptotically normal approximations for tests and confidence intervals. We compare a very simple Wald interval to the nonasymptotic methods in Section 7.7.

For the two-sample binomial problem, the Fisher's exact test with associated melded confidence intervals is a fine default. Slightly better power with a still valid test can be had with the modified Boschloo test (but at the cost of more complex and time-consuming algorithms). For those settings where a method that is not guaranteed to be valid is acceptable, we can use the mid-p version of Fisher's exact test or the closely related mid-p melding intervals and associated p-values.

Finally, in some situations the binary events may be rare or occur over a long period of time (e.g., individuals getting a disease). In that case, it sometimes makes sense to design the study to stop when a certain total number of events have occurred in both samples combined. That design is discussed in Section 7.8.

7.2 Parameters to Measure Effects

Consider the two-sample binomial model with $X_1 \sim Binomial(n_1, \theta_1)$ and $X_2 \sim Binomial(n_1, \theta_2)$. We are often interested in testing in the form of three decision rules (see Section 2.3.4) with null hypothesis $H_0 : \theta_1 = \theta_2$ and two alternative hypotheses $H_{1a} : \theta_1 > \theta_2$ and $H_{1b} : \theta_1 < \theta_2$. Further, regardless of whether we reject H_0, we often want to estimate how much bigger (or smaller) θ_1 is than θ_2. We consider three different effect parameters to operationalize this concept of "how much bigger": the difference in proportions, $\beta_d = \theta_2 - \theta_1$, the ratio of proportions, $\beta_r = \theta_2/\theta_1$, or the odds ratio, $\beta_{or} = \theta_2(1-\theta_1)/\{\theta_1(1-\theta_2)\}$. Then we can equivalently write the three hypotheses as for example, $H_0 : \beta_r = 1$, $H_{1a} : \beta_r < 1$, and $H_{1b} : \beta_r > 1$.

The choice of which of the three effect parameters to use is context dependent. The difference, β_d, is often used because of its simplicity. It has a nice property that we call *symmetry equivariance*: if we switch the group membership and switch the responses then β_d does not change. Returning to Example 7.1, the observed success rate of the treated group is $\hat{\beta}_d = 67/69 - 10/63 = 0.812$ larger than that of the not treated group, and equivalently the observed failure rate of the not treated group is $\hat{\beta}_d = 53/63 - 2/69 = 0.812$ larger than that of the treated group. The ratio does not have symmetry equivariance. The observed success rate of the treated group is $\hat{\beta}_r = (67/69)/(10/63) = 6.1$ times larger than that of the not treated group, but the observed failure rate of the not treated group is $(53/63)/(2/69) = 29.0$ times larger than the treated group. The odds ratio has symmetry equivariance.

The odds ratio seems like an overly complicated effect parameter, but it is a natural parameter to use with Fisher's exact test. Odds ratios will be very important later when we deal with stratification (Section 12.3.3) and modeling of binary effects in logistic regression (see Section 14.3.2). For some scientific problems odds ratios are very useful.

Example 7.2 *Suppose we are interested in the incidence rates of lung cancer in 55–59-year-olds in some population, and we are comparing people who have never regularly smoked cigarettes (Group 1) to those who have smoked regularly (Group 2). Suppose first that we took a simple random sample of n 55–59-year-olds from the population and see how many were in Group 1 (say n_1) and Group 2 (n_2), and see how many developed lung cancer in the previous year in each group, X_1 and X_2 respectively. If we ignore the fact that n_1 and n_2 are not fixed by design, we can treat X_1 and X_2 as binomial with means θ_1 and θ_2. The problem is that in this population in the United States (pooling never smokers and ever smokers), the incidence rate for lung cancers is on the order of 1 per 1,000. So to estimate $\theta_2 - \theta_1$ or θ_2/θ_1 we would typically first get estimates of θ_1 and θ_2, and because the rates are so low, we would need to have n_1 and n_2 be much larger than 1,000. For estimating the odds ratio, $\theta_2(1 - \theta_1)/\{\theta_1(1 - \theta_2)\}$, we do not need to sample that large a number. To see this, consider the data on the ith individual in the population. Let $Y_i = 1$ if that individual developed lung cancer in the previous year and $Y_i = 0$ otherwise, and let $Z_i = 1$ if the ith individual never smoked and $Z_i = 2$ if they did smoke. Then consider if the ith individual was a random draw from the population, $\theta_a = Pr[Y_i = 1 | Z_i = a]$. Using Bayes' rule (Problem P2), we can show that*

$$\frac{\theta_2(1 - \theta_1)}{\theta_1(1 - \theta_2)} = \frac{Pr[Y_i = 1 | Z_i = 2]Pr[Y_i = 0 | Z_i = 1]}{Pr[Y_i = 1 | Z_i = 1]Pr[Y_i = 0 | Z_i = 2]}$$

$$= \frac{Pr[Z_i = 2 | Y_i = 1]Pr[Z_i = 1 | Y_i = 0]}{Pr[Z_i = 1 | Y_i = 1]Pr[Z_i = 2 | Y_i = 0]}. \tag{7.1}$$

So instead of taking a simple random sample from the population, we can get efficient estimates of the odds ratio by taking a much smaller total number of individuals, we take conditional samples of cases (those with $Y_i = 1$) and controls (those with $Y_i = 0$) and estimate the proportion of each smoking group within the cases and controls. This kind of study design is called a case-control study (see Note N1). In the case-control study, we do not need to get good estimates of θ_1 and θ_2 in order to get a good estimate of the odds ratio.

There are some possible results that are essentially uninformative with respect to the parameters β_r and β_{or}. For example, the result $[X_1, X_2] = [0,0]$ is uninformative about β_r because this result is equally likely for the large effect $\beta_r = 100$ and its opposite effect $\beta_r = 1/100$. For example, if $n_1 = n_2 = 10$, $\theta_1 = 0.0001$, and $\theta_2 = 0.01$ then $Pr[X_1 = 0, X_2 = 0] = 0.903$ and $\beta_r = 100$, but if we switch the parameters we still have $Pr[X_1 = 0, X_2 = 0] = 0.903$ but now $\beta_r = 1/100$. Analogously, the points $[0,0]$ and $[n_1, n_2]$ are uninformative about β_{or}. Because of this, we want any test applied to an uninformative result to give a p-value of 1 and fail to reject for any significance level. Note that, although the points $[0,0]$ and $[n_1, n_2]$ are uninformative for β_{or}, they are informative for β_d.

7.3 Conditional Tests

7.3.1 Central Fisher's Exact Test

Let β be the effect parameter that could represent either β_d, β_r, or β_{or}. Write $\beta = \beta(\theta)$, where $\theta = [\theta_1, \theta_2]$ to emphasize the dependence on θ. The p-value is loosely defined as the maximum probability of observing equal or more extreme values under the null hypothesis. So we first need a statistic that we can use to define extremeness with respect to the null hypothesis.

Start with the one-sided null hypothesis, $H_0 : \beta(\theta) \leq \beta_0$. Let $\mathbf{X} = [X_1, X_2]$ and let $T(\mathbf{X})$ be a function of the data where higher values suggest higher values of β. For example, $T(\mathbf{X})$ could be $\hat{\beta}$, an estimator of β calculated by replacing θ_1 and θ_2 with the sample proportions $\hat{\theta}_1 = X_1/n_1$ and $\hat{\theta}_2 = X_2/n_2$. If $T(\mathbf{X})$ is defined for all $\mathbf{X}$, an unconditional p-value for testing $H_0 : \beta(\theta) \leq \beta_0$ maximizes $\Pr_\theta[T(\mathbf{X}) \geq T(\mathbf{x})]$ over all θ in the null hypothesis. This is not straightforward for two reasons. First, it requires searching over all the parameters in the null hypothesis. Second, when $T(\mathbf{X})$ represents $\hat{\beta}_r$ or $\hat{\beta}_{or}$ there are undefined values at $\mathbf{X} = [0,0]$ (for both parameters) and $\mathbf{X} = [n_1, n_2]$ (for $\hat{\beta}_{or}$). We deal with these issues in Section 7.4. An alternative approach is to condition on the total number of events $S = X_1 + X_2$, which we now discuss.

After conditioning on n_1, n_2, and $S = s = x_1 + x_2$, the data are completely defined except for one degree of freedom. This means that after conditioning on $S = s$, we can write $T(\mathbf{X}) = T([s - X_2, X_2])$ in terms of X_2 only. Further, since larger $T(\mathbf{X})$ values suggest larger β, for any $a < b$ we should have $T([s - a, a]) < T([s - b, b])$ for any reasonable ordering function T. For these reasonable T functions, we can order based on X_2 instead of $T(\mathbf{X})$. The distribution of X_2 given $S = s$ is the extended or noncentral hypergeometric distribution,

$$\Pr_{\beta_{or}}[X_2 = x | S = s] = \frac{\dbinom{n_1}{s - x}\dbinom{n_2}{x}(\beta_{or})^x}{\sum_{i=x_{min}}^{x_{max}} \dbinom{n_1}{s - i}\dbinom{n_2}{i}(\beta_{or})^i}, \tag{7.2}$$

where $x_{min} = \max(0, s - n_1)$ and $x_{max} = \min(s, n_2)$. So if $\beta = \beta_{or}$, a one-sided p-value for testing $H_0 : \beta(\theta) \leq \beta_0$ is

$$p_{AG}(\mathbf{x}, \beta_0) = \sup_{\beta \leq \beta_0} \Pr_\beta[X_2 \geq x_2 | S = s] = \Pr_{\beta_0}[X_2 \geq x_2 | S = s], \tag{7.3}$$

(Problem P4), and a one-sided p-value for testing $H_0 : \beta(\theta) \geq \beta_0$ is

$$p_{AL}(\mathbf{x}, \beta_0) = \Pr_{\beta_0}[X_2 \leq x_2 | S = s].$$

A central p-value, $p_c(\mathbf{x}, \beta_0)$ for testing $H_0 : \beta = \beta_0$ is defined in the usual way (see Equation 2.11), and the associated central $100(1 - \alpha)\%$ confidence interval is

$$C_c(\mathbf{x}, 1 - \alpha) = \{b : p_c(\mathbf{x}, b) > \alpha\}. \tag{7.4}$$

The values p_c and C_c are associated with the central Fisher's exact test.

Example 7.3 *Consider the fake data $x_1/n_1 = 2/11$ and $x_2/n_2 = 6/8$. The central Fisher's exact test gives $p_c(\mathbf{x}, 1) = 0.043$ with $\tilde{\beta}_{or} = 11.25$, the maximum likelihood estimate under the extended hypergeometric model (Note N2), and 95% confidence interval $(1.057, 208.70)$.*

It is clear from Equation 7.2 that the odds ratio parameter is the natural parameter to use with conditional exact tests. However, sometimes we use the difference, β_d, or ratio, β_r, for simplicity or because there is a natural reason for the application. We can get confidence intervals for β_d and β_r (as well as for β_{or}) that are compatible with the central Fisher's exact test by using melded confidence intervals. Section 10.11 reviews the general melding method.

Here is the melding method as applied to the two-sample binomial problem. Start with the exact central confidence interval procedures for θ_1 and θ_2 (Section 4.2). Let the $100(1 - \alpha)\%$ exact central interval for θ_1 be $\{\theta_{L1}(\mathbf{x}, 1 - \alpha/2), \theta_{U1}(\mathbf{x}, 1 - \alpha/2)\}$. Define the lower and upper *confidence distribution random variables* for θ_1 as $W_{L1} = \theta_{L1}(\mathbf{x}, U_a)$ and $W_{U1} = \theta_{U1}(\mathbf{x}, U_b)$, where U_a and U_b are independent uniform random variables (Problem P5). Analogously define W_{L2} and W_{U2}. Then the $100(1 - \alpha)\%$ melded confidence interval for $\beta_d = \theta_2 - \theta_1$ is

$$C_{c(d)}(\mathbf{x}, 1 - \alpha) = \{q(\alpha/2, W_{L2} - W_{U1}), q(1 - \alpha/2, W_{U2} - W_{L1})\}, \tag{7.5}$$

where $q(a, W)$ is the ath quantile of the random variable W. To ensure validity we use W_{L2} and W_{U1} for the lower bound and W_{U2} and W_{L1} for the upper bound. The melded confidence interval can be calculated with very high accuracy by Monte Carlo simulation or by numeric integration. Although the melding method did not use any conditioning on the sum, there is an interesting connection to the central Fisher's exact p-value. First, note that $H_0 : \theta_1 = \theta_2$ is equivalent to $H_0 : \beta_d = 0$ and $H_0 : \beta_r = 1$ and $H_0 : \beta_{or} = 1$. To differentiate the different effect parameters, we write the central Fisher's exact p-value in three different ways,

$$p_{c(d)}(\mathbf{x}, 0) = p_{c(r)}(\mathbf{x}, 1) = p_{c(or)}(\mathbf{x}, 1).$$

As an aside, for the difference when $\beta_0 \neq 0$ then $p_{c(d)}(\mathbf{x}, \beta_0)$ is not calculated using Fisher's exact methods because we cannot use the extended hypergeometric distribution as in Equation 7.3. Nevertheless, it turns out (Problem P6) that $p_{c(d)}(\mathbf{x}, 0)$ is compatible with $C_{c(d)}$, meaning $p_{c(d)}(\mathbf{x}, 0) \leq \alpha$ if and only if $0 \notin C_{c(d)}(\mathbf{x}, 1 - \alpha)$.

We can similarly define melded confidence intervals for β_r and β_{or}. The $100(1 - \alpha)\%$ melded confidence interval for $\beta_r = \theta_2/\theta_1$ is

$$\{q(\alpha/2, W_{L2}/W_{U1}), q(1 - \alpha/2, W_{U2}/W_{L1})\}$$

and for $\beta_{or} = \theta_2(1 - \theta_1)/\{\theta_1(1 - \theta_2)\}$ is

$$\left\{ q\left(\frac{\alpha}{2}, \frac{W_{L2}(1 - W_{U1})}{W_{U1}(1 - W_{L2})} \right), q\left(1 - \frac{\alpha}{2}, \frac{W_{U2}(1 - W_{L1})}{W_{L1}(1 - W_{U2})} \right) \right\}.$$

Both of those confidence intervals, as well as $C_{c(d)}$, are compatible with the central Fisher's exact p-value.

Example 7.3 *(continued) Consider the fake data $x_1/n_1 = 2/11$ and $x_2/n_2 = 6/8$. The central Fisher's exact test gives $p_c = 0.043$ for $H_0 : \theta_1 = \theta_2$, and the 95% melded confidence intervals are: $(0.014, 0.881)$ for β_d, $(1.033, 34.46)$ for β_r, and $(1.061, 382.95)$*

for β_{or}. The lower confidence bound for β_{or} is close to the one calculated from the extended hypergeometric distribution, but the two upper confidence bounds for β_{or} are quite different. We expect the two confidence interval methods for β_{or} to be similar for a bound that is close to 1, because both the melded confidence intervals and the extended hypergeometric confidence intervals are compatible with the p-value at the null $\theta_1 = \theta_2$ where $\beta_{or} = 1$.

7.3.2 Fisher–Irwin Test and Blaker Test

In the literature, there are two versions of the two-sided Fisher's exact test, and both are called Fisher's exact test. We will refer to the different versions as the central Fisher's exact test and the Fisher–Irwin test. The latter test was independently proposed by Irwin (1935). It appears that Fisher preferred the central Fisher's exact test over the Fisher–Irwin test (see Yates, 1984, p. 444); nevertheless, many software packages mean the Fisher–Irwin test when they say "Fisher's exact test" (for example, SAS Proc Freq and fisher.test in the stats R package).

The Fisher–Irwin test defines the p-value for testing $H_0 : \theta_1 = \theta_2$ as

$$p_{FI}(\mathbf{x}) = \Pr[f(X_2) \le f(x_2)], \tag{7.6}$$

where X_2 follows a hypergeometric distribution and $f(x)$ is its probability mass function (i.e., $f(x) = \Pr_{\beta_{or}}[X_2 = x | S = s, \beta_{or} = 1]$ from Equation 7.2). To compare the two Fisher's exact p-values, in Figure 7.1 we plot the hypergeometric distribution associated with $x_1/n_1 = 2/11$ and $x_2/n_2 = 6/8$. The central Fisher's exact test p-value is $2 \sum_{x=6}^{8} f(x) = 2 * 0.02155 = 0.0431$. At $x_2 = 6$ the probability mass function (pmf) is $f(6) = 0.0204$ and the pmf values less than that are at $X_2 \in \{0, 7, 8\}$, so the sum $f(6) + f(0) + f(7) + f(8)$ is the Fisher–Irwin p-value of 0.0237. Typically, the Fisher–Irwin p-value will be less than the

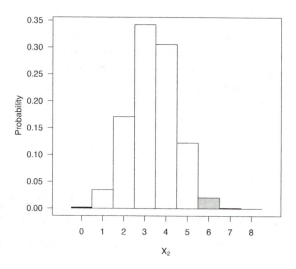

Figure 7.1 Hypergeometric distribution associated with $x_1/n_1 = 2/11$ and $x_2/n_2 = 6/8$. The gray bar is the observed value, $x_2 = 6$, and the black lines at $X_2 = 0, 7, 8$ represent those very small values of $f(X_2)$.

central Fisher's exact p-value, although it is possible for the central Fisher's exact p-value to be less (Problem P7). Blaker (2000) studied a two-sided exact test applied to this problem that always has its p-value less than the central Fisher's exact p-value (Problem P8).

There are two problems with the routine use of the Fisher–Irwin test or the Blaker test. First, there does not exist associated confidence intervals for each of these tests that are compatible with their p-values (see Example 2.5 for data that does not allow compatible confidence intervals using Fisher–Irwin test). Despite this we can get a confidence region associated with the p-value and from that define the matching confidence interval as the minimum and maximum values in the region (see, e.g., the **exact2x2** R package). The second and more common problem is that when we do not use a central p-value, then we cannot automatically perform a three-decision rule test (see Section 2.3.4). In other words, if we reject the two-sided hypothesis, then we cannot use half of that two-sided p-value to give the significance of the appropriate one-sided test. For example, a one-sided Fisher's exact test that is not significant at 0.025, could have a Fisher–Irwin two-sided p-value less than 0.05.

Example 7.4 *PREVAIL II Writing Group (2016): In 2015, the PREVAIL II trial randomized patients in West Africa with Ebola virus disease to ZMapp (an experimental treatment) plus standard of care (SOC) versus standard of care alone. A key analysis considered the subgroup with baseline risk of mortality less than or equal to 50%. In this low risk subgroup, the 28 day mortality rate was 0/24 in ZMapp+SOC arm versus 4/20 in the SOC alone arm. The (two-sided) Fisher–Irwin p-value was 0.0357 which appears significant at the 0.05 level; however the one-sided test Fisher's exact p-value was also 0.0357. Thus, if someone planned to do a two-sided Fishers exact test at the 0.05 level, and used Fisher–Irwin, they would have considered this comparison significant, but if they had planned to use a one-sided Fishers exact test at the 0.025 level, they would not have rejected. No such inconsistency would have arisen with the (two-sided) central Fishers exact test (p = 2*0.0357 = 0.0714): a two-sided rejection decision at level 0.05 with the central Fisher's exact test will always agree with the appropriate one-sided Fisher's exact test rejection decision at the 0.025 level. (PREVAIL II Writing Group (2016) determined that this was a significant difference using a preplanned Bayesian approach.)*

7.4 Unconditional Tests

7.4.1 Overview

There is much discussion in the literature on the choice between the conditional or the unconditional test for the two-sample binomial problem (see, e.g., Yates, 1984). The main issue is with the discreteness of the data and the decisions. If we fix the significance level and condition, then the sample space is much smaller than if we do not condition. The larger sample space allows more places to draw the decision line(s), and therefore the unconditional method tends to be more powerful. However, there is essentially no information about the effect, β, in the sum $s = x_1 + x_2$, and it is standard practice to condition on statistics that do not have any information about the effect of interest, and this suggests the conditional test. Further, some argue that it is arbitrary to set the significance level at 0.05, and we should instead examine p-values, essentially eliminating the advantage of the unconditional tests.

On a practical note, the unconditional tests tend to be more computationally challenging, so conditional tests tend to be used because of that. But this last concern will become less of a problem in the future.

As we have seen in Section 7.3.1, for one-sided tests there is essentially only one ordering of the sample space (see discussion above Equation 7.2), whereas for unconditional tests there are many possibilities for $T(\mathbf{X})$ functions that order the sample space. For example, we can use $T(\mathbf{X}) = \hat{\beta}$ to order the sample space. Unfortunately, some of the more powerful methods use more complicated ordering functions for $T(\mathbf{X})$. Boschloo (1970) proposed using the p-values from the Fisher's exact test as $T(\mathbf{X})$, and the resulting unconditional test is known as Boschloo's test. Martín Andrés et al. (1998) proposed a modification of *Boschloo's test*, that uses the mid p-values from Fisher's exact test as $T(\mathbf{X})$, and we call the resulting generally slightly more powerful test the *modified Boschloo test*. We discuss that test because of its power, not its simplicity!

7.4.2 Modified Boschloo Test

Again, start with the one-sided null hypothesis $H_0 : \beta(\theta) \leq \beta_0$. Let $T(\mathbf{X})$ be a function of the data where higher values suggest higher values of β. Typically, such a function, T, would meet the following reasonable conditions (called *Barnard's convexity conditions*): for any x_1 if $x_2^* > x_2$ then $T([x_1, x_2^*]) \geq T([x_1, x_2])$, and for any x_2 if $x_1^* < x_1$ then $T([x_1^*, x_2]) \geq T([x_1, x_2])$. A good function that meets these conditions and can be used with any β parameter is

$$T(\mathbf{X}) = F(X_2; n_2, n_1, X_1 + X_2) - \frac{1}{2} f(X_2; n_2, n_1, X_1 + X_2), \qquad (7.7)$$

where F and f are the cdf and pmf of the hypergeometric distribution. The unconditional test that orders by $T(\mathbf{X})$ of Equation 7.7 will be called the modified Boschloo test. Confusingly, although the modified Boschloo test is a valid test and is not a mid-p method (see Section 7.5), the T function is the mid p-value of a one-sided Fisher's exact test. Do not be confused, here we are using T as an ordering function, not as a mid p-value. We can use any function to order the sample space to create our p-value, and by using the mid p-value instead of the p-value to order it some more ties in the ordering will be broken, and hence the test will be slightly more powerful. And the resulting unconditional tests will be valid (i.e., exact).

7.4.3 Algorithmic Details

This subsection may be skipped for those interested mostly in the applications. Here is how to calculate one-sided p-values and central confidence intervals for β from any ordering function, T. First, let $\mathcal{X}_I$ be the set of $\mathbf{X}$ with information about β. So $\mathcal{X}_I$ is all possible values of $\mathbf{X}$ when $\beta = \beta_d$, but excludes $\mathbf{X} = [0,0]$ when $\beta = \beta_r$, and excludes $\mathbf{X} = [0,0]$ and $\mathbf{X} = [n_1, n_2]$ when $\beta = \beta_{or}$. For testing any hypothesis, if $\mathbf{x} \notin \mathcal{X}_I$ then $p(\mathbf{x}) = 1$ and the confidence interval includes all values of the parameter. Note, even if $T(\mathbf{x})$ is defined for $\mathbf{x} \notin \mathcal{X}_I$ (e.g., Equation 7.7), we still set $p(\mathbf{x}) = 1$ for $\mathbf{x} \notin \mathcal{X}_I$. When $\mathbf{x} \in \mathcal{X}_I$ then for testing $H_0 : \beta(\theta) \leq \beta_0$ use

$$p_{AG}(\mathbf{x}, \beta_0) = \sup_{\theta : \beta(\theta) \leq \beta_0} \Pr_\theta[T(\mathbf{X}) \geq T(\mathbf{x}) | \mathbf{X} \in \mathcal{X}_I] \Pr_\theta[\mathbf{X} \in \mathcal{X}_I] \tag{7.8}$$

and for testing $H_0 : \beta(\theta) \geq \beta_0$ use

$$p_{AL}(\mathbf{x}, \beta_0) = \sup_{\theta : \beta(\theta) \geq \beta_0} \Pr_\theta[T(\mathbf{X}) \leq T(\mathbf{x}) | \mathbf{X} \in \mathcal{X}_I] \Pr_\theta[\mathbf{X} \in \mathcal{X}_I]. \tag{7.9}$$

This calculation is simplified when T meets Barnard's convexity conditions, and in that case Röhmel and Mansmann (1999) show that we only need to calculate the supremum over the boundary between the hypothesis (i.e., over $\{\theta : \beta(\theta) = \beta_0\}$). We can get lower and upper bounds for the $100(1 - \alpha)\%$ central confidence interval by

$$\beta_L(\mathbf{x}) = \min \{\beta : p_{AG}(\mathbf{x}, \beta) > \alpha/2\} \tag{7.10}$$

and

$$\beta_U(\mathbf{x}) = \max \{\beta : p_{AL}(\mathbf{x}, \beta) > \alpha/2\}. \tag{7.11}$$

A simplication of the confidence limit calculation occurs when T meets Barnard's convexity conditions and $\beta = \beta_d$ so that $\mathcal{X}_I$ is the full sample space. In this case, the one-sided p-values are monotonic in β_0 and we can use a halfing algorithm to search for the confidence bounds.

7.4.4 Other Unconditional Tests

Many other ordering functions, T, have been suggested, each leading to a different unconditional test. We briefly mention several, and state why each is not our default recommendation.

We could order by the simple maximum likelihood estimates of the effect parameters (i.e., plug in the sample proportions into the definition). For example, for testing β_d use $\hat{\beta}_d = \hat{\theta}_2 - \hat{\theta}_1$ where $\hat{\theta}_a$ is the sample proportion for Group a. There are several issues with this. First, for β_d there can be many ties when $n_1 = n_2$. We can break the ties such that we define values as further away from $\hat{\beta}_d = 0$ for larger values of,

$$\frac{\hat{\beta}_d^2}{\widehat{Var}_0(\hat{\beta}_d)} = \frac{(\hat{\theta}_2 - \hat{\theta}_1)^2}{\hat{\theta}(1 - \hat{\theta})/(n_1 + n_2)},$$

where $\hat{\theta} = (x_1 + x_2)/(n_1 + n_2)$. This tie-breaking method gives a reasonably powered test for testing equality ($H_0 : \theta_1 = \theta_2$), but it is not, neral, more powerful than the modified Boschloo test. For β_r, the unconditional exact test based on $\hat{\beta}_r = \hat{\theta}_2/\hat{\theta}_1$ can have very poor power for testing equality even after breaking ties.

In general, the more powerful unconditional tests for testing equality use ordering functions, T, based on score statistics (see Section 10.7). The disadvantage of these score-statistic-based tests is that it is not always possible to create compatible confidence intervals from them. The issue is that the ordering function for the score statistic depends on the null value, β_0, for testing $H_0 : \beta = \beta_0$ when β_0 does not imply equality (e.g., $\beta_0 = 0$ for the difference or $\beta_0 = 1$ for the ratio). This can lead to different orderings for different values of β_0 (which would be used in the creation of the confidence interval).

Wang (2010) and Wang and Shan (2015) developed the smallest valid central confidence intervals for each of the three effect parameters, but they are more complicated than basing the ordering on score statistics. The algorithm for these intervals involves adding one point at a time in such a way as to minimize the length of each of the one-sided confidence interval for that specific nominal level. Computationally, these intervals are very complicated, as the resulting ordering function, T, depends on the nominal level $100(1-\alpha)\%$ and two confidence intervals with slightly different nominal levels may have a different ordering. This can lead to strange intervals, for example, the 97.5% one-sided confidence interval for β_d based on $x_1/n_1 = 2/5$ and $x_2/n_2 = 2/7$ gives $(-1, 0.467)$, but the 98% interval for the same data gives a smaller interval $(-1, 0.442)$. These one-sided intervals are not nested.

Agresti and Min (2001, 2002) recommended noncentral confidence intervals based on unconditional exact tests using score statistics (see Section 10.7). These intervals can be smaller than the central ones, but if directional inferences are important these intervals are not recommended (similar to the recommendation of Section 7.3.2 in the conditional case).

7.4.5 The Berger and Boos Adjustment

Berger and Boos (1994) developed a general method for calculating p-values when there is a nuisance parameter, such as in the two-sample independent binomial case discussed in this chapter. We can add the Berger–Boos adjustment onto any of the unconditional tests previously described.

Here we describe the Berger–Boos adjustment as applied to the one-sided test for $H_0 : \beta_d(\theta) \le \beta_0$. Reparametrize θ as $[\beta_d, \xi]$, where $\xi = \frac{\theta_1 + \theta_2}{2}$. For β_d, Equation 7.8 can be rewritten as

$$p_{AG}(\mathbf{x}, \beta_0) = \sup_{\beta, \xi : \beta \le \beta_0} \Pr_{\beta, \xi}[T(\mathbf{X}) \ge T(\mathbf{x})]. \qquad (7.12)$$

Suppose C_γ is a $100(1 - \gamma)\%$ confidence interval for ξ. Then the Berger–Boos adjusted p-value is

$$p_{AG}^{BB}(\mathbf{x}, \beta_0; \gamma) = \gamma + \sup_{\beta, \xi : \xi \in C_\gamma, \beta \le \beta_0} \Pr_{\beta, \xi}[T(\mathbf{X}) \ge T(\mathbf{x})].$$

Loosely speaking, we restrict the parameter space, searching only over the parameter values likely in the data, but since there is still a γ possibility that there is a parameter in the null space with ξ outside of C_γ, we add on γ to take care of that possibility. This method gives a new valid p-value.

The Berger–Boos adjustment often (but not always) will give a smaller p-value. Further, we can create confidence intervals from Berger–Boos adjusted p-values. For application to the two-sample binomial problem see Lloyd (2008) who develops another adjustment method, the $E + M$ method.

7.5 Mid-p Methods

Similar to how the mid p-values and mid-p confidence intervals were created for the one-sample binary case (see Section 4.3.3), we can create mid-p methods for the two-sample binomial case. We simply adjust the p-value definition, replacing the probability that a test statistic is equal or more extreme than the observed test statistic with the probability that that same test statistic is more extreme plus one-half times the probability that it is equal. This idea can be applied to a one-sample conditional p-value (such as p_{AG} of Equation 7.3), to a one-sample unconditional p-value (e.g., Equation 7.8), or to a two-sided p-value. The resulting test based on a mid p-value will have Type I error rate closer to the nominal α, but the error rate may exceed α for some values of the parameter. We can calculate a compatible confidence interval with the central mid-p Fisher's exact test for β_{or} by inverting the central mid-p Fisher's exact test analogously to Equations 7.10 and 7.11. Similar to the Fisher's exact test, we cannot invert the series of one-sided tests for β_d and β_r with the mid-p Fisher's exact test, because the mid p-value is created using the extended hypergeometric distribution and β_{or}.

However, we can get confidence intervals and p-values for β_d and β_r by using a mid-p modification of the melded confidence interval. These intervals are created by replacing the p-value function with the mid p-value function in the creation of the confidence distribution random variables for θ_1 and θ_2. We can write the mid-p confidence distribution random variable as the weighted sum of the usual lower and upper confidence distribution random variables. For example, for θ_1 we get $W_1 = BW_{1L} + (1 - B)W_{1U}$ where B is a Bernoulli random variable with mean 0.5 (see, e.g., Fay and Brittain, 2016, Appendix). Then for example, for the mid-p melded confidence interval for β_d, in Equation 7.5 we replace W_{1L} and W_{1U} both with W_1 and analogously do the same for the θ_2 confidence distribution random variables. The associated p-values are $p_{AG(d)}(\mathbf{x}, 0) = \Pr[W_2 \leq W_1] = 1 - p_{AL(d)}(\mathbf{x}, 0)$. These p-values will not always equal the mid p-values from Fisher's exact test (see Figure 7.2). The mid-p melding method is not guaranteed to be valid, but like all mid-p methods has closer to nominal Type I error rates (for tests) and closer to nominal coverage (for confidence intervals) averaging over the possible parameter values.

The central Fisher's exact test is so conservative, however, that for its mid-p version there may be very few places (if any) in the null equality parameter space with Type I error rate greater than α. Consequently, although the mid-p version was not designed to be valid, for some sample sizes it turns out to be valid. To show this we calculate the power to reject $H_0 : \theta_2 \leq \theta_1$ for the one-sided Fisher's exact test, the mid-p version of the Fisher's exact test, or the mid-p melded p-value for several sample sizes for different values of $\theta = \theta_1 = \theta_2$ on the boundary between the null and alternative. The results are plotted in Figure 7.2. Thus, the mid-p version of Fisher's exact test may be used if the statistician can check beforehand that the the mid-p version is valid for the designed sample sizes, or if guaranteed validity is not absolutely vital. Figure 7.2 suggests that the mid-p melded method is less likely to be valid than the other two.

We can also get mid-p versions of unconditional exact confidence intervals which like other mid-p methods should be closer to the nominal level. As with other unconditional exact tests, the algorithms for the mid-p version will tend to take much longer than the conditional tests.

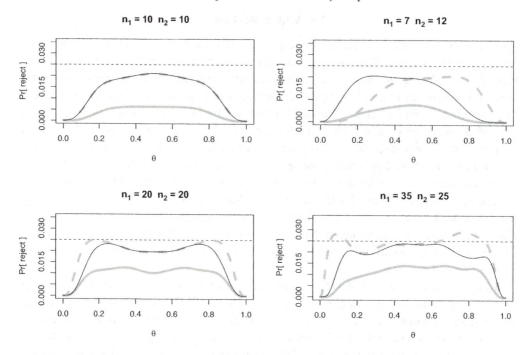

Figure 7.2 Type I error rate for the one-sided Fisher's exact test (thick gray line), the mid-p version of the one-sided Fisher's exact test (thin black line), or the mid-p melded method (thick dashed gray line), all tested at the 0.025 level. In all the cases plotted, the mid-p version of Fisher's exact test has Type I error less than the nominal, even though it is not guaranteed to have that property for all n_1 and n_2.

7.6 Asymptotic Approximations

The differences between all of the previously discussed methods in this chapter will become smaller and smaller as the sample size gets larger. Thus, if you have a small or moderate sample size then any of those previous methods are acceptable; however, very simple asymptotic approximations may not be very good (see Figure 7.4). Because some of the asymptotic approximations are widely used, we briefly review them.

We know from Section 4.3.4 that for large n_a, the sample proportion $\hat{\theta}_a = x_a/n_a$ is approximately normal with mean θ_a and variance $v_a = \theta_a(1 - \theta_a)/n_a$, for $a = 1, 2$. So by properties of sums of independent normal random variables, for large n_1 and n_2, $\hat{\theta}_2 - \hat{\theta}_1$ is approximately normal with mean $\theta_2 - \theta_1$ and variance equal to $v_1 + v_2$. Thus, an approximate $100(1 - \alpha)\%$ confidence interval which we call the *Wald interval* is

$$\hat{\theta}_2 - \hat{\theta}_1 \pm \Phi^{-1}(1 - \alpha/2)\sqrt{\frac{\hat{\theta}_1(1 - \hat{\theta}_1)}{n_1} + \frac{\hat{\theta}_2(1 - \hat{\theta}_2)}{n_2}}. \tag{7.13}$$

There are more accurate asymptotically normal approximations based on score tests (see Section 10.7 and Farrington and Manning (1990)), but since there are nonasymptotic methods that are tractable with fast algorithms (e.g., melded confidence intervals and mid-p melded confidence intervals), we do not explore them further.

There are asymptotic methods for β_r and β_{or} as well. In those cases it is usually useful to perform a transformation on the estimated β and then use the delta method (see Section 10.6) to get confidence intervals, or to use score test methods.

7.7 Comparison of *Binomial* Methods

In Figure 7.3 we illustrate how the methods compare across sample sizes by calculating 95% confidence intervals for β_d when $\hat{\theta}_2 = 0.9$ and $\hat{\theta}_1 = 0.8$ so that $\hat{\beta}_d = 0.1$. We compare the Wald confidence intervals (Equation 7.13), the melded confidence intervals that are compatible with the central Fisher's exact test, and the mid-p melded confidence intervals for $n_1 = n_2 = 10, 100, \ldots, 10^5$. We additionally tried the confidence intervals associated with the modified Boschloo test, but that method is very computationally involved (the $n_1 = n_2 = 1{,}000$ case took over five hours, while all other methods took under a second), so we did not calculate it for $n_a > 1{,}000$. For $n_1 = n_2 = 100$ or larger the mid-p melded CIs and the melded CIs appear similar and close to those of Wald, and all three CIs are essentially equal to the Wald CIs for $n_1 = n_2 = 1{,}000$.

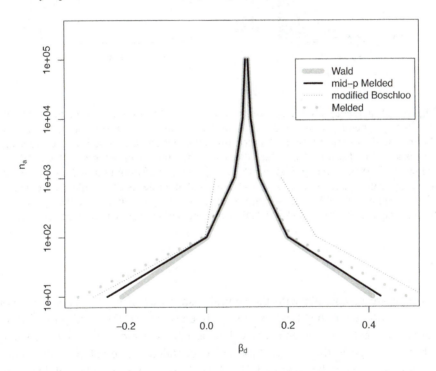

Figure 7.3 95% confidence intervals for β_d by modified Boschloo CI (black small dotted), Wald CI (gray solid), melded CI (gray dotted), or mid-p melded (black solid). Each confidence interval is based on $\hat{\theta}_2 = 0.9$ and $\hat{\theta}_1 = 0.8$ but with different sample sizes, $n_1 = n_2 = 10, 100, \ldots, 10^5$.

We plot the coverage of some of those 97.5% confidence intervals in Figure 7.4. First, we notice that the simple Wald interval is under covered throughout much of the parameter space, while some portion of the parameter space is conservative. The melded confidence

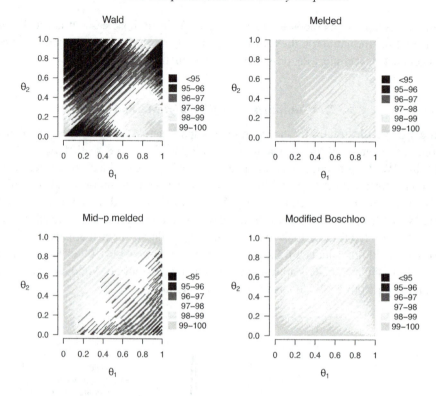

Figure 7.4 Percentage coverage of the one-sided 97.5% lower confidence interval (CI) for β_d when $n_1 = n_2 = 15$ by Wald CIs, melded CIs, mid-p melded CIs, and the modified Boschloo CIs (i.e., unconditional exact test CIs based on the ordering by Fisher's exact test mid p-value). We calculated the coverage at the 100×100 θ values in $\{[0.005, 0.005], [0.005, 0.015], \ldots, [0.995, 0.995]\}$ and plotted the coverage as a gray scale square. The colors are chosen so that lighter is better: ideal coverage is 97.5% (white is 97–98%), conservative coverage is better than under coverage so the lighter gray are 98–99% and 99–100%, while the darker gray denote under coverage.

intervals (that are compatible with the central Fisher's exact test) have at least nominal coverage everywhere but they tend to be slightly conservative. The mid-p melded interval has closer to nominal coverage on average, but has places where there is under coverage.

The best coverage is the modified Boschloo confidence interval (i.e., the confidence interval associated with the modified Boschloo test); it always gives at least nominal coverage but is less conservative than the melded interval. For large sample sizes, the unconditional exact interval can become very time consuming (although future algorithms may alleviate this problem).

There is no simple rule for when the asymptotic confidence interval converges to the exact. The asymptotic method can provide invalid inference for small and moderate sample sizes, but with increased sample size, all tests provide equivalent conclusions. Thus, a simple, and robust strategy is to use an exact method routinely.

7.8 Other Study Designs

The chapter deals with two-sample tests with binary responses, whose results can be summarized into a 2×2 table such as the following:

	Group 1	Group 2	
events	x_1	x_2	s
No events	$n_1 - x_1$	$n_2 - x_2$	$n - s$
	n_1	n_2	n

In the two-sample binomial problem, the bottom marginal totals (n_1 and n_2) are fixed, but the right-hand marginal totals (s and $n-s$) are random by design. In Section 7.3 we condition on $X_1 + X_2 = s$, but that is a computational technique that is motivated by the near-ancillarity of that marginal total on the parameters of interest (β_d, β_r, and β_{or}). When a conditional test such as Fisher's exact test is applied, fixing both marginals is used in calculations, but the use of the conditional test does not require that both marginals are fixed by design. In fact, Fisher's exact test may be applied when neither marginal is fixed by design (e.g., a simple random sample from a population where we test for independence of two binary variables measured on each individual in the sample). For that case, see also the goodness-of-fit tests in Section 19.2.3, Equation 19.2. The focus of this subsection is the case where both marginals are fixed by design.

The motivating problem for this section is a vaccine trial for COVID-19. A study design that fits a two-sample binomial model would be the following. Quickly accrue n individuals into the study, randomize $n_2 = n/2$ individuals to get the new vaccine and $n_1 = n/2$ to get a sham (i.e., placebo) vaccine, and then follow all n individuals for six months. At the end of six months, count the number of cases in the vaccine arm, say X_2, and in the placebo arm, say X_1. The two-sample binomial model fits this design well, where $X_a \sim Binomial(n_a, \theta_a)$ for $a = 1, 2$. The problem is that at the beginning of the COVID-19 pandemic, it is not clear how many people in the placebo group would get COVID-19 within the six months of the study. If too many people in the placebo group got COVID-19, then the study will have been longer than it needed to be to find out of the vaccine worked, while conversely, if not enough people in the placebo group got COVID-19 then the study may not be long enough to really know with precision whether the vaccine worked. A solution is to stop the study after s people, combined across groups, have gotten COVID-19, where s is chosen to be large enough to have sufficient power to show that the vaccine works.

Consider the following probability model to go with that design. Suppose there are n_a individuals in arm $a = 1, 2$. Let $X_a(t)$ be the number of events in arm a at time t and assume $X_a(t) \sim Poisson(n_a \lambda_a t)$. Under this simple model, the number of events in each arm is proportional to the number of people in that arm times the time on study (we can relax these assumptions later). We stop the study at the time when $X_1(t) + X_2(t) = s$. We are interested in the ratio of the two rates, $\beta = \lambda_2/\lambda_1$, and want to test the one-sided hypotheses, $H_0 : \beta \geq \beta_0$ versus $H_1 : \beta < \beta_0$. We use the following (Problem P9),

$$X_2(t) | X_1(t) + X_2(t) = s \sim Binomial\,(s, \pi)\,,$$

where

$$\pi = \frac{n_2\lambda_2}{n_1\lambda_1 + n_2\lambda_2} = \frac{\beta}{(n_1/n_2) + \beta}. \tag{7.14}$$

Thus, we can use methods for the single binomial random variable for inferences, such as the exact central confidence interval (see Section 4.2). The following hypotheses are equivalent:

	Original	Tranformed
	$H_0 : \beta \geq \beta_0$	$H_0 : \pi \geq \pi_0$
	$H_1 : \beta < \beta_0$	$H_1 : \pi < \pi_0,$

where

$$\pi_0 = \frac{\beta_0}{(n_1/n_2) + \beta_0}.$$

There are some modifications that can be made to relax assumptions. Instead of assuming $X_a(t) \sim \text{Poisson}(n_a\lambda_a t)$, we can assume $X_a(t) \sim \text{Poisson}(\lambda_a t_a)$, where t_a is the total person-years at risk for Group a. In this case, just replace n_1/n_2 with t_a/t_2 in the equations.

Example 7.5 Sadoff et al. (2021), Ad26.COV2.S Vaccine against COVID-19 (Johnson and Johnson vaccine): *For the J & J vaccine against COVID-19, there were $n_2 = 19,514$ an $n_1 = 19,544$ randomized and at-risk people in the vaccine and placebo arms, respectively, with $x_2 = 116$ and $x_1 = 348$ moderate or more severe COVID-19 cases in the two arms. The proportion in the vaccine group is $\hat{\pi} = 0.250 = 116/(348 + 116)$, and the 95% exact central confidence interval is $(0.211, 0.292)$. The number of person-years at risk is very similar but slightly more in the vaccine arm (because there is on average a longer time at risk per participant since fewer have the event and are taken out of the risk set after that), $t_2 = 3,116.6$, compared to the placebo arm, $t_1 = 3,096.1$. Vaccine efficacy (VE) is defined as $1 - \beta$, so translating the results into VE, we get $VE = 66.9\%$ with 95% confidence interval $(59.0\%, 73.4\%)$. Vaccine trials test $H_0 : VE \leq 0.30$ versus $H_1 : VE > 0.30$, and in this case the test is highly significant ($p < 0.0001$). The p-value comes from the binomial test $H_0 : \pi \geq \pi_0$ versus $H_1 : \pi < \pi_0$, where $\pi_0 = (1 - 0.30)/((t_1/t_2) + (1 - 0.30))$.*

7.9 Summary

This chapter primarily deals with the two-sample binomial problem. There are three commonly used parameters to measure effects when comparing proportions from two groups: difference in proportions; the ratio of proportions; and odds ratio. The difference in proportions and ratio of proportions are both simple measures. However, unlike the ratio of proportions, the difference of proportions has the nice property of symmetry equivariance: if the group membership is switched and the responses are also switched, the parameter does not change. While the odds ratio is much less intuitive than the other two, it is the natural metric corresponding to some key analytic methods including Fishers exact test and logistic regression. The odds ratio, like the difference in proportions, also has the advantage of symmetry equivariance. The choice between the parameters will often be driven by the context. For example, the ratio of two extremely small proportions may be more meaningful than the difference of these proportions. Also, some research arenas have traditionally used one effect parameter, and as long as the traditional parameter is reasonable, it usually makes sense to continue with that practice.

Fisher's exact test is a commonly used conditional exact test, however it is not well recognized that there are two versions of the two-sided Fisher's exact test. We recommend the central Fishers exact test, which is compatible with melded confidence intervals for any of the three parameters. The alternate version is the Fisher–Irwin test, which is often the default Fishers exact test in software packages. We do not favor the Fisher–Irwin for two reasons. First, there is no existing compatible confidence intervals. Second, there is an unappealing relationship between one- and two-sided testing, that is, if we reject at 0.05 level in a two-sided test, we do not necessarily reject at 0.025 level in a one-sided test. In those settings where strict control of the Type I error is not essential; a mid-p version of the central Fisher's exact test and the associated melding procedure could be considered for greater power.

Another class of exact tests are unconditional, that is, unlike Fishers exact test, these procedures do not condition on the margins of the 2×2 tables. These tests are generally more powerful than the Fishers exact test. However, they can be very computationally intensive. In addition, they are less well known, which might be a consideration when working with nonstatistical colleagues. However, their use and their inclusion in statistical packages may continue to increase as the corresponding improvement in power becomes more widely recognized. In the past, there have been philosophical disagreements about conditional versus unconditional tests, but there seems to be less controversy about this question in the contemporary statistical literature.

Some researchers still use asymptotic approaches to provide an approximate comparison of two proportions. This made sense prior to the modern computational era, but, in our view, the prudent approach now is routine use of exact methods since for most moderate sample sizes those methods are tractable. Asymptotic approximations may be useful when calculation speed is an issue (i.e., when testing millions of hypotheses).

7.10 Extensions and Bibliographic Notes

Hirji (2006) is a very comprehensive book on exact hypothesis tests. There are good references to the historical Fisher exact test and the Irwin test, and good descriptions of some algorithms for calculating exact tests. Fay and Hunsberger (2021) give a review of the two-sample binomial problem with a similar perspective as this chapter but with more details. Yates (1984) gives a nice perspective on the historical debates about tests on 2×2 tables (of which the two-sample binomial problem is a special case).

For good score test-based methods see Farrington and Manning (1990) for β_d, Miettinen and Nurminen (1985) for β_r, and Agresti and Min (2002) for β_{or}. For details on the delta method applied to the two-sample binomial problem see section 3.1 of Agresti (2013). Agresti and Min (2001) gives a nice summary of how noncentral score-based asymptotic confidence intervals can give overall coverage closer to the nominal level than many exact methods; however, these confidence intervals and associated tests have no advantage over central methods when used with three-sided tests, so if directional inferences are important, they should not be used (see Section 2.3.4).

7.11 R Functions and Packages

The exact2x2 R package does many of the tests in this chapter. The central Fisher's exact test (tsmethod = "central") and the Fisher–Irwin (tsmethod = "minlike") give p-values and associated confidence intervals in the fisher.exact function of exact2x2. The melding method is calculated by the binomMeld.test function. There are many versions of the unconditional exact test in the uncondExact2x2 function (including Boschloo test, modified Boschloo test [using method = "FisherAdj"], and ordering functions based on score tests), and uncondExact2x2 has options for the Berger–Boos and $E + M$ adjustments. The test functions of exact2x2 have a mid-p argument to give mid-p versions. There are also functions to calculate sample size and power.

The fisher.test function in base R calculates the two-sided p-value using the Fisher–Irwin test but the confidence interval as an inversion of the central Fisher's exact test (see Fay, 2010a).

The Exact R package gives p-values for Barnard's CSM test (an unconditional exact test, see, e.g., Fay and Hunsberger, 2021) and Boschloo's test. The ExactCIdiff R package calculates the smallest length confidence interval for β_d of Wang et al. (2010).

7.12 Notes

N1 **Designing a case-control study:** In Example 7.2 we presented the example of determining the odds of lung cancer incidence in 55–59-year-old people who have smoked over the odds of that population who have never smoked. That example was unrealistic in two ways. First, we are rarely only interested in only 55–59-year-olds. Second, even if we condition on cases and controls, for actual studies we rarely take a simple random sample from those conditional populations. In an actual case-control study, we can select our controls to match to cases of similar age or a set of variables that are thought to be related to the risk for the disease but not the exposure variable. Wacholder et al. (1992) in the first in a series of three papers discuss the principles of designing a case-control study, and in two companion papers give more practical details. Breslow (1996) gives an account of the mathematics behind the case-control design.

N2 **Estimating odds ratios:** For $X_a \sim Binomial(n_a, \theta_a)$ with $a = 1, 2$, the odds ratio is $\beta_{or} = \theta_2(1 - \theta_1)/\{\theta_1(1 - \theta_2)\}$. Two common ways to estimate β_{or} are either the maximum likelihood estimate (MLE) under the unconditional likelihood (two independent binomial distributions), or the MLE under the conditional likelihood (the extended hypergeometric distribution). For example, if $x_1/n_1 = 2/11$ and $x_2/n_2 = 6/8$, then the MLE under the unconditional likelihood is $\hat{\beta}_{or} = \hat{\theta}_2(1 - \hat{\theta}_1)/\{\hat{\theta}_1(1 - \hat{\theta}_2)\} = 13.5$, where $\hat{\theta}_a = x_a/n_a$ for $a = 1, 2$. The MLE under the conditional likelihood is $\tilde{\beta}_{or} = 11.25$.

7.13 Problems

P1 Calculate the conditional exact test using exact2x2 with all three options of the tsmethod argument on the following fake data,

```
      Succ Fail
Drug   240  120
Plac    90   30
```

Explain why the three methods give different p-values. Why does the "minlike" method have a larger p-value?

P2 Prove Equation 7.1 using Bayes' rule (see, e.g., Casella and Berger, 2002),

$$\Pr(A|B) = \frac{\Pr(B|A)\Pr(A)}{\Pr(B)},$$

where A and B are some events (e.g., $Y_1 = 1$ or $Z_i = 2$), and assume $\Pr(A) > 0$ and $\Pr(B) > 0$.

P3 Derive the extended hypergeometric distribution (Equation 7.2).

P4 Show that the second equality in Expression 7.3 is true. Hint: show that the cumulative distribution function of the extended hypergeometric distribution is a decreasing function of β_{or} (see Mehta et al., 1985, Appendix).

P5 Show that the lower and upper confidence distribution random variables for $X \sim Binomial(n, \theta)$ are $Beta(x, n - x + 1)$ and $Beta(x + 1, n - x)$. You need to extend the traditional definition of a beta random variable so that $Beta(0, b) = \lim_{a \to 0} Beta(a, b)$ is a point mass at 0 and $Beta(a, 0) = \lim_{b \to 0} Beta(a, b)$ is a point mass at 1. Hint: use integration by parts to relate the cumulative distribution of beta to that of a binomial. Then use the probability integral transformation.

P6 Difficult: Show that the central Fisher's exact p-value is compatible with the melded confidence interval for β_d. You may use the fact that the melded interval is compatible with one-sided p-values associated with it, $p_{AG(d)}(\mathbf{x}, 0) = \Pr[W_{L2} \le W_{U1}]$ and $p_{AL(d)}(\mathbf{x}, 0) = \Pr[W_{L1} \le W_{U2}]$. Hints: use results from exercise P5. Write the one-sided melded p-values that are not degenerate as functions of the cumulative distribution function for a beta-binomial distribution. Then use the relationship between the beta binomial and the hypergeometric. (For the last step, see (Johnson et al., 2005, Equation 6.64); but there is a typo, $F_h(n - k - 1)$ should be $F_h(n - k)$.)

P7 Calculate the two-sided p-value for both versions of the Fisher's exact test for $x_1/n_1 = 0/10$ and $x_2/n_2 = 22/100$. Explain why the central Fisher's exact p-value is less than the Fisher–Irwin one. Find another example where that inequality holds.

P8 Blaker's exact test applied to the two-sample binomial problem is a conditional test like the Fisher's exact tests. It conditions on the sum $S = x_1 + x_2$ and uses the fact that $X_2|S = x_1 + x_2$ is an extended hypergeometric distribution (see Equation 7.2). Let $F_\beta(x) = \Pr_\beta[X_2 \le x|S = x_1 + x_2]$ and $\bar{F}_\beta(x) = \Pr_\beta[X_2 \ge x|S = x_1 + x_2]$, where here β is the odds ratio parameter. Let $T_{Blaker}(X, \beta) = \min\{F_\beta(X), \bar{F}_\beta(X)\}$. Blaker's two-sided p-value for $H_0 : \beta = \beta_0$ is $p_B(\mathbf{x}, \beta) = \Pr_\beta[T_B(X_2, \beta) \le T_B(x_2, \beta)]$. Its associated $100(1 - \alpha)\%$ confidence interval is the smallest interval that contains the region $\{\beta : p_B(\mathbf{x}, \beta) > \alpha\}$. Show that $p_B(\mathbf{x}, \beta_0) \le p_c(\mathbf{x}, \beta_0)$ for all $\mathbf{x}$ and β_0, where p_c is the central Fisher's exact p-value.

P9 **Poisson conditional on the sum of two Poissons:** Show that if $X_a \sim Poisson(\theta_a)$ for $a = 1, 2$ and X_1 and X_2 are independent, then

$$X_2|X_1 + X_2 = m \sim Binomial\left(m, \frac{\theta_2}{\theta_1 + \theta_2}\right).$$

Hint: $X_1 + X_2 \sim Poisson(\theta_1 + \theta_2)$.

8

Assumptions and Hypothesis Tests

8.1 Overview

This chapter illustrates various subtle issues about assumptions and the choice of a test. It is quite abstract and can be omitted by the more applied reader. In Chapter 2 we described how a hypothesis test is a decision rule paired with a set of assumptions that define null and alternative hypotheses. Section 8.2 provides a framework for saying which of two hypothesis tests has fewer assumptions. The test with fewer assumptions is called *less restrictive*.

Although comparing the amount of assumptions in tests is useful for comparing the breadth of applicability of two hypothesis tests, the applied statistician needs more information to decide on a decision rule. Suppose two hypothesis tests are being considered and the researchers decide that the assumptions of the first of those two tests do not hold in the application, while the assumptions of the second test are plausible. Then it would seem clear that the second test is more appropriate. However, what is not so clear is whether the decision rule associated with the first test can be applied; that is, whether it may be valid or approximately valid under a different set of plausible assumptions. For example, what if a test for normality for the one-sample t-test shows that the normality assumption is very unlikely to hold. In that case, it is useful to know that there are other assumptions associated with the decision rule of the t-test that are asymptotically valid. Thus, contrary to the advice in some elementary statistics texts, the t-test decision rule (and hence the p-value associated with the t-test) may still be used in an application when the normality assumption does not hold. Section 8.3 gives us terminology and notation to talk about this idea that decision rules may be valid or asymptotically valid for several different sets of assumptions.

8.2 Which Test Is More Restrictive?

Recall from Chapter 2 that our formal definition of a hypothesis test consists of three parts: (1) the sample space, $\mathcal{X}$; (2) a set of probability models partitioned into the null hypothesis set, $\mathcal{P}_0$, and the alternative hypothesis set $\mathcal{P}_1$; and (3) a decision rule, δ, which is a function that depends on the data, X, and a significance level, α, and returns a value of 1 (for reject the null hypothesis) or 0 (for fail to reject the null hypothesis). In this chapter we will group together all the assumptions and denote them, $A = (\mathcal{X}, \mathcal{P}_0, \mathcal{P}_1)$; the following example in this section illustrates this notation. Under this notation, a hypothesis test is formally the pair, (δ, A).

In most mathematical statistics books, the choice of hypothesis test is often presented as a choice between two hypothesis tests with the same set of assumptions but different choices of decision rules, in other words, a choice between (δ_1, A) and (δ_2, A). In this situation, typically the first priority is choosing the hypothesis test that is valid. If both hypothesis tests are valid, then the second priority is determining which test has greater power under a specific alternative probability model of interest. This chapter is about a different problem.

For real applications, a major challenge is deciding on which assumptions apply to the situation at hand. If you are comparing two sets of assumptions and one set of assumptions can be thought of as a subset of the other set, then the bigger set is less restrictive and can be applied more widely. Here is a formal definition:

Definition 8.1 Let $A_1 = (\mathcal{X}_1, \mathcal{P}_0^{(1)}, \mathcal{P}_1^{(1)})$ and $A_2 = (\mathcal{X}_2, \mathcal{P}_0^{(2)}, \mathcal{P}_1^{(2)})$ be two sets of assumptions. We say that A_1 is more restrictive than A_2 (or equivalently, A_2 is less restrictive than A_1) if $\mathcal{X}_1 \subseteq \mathcal{X}_2$, $\mathcal{P}_0^{(1)} \subseteq \mathcal{P}_0^{(2)}$, $\mathcal{P}_1^{(1)} \subseteq \mathcal{P}_1^{(2)}$, and either $\mathcal{P}_0^{(1)} \cap \mathcal{P}_0^{(2)} \neq \mathcal{P}_0^{(2)}$ or $\mathcal{P}_1^{(1)} \cap \mathcal{P}_1^{(2)} \neq \mathcal{P}_1^{(2)}$, and we denote this as $A_1 \sqsubset A_2$.

Here is an example to demonstrate this concept:

Example 8.1 An example where one test has more restrictive assumptions than a competing test: *Consider two one-sample tests from Chapter 5, the one-sample t-test assuming normality and the exact test on the median. Although we typically do not think of the t-test as testing the median, recall that the mean equals the median for a symmetric distribution such as the normal distribution. So under the assumption of normality, a test on the mean is equivalent to a test on the median. Suppose $X_1, \ldots, X_n$ all come from a distribution F. Let θ be the median of F. Let $\mathcal{X} \equiv \mathcal{X}_1 = \mathcal{X}_2$ be the set of real numbers. Let the one-sample t-test be (δ_1, A_1), where $A_1 = (\mathcal{X}, \mathcal{P}_0^{(1)}, \mathcal{P}_1^{(1)})$ with $\mathcal{P}_0^{(1)}$ and $\mathcal{P}_1^{(1)}$ associated with the hypotheses*

$$H_0^{(1)} : \theta \leq \theta_B, \text{ and } F \in \Phi_N$$
$$H_1^{(1)} : \theta > \theta_B, \text{ and } F \in \Phi_N,$$

where Φ_N is the set of all normal distributions. Let δ_2 be the exact test on the median, and let its associated hypotheses be A_2, where A_2 has null and alternative hypotheses:

$$H_0^{(2)} : \theta \leq \theta_B, \text{ and } F \in \Phi_C$$
$$H_1^{(2)} : \theta > \theta_B, \text{ and } F \in \Phi_C,$$

with Φ_C being the set of all continuous distributions with support on the real line. Since $\Phi_N \subset \Phi_C$, we have $A_1 \sqsubset A_2$. Therefore, if the normality assumption does not hold for our data, then the exact median test, (δ_2, A_2), is more appropriate than the t-test assuming normality, (δ_1, A_1). However, in Section 8.3, we show how the t-test decision rule, δ_1, may be associated with a different set of assumptions.

It is not always obvious which test has more restrictive assumptions. Sometimes a test associated with a model, although it appears more restrictive, is actually less restrictive.

Example 8.2 An example where one test might appear to have more assumptions than a competing test but is actually less restrictive: *Consider a two-sample test with a baseline* *covariate. Suppose the data on the ith individual are the response, y_i, a baseline covariate, x_i, and a treatment indicator, z_i, that equals 1 if the ith individual got treatment and 0 if that individual got control. Consider the linear model, where we model Y_i, the random variable (RV) associated with y_i, as*

$$Y_i = \beta_0 + \beta_1 x_i + \beta_2 z_i + \epsilon_i,$$

where we assume that $\epsilon_1, \dots, \epsilon_n$ are independent and normally distributed with mean 0 and variance σ^2. Let A_3 be the assumptions associated with this model and partition the null and alternative hypotheses into

$$H_0^{(3)} : \beta_2 = 0,$$
$$H_1^{(3)} : \beta_2 \neq 0.$$

Let δ_3 be the valid decision rule for testing between these hypotheses using a linear model (see Chapter 14). Consider another set of assumptions, A_4, where if $z_i = 0$ then $Y_i \sim N(\mu_0, \sigma^2)$ and if $z_i = 1$ then $Y_i \sim N(\mu_1, \sigma^2)$, and $N(\mu, \sigma^2)$ represents a normal distribution with mean μ and variance σ^2. For A_4 we partition the null and alternative as

$$H_0^{(4)} : \mu_0 = \mu_1,$$
$$H_1^{(4)} : \mu_0 \neq \mu_1.$$

Let δ_4 be the two-sample t-test using a pooled variance estimator that is valid for testing A_4 (see Chapter 9). On first glance it may seem that the A_3 is more restrictive than A_4, because A_3 assumes a model while A_4 does not appear to assume as much. In fact, A_4 is just A_3 with the restriction that $\beta_1 = 0$ and renaming $\beta_0 = \mu_0$, and $\beta_2 = \mu_1 - \mu_0$. Thus, $A_4 \sqsubset A_3$.

If the issue is deciding between two decision rules, then there is less clarity, because each decision rule may be associated with more than one set of assumptions, as we discuss in Section 8.3.

8.3 Multiple Perspective Decision Rules

Suppose a decision rule, δ, is valid or asymptotically valid under assumptions A_1 as well as under assumptions $A_2, \dots, A_k$. Then we call the set $(\delta, A_1, \dots, A_k)$ a *multiple perspective decision rule*. We call each of the k hypothesis tests, (δ, A_i), $i = 1, \dots, k$ a *perspective* on that decision rule, since the decision rule may be interpreted under the perspective of any of the set of assumptions out of the k possible. Note that the p-value is a function of the data and the decision rule only, so that the same p-value may be used for any perspective in the multiple perspective decision rule. The properties of that p-value (e.g., validity or uniform asymptotically validity) depend on the associated assumptions, so the same p-value may have different properties under different perspectives.

In the context of multiple perspective decision rules, to avoid confusion we define a test as a pair, (δ, A_i), and a decision rule as δ only. This is in contrast to standard usage, because in the standard usage only one set of assumptions is used at a time, so assumptions are often left implied. This definition will lead to awkward phrases such as "the t-test decision rule" for this section.

Why is the concept of multiple perspective decision rules helpful? One case is when a more restrictive perspective shows some good properties, while a less restrictive perspective allows the decision rule to be applied more widely.

Example 8.3 Example where the *p*-value from a given test can be interpreted differently under different assumptions: *Consider the one-sample t-test decision rule under the normality assumption (see Section 5.3.2). That test is valid under the normality assumption, say A_1, and that is usually the set of assumptions presented when the test is introduced in statistics textbooks (for example, see Chapter 5, Note N4). But that same decision rule is uniformly asymptotically valid when we make no distributional assumptions except that the variance is assumed greater than 0 and $E(X^4)$ is bounded away from both 0 and ∞, say A_2 (see Chapter 5, Note N5). The p-value is a function of the decision rule only, and not a function of the assumption set. Thus, the p-value may be interpreted as exact under A_1 and approximately valid for large samples under A_2. Applied statisticians should keep both sets of assumptions in mind when interpreting the p-value from that test. For example, if we test the normality assumption and reject it, we can still use the p-value from the t-test as an approximation if the less restrictive assumptions are reasonable and the sample size is large.*

Another case to show the usefulness of the MPDR framework is when the less restrictive perspective allows the decision rule to be applied with minimal assumptions, but the more restrictive perspective allows us to achieve matching inferences for a confidence interval (CI).

Example 8.4 Example where a more restrictive set of assumptions for a test facilitates the construction of a matching CIs: *Consider the two-sample situation of Section 2.2.2 and Chapter 9. Let the data be $\mathbf{x} = [\mathbf{y}, \mathbf{z}]$, where $\mathbf{y}$ is an n vector of responses, and $\mathbf{z}$ is an n vector of associated group indicators, and suppose there are no ties in $\mathbf{y}$. Consider an exact version of the Wilcoxon–Mann–Whitney decision rule (this is the decision rule that is associated with the Wilcoxon rank sum test and equivalently with the Mann–Whitney U test). Consider two sets of assumptions that are both valid for this decision rule. Both assumption sets have that $Y_1, \ldots, Y_n$ are independent, $\mathcal{X}$ states that each Y_i comes from the set of real numbers, $Y_i \sim F$ when $z_i = 1$ and $Y_i \sim G$ when $z_i = 0$, and F, G represent continuous distributions. Let A_1 be a not very restrictive set of assumptions that defines the null and alternative as*

$$H_0^{(1)} : F = G$$
$$H_1^{(1)} : F \neq G.$$

The problem with this set of assumptions is that there is no finite dimensional way to describe the treatment effect. If we impose a more restrictive set of assumptions then we can represent the treatment effect as a location shift, Δ. Let A_2 define the null and alternative as

$$H_0^{(2)} : F = G$$
$$H_1^{(2)} : G(y) = F(y + \Delta); \Delta \neq 0.$$

Under A_2 if we reject the null hypothesis of $F = G$, then we conclude that the location shift Δ is not equal to 0. This is important when presenting the p-value of the Wilcoxon–Mann–Whitney decision rule together with a confidence interval on Δ. Recall confidence intervals *can be derived through a series of decision rules, say $\delta^*_{\Delta_0}$, that each are valid for one of a series of assumptions, say $A^*_2(\Delta_0)$. Suppose $A_2 = A^*_2(0)$ (i.e., A_2 can be considered one of a series of hypothesis tests used to create a confidence interval) where $A^*_2(\Delta_0)$ has*

$$H_0^{*(2)} : G(y) = F(y + \Delta); \Delta = \Delta_0$$
$$H_1^{*(2)} : G(y) = F(y + \Delta); \Delta \neq \Delta_0.$$

Then using the series, we can create a confidence interval for Δ. Further, we can create matching inferences for the Wilcoxon–Mann–Whitney decision rule. For example, the 95% confidence interval does not contain 0 if and only if the p-value is less than or equal to 0.05 (see Fay and Malinovsky (2018) or Section 9.5 for details on the Wilcoxon–Mann–Whitney test and compatible confidence intervals).

8.4 Summary

This chapter illustrates subtleties about tests and their associated assumptions. Examples include: a) a linear model with a baseline term is actually assuming less than a t-test, b) a one-sample t-test can yield an approximately valid p-value even if the data are clearly not from a normal distribution if the sample size is large, and c) one might choose to make more assumptions than necessary for a given test, if it facilitates the construction of matching confidence intervals.

8.5 Extensions and Bibliographic Notes

Fay and Proschan (2010) introduced the multiple perspective decision rule. They focused on decision rules that determine which of two samples tends to have larger responses, such as the t-test decision rule and the Wilcoxon–Mann–Whitney decision rule. They additionally discuss an interesting genetics example where the multiple perspectives for a decision rule are vital to the proper interpretation of the associated p-value. Hand (1994, see section 4.2) discusses choosing different hypotheses under different sets of assumptions, where the assumptions are called *statistical models*.

9

Two-Sample Studies with Ordinal or
Numeric Responses

9.1 Overview

9.1.1 Response Type: Ordinal or Numeric

In this chapter we consider two-sample studies with ordinal or numeric responses. Through-out this chapter we will assume that the responses are independent. If the observations are paired, see Chapter 6.

Recall (see Chapter 5), ordinal responses are categorical responses, where the categories are ordered and the distance between responses is not known. For example, the responses could be a food preference survey with possible responses: *Extremely Dislike, Moderately Dislike, No Preference, Moderately Favorable*, and *Extremely Favorable*. In that case, the difference in preference between adjacent categories is not necessarily the same for all pairs of categories. Numeric data, on the other hand, has numeric values associated with each response and the difference between responses is clearly defined. We can treat ordinal data as numeric by assigning numeric scores to each category of response. This scoring will impose added structure to the data and may be helpful for interpretation for some applications. Additionally, we can treat numeric data as ordinal data by only performing statistics on the ranks. Two-sample tests of equality of the distributions from each sample (i.e., $H_0 : F_1 = F_2$) performed on the ranks will be invariant to monotonic transformations on the responses such as the log transformation. For example, if we are measuring antibodies in the blood, it is not necessarily clear if we should be comparing difference effects on the original arithmetic scale (so that the effect is additive, and the difference in response between 2 and 4 is similar in effect to the difference between 10 and 12), or comparing difference effects on the log scale (so that the effect is multiplicative, and the doubling from 2 to 4 is similar in effect to doubling from 10 to 20). By performing tests on the ranks, we get the same probability of rejection of $H_0 : F_1 = F_2$ regardless of whether the responses are log transformed or not. However, this type of invariance does not apply when the null hypothesis is not $F_1 = F_2$, such as testing related to calculating confidence intervals on the difference in medians (see Section 9.4.1). Nevertheless, because the ordinal may be treated as numeric and the numeric as ordinal, we tackle both types of responses together in one chapter.

9.1.2 Ordinal Responses

We first consider ordinal responses with k possible categories. As in Chapter 5, we write the random variable for the ordinal response from the ith individual, Y_i^* a $k \times 1$ vector of indicators, as a discrete variable with possible values $\{1, 2, \ldots, k\}$ and denote it Y_i

(see Equation 5.1). Suppose there are $n = n_1 + n_2$ total responses, and let $\mathbf{y}$ and $\mathbf{z}$ be the $n \times 1$ vectors of observed responses and group labels. Let the response distributions be $Y_i|z_i = 1 \sim F_1$ and $Y_i|z_i = 2 \sim F_2$. In this case the two group distributions F_1 and F_2 are each completely specified with $k - 1$ parameters. It is not immediately clear how to take this problem with $2k - 2$ parameters and define an effect of interest for comparing the two groups. One effect parameter that is useful is the proportional odds parameter from the *proportional odds model* (also known as the cumulative logit model, or ordinal logistic regression). This model imposes some structure on the F_1 and F_2 forcing the following relationship,

$$\frac{F_1(x)}{1 - F_1(x)} = \frac{\psi F_2(x)}{1 - F_2(x)} \text{ for all } x \in \{1, \ldots, k - 1\}. \tag{9.1}$$

For convenience let W_1 and W_2 be an arbitrary response from F_1 and F_2, respectively. Then the proportional odds model has $k - 1$ nuisance parameters plus ψ, which represents either of the two odds ratios,

$$\psi = \frac{Odds(W_1 \leq j)}{Odds(W_2 \leq j)} = \frac{Odds(W_2 > j)}{Odds(W_1 > j)} \tag{9.2}$$

for any $j \in \{1, \ldots, k - 1\}$, where $Odds(A) = \Pr[A]/(1 - \Pr[A])$ for an event A. The model is palindromically equivariant meaning that the inferences do not change if you reverse the order of the responses and switch group labels. Another important model that is similar but that is not palindromically equivariant is the proportional hazards model. We delay talking about that model until Chapter 16, since it is an important model in time-to-event data. Inferences related to the proportional odds parameter are discussed in Section 9.2.

Another important effect parameter for the two-sample ordinal response problem is the Mann–Whitney parameter, ϕ, that we define using the *Mann–Whitney functional*, $h_{MW}(\cdot, \cdot)$, (a *functional* is a function that takes functions [in this case cumulative distribution functions] as its input),

$$\phi \equiv h_{MW}(F_1, F_2) = \Pr[W_1 < W_2] + \frac{1}{2}\Pr[W_1 = W_2]. \tag{9.3}$$

This parameter is very closely related to the Wilcoxon–Mann–Whitney (WMW) test (also called the Wilcoxon rank sum test or the Mann–Whitney U test). The *Mann–Whitney parameter*, ϕ, is known by many other names: the *probabilistic index* applied to the two-sample problem (Thas et al., 2012), the area under a receiver operating characteristic (ROC) curve when the response is some diagnostic variable and the two groups are those with the disease of interest or those without (Hanley and McNeil, 1982), or the *concordance index*, or c-index when the response is some predictor of death and the two groups are those that died and those that survived (Harrell et al., 1996) (although with censoring, the c-index does not estimate ϕ, see Koziol and Jia (2009)). For the continuous case many just define $X \sim F_X$ and $Y \sim F_Y$ and call $h_{MW}(F_X, F_Y)$ by $\Pr[X < Y]$ (Mee, 1990).

Importantly, $h_{MW}(F_1, F_2)$ does not require imposing additional structure to the problem as in the proportional odds model; however, when we use $h_{MW}(F_1, F_2)$ without additional structure it has several problems. First, the functional may be intransitive (see Section 9.5.2), meaning that if you have three groups with $W_1 \sim F_1$, $W_2 \sim F_2$ and $W_3 \sim F_3$, you can have $h_{MW}(F_1, F_2) > 0.5$ (implying W_2 tends to be larger than W_1), $h_{MW}(F_2, F_3) > 0.5$ (implying W_3 tends to be larger than W_2), *and* $h_{MW}(F_3, F_1) > 0.5$ (implying W_1 tends to be larger than W_3). Second, unlike other parameters, the Mann–Whitney parameter is not automatically easily interpretable as a causal parameter. The Mann–Whitney parameter does have a causal interpretation, but it is often confused with a different closely related causal parameter. We discuss these two parameters and the subtle distinction between them in Section 9.5.3. Finally, as with many effect parameters it is difficult to get exact confidence intervals for $h_{MW}(F_1, F_2)$ (see Section 9.5.5).

The Mann–Whitney parameter is closely related to the WMW test which we discuss in Section 9.5.1. Actually, *the* WMW test is a vague term, because there are exact and asymptotic versions of the WMW decision rule, and each version can be interpreted under many different assumptions (see Chapter 8). When we use the term WMW test, it is relevant to any version, and when we discuss a specific version, we will specify that particular version. Unless stated otherwise, we will always assume the WMW test has $H_0 : F_1 = F_2$, so that the Type I error rate is theoretically correct, either exactly or asymptotically depending on the version. Because the WMW test is often applied to numeric responses, we place Section 9.5 after sections on inferences for numeric responses.

9.1.3 Numeric Responses

Now consider numeric responses. We consider primarily four parameters to compare the two groups: the difference in means, the ratio of geometric means, the difference in medians, or the ratio of medians. To frame the discussion about these four parameters, consider the following example:

Example 9.1 *Mullen et al. (2008): People living in the United States who had never been exposed to malaria are randomized to three groups to test their response to an experimental malaria vaccine. The three groups are: high dose of antigen (the antigen is the AMA1 protein that is on the outside of the malaria parasite during the blood stage), low dose of antigen plus adjuvant (an adjuvant is something added to a vaccine to stimulate an immune response), and high dose of antigen plus adjuvant. The study tests for many effects. We focus on comparing the day 70 antibody response between the low dose plus adjuvant group and the high dose plus adjuvant group.*

The data for the example are plotted in Figure 9.1. In general, we want to know if giving a higher dose will increase the antibody level. The difference in means has a straightforward interpretation to answer this question. For these data the low dose has slightly higher mean, with difference in means: $\hat{\mu}_2 - \hat{\mu}_1 = 2{,}950.8 - 3{,}058.3 = -107.5$. For positive numeric responses, the difference in geometric means is another possibility, giving $\hat{u}_2 - \hat{u}_1 = 2{,}709.4 - 2{,}721.7 = -12.3$. (Note, the geometric mean parameters, u_1 and u_2, are different from the usual means, μ_1 and μ_2.) For antibody responses, because we usually do not want a few very large observations to overly influence the results, we often use geometric means (the back-transform of the mean of the log-transformed data, see part (b)). If there are a few very low responses, then those low values may overly influence the geometric means compared to the arithmetic mean. See the response of 879 in the low group, where part (b) shows a much larger separation from the bulk of the data than part (a). Another possibility for an effect parameter is the ratio of geometric means, estimated for this example as $\hat{u}_2/\hat{u}_1 = 0.995$. Alternatively, we could use the difference in medians, $2{,}476.0 - 3{,}059.5 = -583.5$, or the

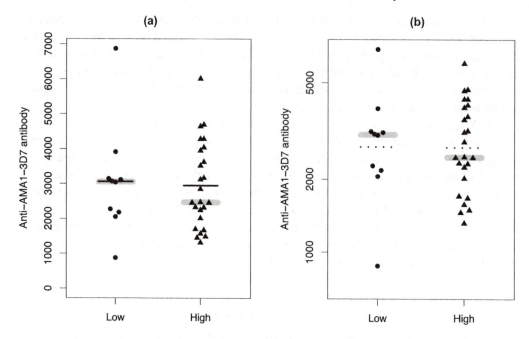

Figure 9.1 Antibody response to the AMA1 protein from the 3d7 strain of the malaria parasite in the low dose+adjuvant and high dose+adjuvant groups. Part (a) is on the natural scale, and part (b) is on the log scale. The black solid line is the arithmetic mean, the black dotted line is the geometric mean, and the gray lines are the medians. Points are jittered so they do not overlap. Antibody is measured in µg/ml.

ratio of medians, $2,476.0/3,059.5 = 0.809$. Other choices are possible, but we focus on four of these choices because the inferences are common and tractable (one choice, the difference in geometric means is rarely used).

9.1.4 Outline of Chapter

The chapter is organized primarily on the effect parameters. We discuss methods to use when the focus is on obtaining confidence intervals for the effect parameter of interest, but we also discuss methods to use when the focus is primarily on power to detect differences between the groups. Section 9.2 discusses inferences on proportional odds. In Section 9.3 we discuss differences in means and ratio of geometric means since the latter can be calculated by back-transforming the differences in the means of the log-transformed data. Section 9.4 discusses differences in medians or ratios of medians. In Section 9.5 we discuss the Mann–Whitney parameter. This is very closely related to the WMW test, since the WMW test is consistent (i.e., the power goes to 1 as the sample size goes to infinity) whenever the Mann–Whitney parameter is not equal to 1/2. Finally, we discuss in Section 9.6 on choosing between tests when the choice of effect parameter is less important than the power to show superiority of one group. For example, consider the vague scientific hypothesis that the new treatment will give (or tend to give) larger responses than the standard treatment, where larger is better.

The researcher may be happy with any of the effect parameters (e.g., difference in means, ratio of medians), since each can show in its own manner if the new treatment tends to lead to larger responses than the standard treatment. The researcher may care more about the power under different scenarios than the specific effect parameters tested.

9.2 Inferences on Proportional Odds

We rewrite the proportional odds model of Equation 9.2 in terms of the log of the odds that $Y_i \leq j$ given the group is z_i as

$$\log\left(Odds\left\{\Pr[Y_i \leq j|z_i]\right\}\right) = \log\left(\frac{F(j|z_i)}{1 - F(j|z_i)}\right) = \gamma_j - \beta I(z_i = 2), \qquad (9.4)$$

where $F(j|z_i) = \Pr[Y_i \leq j|z_i]$ and I is the indicator function: $I(A) = 1$ when A is true and 0 otherwise. The negative before the β is a convention (not always used) that ensures that larger values of β predict higher responses. The model is written as the log odds that Y_i gives a smaller response, and the proportional odds assumption ensures that the odds ratios are the same for $j = 1, \ldots, k - 1$. In other words, the model says that the odds ratio does not depend on j:

$$\frac{Odds(Y_i \leq j|z_i = 2)}{Odds(Y_i \leq j|z_i = 1)} = \frac{\exp(\gamma_j - \beta)}{\exp(\gamma_j)} = \exp(-\beta).$$

We can alternatively express the odds ratio using the odds that $Y_i > j$:

$$\psi = \frac{Odds(Y_i > j|z_i = 2)}{Odds(Y_i > j|z_i = 1)} = \frac{\exp(-\gamma_j + \beta)}{\exp(-\gamma_j)} = \exp(\beta), \qquad (9.5)$$

which matches the earlier definition (Equation 9.2), and in this form larger values increase Y_i for $z_i = 2$ compared to $z_i = 1$. This latter expression is more common for expressing odds ratios for binary responses (i.e., $k = 2$).

To estimate β we typically use maximum likelihood estimation. When the proportional odds model is true, then we can use the usual asymptotic methods for likelihoods (see Chapter 10). Generally, the best of these methods is the likelihood ratio method, although the Wald method is the easiest to calculate. The score method can give good local power, and in this case the permutation test on the score statistics is equivalent to the WMW test (see, e.g., McCullagh and Nelder, 1989, p. 188). The first two likelihood-based methods estimate the variance using Fisher's information, which requires that the model be correctly specified. When the model is misspecified then β is interpreted as a kind of average odds ratio effect over all the possible response values, and we can get an asymptotically valid test of this average odds ratio effect using a sandwich estimator of the variance. The relationship to the Mann–Whitney parameter is discussed in Section 9.5.6.

9.3 Differences in Means

9.3.1 Overview

In Section 5.3.1 we showed why there cannot be a useful valid nonparametric test on the mean. Similar arguments for the difference in means show that there does not exist a valid nonparametric test on the difference in means that has enough power to be at all useful (i.e., the power under any alternative cannot be greater than the significance level, α). So to perform tests and create confidence intervals on the difference in means, say $\Delta = \mu_2 - \mu_1$, we need to make some assumptions about the distributions F_1 and F_2.

A two-sample permutation test can test $H_0 : F_1 = F_2$ versus $H_1 : F_1 \neq F_2$ without any assumptions about the distributions, but it does not help in creating confidence intervals for the difference in means. We can get confidence intervals by additionally making the shift assumption (i.e., assume that $F_2(y) = F_1(y - \Delta)$), then inverting a series of null hypotheses on the shift parameter to get confidence intervals (Section 9.3.2).

Another approach is to make parametric assumptions, and the most common of these is the normality assumption. This leads to t-tests and Behrens–Fisher tests as discussed in Section 9.3.3. In Section 9.3.4 we show that for large samples, t-tests work well even for nonnormal data.

9.3.2 Permutation Test and Associated Confidence Intervals under the Shift Assumption

In this section, we consider the permutation test on the difference in means. See Section 10.9 for a general procedure for creating permutation tests.

A two-sample permutation test is valid for testing the null hypothesis $H_0 : F_1 = F_2$. First, consider the most general form of the hypothesis test, where we make no assumptions on F_1 and F_2, and the alternative is $H_1 : F_1 \neq F_2$. For the difference in means effect, we can just permute the estimate of the difference in means. Write the sample difference in means as

$$\hat{\Delta}(\mathbf{y}, \mathbf{z}) = \left(\frac{1}{n_2} \sum_{i=1}^{n} y_i I(z_i = 2) \right) - \left(\frac{1}{n_1} \sum_{i=1}^{n} y_i I(z_i = 1) \right) = \bar{y}_2 - \bar{y}_1. \qquad (9.6)$$

There are $N = \binom{n}{n_1}$ unique permutations of the group labels (i.e., there are N ways of assigning n_1 individuals to Group 1 and $n_2 = n - n_1$ to Group 2). Let $\hat{\Delta}_{obs}$ be the observed difference in means, and $\hat{\Delta}_1, \ldots, \hat{\Delta}_N$ be the difference in means for each of the N permutations. Then a two-sided p-value for testing $H_0 : F_1 = F_2$ versus $H_1 : F_1 \neq F_2$ is

$$p_{abs}(0, \mathbf{y}, \mathbf{z}) = \frac{\sum_{j=1}^{N} I\left(|\hat{\Delta}_j| \geq |\hat{\Delta}_{obs}| \right)}{N}, \qquad (9.7)$$

where the 0 represents the null value of Δ. The p-value may be approximated by Monte Carlo replication if necessary (see Section 10.9.5). We can think of the p-value as the probability of observing equal or more extreme values of $\hat{\Delta}$ under the permutation distribution, where more extreme is larger values of $|\hat{\Delta}_j|$. One might think that if $\hat{\Delta}_{obs} = \bar{y}_2 - \bar{y}_1 > 0$, and $|\hat{\Delta}_{obs}|$, is large so that we reject $H_0 : F_1 = F_2$, this implies that we can conclude that $\mu_2 - \mu_1 > 0$. However, we cannot draw that conclusion without additional assumptions, that is, the alternative accepted after rejecting $F_1 = F_2$ is only $F_1 \neq F_2$. A scenario that illustrates why we cannot conclude anything about the direction of the difference in means without additional assumptions occurs when a small proportion of the distribution is extremely small (as exemplified by Example 9.2 which we describe shortly).

In order to make inferences about the direction of the difference in means (i.e., whether $\Delta > $ or $\Delta < 0$), we should formalize hypotheses about Δ, for example, test the null hypothesis, $H_0 : \Delta \leq 0$. First, we show that we cannot create a "one-sided" permutation p-value associated with that hypothesis. Consider a permutation p-value based on rejecting when $\hat{\Delta}_{obs}$ is large,

$$p_L(0, \mathbf{y}, \mathbf{z}) = \frac{\sum_{j=1}^{N} I\left(\hat{\Delta}_j \geq \hat{\Delta}_{obs}\right)}{N}. \tag{9.8}$$

Although this p-value has small values when $\hat{\Delta}_{obs}$ is large, it does not have proper Type I error for testing the null hypothesis, $H_0 : \Delta \leq 0$. (We use the subscript L because inverting that type of p-value can give a one-sided confidence interval with a lower limit; that will be discussed on page 164.) However, $p_L(0, \mathbf{y}, \mathbf{z})$ is valid for testing the null $H_0 : F_1 = F_2$ versus $H_1 : F_1 \neq F_2$. Consider the following hypothetical example:

Example 9.2   *Imagine a strain of lab-raised mice that has a mean lifespan of 989 days. Suppose a new additive to the standard diet increases the mean lifespan to 1,088 days in 90% of the mice, but 10% of the mice have a bad reaction to the additive that has a drastic fatal effect (mean lifespan 30 days). Figure 9.2 shows the distributions. Suppose we plan a randomized study with $n_1 = n_2 = 5$ in the two groups (Group 1 is standard diet, Group 2 is standard diet plus additive), and we plan to use a permutation test with the difference in lifetime as the test statistic, that rejects when $\bar{y}_2 - \bar{y}_1$ is large (i.e., use Equation 9.8). By simulation, the power to reject (at the 5% significance level) is 58.5%. This occurs because*

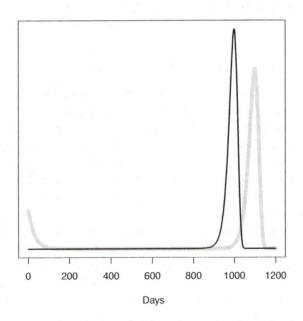

Figure 9.2 Distributions for Example 9.2. The black line is the distribution (probability density function) for Group 1, Weibull(50, 1000), in other words, a Weibull with shape = 50 and scale = 1000. The gray line is the distribution for Group 2, a mixture of two Weibulls, 10% Weibull(1, 30) and 90% Weibull(50, 1, 100).

59% of the time none of the five mice in Group 2 have the bad reaction, and the power conditional on no bad reactions is nearly 100%. However, even though those rejections have $\bar{y}_2 > \bar{y}_1$, the true mean lifespan for Group 2 is 982 days, which is less than that for Group 1 (989 days).

Thus, we must make an additional assumption on the distributions in order to get confidence intervals on the difference in means. One simple assumption is the shift assumption: we assume that both distributions have the same shape but are shifted by Δ. In other words, $F_2(x) = F_1(x - \Delta)$ for all x. Once we make the shift assumption, then if we reject $F_1 = F_2$, then we can declare that $\Delta \neq 0$ with a significance level of α. Additionally, with the shift assumption we can test

$$H_0 : \Delta = \Delta_0$$
$$H_1 : \Delta \neq \Delta_0,$$

by using a permutation test. Specifically, we perform the permutation test in two steps. First, we subtract Δ_0 from all the responses from Group 2 to create shifted responses,

$$y_i^{shift} = y_i - I(z_i = 2)\Delta_0. \tag{9.9}$$

Then under the null hypothesis of $H_0 : \Delta = \Delta_0$, all Y_i^{shift} random variables have the same distribution. Thus, a permutation test p-value for that hypothesis is $p_{abs}(0, \mathbf{y}^{shift}, \mathbf{z})$, in other words, replacing y_i in Equation 9.7 with y_i^{shift}. We define the notation for p_{abs} to explicitly relate the p-value to the hypothesis $H_0 : \Delta = \Delta_0$. Then $p_{abs}(\Delta_0, \mathbf{y}, \mathbf{z}) \equiv p_{abs}(0, \mathbf{y}^{shift}, \mathbf{z})$. Then we can create a confidence interval for Δ by inverting a series of hypothesis tests for different values of Δ_0. Specifically, a $100(1 - \alpha)\%$ confidence interval for Δ (under the shift assumption) is

$$C(1 - \alpha, \mathbf{y}, \mathbf{z}) = \{\Delta_0 : p_{abs}(\Delta_0, \mathbf{y}, \mathbf{z}) > \alpha\}. \tag{9.10}$$

We make inferences about the differences in means of two groups by using that permutation method as follows. Test $H_0 : F_1 = F_2$ and reject if $p_{abs}(0, \mathbf{y}, \mathbf{z}) \leq \alpha$. If we reject, then we conclude that F_1 and F_2 are significantly different at level α. Then adding the additional shift assumption, we can construct the $100(1 - \alpha)\%$ confidence interval (Equation 9.10). With the shift assumption, if the confidence interval excludes $\Delta = 0$ and all values in the interval are larger than 0, that means that we can conclude that the difference in means is significantly larger than 0 at level α. The latter conclusion requires the shift assumption, while conclusion of rejecting $H_0 : F_1 = F_2$ at level α does not require the shift assumption. This is often true with hypothesis tests: we can conclude that two distributions are different without making many assumptions, but in order to get confidence intervals and make directional statements about the effects, we must make additional assumptions.

If we care about directional inferences (i.e., we are not just interested in showing $\Delta \neq 0$, we want to show one of $\Delta > 0$ or $\Delta < 0$), then we can test one-sided hypotheses using permutation tests and the shift assumption. Suppose we want to test $H_0 : \Delta \leq \Delta_0$ versus $H_1 : \Delta > \Delta_0$. With the shift assumption, the one-sided-type p-value of Equation 9.8 tests that one-sided hypothesis. Analogously to the expansion of the definition of p_{abs}, let $p_L(\Delta_0, \mathbf{y}, \mathbf{z}) \equiv p_L(\mathbf{y}^{shift}, \mathbf{z})$, where the latter comes from replacing $\mathbf{y}$ with $\mathbf{y}^{shift}$ in Equation 9.8 and $\mathbf{y}^{shift}$ is defined in Equation 9.9. To test $H_0 : \Delta \geq \Delta_0$ versus $H_1 : \Delta < \Delta_0$, the

p-value is $p_U(\Delta_0, \mathbf{y}, \mathbf{z})$ defined by replacing the $\geq$ in Equation 9.8 with $\leq$. Then a two-sided central p-value is $p_c(\Delta_0) = \min(1, 2 * p_L(\Delta_0), 2 * p_U(\Delta_0))$, and the $100(1 - \alpha)\%$ central confidence interval is the set of Δ_0 for which $p_c(\Delta_0) > \alpha$ (analogous to Equation 9.10). Under the shift assumption, if $p_c(0) \leq \alpha$ and the $100(1 - \alpha)\%$ confidence interval excludes $\Delta = 0$ and all values in the interval are larger than 0, that means that $p_L(0) \leq \alpha/2$ and we can conclude that the difference in means is significantly larger than 0 at level $\alpha/2$.

Example 9.2 *(continued) Suppose we performed the mouse additive study on 10 mice and observed lifetimes of $\{997, 1026, 953, 967, 996\}$ days in the standard diet group (Group 1), and $\{1112, 1115, 1113, 1090, 1073\}$ in the diet with additive group (Group 2). The permutation test gives $p_L(0) = 1/252 = 0.004$, since we observed the largest value of $|\hat{\Delta}_0|$ of the $\binom{10}{5} = 252$ permutations. So, without any assumptions we can conclude that the additive is significantly affecting the lifetimes. Making the shift assumption, the 95% confidence interval on $\Delta = \mu_2 - \mu_1$ is $(77.0, 152.5)$, and we can conclude that the mean lifetime is significantly larger (up to level $p_L(0)$) when the dietary additive is used. In other words, without the shift assumption, we conclude that there are significant differences in the distributions at level 0.05 (since $p_c(0) < 0.05$), and with the shift assumption we conclude that the mean lifetime is significantly larger (p-value = 0.004) in the additive group than in the standard diet group. If the shift assumption is wrong (e.g., if Figure 9.2 reflects the true distributions), then the latter conclusion is wrong. Of course, with real data we never know with certainty how closely our assumptions match up with reality, although inspection of the data may provide some clues.*

If the responses are all positive, we can take a log transformation, and perform the difference in means on the log-transformed data. The antilog of this is the ratio of geometric means.

Example 9.2 *(continued) Taking the $\log_{10}$ transformation and performing the difference in means permutation method, we again get $p_L(0) = 1/252$. Then assuming a shift assumption on the distribution of the log-transformed responses, a 95% central confidence interval on the difference in the expected $\log_{10}$ transformed values is $(0.03235, 0.06404)$. So the 95% central confidence interval on the ratio of geometric mean lifetimes (Group 2/Group 1) is $(10^{0.03235}, 10^{0.06404}) = (1.077, 1.159)$.*

There are many algorithms to calculate permutation tests. For sample sizes like $n_1 = n_2 = 5$ we can easily completely enumerate all the permutations. But for large sample sizes $\binom{n}{n_1}$ quickly makes complete enumeration intractable. In Section 10.9.4 we show that for many permutation tests there is a central limit theorem. The application to this problem is that under the shift assumption, for large enough sample sizes, and under some regularity conditions on F_1 (see Section 10.9.4 for details), we can treat T_{perm} as a standard normal random variable, where

$$T_{perm}(\Delta_0) \equiv \frac{\hat{\Delta} - \Delta_0}{\sqrt{V_{perm}}} = \frac{(\bar{y}_2 - \bar{y}_1) - \Delta}{\sqrt{\left\{\frac{1}{n-1}\sum_{i=1}^{n}(y_i^{shift} - \bar{y}^{shift})^2\right\}\left(\frac{1}{n_1} + \frac{1}{n_2}\right)}}, \qquad (9.11)$$

where $y_i^{shift} = y_i - I(z_i = 2)\Delta_0$ and $\bar{y}^{shift} = n^{-1} \sum_{i=1}^{n} y_i^{shift}$. We can use that approximation to get p-values and confidence intervals. One can show (Problem P2) that a $100(1-\alpha)\%$ (asymptotic permutation shift) confidence interval for Δ is

$$\hat{\Delta} \pm c_{\alpha/2} \sqrt{\frac{n}{n - c_{\alpha/2}^2 - 1} \left(\frac{\tilde{\sigma}_1^2}{n_2} + \frac{\tilde{\sigma}_2^2}{n_1} \right)}, \tag{9.12}$$

where $c_{\alpha/2} = \Phi^{-1}(1 - \alpha/2)$ is the $(1 - \alpha/2)$th quantile of the standard normal distribution, and

$$\tilde{\sigma}_a^2 = \frac{1}{n_a} \sum_{i:z_i=a} (y_i - \bar{y}_a)^2$$

and $\bar{y}_a$ is the mean for Group a. Note, in the definition of $\tilde{\sigma}_1^2$ we divide by n_1 instead of $(n_1 - 1)$ as in the usual unbiased estimate of the variance, $\hat{\sigma}_1^2$ (see Equation 9.13). Also in the confidence interval expression, $\tilde{\sigma}_1^2$ is divided by n_2 not n_1 as one might expect. For the confidence interval of Example 9.2, the exact 95% confidence interval was $(77.0, 152.5)$ and the approximation by equation 9.12 gives $(75.7, 149.9)$.

9.3.3 Difference in Means between Normals: t-Tests and Behrens–Fisher Test

A very common and historically important way that difference in means have been tested is by assuming F_1 and F_2 are both normally distributed. Let F_a be a normal distribution with mean μ_a and variance σ_a^2. If we can further assume that $\sigma_1 = \sigma_2$, then the uniformly most powerful unbiased test is Student's two-sample t-test. Let

$$T_t(\Delta_0) = \frac{\hat{\Delta} - \Delta_0}{\sqrt{\hat{\sigma}_p^2 \left(\frac{1}{n_1} + \frac{1}{n_2} \right)}},$$

where $\hat{\sigma}_p^2$ is the pooled variance estimate,

$$\hat{\sigma}_p^2 = \frac{1}{n - 2} \left((n_1 - 1)\hat{\sigma}_1^2 + (n_2 - 1)\hat{\sigma}_2^2 \right),$$

where

$$\hat{\sigma}_a^2 = \frac{1}{n_a - 1} \sum_{i:z_i=1} (y_i - \bar{y}_a)^2 \text{ for } a = 1, 2. \tag{9.13}$$

Under the null hypothesis $H_0 : \Delta = \Delta_0$ (and $\sigma_1 = \sigma_2$), then $T_t(\Delta_0)$ is distributed as a t-distribution with $n - 2$ degrees of freedom, so the (two-sided central) p-value using Student's two-sample t-test is

$$p_c(\Delta_0) = 2F_{t,n-2}(-|T_t(\Delta_0)|), \tag{9.14}$$

where $F_{t,df}(t)$ is the cumulative distribution for a central t-distribution with df degrees of freedom. The $100(1 - \alpha)\%$ confidence interval associated with that p-value is

$$\hat{\Delta} \pm F_{t,n-2}^{-1}(1 - \alpha/2) \sqrt{\hat{\sigma}_p^2 \left(\frac{1}{n_1} + \frac{1}{n_2} \right)}. \tag{9.15}$$

As the sample size gets large, then $F_{t,n-2}^{-1}(1-\alpha/2)$ approaches $c_{\alpha/2}$, the $(1-\alpha/2)$th quantile of the standard normal. So Student's t-test confidence interval (Expression 9.15) is similar to the permutation shift confidence interval (Expression 9.12), especially when $n_1 = n_2$. This makes sense, since with equal variances, then we have a shift relationship between the two normal distributions, $F_2(x) = F_1(x - \Delta)$.

Now assume F_1 and F_2 are normal but allow $\sigma_1 \neq \sigma_2$. This is known as the *Behrens–Fisher problem*. In this case, there is no uniformly most powerful unbiased test. Let

$$T_{BF}(\Delta_0) = \frac{\hat{\Delta} - \Delta_0}{\sqrt{\left(\frac{\hat{\sigma}_1^2}{n_1} + \frac{\hat{\sigma}_2^2}{n_2}\right)}}.$$

Under $H_0 : \Delta = \Delta_0$ then the distribution of $T_{BF}(\Delta_0)$ depends on not only n_1 and n_2 but also on the ratio of variances, $\gamma = \sigma_1^2/\sigma_2^2$ (Nel et al., 1990). There are several methods that use $T_{BF}(\Delta_0)$ to test and create confidence intervals. All the $100(1-\alpha)\%$ confidence intervals can be put into the form

$$\hat{\Delta} \pm c \sqrt{\frac{\hat{\sigma}_1^2}{n_1} + \frac{\hat{\sigma}_2^2}{n_2}}, \tag{9.16}$$

where c is a different value that depends on the method. One valid method is Hsu's method, that uses $c_H = F_{t, \min(n_1,n_2)-1}^{-1}(1-\alpha/2)$. An approximate method is Welch's method that uses $c_W = F_{t,d_w}^{-1}(1-\alpha/2)$, with the degrees of freedom estimated with Satterthwaite's approximation,

$$d_w = \frac{\left(\frac{\hat{\sigma}_1^2}{n_1} + \frac{\hat{\sigma}_2^2}{n_2}\right)^2}{\frac{\hat{\sigma}_1^4}{n_1^2(n_1-1)} + \frac{\hat{\sigma}_2^4}{n_2^2(n_2-1)}} \tag{9.17}$$

(see Problem P3). It can be shown that $c_W \leq c_H$ (see, e.g., Dudewicz and Mishra, 1988, p. 502), so that the Welch intervals are always less than or equal to those of Hsu.

A third method that is valid (although validity has been shown only with simulations) is the Behrens–Fisher method, with c_{BF} equal to the $(1-\alpha/2)$th quantile of the Behrens–Fisher distribution, which is a distribution whose random variable is distributed the same as

$$W_{n_1-1} \cos(R) + W_{n_2-1} \sin(R),$$

where W_{n_1-1} and W_{n_2-1} are independent t random variables with $n_1 - 1$ and $n_2 - 1$ degrees of freedom, and $R = \tan^{-1}\left(\frac{s_2/\sqrt{n_2}}{s_1/\sqrt{n_1}}\right)$.

Our general recommendation is to not assume equal variances unless there is a reason to do so. We do not recommend testing for equal variances, then assuming equal variances only if you cannot reject, as this two-step procedure will inflate the Type I error rate (Moser et al., 1989). If bounding the Type I error rate is needed, then the Behrens–Fisher test generally has better power than Hsu's test. If bounding the Type I error rate is not vital, then Welch's t-test gives good power and usually does not inflate the Type I error rate by much if at all (Mehta and Srinivasan, 1970). As an example, consider the following simulation using 10^4 replications, with $n_1 = 20, n_2 = 10$, and $\sigma_2^2/\sigma_1^2 = 0.5$. The simulated rejection rates under the null hypothesis of $\Delta = 0$ for the four methods is ($pow_t = 3.1\%$, $pow_W = 5.1\%$, $pow_H = 3.3\%$, and $pow_{BF} = 4.2\%$) and under the alternative of $\Delta = 1$ is ($pow_t = 79.9\%$, $pow_W = 85.4\%$, $pow_H = 80.8\%$, and $pow_{BF} = 83.1\%$).

9.3.4 Comparison between Tests and Robustness of Tests

Some textbooks focus on the normality assumption for t-tests. They might state that if we test for normality and reject it, then we should not use a t-test (in any of the forms), and if we fail to reject the hypothesis of normality, then it is okay to use t-tests. There are two problems with this approach. First, failing to reject a null hypothesis does not mean that we have shown that the null hypothesis is true, it could be that we just do not have enough information (due to perhaps a small sample size) to reject the null hypothesis. Second, and more importantly, even if the data generating process is not normal, often the t-tests are approximately valid. For example, even though binary responses are nowhere close to normally distributed, we can use a t-test to test for differences in means of binary responses and it will be approximately valid if the sample size is large enough because of the central limit theorem. In this section we review some of that approximate validity.

It is useful to think of the decision rules on difference in means as multiple perspective decision rules (see Chapter 8). For example, the usual t-test p-value (see Equation 9.14) can be thought of as testing several different hypotheses. It tests the null hypothesis that responses from both groups come from the same normal distribution. If that null hypothesis is true, then the t-test is the uniformly most powerful unbiased test. But the t-test p-value also can test that both groups come from the same distribution that is unknown except it has finite variance. In that case the test is (pointwise) asymptotically valid, meaning that for any specific distribution in the null hypothesis, the power to reject approaches the nominal significance level as the sample size get large. If the data come from a Cauchy distribution (a t-distribution with one degree of freedom), then the variance is infinite, and the asymptotic validity does not hold. Unfortunately, even with finite variance distributions it is hard to get precise about what sample size is large enough, because it depends on how nonnormal the distribution is and the nature of the nonnormality, and we will not attempt to quantify that. However, as long as we can assume that the two distributions are equal (as opposed to assuming just the means are equal, for example) under the null hypothesis, the permutation t-test will be valid for any sample size.

On the other hand, once we assume normality, even if we have two different variances under the null hypothesis, then the Behrens–Fisher test will be valid, and Welch's t-test will be approximately valid. We expect they will be approximately valid for nonnormal data as well as long as the sample size is large enough and the distributions have finite variances. The tests sometimes even are approximately valid for small samples. To demonstrate this, we simulate with 10^4 replicates a two-group test with $n_1 = 20$ and $n_2 = 10$ with F_1 a Poisson with mean 10 and F_2 a distribution of $Y = 5 + W$, where W is Poisson with mean 5. These two distributions have very different variances (Group 1 has variance equal 10, Group 2 has variance equal 5), but the same means. The probability of rejecting at the $\alpha = 0.05$ level using the two-sided t-tests or Behrens–Fisher test are: $pow_t = 3.1\%$, $pow_W = 5.0\%$, and $pow_{BF} = 4.1\%$. So we appear to have approximate validity for all three tests in this case. Repeating the simulation except changing F_2 to the distribution of $Y = 2.5 + W$, where W is Poisson with mean 5, we get powers: $pow_t = 59.4\%, pow_W = 67.2\%$, and $pow_{BF} = 64.0\%$.

Because we can rarely assume exact normality, and we may not wish to assume equal distributions, Welch's *t*-test is a reasonable default test for testing differences in means. However, note that in Section 9.6 we generally recommend the WMW test when the primary goal is to determine if one groups values are generally higher than the other as opposed to estimating a specific effect parameter.

9.4 Differences in Medians

9.4.1 Overview

For continuous data, one of the pleasing properties of medians is that for monotonic transformations the median of the transformed data is equal to the transformation of the median. In other words, suppose $g(\cdot)$ is a strictly increasing transformation and let $Y \sim F$. Let a transformed response be $X = g(Y)$. Then $median(X) = g(median(Y))$. For discrete numeric data, the property holds, but its statement requires more care (see Note N1). For the sample median, the property holds when there is an odd number of observations, but for an even number of observations the default estimator that averages the two middle observations will change with transformations (Problem P4). For example, with an odd number of observations, if you find the sample median of the log-transformed data and exponentiate it, this is equal to the median of the untransformed data. An analogous property does not hold with the difference in medians. The difference in medians of two groups is not equal to the exponential of the difference in the medians of the log-transformed data. So when dealing with difference in medians an important choice for positive responses is whether we should log transform the data or not.

Suppose the responses are positive. Let θ_1 and θ_2 be the medians for the two groups. Then the difference in medians, $\theta_2 - \theta_1$, is quite different from the exponential of the difference in log medians,

$$\exp\{\log(\theta_2) - \log(\theta_1)\} = \frac{\theta_2}{\theta_1}.$$

So for positive responses, we must decide whether we care more about the difference in medians or the ratio of medians. But after we decide, we can use essentially the same algorithms on either the untransformed data (for the difference) or the log-transformed data (for the ratio).

There are primarily three main approaches for making inferences on the difference in medians. We can meld together the one-sample confidence intervals of the median (see Section 10.11 for a more general discussion of melding), we can bootstrap (see Section 10.10), or we can assume a shift model and use a nonparametric test on the shifted data. We discuss the melding first, since it requires the fewest assumptions and it is our default recommendation. But since melding is a relatively new method and gives larger confidence interval lengths, we also discuss the other two methods.

9.4.2 Melded Confidence Interval

In Section 5.2 we explored exact confidence intervals on the median in the one-sample case. We can use melding of confidence intervals to combine the one-sample exact confidence intervals on the median to create two-sample "exact" confidence intervals on the difference in medians. See Section 10.11 for the details of the motivation and algorithm for melding. Here we note that the melded confidence interval applied to this situation is designed to guarantee coverage for any sample size without making any assumptions about continuity (i.e., it is valid for discrete data as well as continuous data), and without making any assumptions about the relationship between the two distributions. Another nice property for the difference in median test is that the p-values for one-sided tests $H_0 : \theta_2 \leq \theta_1$ or $H_0 : \theta_2 \geq \theta_1$ are invariant to monotonic transformations on the responses. The disadvantage of the melding approach is that for very small sample sizes it may not be possible to get finite confidence intervals. For example, for continuous responses if both $n_1 < 7$ and $n_2 < 7$, the 95% confidence intervals on the difference in medians will always be $(-\infty, \infty)$ regardless of the data. This is the price for making very few assumptions, or rather, without substantial assumptions, there is insufficient information to draw conclusions about the differences in medians with very small sample sizes.

Example 9.1 *(continued) Returning to the example of Figure 9.1, using the melding method, we get estimates and 95% central confidence intervals on the difference in medians (median from the high dose+Adjuvant minus median of the low dose+Adjuvant) of* $-583.5(95\%CI : -1455.0, 1444.0)$*, and on the ratio of medians (high/low) of* $0.809(95\%CI : 0.596, 1.661)$*. The one-sided p-values associated with this test are invariant to monotonic transformations on the data, so we get the same one-sided p-value of* 0.577 *for testing the alternative* $H_1 : \theta_2 - \theta_1 < 0$ *and the alternative* $H_1 : \theta_2/\theta_1 < 1$*.*

9.4.3 Other Approaches

Bootstrapping is a very general method that will work reasonably well for the difference in medians. Like the melding, we do not need a continuity assumption or a shift assumption. But unlike melding, the bootstrap method technically requires a large sample size (i.e., its properties are based on asymptotics) to be approximately valid, although it may give reasonable results for small samples.

An alternative approach assumes continuous responses such that the distributions of the two groups are equal except a location shift (it is analogous to the location shift approach for means, Section 9.3.2). Suppose that if $z_i = a$ then $Y_i \sim F_a$ with median θ_a for $a = 1, 2$. Let $\Delta = \theta_2 - \theta_1$. The shift assumption says that if $z_i = 2$, then $(Y_i - \Delta) \sim F_1$. Thus, under the shift assumption, we can test $H_0 : \Delta = \Delta_0$ using the responses, shifted in Group 2 only. In other words, let $Y_i^{shift}(\Delta_0) = Y_i - \Delta_0 I(z_i = 2)$, where $I(z_i = 2) = 1$ for Group 2 responses and 0 otherwise. Then we can use any test procedure where the null hypothesis is $H_0 : F_1 = F_2$, except apply it to $\mathbf{Y}^{shift}(\Delta_0) = [Y_1^{shift}(\Delta_0), \ldots, Y_n^{shift}(\Delta_0)]$. For example, we could use a WMW test (see Section 9.5.1) on $Y^{shift}(\Delta_0)$ to test $H_0 : \Delta = \Delta_0$ and define the p-value as $p(\mathbf{y}, \Delta_0)$. Then we can create a $100(1 - \alpha)\%$ confidence interval by inverting the series of p-values as

$$C(\mathbf{y}, 1 - \alpha) = \{\Delta : p(\mathbf{y}, \Delta) > \alpha\}.$$

We call this the shift confidence interval associated with the WMW test. The advantage of this confidence interval is that it is compatible with the WMW test, so that if we reject the null $H_0 : \Delta = 0$ (i.e., if $p(\mathbf{y}, 0) \leq \alpha$), then $\Delta = 0$ is not in the $100(1 - \alpha)\%$ confidence interval. We can get an estimator that is associated with the WMW test by choosing the middle of the set $\{\Delta : p(\mathbf{y}, \Delta) = 1\}$, this is known as the *Hodges–Lehmann estimator* associated with the WMW test.

Example 9.1 *(continued) We get Hodges–Lehmann estimator and location shift confidence intervals based on the WMW test on the log-transformed responses. This gives an estimator of θ_2/θ_1 (median(high dose+adjuvant)/median(low dose+adjuvant)) of* $1.013(95\%CI : 0.684, 1.394)$*. Notice that the estimator of the ratio is not equal to the ratio of sample medians, and in fact is greater than 1, while the ratio of sample medians is equal to 0.809.*

To highlight the differences between the three methods, we reproduce part of a simulation from (Fay et al., 2015). Suppose Group 1 and 2 are distributed Poisson with means 2.6 and 2.7, respectively. This means that the medians are 2 (Group 1) and 3 (Group 2). Simulating 10^4 data sets, the simulated coverage of 95% confidence intervals in percent are: for location shift method ($n = 20$, coverage = 87.5%; $n = 100$, coverage = 57.0%), for the nonparametric percentile bootstrap ($n = 20$, coverage = 90.3%; $n = 100$, coverage = 91.8%), and for the melding method ($n = 20$, coverage = 100.0%; $n = 100$, coverage = 100.0%). The location shift method fails because the Poisson case is not continuous nor a shift. While closer to nominal coverage, the bootstrap still suffers from considerable undercoverage perhaps because of the many ties and smaller sample size. In contrast, the melding method has conservative coverage, and the ratio of median confidence interval lengths are two times larger than the bootstrap. If the continuity and shift assumptions hold, then both the location shift method and the melded method are valid, and the location shift method appears to be more efficient (see Fay et al., 2015, table 1).

9.5 Mann–Whitney Parameter

9.5.1 Relationship to Wilcoxon–Mann–Whitney Test

Wilcoxon (1945) proposed a two-sample permutation test based on the sum of the midranks (ranks that control for ties, see Expression 9.18) in one group, and Mann and Whitney (1947) more fully studied its properties in the continuous case. There is an exact version and some asymptotic versions of the test, and we will call any version a Wilcoxon–Mann–Whitney (WMW) test. The *exact WMW test* is simply an exact permutation test on the difference in means after converting the responses to their overall rank among both groups. The test is also known as the Wilcoxon rank sum test or the Mann–Whitney U test. The WMW test can be used with the shift assumption to test for a difference in medians (see Section 9.4.3). But in this section, we will formulate the test as a permutation test on $h_{MW}(\hat{F}_1, \hat{F}_2)$, where $\hat{F}_a$ is the empirical distribution of Group a. This formulation is equivalent to the permutation test on the difference in mean ranks, but allows us to relate the test to estimating the effect parameter, $h_{MW}(F_1, F_2)$.

Let n_1 and n_2 be the sample sizes in the two groups, and $n = n_1 + n_2$. The sum of the midranks in Group 2, is equal to

$$\frac{n_2(n_2 + 1)}{2} + n_1 n_2 h_{MW}(\hat{F}_1, \hat{F}_2) \tag{9.18}$$

(see Problem P5). Then we can show (see Problem P6) that the difference in the mean ranks between the two groups is $n\{h_{MW}(\hat{F}_1, \hat{F}_2) - \frac{1}{2}\}$. The traditional way to do the test gives a two-sided p-value that is equivalent to performing a permutation test using $|h_{MW}(\hat{F}_1, \hat{F}_2) - \frac{1}{2}|$ as the test statistic. (See Section 10.9 for general permutation tests.) Similar to Section 9.3.2 for a two-sample permutation test, let $\hat{\phi}_{obs} = h_{MW}(\hat{F}_1, \hat{F}_2)$ and let $\hat{\phi}_1, \ldots, \hat{\phi}_N$ be the $N = \binom{n}{n_1}$ values of the $h_{MW}(\hat{F}_1, \hat{F}_2)$ after permuting the treatment assignment. A two-sided p-value is for testing $H_0 : F_1 = F_2$ versus $H_1 : F_1 \neq F_2$ is

$$p_{abs}(0.5, \mathbf{y}, \mathbf{z}) = \frac{\sum_{j=1}^{N} I\left(|\hat{\phi}_j - \frac{1}{2}| \geq |\hat{\phi}_{obs} - \frac{1}{2}|\right)}{N}, \tag{9.19}$$

where 0.5 for the first argument in p_{abs} represents the value of $h_{MW}(F_1, F_2)$ under the null hypothesis when $F_1 = F_2$.

Since the empirical distributions consistently estimate the true distributions, the WMW test will be consistent (meaning its power to reject will go to 1 as the sample sizes get larger, specifically, as $n_1/(n_1 + n_2) \to \lambda$ with $0 < \lambda < 1$ and $n_1 \to \infty$) as long as $h_{MW}(F_1, F_2) \neq 0.5$.

For large sample sizes we can use an asymptotic approximation. We take $\hat{\phi}_{obs}$, subtract the mean under the null hypothesis of $F_1 = F_2$ (which is $1/2$), add a small continuity correction, c (to make the approximation better with small samples), and divide by the variance of $\hat{\phi}$ under the permutation distribution. This gives

$$Z_{perm} = \frac{\hat{\phi}_{obs} - \frac{1}{2} + c}{\sqrt{V_{perm}}}, \tag{9.20}$$

where $c = \text{sign}(\hat{\phi}_{obs} - \frac{1}{2})/(n_1 n_2)$ and

$$V_{perm} = \frac{N+1}{12mn}\left(1 - \frac{\sum_{j=1}^{k}(d_j^3 - d_j)}{N^3 - N}\right) = \frac{(N+1)t}{12mn}, \tag{9.21}$$

where $d_1, \ldots, d_k$ are the number of responses from both groups in each of the k unique responses, and

$$t = 1 - \frac{\sum_{j=1}^{k}(d_j^3 - d_j)}{N^3 - N} \tag{9.22}$$

is the "tie adjustment factor." If the responses are continuous, there are no ties and $t = 1$. We treat Z_{perm} as standard normal, so for example a two-sided p-value is $2(1 - \Phi(|Z_{perm}|))$, where Φ is the cumulative distribution of the standard normal.

Before we explore getting confidence intervals for $h_{MW}(F_1, F_2)$, we discuss some of the important properties of the Mann–Whitney parameter.

9.5.2 Intransitivity of the Mann–Whitney Functional

One problem with the Mann–Whitney functional is that it is not transitive. This is best shown by a simple example.

Example 9.3 *Let X, Y, and Z be three discrete random variables such that each have three possible values that are realized with equal probability: $X \in \{1, 5, 9\}$, $Y \in \{2, 6, 7\}$, and $Z \in \{3, 4, 8\}$. Since the three sets are disjoint, we can ignore ties. By simple calculations, $\Pr[Z > Y] = 5/9$ and $\Pr[Y > X] = 5/9$. Since these probabilities are greater than $1/2$, we can say that Z tends to be larger than Y, and Y tends to be larger than X, and we might expect that Z value will tend to be larger than X, but this is not so: $\Pr[Z > X] = 4/9$.*

9.5.3 Causal Interpretation of the Mann–Whitney Functional

We discuss a causal interpretation of the Mann–Whitney functional and its relationship to a closely related parameter. See Section 3.2.2 and Chapter 15 for more details on causal estimands and potential outcomes.

In this section, consider a randomized trial of control versus treatment. The ith individual is randomized to one of two interventions represented by z_i, with $z_i = 1$ being the control and $z_i = 2$ being treatment. Suppose that some basic causal assumptions apply: every individual gets the intervention they are randomized to, and the intervention and outcome of any individual does not affect the outcome of any other individual. Imagine before the study starts that each individual has two unknown potential outcomes that would be observed if the individual is randomized to each intervention, and denote these two potential outcomes as $\mathbf{Y}_i = [Y_i(1), Y_i(2)]$, where $Y_i(1)$ is the outcome observed if $z_i = 1$ and $Y_i(2)$ is the outcome observed if $z_i = 2$. Suppose the potential outcome vectors are independent and $\mathbf{Y}_i \sim F_{12}$, where F_{12} is a bivariate distribution. Importantly, although the bivariate $\mathbf{Y}_i$ are independent, it is very likely there will be correlation between $Y_i(1)$ and $Y_i(2)$ because they both come from individual i. By randomization $\mathbf{Y}_i$ is independent of Z_i. In a randomized trial we only observe one potential outcome for each individual. Let the marginal distributions be $Y_i(1) \sim F_1$ and $Y_i(2) \sim F_2$. Since we only observe one of the potential outcomes for each individual, we cannot identify F_{12}, but we can only identify F_1 and F_2. So for example, although $Y_i(1)$ and $Y_i(2)$ are likely to be correlated, there is no way to measure that correlation in the simple randomized trial.

For causal inferences we want to compare the causal effect of getting intervention $z_i = 2$ instead of $z_i = 1$, so we want to compare the potential outcomes $Y_i(2)$ and $Y_i(1)$ on the same individual. A simple unit-level causal effect for the ith unit is the difference $Y_i(2) - Y_i(1)$, but for the WMW test we want a rank-based unit-level causal effect, say $R\{Y_i(2)\} - R\{Y_i(1)\}$, where R is a ranking function. Even better than a ranking function is a function that transforms the ranks onto the unit interval, $[0, 1]$, so that the resulting values do not depend so much on the sample size. In other words, we want a distribution-like function. Let this function be $G(y) = \Pr[W \le y] + \frac{1}{2}\Pr[W = y]$, where $W = BY(1) + (1 - B)Y(2)$, $Y(1) \sim F_1$, $Y(2) \sim F_2$, and B is a Bernoulli random variable with parameter 0.5. So for a randomized trial where $z_i = 1$ and $z_i = 2$ are equally likely, W represents a random observed outcome from the trial. Then $G(Y_i(z_i))$ represents the quantile level of the observed response for the ith individual, where the quantile level is corrected for ties (so can be used with discrete responses), and the level refers to the population of all potential outcomes. Our rank-like individual causal effect is the difference in these quantile levels, $D_i = G\{Y_i(2)\} - G\{Y_i(1)\}$. Then it can be shown that (Problem P8),

$$\phi = h_{MW}(F_1, F_2) = \frac{1}{2} + \mathrm{E}(D_i). \tag{9.23}$$

A noncausal interpretation of the Mann–Whitney parameter is

$$\phi = h_{MW}(F_1, F_2) = \mathrm{Pr}[Y_i(1) < Y_j(2)] + \frac{1}{2}\mathrm{Pr}[Y_i(1) = Y_j(2)], \tag{9.24}$$

where $Y_i(1)$ represents a randomly selected outcome from the arm randomized to control ($z_i = 1$) and independently $Y_j(2)$ represents a randomly selected outcome from the arm randomized to treatment ($z_j = 2$). Notice the indices for $Y_i(1)$ and $Y_j(2)$ are different, denoting different individuals. This is not the same as

$$\eta = h_{PBT}(F_{12}) = \mathrm{Pr}[Y_i(1) < Y_i(2)] + \frac{1}{2}\mathrm{Pr}[Y_i(1) = Y_i(2)], \tag{9.25}$$

where here $Y_i(1)$ and $Y_i(2)$ have the same index, and represent responses from the same randomly selected individual. The parameter, $\eta = h_{PBT}(F_{12})$, is the proportion of individuals from the population that would respond with larger values on intervention 2 than on intervention 1 (plus a correction for ties). In the scenario of this section, $\eta = h_{PBT}(F_{12})$ is the proportion of the population that benefits from treatment compared to control (corrected for ties), motivating the h_{PBT} notation. Unfortunately, since F_{12} is not identifiable, we cannot estimate $h_{PBT}(F_{12})$ from the trial, we can only estimate functionals of the marginal distributions such as the Mann–Whitney parameter, $h_{MW}(F_1, F_2)$ (see Equation 9.3). Although Equations 9.24 and 9.25 look similar, they can be very different.

Example 9.4 *Consider a randomized trial where patients are randomized to get either a new treatment ($z = 2$) or a standard treatment ($z = 1$). Suppose larger values are better and the responses are continuous, so that $h_{PBT}(F_{12})$ represents the proportion of patients that are better off on the new treatment than on the standard treatment. In Figure 9.3 we plot distributions from a hypothetical example where the new treatment only helps about 35% of individuals (although with substantial improvement), but the other 65% are slightly worse off on the new treatment (for details see Problem P11). The value of $\eta = h_{PBT}(F_{12})$ for this example is 0.358, properly showing that in truth only 35.8% of patients would do better on the new treatment. The value of $\phi = h_{MW}(F_1, F_2) = 0.654$ is the probability that a response from an arbitrary individual who got the new treatment is larger than a response from an arbitrary individual who got the standard treatment. Alternatively, $\phi - 0.5 = 0.154$ is the expected quantile level change for an individual who got treatment rather than control. So by $h_{MW}(F_1, F_2)$ the new treatment appears better, but by $h_{PBT}(F_{12})$ the standard treatment appears better (Problem P12).*

Fay et al. (2018a) studied the causal interpretation of $\phi = h_{MW}(F_1, F_2)$ and explored the relationship to $h_{BPT}(F_{12})$. They reviewed bounds on $h_{BPT}(F_{12})$ given estimates of F_1 and F_2.

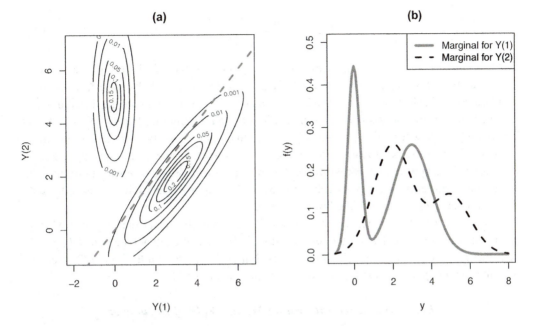

Figure 9.3 Bivariate distribution (a) and marginal distributions (a) for Example 9.4. The dotted gray line in part (a) is the line of equality for reference. See Problem P11 for a full specification of the model.

9.5.4 Testing the Mann–Whitney Parameter

Suppose we wanted to test,

$$H_0 : h_{MW}(F_1, F_2) = \phi_0$$
$$H_1 : h_{MW}(F_1, F_2) \neq \phi_0,$$

without making any assumptions about the distributions F_1 and F_2. This is known as the *nonparametric Behrens–Fisher problem* (Brunner and Munzel, 2000). Typically, we want to test $\phi_0 = 0.5$ since $h_{MW}(F, F) = 0.5$ for any F (even discrete distributions). First, the WMW test applied to that hypothesis will not be valid (i.e., will not retain the Type I error rate). Suppose F_1 and F_2 are both normal with $F_1 = N(0, 1)$ and $F_2 = N(0, 16)$ (so that $h_{MW}(F_1, F_2) = 1/2$) and let $n_1 = 5,000$ and $n_2 = 1,000$. This sample size is large enough that we can use an asymptotic approximation of the permutation test (Equation 9.20). Simulating 10,000 data sets, two-sided significance level of 5%, we get a simulated Type I error rate of 15.9% (Fay and Malinovsky, 2018); that is, the WMW test of $H_0 : h_{MW}(F_1, F_2) = 1/2$ is not even asymptotically valid. If we were testing $H_0 : F_1 = F_2$, then the variances would be equal under the null, and the WMW test would be valid.

The null hypothesis of $H_0 : h_{MW}(F_1, F_2) = 1/2$ contains distributions with $F_1 \neq F_2$, so as we have just shown, not all permutation tests are valid (recall the WMW test is a permutation test). However, there are some permutation tests that are asymptotically valid if we choose the test statistic carefully. Brunner and Munzel (2000) showed the following test statistic is asymptotically valid (specifically, pointwise asymptotically valid (PAV), see Note N2) under $H_0 : \phi = \phi_0$,

$$T \equiv T(\mathbf{y}, \mathbf{z}) = \frac{h_{MW}(\hat{F}_1, \hat{F}_2) - \phi_0}{\sqrt{V}} \qquad (9.26)$$

and V is an estimator of the variance of $h_{MW}(\hat{F}_1, \hat{F}_2)$ (see Problem P13). The statistic $T(\mathbf{y}, \mathbf{z})$ is a studentized test statistic, because we divide $h_{MW}(\hat{F}_1, \hat{F}_2) - \phi_0$ by an estimator of its standard error. Further, we can use Student's t-distribution to approximate the distribution of T under the null hypothesis if we add the additional assumption that $\mathrm{Var}(F_1(W_2)) > 0$ and $\mathrm{Var}(F_2(W_1)) > 0$, where $W_1 \sim F_1$ and $W_2 \sim F_2$. This is not an overly restrictive assumption, and just eliminates some strange zero variance cases. With this added assumption, $T(\mathbf{y}, \mathbf{z})$ of Equation 9.26 is approximately distributed t with degrees of freedom using an approximation analogous to Satterthwaite's (see Problem P13), and the approximation becomes better with larger sample sizes. We call this a *Brunner–Munzel test*.

When $\phi_0 = 1/2$, i.e., for testing $H_0 : \phi = 1/2$, we can perform an exact Brunner–Munzel test using a permutation test on the test statistic T. This has the advantage that *if* $F_1 = F_2$, then the p-value is valid for all sample sizes, but if $F_1 \neq F_2$ while $\phi = 1/2$, then the permutation test p-value is still (pointwise) asymptotically valid.

9.5.5 Confidence Interval on Mann–Whitney Parameter

There are many methods for calculating a confidence interval for $h_{MW}(F_1, F_2)$. We will discuss two useful ones, a robust method that needs few assumptions about F_1 and F_2, and a method that works reasonably well with the added proportional odds assumption.

First, consider a robust way to calculate a confidence interval for $h_{MW}(F_1, F_2)$. We can invert a series of nonparametric Behrens–Fisher tests for different values of ϕ_0. Let $p(\mathbf{x}, \phi_0)$ where $\mathbf{x} = [\mathbf{y}, \mathbf{z}]$ is the p-value calculated using the t-distribution approximation for Equation 9.26. Then we invert that p-value procedure to get a Brunner–Munzel $100(1 - \alpha)\%$ confidence interval,

$$C(\mathbf{x}, 1 - \alpha) = \{\phi_0 : p(\mathbf{x}, \phi_0) > \alpha\},$$

which can be written as

$$\hat{\phi} \pm q_t(1 - \alpha/2, d_{nBF})\sqrt{V}, \qquad (9.27)$$

where $q_t(a, d)$ is the ath quantile of a t-distribution with d degrees of freedom, and V and d_{nBF} are given in Problem P13.

An issue with the robust confidence interval of Expression 9.27 is that the confidence interval is not compatible with the WMW test. This follows from the fact that the WMW test is not valid for all cases with $\phi = 1/2$ but $F_1 \neq F_2$. In order to have a confidence interval compatible with the WMW test, we need to add additional assumptions. An example is adding the assumption that F_1 and F_2 are related by proportional odds (see Equation 9.1). This leads us to our second method for calculating a confidence interval on ϕ, a proportional odds model confidence interval on ϕ. The advantage of this interval is that is compatible with the usual asymptotic version of the WMW test (see Equation 9.20), meaning the two-sided test rejects at level α if and only if the $100(1 - \alpha)\%$ confidence interval excludes $1/2$.

The key to the proportional odds model confidence interval on ϕ is that we approximate the variance of ϕ with a function that depends on ϕ, and when $\phi = 1/2$ that function reduces to V_{perm} (see Equation 9.21). That variance function is

$$V(\phi) = \frac{t\phi(1 - \phi)}{mn}\left\{1 + \frac{N - 2}{2}\left(\frac{\phi}{1 + \phi} + \frac{(1 - \phi)}{2 - \phi}\right)\right\},$$

where t is the tie adjustment factor of Equation 9.22. The motivation of $V(\phi)$ is fairly involved, but it can be thought of as an approximate variance estimate under a continuous proportional odds model (a shift in logistic distributions) with a tie correction (see Fay and Malinovsky, 2018). We create a Z statistic with continuity correction, as

$$Z(\phi) = \frac{\hat{\phi} - \phi + c}{\sqrt{V(\phi)}},$$

where $c = sign(\hat{\phi} - \phi)/(n_1 n_2)$ is the same continuity correction used with the asymptotic WMW test (see Equation 9.20). Then we treat $Z(\phi)$ as standard normal when $h_{MW}(F_1, F_2) = \phi$. We create a $100(1 - \alpha)$% confidence interval by solving for the two ϕ values that just reject at the $\alpha/2$ level (except when $\hat{\phi}$ are extreme). Specifically, the $100(1 - \alpha)$% confidence interval is (ϕ_L, ϕ_U), where $\phi_L = 0$ if $\hat{\phi} = 0$, otherwise it is the value of ϕ that solves $Z(\phi) = \Phi(\alpha/2)$ and $\phi_U = 1$ if $\hat{\phi} = 1$, otherwise it is the value of ϕ that solves $Z(\phi) = \Phi(1 - \alpha/2)$.

Since the proportional odds model confidence interval on ϕ is compatible with the asymptotic WMW test, it can be used in the following way. First test $H_0 : F_1 = F_2$ versus $H_1 : F_1 \neq F_2$ with the asymptotic WMW test (see Equation 9.20). Then give a $100(1 - \alpha)$% proportional odds model confidence interval on ϕ. That interval will have approximately nominal coverage if we can add the additional assumption of proportional odds. The coverage may be poor if the proportional odds assumption is not approximately true. There is also a proportional odds model confidence interval that is compatible with the exact implementations of the WMW test, but it is intractable except for small sample sizes (for details see Fay and Malinovsky, 2018).

9.5.6 Latent Continuous Models and the Mann–Whitney Functional

Imagine a continuous outcome used in two different studies on the same population that both estimate the same Mann–Whitney parameter (ϕ_ℓ) such that we can compare the two estimates of ϕ_ℓ. Now suppose that one of the studies took the continuous data and partitioned it into k ordered categories. The Mann–Whitney parameter from the partitioned responses, say ϕ, is estimating a different value that would be pulled closer to $1/2$ than ϕ_ℓ because of the ties. In this section, we show how to use a proportional odds model to estimate ϕ_ℓ (the latent continuous Mann–Whitney parameter) from the study that only observes the partitioned categorical outcomes.

For this subsection, assume we have a proportional odds model with odds ratio parameter ψ. For this section, we denote the Mann–Whitney parameter from the ordinal data as ϕ, and the Mann–Whitney parameter from the latent (i.e., unobserved) continuous responses as ϕ_ℓ. The parameter, ϕ_ℓ, is a function of ψ only, but ϕ is a function of ψ and a vector of partitioning parameters, γ.

First, suppose we have latent continuous responses $(X_1^*, \ldots, X_n^*)$ and associated group indicators $(z_1, \ldots, z_n)$, with $z_i \in \{1, 2\}$. When $z_i = 1$ then $X_i^* \sim F_1^*$ and when $z_i = 2$ then $X_i^* \sim F_2^*$. We can show that for any proportional odds model on continuous responses there is some monotonic transformation of the responses such that the transformed responses are a location shift on a standard logistic distribution. In other words, there exists a transformation function, say $h(\cdot)$, such that $X_1 = h(X_1^*), \ldots, X_n = h(X_n^*)$, and letting $X_i \sim F_a$ when $z_i = a$ for $a = 1,2$, then F_1 is the cumulative distribution function (cdf) of a standard logistic distribution and $F_2(x) = F_1(x - \beta)$ for all x (see Problem P14). Following Equation 9.5 we define $\psi = \exp(\beta)$.

Because of the monotonicity of the transformation, $h_{MW}(F_1^*, F_2^*) = h_{MW}(F_1, F_2) = \phi_\ell$. We can solve for ϕ_ℓ as a function of ψ using the shift in logistic model (Problem P15),

$$\phi_\ell = \varphi_\ell(\psi) = \begin{cases} 0 & \text{if } \psi = 0 \\ \frac{1}{2} & \text{if } \psi = 1 \\ 1 & \text{if } \psi = \infty \\ \frac{\psi\{\psi - 1 - \log(\psi)\}}{(\psi - 1)^2} & \text{otherwise .} \end{cases} \tag{9.28}$$

Now consider the observed ordered categorical responses $(Y_1, \ldots, Y_n)$ each of which has one of k ordered categories. We call this the *coarsened data*. We can think of an unobserved process where the latent responses X_i^* are binned into k categories, meaning $Y_i = j$ if $X_i^* \in (\xi_{j-1}, \xi_j]$, for $j = 1, \ldots, k$ and $-\infty \equiv \xi_0 < \xi_1 < \cdots < \xi_{k-1} < \xi_k \equiv \infty$. We can get the distribution of the coarsened responses using $\beta = \log(\psi)$ and $\gamma = [\gamma_1, \ldots, \gamma_{k-1}]$ (also called the *cutpoint parameters*) according to Equation 9.4:

$$F(j|z_i) = \Pr[Y_i \le j|z_i] = \frac{1}{1 + \exp\left(-\{\gamma_j - \beta I(z_i = 2)\}\right)}. \tag{9.29}$$

Let $F_1(j) = F(j|1)$, $F_2(j) = F(j|2)$, and let the associated probability mass functions be $f_a(j) = F_a(j) - F_a(j-1)$ for $a = 1,2$. Then ϕ as a function of ψ and γ is

$$\phi = \varphi(\psi, \gamma) = \sum_{j=2}^{k} F_1(j-1) f_2(j) + \frac{1}{2} \sum_{j=1}^{k} f_1(j) f_2(j).$$

Table 9.1 gives some values of ϕ_ℓ and ϕ for different values of ψ and γ. Notice how the coarsening pulls the ϕ_ℓ toward 0.5.

ψ	$\varphi_\ell(\psi)$	$\varphi(\psi, [-1.1, -0.7, 0.3, 2.3])$	$\varphi(\psi, [-1.5, 0.9])$
0.25	0.283	0.301	0.317
0.50	0.386	0.394	0.406
1.00	0.500	0.500	0.500
2.00	0.614	0.605	0.594
4.00	0.717	0.699	0.681

Table 9.1 *Different values of $\phi_\ell = \varphi(\psi)$ and $\phi = \varphi(\psi, \gamma)$*

For a data analysis, these equations give us another way of interpreting the proportional odds parameter for the two-sample case. Suppose we ran a proportional odds model on a two-sample ordinal response problem and got estimates $\hat{\beta}$ and the vector $\hat{\gamma}$, with 95% confidence interval on β as (β_L, β_U). The odds ratio estimate is $\hat{\psi} = \exp(\hat{\beta})$ with 95% confidence interval equal to $(e^{\beta_L}, e^{\beta_U})$ which we write as (ψ_L, ψ_U). Then our estimate of the latent

Mann–Whitney parameter, ϕ_ℓ, is $\varphi_\ell(\hat{\psi})$ with 95% confidence interval $(\varphi_\ell(\psi_L), \varphi_\ell(\psi_U))$. A plug-in estimate of the usual Mann–Whitney parameter, ϕ, is $\varphi(\hat{\psi}, \hat{\gamma})$. Because small changes in the values of the γ parameters do not change ϕ much, a simple 95% confidence interval for ϕ ignores the variability of the γ parameters, giving $\left(\varphi(\psi_L, \hat{\gamma}), \varphi(\psi_U, \hat{\gamma})\right)$.

9.6 Choosing a Test when Power Concerns Dominate

In this section we compare some tests (like the t-test and the WMW test), and show that even when the data appear normally distributed often the WMW test can be more powerful. Although this section focuses on power, in practice other aspects of the study may be more important (see, e.g., Section 3.3.2).

Suppose a statistician is designing a study comparing two groups and the primary endpoint is ordinal or numeric where larger values denote a better response. Further, suppose that in designing the study, the primary concern of the research is showing that one group is better or worse than the other, and the choice of effect (i.e., the metric of the treatment effect) is secondary. We address this in two cases separately: first the case when the data are ordinal, and then the case when the data are numeric.

When the data are ordinal, we have been writing the responses as $Y_i = j$ for $j \in \{1, \ldots, k\}$, but we only consider tests that are invariant to monotonic transformations of the Y_i. We could define different scores besides $\{1, \ldots, k\}$ for the k categories, such as scores based on the overall midranks or scores based on the subject matter context; however, such scoring would essentially treat the responses as numeric, so we address the power consideration for scores in the numeric case. The main ordinal tests are tests on proportional odds or the Mann–Whitney parameter. The first point is that if we can assume proportional odds and create a permutation test on the efficient score (see Section 10.7), this is the WMW test. In this case, the WMW test has some good power properties for alternative models close to the null.[1] So under the proportional odds assumption the WMW test is a good choice, and we can use the proportional odds model confidence interval on ϕ with that. If the proportional odds assumption does not fit the application, then the Brunner–Munzel confidence intervals of Section 9.5.4 could be used.

Now consider when the data are numeric. For example, consider a study similar to Example 9.2, where two diets are being tested in some lab mice and the endpoint is time to natural death in days. Suppose in designing this study, any of the effects (e.g., difference in means, difference in medians, Mann–Whitney parameter) would be reasonable. In this case, the statistician could choose the test based on power concerns.

Suppose the researcher has data on a sample of the lifetimes of 112 mice on standard diet from a different lab a few years earlier. The researcher says that they expect a similar distribution on the new diet except shift up by about a week. (They say there is no way the diet is going to kill a small proportion of the mice as in Figure 9.2.)

We will consider two scenarios. First, suppose that the researcher shows you histogram (a) of Figure 9.4. That data looks approximately normal and has mean $1,000.0$ (it is fake data after all!), and standard deviation 9.1. The uniformly most powerful unbiased test of a shift in

---

[1] Specifically, it is the locally most powerful similar test for one-sided alternatives (McCullagh, 1980). See Lehmann and Romano (2005, p. 115) for the definition of a similar test.

normal distributions is the t-test. Further, for large samples we do not need the unbiasedness condition, and the asymptotically most powerful test is the t-test. But how does the power of the WMW test compare? Or the power of the difference in medians test? Further, what if the histogram was (b) of Figure 9.4 that shows some apparent skewness?

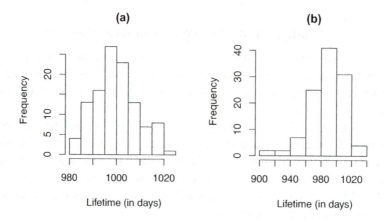

Figure 9.4 Two histogram of 112 mice lifetimes on standard diet. Both data (a) and (b) are fake data.

One way to compare two tests is by using relative efficiency, which is a ratio of sample sizes needed to get a certain power. Suppose we have an assumed data generating model under the alternative (e.g., a normal shift of 7 with standard deviation of 9). Let $N_A(\alpha, 1 - \gamma) \equiv N_A$ be the expected total sample size needed using Test A to reject a one-sided test at significance level α and get a power of $1 - \gamma$, where we do not force N_A to be an integer. Analogously define $N_B = N_B(\alpha, 1 - \gamma)$. The relative efficiency of Test A with respect to Test B is defined as the ratio N_B/N_A. So if Test A needs a larger sample size than Test B to get the same power, the relative efficiency of Test A with respect to Test B will be less than 1.

We compare the asymptotic WMW test, Welch's t-test, Student's t-test, and the melded difference in medians test. We simulate tests under the normal shift model with standard deviation 9 and shift 7 and equal sample sizes in each group, and try all sample sizes (per group) between 30 and 40. First, using only 10^3 simulations for the melded difference in median test, none of the simulated powers were greater than 70%, so that will not be considered further. For the other tests using 10^5 simulations so that the simulated powers are sufficiently smooth and are increasing with sample size, we linearly interpolate the simulated powers between the sample size less than 90% and the one greater than 90% to get the expected sample size that equals 90%. We get simulated total sample sizes of $\hat{N}_{st} = 2 * 35.71$ (for Student's t-test), $\hat{N}_{wt} = 2 * 35.73$ (for the Welch t-test), and $\hat{N}_{wmw} = 2 * 37.87$ (for the WMW test). Ideally, we would simulate using the exact WMW test, but that would take too long, and we expect that the asymptotic WMW test will be close to valid. The simulated efficiency of the asymptotic WMW test relative to Student's t-test is $\hat{N}_{st}/\hat{N}_{wmw} = 94.3\%$, and the simulated efficiency of Welch's t-test relative to Student's t-test is $\hat{N}_{st}/\hat{N}_{wt} = 99.9\%$. So in terms of power, there appears to be very little reason to not use Welch's t-test instead of Student's t-test, since it is better if the two normal distributions have different variances and it has almost the same power as Student's t-test. Even the WMW test is about 94% as efficient as Student's t-test.

There is a way to calculate an asymptotic relative efficiency (ARE) of Test A relative to Test B, where we let the difference in means approach 0 at a rate such that the ratio N_B/N_A approaches a constant. Nicely, for the location shift models where the tests can be based on asymptotically normal test statistics, the ARE does not depend on the significance level or the power (see Lehmann, 1999, section 3.4). The ARE of the WMW test relative to Student's t-test for the normal location shift is $3/\pi = 95.5\%$, close to our simulated value of 94.3%.

We can use the ARE machinery to make some statements about location shift models. First, to study distributions with large tails, we consider t-distributions. The t-distributions range from degrees of freedom equal to $df = 1$ (also called a Cauchy distribution) that have the largest tails (and have infinite variance), to the t-distribution with $df \to \infty$ (giving a normal distribution). We calculate the ARE of the WMW test with respect to Student's t test for a location shift in a t-distribution with df degrees of freedom. We solve for the df that gives ARE = 1, which is $df = 18.76$. We plot that distribution (scaled to have variance = 1) compared to a standard normal in Figure 9.5 (a). We see that the amount that the tails of that t-distribution are larger than the standard normal is almost imperceivable. For any t-distribution with larger tails ($df < 18.76$), the WMW test is asymptotically more efficient than Student's t-test. With most reasonably sized data sets we would not be able to detect if the tails were larger than the t-distribution with $df = 18.76$ compared to a normal. A study of many high-quality data sets with normal looking distributions, found that t-distributions with df = 3 to 10 fit better than normal distributions (Hampel et al., 1986, p. 23). Thus, if power concerns dominate, it may make sense to default to the WMW test over a t-test even for normal looking data.

Now consider skewed distributions. To study this, we consider the ARE of the WMW test with respect to Student's t-test for a location shift in a log-gamma distribution. We solve for the shape parameter that gives ARE = 1 and find $a = 5.55$. A distribution with this shape parameter, scaled to have a variance of 1, is plotted in part (b) of Figure 9.5. We see that the skewness is a little more perceivable than the large tail example of part (a), but for reasonably sized data sets we may not be able to detect that amount of skewness compared to normal (which is symmetric of course). It is reasonable to conclude that if there is perceivable skewness, then the WMW test is likely to be more efficient than a t-test.

So for reasonably large sample sizes, even when the historic data look close to normal, it is quite possible that the WMW test will be more powerful. Further, if the historic data appear nonnormal then we recommend the WMW test over the t-tests, if the primary concern is power.

Now consider very small sample sizes. If we are testing $n < 4$ in each group at the two-sided $\alpha = 0.05$ level, then regardless of the responses, the WMW test cannot reject. So if a researcher is in that situation, the t-test is the better option. But with sample sizes this small, even the inferential values of the t-test may need to be treated as descriptive. Consider a fake randomized clinical trial with $n = 2$ in each group. Suppose that the data are 9.4, 10.2 in the control group and 16.6, 15.3 in the treatment group. Welch's t-test gives a two-sided $p = 0.0251$. The problem is that even though we randomized, there could be some unobserved (or unrecorded) binary covariate that completely explains the difference. For example, the effect could be entirely due to sex differences, and all the females were in

(a)

Scaled t distribution (df = 18.76) vs standard normal

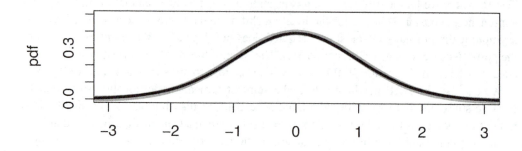

(b)

log gamma (a = 5.55) vs standard normal

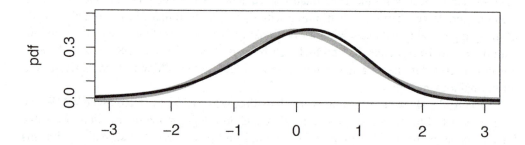

Figure 9.5 Part (a) is the probability density function (pdf) of a central t-distribution with degrees of freedom, df = 18.76, scaled to have variance 1 (black) compared to the pdf of a standard normal distribution (gray). t-distributions with smaller degrees of freedom have larger tails and have the WMW test has greater asymptotic efficiency for a location shift than the t-test. Part (b) is the pdf of a log-gamma distribution with shape parameter $a = 5.55$ (black) and the standard normal pdf (gray). In log-gamma distributions that are more skewed ($a < 5.55$), the WMW test has greater asymptotic efficiency for a location shift than the t-test.

one arm of the study and all males were in the other. If the two values of the covariate are evenly distributed in the population and we take a simple random sample, there is a 12.5% chance that both individuals in one experimental arm will have one value of the covariate (e.g., be all females or all males) and both in the other arm have the other value. So even though the $p = 0.0251$ under the assumption of a shift in normal distributions due to the intervention group, if the true distribution is a shift in normal distributions due to some other unobserved covariate (e.g., sex of the individual), then the low p-value could be entirely due to the unobserved covariate. Thus, when $n < 4$ we would presumably use a Welch's t-test, but we should interpret it cautiously.

9.7 Summary

The chapter first focuses on tests and confidence intervals for particular group effect metrics and then focuses on test selection when the effect metric is a secondary consideration to power and validity. Table 9.2 gives a tabular summary of the main methods.

Response type	Primary goal	Sect	Method for test and confidence interval	Assumptions	Comments
Ordinal or numeric	Estimate odds ratio	9.2	Proportional odds with maximum likelihood (i.e., ordinal logistic regression)	Proportional odds likelihood regularity, Ch 10: Note N3	Number of response categories fixed (does not depend on n)
	Estimate Mann–Whitney parameter	9.5.4 9.5.5	Brunner–Munzel	Largely assumption free	CI requires large sample, incompatible with WMW test
		9.5.4 9.5.5	WMW test with proportional odds model CI	Test: $F_1 = F_2$ CI: proportional odds	CI compatible with WMW test
	Power to compare arms	9.4.3	WMW test		WMW is default for ordinal data
Requires numeric data	Estimate difference in means	9.3.2	Permutation test with shift CI	Location shift	Location shift fails for discrete responses
		9.3.3	Classical t-test with CIs	Normal with equal variances	Not recommend as default
		9.3.3	Behrens–Fisher or Welch t-test with CIs	Normal, allow different variances	Behrens–Fisher is valid, Welch t-test is approximately valid
	Estimate difference in medians	9.4.2	Melding of exact CIs on median	Largely assumption free large samples	Can be conservative with small samples
		9.4.3	Nonparametric bootstrap		Can be anti-conservative
		9.4.3	WMW test with shift CI and associated Hodges–Lehmann estimator	Location shift	Location shift fails for discrete responses
			Classical t-test with CIs	Location shift	Better to use WMW test with shift, Sec 9.4.3
	Power to compare arms	9.6	WMW test		WMW test is robust with respect to power
		9.6	t-tests		Slightly more powerful if data normal

Table 9.2 *Summary of methods and assumptions*

Ordinal Data

When data are truly ordinal such as a five-point scale that goes from *Agree Strongly* to *Disagree Strongly*, there are two main parameters to quantify the group effect. First is the proportional odds parameter, ψ, which assumes that the odds that a response in Group 1 is $\leq j$ divided by the same odds in Group 2 equals ψ for any value $j < 5$ on the five-point ordinal scale. Second is an assumption-free parameter, the Mann–Whitney parameter ϕ, which represents the probability that a random observation in Group 2 is larger than a random observation in Group 1 plus half the probability that they are equal.

When a proportional odds assumption seems reasonable, the number of categories for the ordinal responses does not depend on n, the sample size is large, and the primary focus is on estimation of the proportional odds parameter, then maximum likelihood estimation and inferences can be used. Alternatively, if the proportional odds assumption is not tenable, then the Mann–Whitney parameter is another option, and confidence intervals about that parameter can be made with large samples using the Brunner–Munzel method, which uses a t-distribution based on a t-statistic, the sample Mann–Whitney parameter minus its true value divided by its variance. For small samples there is an exact Brunner–Munzel method, a permutation test based on the t-statistic, but there are no existing compatible confidence intervals for ψ with that permutation test.

The important Wilcoxon–Mann–Whitney (WMW) test has many derivations. Historically, it is a permutation test on the difference in mean ranks of the two groups, but it is also the permutation test based on the sample Mann–Whitney parameter. The WMW test has both small sample and asymptotic implementations, with each of those implementations having compatible confidence intervals, and the confidence intervals require the proportional odds model. The WMW test may be derived as a permutation test on the score statistic for the proportional odds model. Because of that relationship, the WMW test will have inferential results similar to the likelihood methods on the proportional odds model.

Numeric Data

When a confidence interval of the difference in means is the main goal, we discussed two choices. For small sample sizes with data that look far from normal, one option is a shift confidence interval based on a permutation test that uses a shift assumption (i.e., assumes that $F_2(y) = F_1(y - \Delta)$). However, if the shift assumption is not reasonable, and the sample size is large or the data look close to normally distributed, then a Welch t-test is recommended, with its corresponding confidence interval for difference in means. Either of the two approaches may be used for obtaining a confidence interval on the ratio of geometric means by computing a confidence interval on the difference in means of the log-transformed data, and then taking antilogs of this to get an interval on the ratio of geometric means.

When a confidence interval of the difference of medians is the main goal there are three possible approaches: melded, bootstrap, and shift confidence intervals. When the shift assumption does not hold, the shift confidence interval can have very poor coverage. Further, the bootstrap confidence interval can also be quite anti-conservative in some situations, whereas the melded confidence interval may be very conservative. When the shift assumption holds, the shift confidence interval is preferred for estimating the difference in medians.

Even with numeric data, one might find the Mann–Whitney parameter the most useful metric. In this case, the options are the Brunner–Munzel method which requires few assumptions, or the WMW test and its compatible proportional odds model confidence intervals.

If the primary research objective is a powerful and robust test, we generally recommend the WMW test, even for numeric data. When data are normally distributed, the asymptotic relative efficiency of WMW versus the t-test is 95.5%. Conversely, when data are distributed with modestly larger tails than a normal distribution, which one study found is common in applications, the WMW will have better power than the t-test even asymptotically. Finally, the WMW is safe to use when the data are clearly not normally distributed, and when the sample sizes are small and distributions are poorly understood. Of course, in a setting where there is much confidence about the normality of the data, Welch's t-test would be the natural choice.

In closing, we note the Mann–Whitney parameter in a randomized trial requires care to be viewed from a causal perspective, since it is easily mistaken for the proportion of the population with larger responses under the second group.

9.8 Extensions and Bibliographic Notes

See Fay and Proschan (2010) for a comprehensive discussion of multiple perspective decision rules, especially as it applies to the tests in this chapter. Their paper also gives more precise statements about asymptotic validity. In particular, it differentiates between tests that are pointwise asymptotically valid (PAV) (for any fixed distribution under the null, the power to reject is asymptotically less than or equal α), and those that additionally are uniformly asymptotically valid (UAV) (the size of the test is asymptotically less than or equal α). For example, if the distribution under the null hypothesis has finite variance, then the different versions of the t-test are all PAV, but if additionally $E(Y^4)$ is finite then the t-tests are UAV. Table 1 of Fay and Proschan (2010) gives validity and consistency of 8 decision rules (including rules related different versions to t-tests, WMW tests, and Brunner–Munzel tests), under 14 different perspectives. Additionally, much of Section 9.6 is modeled after parts of Fay and Proschan (2010).

Brown and Hettmansperger (2002) discuss the intransitivity of the Mann–Whitney functional. Fay et al. (2018a) discuss causality and the Mann–Whitney functional in a randomized experiment. They focus on the comparing the proportion who benefit from treatment (plus tie correction) (i.e., $h_{PBT}(F_{12})$) and the Mann–Whitney functional.

Pratt (1964) discusses the lack of validity of the Mann–Whitney test when $\phi = 1/2$ but $F_1 \neq F_2$. For details on the nonparametric Behrens–Fisher tests (i.e., test on the Mann–Whitney parameter) and the associated confidence intervals on ϕ see Brunner and Munzel (2000), Neubert and Brunner (2007), Chung and Romano (2016), and Pauly et al. (2016). For the proportional odds model confidence intervals of ϕ that are compatible with implementations of WMW tests see Fay and Malinovsky (2018), where the asymptotic version is closely related to a confidence interval in Newcombe (2006).

9.9 R Functions and Packages

In R, let y1 and y2 be vectors of responses. Then the difference in means (Grp 1 - Grp 2) using Welch's t test uses t.test(y1,y2). If y1 and y2 are positive, then we can get confidence intervals on the ratio of geometric means (Grp 1/Grp 2) as `exp( t.test(log(y1), log(y2))$conf.int )`. For the difference or ratio in medians by melding use the `mdiffmedian.test` function in the `bpcp` R package. For a proportional odds model use the `polr` function in the `MASS` R package. For exact WMW tests use the `coin` R package. For the location shift confidence intervals associated with the WMW test use `wilcox.test`, and for the proportional odds model confidence interval use the `wmwTest` function in the `asht` R package. For the Brunner–Munzel approach to the nonparametric Behrens–Fisher problem use the `npar.t.test` function of the `nparcomp` R package.

9.10 Notes

N1 **Transformations and medians:** Consider a discrete numeric random variable, X, with distribution, F, and a strictly increasing transformation function, g. A median of X is a value θ such that $F(\theta) = \Pr[X \le \theta] \ge 0.5$ and $\bar{F}(\theta) = \Pr[X \ge \theta] \ge 0.5$ (see Section 5.2). Let Θ_x be the set of all medians of X. Let $g(X) \equiv Y$ and let Θ_y be the set of all medians of Y. Then for any median of X, say θ such that $\theta \in \Theta_x$, then $g(\theta)$ is a median of Y, i.e., $g(\theta) \in \Theta_y$.

N2 **Nonparametric Behrens–Fisher test and asymptotic validity:** If we perform a permutation test using the test statistic in Equation 9.26, the test is pointwise asymptotically valid but not uniformly asymptotically valid. Lehmann and Romano (2005, p. 422) discuss the differences between the two types of asymptotic validity. Roughly, pointwise asymptotically valid means that for any probability model in the null hypothesis the test is asymptotically valid, while uniformly asymptotically valid allows the probability model chosen within the null hypothesis to change with the sample size (e.g., get more and more extreme in the null hypothesis space as the sample size changes).

9.11 Problems

P1 Show that any valid test of the difference in means, Δ, of the form $H_0 : \Delta = \Delta_0$ versus $\Delta \ne \Delta_0$ (with no assumptions about F_1 and F_2) cannot have power greater than the significance level, α. Hint: use mixture distributions (see, e.g., Fay and Proschan, 2010, for answer).

P2 Show that the confidence interval by the location shift method for the difference in means using the permutational central limit theorem is given by Equation 9.12. Hints: let $\hat{\Delta} - \Delta = \sum_{i=1}^{n} y_i b_i$, where $b_i = -1/n_1$ if $z_i = 1$ and $b_i = 1/n_2$ if $z_i = 2$. Remember, V_{perm} depends on Δ. This is a large sample result, so assume that $n - c_{\alpha/2}^2 - 1$ is positive. Final hint, there is a lot of algebra.

P3 The value d_w (Equation 9.17) is known as Satterthwaite's approximation. Show that Equation 9.17 can be derived by assuming that $d_w \hat{V}/V$ is approximately distributed χ^2 with d_w degrees of freedom, where $V = (\sigma_1^2/n_1 + \sigma_2^2/n_2)$ and $\hat{V}$ replaces σ_a with $\hat{\sigma}_a$. Hint: start with setting $\mathrm{Var}(d_w \hat{V}/V)$ to $2d_w$.

P4 Calculate the sample median of the first nine positive integers. Then calculate the exponentiation of the sample median of the log transformation of those same integers. Are they the same? Repeat for the first 10 positive integers. Are they the same? If not, why not?

P5 Show that the sum of the midranks in Group 2 is equal to Expression 9.18. Hints: write the midrank as a function of the empirical distribution. Write the empirical distribution as a weighted sum of the empirical distribution from each group. Define $s_1, \ldots, s_k$ as the unique set of y_i values from both groups, and use the fact that for any function of y_i, say $G(y_i)$, we have the equality $\sum_{i=1}^{n} I(z_i = 2)G(y_i) = \sum_{j=1}^{k} n_2 \hat{f}_2(s_j)G(s_j)$, where $\hat{f}_2(s)$ is the empirical probability mass function at s.

P6 Show that the difference in mean ranks is equal to $n\left\{h_{MW}(\hat{F}_1, \hat{F}_2) - \frac{1}{2}\right\}$. Hint: use Expression 9.18 and the fact that the sum of the ranks of both groups is $n(n+1)/2$.

P7 **Efron dice:** Define three dice named *A*, *B*, and *C* where the numbers 1 through 18 are each put on one of the unique sides of the three dice. Imagine a two-person game where the first person picks a die, and the second person picks from the remaining two dice, the two people roll their chosen die, and the person with the highest roll on their die wins the game. Define the dice such that if you go second, you can always have an advantage. A set of such dice are called *Efron dice* (Brown and Hettmansperger, 2002).

P8 Show that Equation P8 holds. Hint: prove it separately for continuous and discrete distributions (see Fay et al., 2018a).

P9 Suppose $[Y(1), Y(2)]$ is bivariate normal with mean $[\mu_1, \mu_2]$ and covariance matrix,

$$\begin{bmatrix} v_1 & \rho\sqrt{v_1 v_2} \\ \rho\sqrt{v_1 v_2} & v_2 \end{bmatrix}.$$

Show that $\eta = h_{PBT}(F_{12}) = \Pr[Y(1) < Y(2)] + \frac{1}{2}\Pr[Y(1) = Y(2)]$ equals $\Phi\left(\frac{\mu_1 - \mu_2}{\sqrt{v_1 + v_2 - 2\rho\sqrt{v_1 v_2}}}\right)$, where $\Phi(z) = \Pr[Z \le z]$ and $Z \sim N(0, 1)$. What is $h_{MW}(F_1, F_2)$, where F_1 and F_2 are the marginal distributions of F_{12}? Hint: for any bivariate normal random variable $Y = [Y_1, Y_2]$ with mean μ and covariance Σ, then for any vector $w = [w_1, w_2]$ the weighted sum, $w_1 Y_1 + w_2 Y_2$ is normal with mean $w\mu$ and covariance $w\Sigma w'$.

P10 Suppose $\mathbf{Y} = [Y(1), Y(2)]$ has a bivariate distribution, F_{12}, with marginal distributions, F_1 and F_2. Assume F_{12} is a mixture of two bivariate normal distributions, with $\mathbf{Y} = \lambda \mathbf{Y}_G + (1 - \lambda)\mathbf{Y}_H$ and $\mathbf{Y}_G \sim G_{12}$ and $\mathbf{Y}_H \sim H_{12}$. Let the marginal distributions be $G_1, G_2, H_1,$ and H_2. Show that $h_{PBT}(F_{12}) = \lambda h_{PBT}(G_{12}) + (1 - \lambda)h_{PBT}(H_{12})$, and $h_{MW}(F_1, F_2) = \lambda^2 h_{MW}(G_1, G_2) + \lambda(1 - \lambda)h_{MW}(G_1, H_2) + \lambda(1 - \lambda)h_{MW}(H_1, G_2) + (1 - \lambda)^2 h_{MW}(H_1, H_2)$.

P11 The model of Figure 9.3 is a mixture of two bivariate normal distributions, one (with weight 0.35) has mean $[0, 5]$ and covariance matrix

$$\begin{bmatrix} 0.1 & 0 \\ 0 & 1 \end{bmatrix},$$

and the other (with weight 0.65) has mean $[3, 2]$ with covariance matrix

$$\begin{bmatrix} 1 & 0.9 \\ 0.9 & 1 \end{bmatrix}.$$

Write an R function to calculate or approximate $h_{PBT}(F_{12})$ and $h_{MW}(F_1, F_2)$ for that model. Explore different parameter values for the mixture bivariate normal model to see when $h_{PBT}(F_{12})$ and $h_{MW}(F_1, F_2)$ are similar or not.

P12 **Hand's paradox:** Hand (1992) studied the case when $h_{MW}(F_1, F_2)$ and $h_{PBT}(F_{12})$ imply differing directional effects, and Fay et al. (2018a) call this situation *Hand's paradox*, since by comparing Equation 9.24 and Equation 9.25 it seems like $h_{MW}(F_1, F_2)$ and $h_{PBT}(F_{12})$ are measuring the same thing, when they are not. Hand's paradox can exist in very simple situations. Give a bivariate distribution, F_{12}, with three equally likely potential outcome vectors, $[Y_i(1), Y_i(2)]$, that results in Hand's paradox.

P13 The variance estimate of $h_{MW}(\hat{F}_1, \hat{F}_2)$ is

$$V = \frac{\hat{\tau}_1^2}{n_1} + \frac{\hat{\tau}_2^2}{n_2},$$

where

$$\hat{\tau}_1^2 = \frac{1}{n_1 - 1} \sum_{i=1}^{n} I(z_i = 1) \left\{ \hat{F}_2(y_i) - \frac{1}{n_1} \sum_{h=1}^{n} I(z_h = 1) \hat{F}_2(y_h) \right\}^2$$

where $\hat{F}_2$ is the empirical distribution for the n_2 observations with $z_i = 2$, and $\hat{\tau}_2^2$ is defined analogously (see Chung and Romano, 2016, p. 102). Exercise: reexpress V as a function of the overall midranks, and the midranks within each group. (Answer: See Appendix A of Fay and Proschan (2010) (however, there is a typo: there should not be a square root in the definition of V_J) or section 2.1 of Pauly et al. (2016).) The degrees of freedom approximation, d_{nBF}, uses Satterthwaite approximation, Equation 9.17, except with s_1 and s_2 replaced by $\hat{\tau}_1$ and $\hat{\tau}_2$.

P14 Show that for any continuous response proportional odds model for the two-sample problem, there exists a monotonic transformation that transforms the responses to a location shift model on the standard logistic distribution. Hint: use the probability integral transformation. See Fay and Malinovsky (2018, Appendix B).

P15 Derive Equation 9.28. Hint: there is a difficult integral, use Mathematica or some computer algebra software to calculate it.

10

General Methods for Frequentist Inferences

10.1 Overview

In Chapter 2 we presented a general theory for creating hypothesis tests. There, as here, we focus on non-randomized decision rules, $\delta(\mathbf{x}, \alpha)$ that return either a 1 (reject the null hypothesis) or 0 (fail to reject the null). We showed the relationship of the decision rule function to the p-value function (Equation 2.2) and the confidence set procedure (see, e.g., Equation 2.5). Because of this relationship, informally we can define hypothesis tests by the assumptions (see Chapter 8) together with either the decision rule, δ, the p-value function, p, or the confidence set procedure, C_S. For scalar parameters, usually the confidence sets will be confidence intervals. Whereas in Chapter 2 we focused on properties that resulted once we had found a valid (or approximately valid) decision rule, in this chapter we focus on how we pick or create a valid decision rule (or p-value or confidence interval) in the first place.

We emphasize that nearly every section of this chapter deserves a book unto itself, and in fact such books exist (see Section 10.15). Our purpose is to give a brief introduction to methods, to give the user some idea of the assumptions needed for each method (or some of the most useful versions of the method), and to give some examples of when those assumptions are met or not met in real or simple made up examples.

This chapter is purposefully general because sometimes applied problems do not fit into a predefined simple model like the two-sample binary response model. The applied statistician should have knowledge and understanding of methods that are very flexible that can be applied to all kinds of situations, hence the generality, and unfortunately, the higher level of abstraction.

Much of this chapter is about problems where the probability models are described by the scalar parameter of interest, and a set of nuisance parameters that may be high dimensional (or even infinite dimensional). Many of the methods of this chapter outline different ways of dealing with those nuisance parameters.

10.2 One-Dimensional Parameter

A parametric model is a model where θ has a finite number of parameters. The simplest case is when θ is one-dimensional and the null and alternative are simple hypotheses (i.e., only one parameter value). In this case the Neyman–Pearson lemma (see Note N1) shows that the most powerful valid test for this situation rejects when the likelihood ratio is large. We do not emphasize that lemma for two reasons. First, the application of the lemma is for simple null and simple alternative hypotheses which are rarely used in applications since we often define the null or alternative hypotheses as a range of parameter values. Second, the application of the lemma to discrete data results in randomized decision rules which are rarely used in applications (see Problem P1).

We take the main idea of the Neyman–Pearson lemma (base decision rules on the likelihood ratio) and apply it to some examples. Our approach is to propose a three-step process that will give valid, nonrandomized tests for many one-sided composite hypotheses, although it will not work for all situations. The three steps are:

(1) formulate a likelihood;
(2) check that the likelihood ratio is monotone (i.e., monotonic) in some statistic, and
(3) calculate the critical value or p-value based on the null hypothesis distribution of that statistic.

If the likelihood ratio is not monotone in some statistic, then the p-value can still be calculated, but the power of the test may not be optimal or even good. Luckily, many practical likelihoods do have monotone likelihood ratios.

We go through these three steps for the simple case of n independent binary observations with the same probability of success, θ. Let $X_1, \ldots, X_n$ be the random variables for the binary observations. For discrete data, the likelihood is the probability of observing the data. For independent data, we multiply the probability of each observation, so the likelihood is

$$f(\mathbf{x}; \theta) = \Pr_\theta(X_1 = x_1, X_2 = x_2, \ldots, X_n = x_n)$$

$$= \prod_{i=1}^{n} \theta^{x_i} (1 - \theta)^{1-x_i} = \theta^s (1 - \theta)^{n-s},$$

where we let $s = \sum_{i=1}^{n} x_i$ and $s \in \{0, \ldots, n\}$. Suppose we are testing the one-sided hypothesis $H_0 : \theta \leq \theta_0$ versus $H_1 : \theta > \theta_0$. Let $\theta_a \in \Theta_0$ and $\theta_b \in \Theta_1$ be parameter values in the null or alternative space, so that $\theta_a < \theta_b$. The likelihood ratio is the likelihood under the alternative over the likelihood under the null. If we test the simple hypothesis, $H_0 : \theta = \theta_a$ versus $H_1 : \theta = \theta_b$, then the likelihood ratio is

$$R(x; \theta_a, \theta_b) = \frac{\theta_b^s (1 - \theta_b)^{n-s}}{\theta_a^s (1 - \theta_a)^{n-s}}$$

$$= \left(\frac{\theta_b(1 - \theta_a)}{\theta_a(1 - \theta_b)} \right)^s \left(\frac{1 - \theta_b}{1 - \theta_a} \right)^n.$$

For simple hypotheses on discrete data, the likelihood ratio represents the probability of observing the data under the alternative model over that probability under the null model. Now we check the monontonicity of $R(x; \theta_a, \theta_b)$ with respect to the only statistic that is not fixed, s. Since $\theta_a < \theta_b$, we know that $\frac{\theta_b(1-\theta_a)}{\theta_a(1-\theta_b)} > 1$ and $R(s; \theta_a, \theta_b)$ is increasing in s. So we can base our decision rule on s, larger values of s give larger likelihood ratio values than smaller values of s. So we reject for larger values of s, since these are more likely under the alternative. To define the decision rule we need the distribution of S, the random variable associated with s. We get this distribution by summing over all $\binom{n}{s}$ possible values of $\mathbf{X}$ that result in $S = s$, each of which is equally likely, leading to (the probability mass function of) the binomial distribution,

$$f(s;n,\theta) = \binom{n}{s} \theta^s (1-\theta)^{n-s}.$$

So we can define the decision rule for $H_0 : \theta \leq \theta_0$ versus $H_1 : \theta > \theta_0$ as $\delta(s,\alpha) = I(s \geq c(\alpha))$, where $c(\alpha)$ is the smallest value that gives a valid test,

$$\sup_{\theta \leq \theta_0} \Pr_\theta[S \geq c(\alpha)] \leq \alpha.$$

In terms of a *p*-value, this is

$$p(s,\Theta_0) = \sup_{\theta \leq \theta_0} \Pr_\theta[S \geq s].$$

Since $\Pr_\theta[S \geq s]$ is monotonic in θ, then this simplifies to

$$p(s,\Theta_0) = \Pr_{\theta_0}[S \geq s].$$

The creation of *p*-values and decision rules may not be as straightforward for two-sided hypotheses, but for applied tests we usually care more about hypotheses with three decisions (Section 2.3.4). Also, the likelihood ratio is not guaranteed to be monotone, but it is monotone in most practical examples with a one-dimensional parameter.

Now consider a different design where we have independent binary observations, but instead of sampling n observations regardless of the responses, we keep sampling until we observe s successes:

$$f\left(\mathbf{x}; \sum_{i=1}^N x_i = s, x_N = 1, \theta\right) = \Pr_\theta\left(X_1 = x_1, X_2 = x_2, \ldots, X_N = 1 \middle| \sum_{i=1}^N X_i = s, X_N = 1\right)$$

$$= \theta \prod_{i=1}^{N-1} \theta^{x_i}(1-\theta)^{1-x_i} = \theta^s (1-\theta)^{(N-1)-(s-1)}$$

$$= \theta^s (1-\theta)^{N-s} = \theta^s (1-\theta)^M,$$

where s is fixed and $N \in \{s, s+1, \ldots\}$ or alternatively, $M \in \{0, 1, \ldots\}$ are random. Summing over the $\binom{N-1}{s-1} = \binom{M+s-1}{s-1}$ ways that the first $N-1$ observations can have exactly $s-1$ successes, we get a negative binomial distribution on M,

$$f(m;s,\theta) = \binom{m+s-1}{s-1} \theta^s (1-\theta)^m.$$

The likelihood ratio of $H_0 : \theta = \theta_a$ versus $H_1 : \theta = \theta_b$ for $\theta_a < \theta_b$ is decreasing in m, so we reject for smaller values of m. And the one-sided *p*-value for testing $H_0 : \theta \leq \theta_0$ versus $H_1 : \theta > \theta_0$ is

$$p(m,\Theta_0) = \sup_{\theta \leq \theta_0} \Pr_\theta[M \leq m] = \Pr_{\theta_0}[M \leq m],$$

where the last step comes from the monotonicity of $\Pr_\theta[M \leq m]$ in θ.

The Poisson example is Exercise P2. Even though for continuous data $\Pr[X_i = x] = 0$ for all x, we can still use the likelihood ratio in order to create a test through a decision rule or *p*-value. The most commonly used continuous distributions have θ more than one dimension, so we do not discuss them in this Section.

10.3 Pivots

Now consider the case when θ is a vector, and we reparametrize it into $[\beta, \xi]$, where β is a scalar parameter of interest, and ξ is vector nuisance parameters. There are several ways to create tests about β. One important way is through a pivotal quantity. First, we define a pivot.

Definition 10.1 Pivot: *The quantity $T(\mathbf{X}, \theta)$ is a pivot if its distribution does not depend on θ, and the distribution of $\mathbf{X}$ is a function of θ alone.*

There is not a theoretical way to find a pivot in every situation, and each use of the pivot is almost its own special case. But there are a few special cases that are important and useful for the applied researcher.

The *t*-test for a test of a normal mean is the classic example (see Section 5.3.2). In this case, the parameter of interest is $\beta \equiv \mu$ the mean of the normal distribution, and the nuisance parameter is $\xi \equiv \sigma$ the standard deviation. In this case, we can get a pivot using a ratio statistic.

Example 10.1 *Suppose $X_1, \ldots, X_n$ are independent with $X_i \sim N(\mu, \sigma^2)$. The likelihood can be written as a function of the sample mean, $\bar{x}$, and the unbiased sample variance s^2, with associated random variables, $\bar{X}$ and S^2. It can be shown that $\bar{X}$ and S^2 are independent, and $\bar{X} \sim N(\mu, \sigma^2/n)$ and $(n-1)S^2/\sigma^2 \sim \chi^2_{n-1}$, where χ^2_{n-1} is a chi-squared distribution with $n-1$ degrees of freedom. Also, the distribution of a standard normal divided by the square root of an independent chi-squared divided by its degrees of freedom is a t-distribution with the same degrees of freedom. Therefore,*

$$T(\mathbf{X}, \mu, \sigma) = \frac{\frac{\bar{X} - \mu}{\frac{\sigma}{\sqrt{n}}}}{\sqrt{\frac{\left(\frac{(n-1)S^2}{\sigma^2}\right)}{n-1}}} = \frac{\bar{X} - \mu}{\frac{S}{\sqrt{n}}}$$

is distributed t with $n-1$ degrees of freedom, and we see that this distribution is not a function of either parameter. Letting $F^{-1}_{t,df}(a)$ be the ath quantile of a t-distribution with $n-1$ degrees of freedom, we have

$$\Pr\left[F^{-1}_{t,n-1}\left(\frac{\alpha}{2}\right) \leq \frac{\bar{X} - \mu}{\frac{S}{\sqrt{n}}} \leq F^{-1}_{t,n-1}\left(1 - \frac{\alpha}{2}\right)\right] = 1 - \alpha$$

$$\Rightarrow \Pr\left[\bar{X} - F^{-1}_{t,n-1}\left(1 - \frac{\alpha}{2}\right)\frac{S}{\sqrt{n}} \leq \mu \leq \bar{X} - F^{-1}_{t,n-1}\left(\frac{\alpha}{2}\right)\frac{S}{\sqrt{n}}\right] = 1 - \alpha$$

and the two extremes within the probability statement represent the usual $100(1 - \alpha)\%$ confidence interval for μ (note $= F^{-1}_{t,n-1}\left(\frac{\alpha}{2}\right) = -F^{-1}_{t,n-1}\left(1 - \frac{\alpha}{2}\right)$ and compare to Equation 5.6).

Another important pivot is known as the *probability integral transformation*.

Example 10.2 *Suppose X is a continuous random variable with cumulative distribution function F. Then $F(X)$ is uniformly distributed for any F.*

10.4 Multivariate Normal Inferences

Because asymptotic normality is an important tool for statistical inference, it is useful to know some basic facts about the multivariate normal distribution. For this subsection, let θ be a $k \times 1$ parameter vector and suppose

$$\hat{\theta} \sim N(\theta, \Sigma),$$

where Σ is a $k \times k$ variance-covariance matrix, and we assume that Σ^{-1} exists. Then

$$\left(\hat{\theta} - \theta\right)^{\top} \Sigma^{-1} \left(\hat{\theta} - \theta\right) \sim \chi_k^2,$$

where χ_k^2 is a chi-squared distribution with k degrees of freedom. If C is a $k \times h$ matrix and b is an $h \times 1$ vector of constants, then

$$C^{\top}\hat{\theta} + \mathbf{b} \sim N\left(C^{\top}\theta + \mathbf{b}, C^{\top}\Sigma C\right).$$

For example, if $\theta^{\top} = [\theta_1, \theta_2]$ then letting $C^{\top} = [1, -1]$ we get $\hat{\theta}_2 - \hat{\theta}_1 \sim N(\theta_2 - \theta_1, \sigma_1^2 + \sigma_2^2 - 2\sigma_{12})$, where

$$\Sigma = \begin{bmatrix} \sigma_1^2 & \sigma_{12} \\ \sigma_{12} & \sigma_2^2 \end{bmatrix}.$$

In a more general case, partition θ such that $\theta = [\beta, \xi]$ where β is an $r \times 1$ vector, and similarly partition $\hat{\theta}$. Write Σ as

$$\Sigma = \begin{bmatrix} \Sigma_{\beta} & \Sigma_{\beta\xi} \\ \Sigma_{\beta\xi}^{\top} & \Sigma_{\xi} \end{bmatrix}.$$

Then $\hat{\beta} \sim N(\beta, \Sigma_{\beta})$. If $r = 1$ then $(\hat{\beta} - \beta_0)/\sqrt{\Sigma_{\beta}} \equiv Z \sim N(0, 1)$, and a two-sided p-value for testing $H_0 : \beta = \beta_0$ versus $H_1 : \beta \neq \beta_0$ is $p(\hat{\beta}; \beta_0) = 2\{1 - \Phi(|Z|)\}$ and a $100(1 - \alpha)\%$ confidence interval for β is $\hat{\beta} \pm \sqrt{\Sigma_{\beta}}\Phi^{-1}(1 - \alpha/2)$. When $r > 1$, let

$$\Sigma^{-1} = \begin{bmatrix} \Psi_{\beta} & \Psi_{\beta\xi} \\ \Psi_{\beta\xi}^{\top} & \Psi_{\xi} \end{bmatrix}.$$

Then

$$Q(\hat{\beta}, \beta) \equiv \left(\hat{\beta} - \beta\right)^{\top} \Psi_{\beta} \left(\hat{\beta} - \beta\right) \sim \chi_r^2.$$

If Σ_{β}^{-1} and Σ_{ξ}^{-1} exist, then

$$\Psi_{\beta} = \left(\Sigma_{\beta} - \Sigma_{\beta\xi}\Sigma_{\xi}^{-1}\Sigma_{\beta\xi}^{\top}\right)^{-1}.$$

Then a p-value for testing $H_0 : \beta = \beta_0$ versus $H_1 : \beta \neq \beta_0$ is $p(\hat{\beta}; \beta_0) = 1 - F_{\chi_r^2}\{Q(\hat{\beta}, \beta_0)\}$, where $F_{\chi_r^2}(x)$ is the cdf of a chi-squared distribution with r degrees of freedom, and a $100(1 - \alpha)\%$ confidence interval for β is $C(\hat{\beta}, 1 - \alpha) = \{\beta_0 : p(\hat{\beta}; \beta_0) > \alpha\}$.

10.5 Asymptotic Normality and the Central Limit Theorem

In some cases, the distribution of a statistic is not known or is so complicated that we must approximate it. A very useful method is to approximate the distribution with a limiting distribution as the sample size goes to infinity. We then use the simpler limiting distribution when the sample size is "large" (it is not always easy to determine how large the sample needs to be for the approximation to work well). We call these types of large sample methods *asymptotics*. The most important case is asymptotic normality that occurs often when we use statistics that can be written as means. We present the Wald test, an example of this approach, in Chapter 7. Before we discuss more applications, we review some definitions.

Definition 10.2   *Convergence in Distribution (also called convergence in law): Let $X_i \sim F_i$ for $i = 1, 2, \ldots$ and let $X \sim F$. Then if*

$$\lim_{n \to \infty} F_n(x) = F(x) \quad \text{for each } x \text{ at which } F \text{ is continuous,}$$

then we say X_n converges in distribution to X, and write $X_n \overset{d}{\to} X$.

Definition 10.3   *Convergence in Probability: Using notation from the previous definition, if $X_n \overset{d}{\to} X$ with $\Pr[X = c] = 1$, where c is a constant, then we say X_n converges in probability to c, and write $X_n \overset{P}{\to} c$ (see Lehmann, 1999, p. 66, Example 2.3.5). An equivalent more standard definition is, $X_n \overset{P}{\to} c$ if for each $\epsilon > 0$,*

$$\lim_{n \to \infty} \Pr[|X_n - X| < \epsilon] = 1.$$

For using convergence in probability in practice it is useful to know the following theorem.

Theorem 10.1   *If $X_n \overset{P}{\to} c$ and f is a continuous function at c, then $f(X_n) \overset{P}{\to} f(c)$. (see, e.g., Lehmann, 1999, Theorem 2.1.4)*

Here is the main theorem.

Theorem 10.2   *Classical Central Limit Theorem: Let $X_1, \ldots, X_n$ be independent and come from the same distribution with $\mathrm{E}(X_i) = \mu$ and $\mathrm{Var}(X_i) = \sigma^2 < \infty$. Let $\bar{X}$ be the mean. Then*

$$\frac{\sqrt{n}\,(\bar{X} - \mu)}{\sigma} \overset{d}{\to} N(0, 1).$$

For most applications, we do not know σ, but have an estimator, $\hat{\sigma}^2 \to \sigma^2$. So for the central limit theorem to be useful in this case we need two more theorems.

Theorem 10.3 *Slutsky: If $X_n \xrightarrow{d} X$, $A_n \xrightarrow{P} a$ and $B_n \xrightarrow{P} b$, where a and b are constants, then*

$$A_n + B_n X_n \xrightarrow{d} a + bX.$$

Combining the central limit theorem with $\hat{\sigma}^2 \xrightarrow{P} \sigma^2$, Theorem 10.1 and Slutsky's theorem, we get

$$\frac{\sqrt{n}\,(\bar{X} - \mu)}{\hat{\sigma}} \xrightarrow{d} N(0, 1).$$

For applications it is often useful to replace the standard normal with the central t-distribution with degrees of freedom related to the variableness of $\hat{\sigma}/\sqrt{n}$. We rewrite the derivation of the t-distribution from Example 10.1 to be more general.

Example 10.3 Satterthwaite-like approximation: *Suppose $\hat{\beta} \sim N(\beta, \tau^2)$ and $\hat{\tau}^2 \sim Gamma(d/2, 2\tau^2/d)$, which is gamma with mean $= \tau^2$ and variance $= 2\tau^4/d$. Further, suppose $\hat{\beta}$ and $\hat{\tau}$ are independent. Then $(\hat{\beta} - \beta)/\hat{\tau}$ is distributed central t with d degrees of freedom. So if $\hat{v} \approx \mathrm{Var}(\hat{\tau}^2)$, then $\hat{d} = 2\hat{\tau}^4/\hat{v}$ is an approximate degrees of freedom and an approximate $100(1 - \alpha)\%$ confidence interval for β is $\hat{\beta} \pm F_{t,\hat{d}}^{-1}(1 - \alpha/2)\hat{\tau}$. This often has better coverage than the normal confidence interval which is like treating $\hat{d}$ as ∞, since $\lim_{d\to\infty} t_d$ is a standard normal. Even when $\hat{\tau}^2$ is only approximately distributed Gamma, this method can be very useful. For example, Welch's-t test (page 139) is this method applied to the two-sample heterogeneous difference in normal means problem, and that application is more closely related to Satterthwaite's original formula.*

If the data are a vector, we can still get a central limit theorem.

Theorem 10.4 *Multivariate Central Limit Theorem: Let $X_1, \ldots, X_n$ be independent $p \times 1$ random vectors that come from the same distribution with $\mathrm{E}(X_i) = \mu$ and $\mathrm{Cov}(X_i) = \Sigma$, where μ is a $p \times 1$ vector and Σ is a $p \times p$ covariance matrix. Let $\bar{X}$ be the mean. Then*

$$\sqrt{n}\,(\bar{X} - \mu) \xrightarrow{d} N(0, \Sigma). \tag{10.1}$$

For random variables that converge to a normal (possibly multivariate), then we say they are asymptotically normal, and use the notation AN. For example, we can restate Expression 10.1 as

$$\bar{X} \text{ is } AN\left(\mu, \frac{1}{n}\Sigma\right). \tag{10.2}$$

For making approximate large sample inferences, we can act as if $\bar{X} \sim N\left(\mu, \frac{1}{n}\Sigma\right)$ and use the methods of Section 10.4.

There are many versions of the central limit theorem; we mention one more. The Liapounov central limit theorem allows that the means and variances of X_i need not be identical; under certain conditions (see Note N2) $\bar{X}$ is asymptotically normal with mean equal to the average of the means of the X_i and variance equal to the average of the variances.

10.6 Delta Method

Suppose a parameter estimate is asymptotically normal but the convergence is slow (i.e., it takes very large sample sizes for the normal approximation to work well). Often we can improve the approximation by taking a transformation, where the transformed estimate may more quickly approach the normal distribution asymptotically than the untransformed estimate. The delta method allows us to accomplish this.

Theorem 10.5 Delta method: *If $\sqrt{n}(\hat{\theta} - \theta) \xrightarrow{d} N(0, \tau^2)$ then $\sqrt{n}(h(\hat{\theta}) - h(\theta)) \xrightarrow{d}$ $N(0, \{h'(\theta)\}^2\tau^2)$, for any function h whose derivative evaluated at θ, $h'(\theta)$, exists and is not 0.*

Example 10.4   *Suppose we have n independent binary observations, $Y_1, \ldots, Y_n$, each with mean θ and variance $\theta(1 - \theta)$. Then the sample proportion, $\hat{\theta}$, is the sample mean, $\bar{y}$. So as long as θ is not too close to 0 or 1 we could use the central limit theorem to approximate the distribution of $\hat{\theta}$ as $AN(\theta, \theta(1 - \theta)/n)$. A problem is that $\theta \in [0, 1]$ but the support of the normal distribution is the whole real line. So it might make sense to use a transformation that takes values in $(0, 1)$ and transforms them to numbers on the real line. We could use the logit transformation, $h(t) = \log(t/(1 - t))$. Then $h'(t) = \frac{1}{t(1-t)}$. So the delta method says that*

$$h(\hat{\theta}) \text{ is } AN\left(h(\theta), \frac{1}{n\theta(1 - \theta)}\right).$$

Then an approximate $100(1 - \alpha)\%$ confidence interval for θ is

$$h^{-1}\left\{h(\hat{\theta}) \pm \Phi^{-1}(1 - \alpha/2)\frac{1}{\sqrt{n\hat{\theta}(1 - \hat{\theta})}}\right\},$$

where $h^{-1}(h(t)) = 1/(1 + exp(-h(t))) = t$. When $\hat{\theta} = 0.10$ and $n = 100$ then the 95% confidence interval by the Wald method is $(0.041, 0.159)$ while the confidence interval using the logit transformation with the delta method is $(0.055, 0.176)$, which is closer to the exact central interval, $(0.049, 0.176)$.

Theorem 10.6 Multivariate delta method: *Let $\hat{\theta}$ be a $p \times 1$ vector estimate of the vector θ with $\sqrt{n}(\hat{\theta} - \theta) \overset{d}{\to} N(0, \Sigma)$. Let $h(t)$ be a function that inputs a $p \times 1$ vector $t = [t_1, \ldots, t_k]$ and outputs a scalar. Suppose $h(t)$ is continuous at θ, and let*

$$h'(\theta)^\top = \left[\frac{\partial h(t)}{\partial t_1}, \frac{\partial h(t)}{\partial t_2}, \cdots, \frac{\partial h(t)}{\partial t_p} \right]_{t=\theta},$$

where $h'(\theta)$ exists and is not equal to 0. Then

$$\sqrt{n} \left(h(\hat{\theta}) - h(\theta) \right) \overset{d}{\to} N \left(0, h'(\theta)^\top \Sigma h'(\theta) \right).$$

As an example, consider the ratio of two parameters that are asymptotically normal, where the mean of the distribution of the denominator is positive and not too close to 0 (with respect to its standard error).

Example 10.5 Delta method on a ratio: *Suppose $\sqrt{n}(\hat{\theta} - \theta) \overset{d}{\to} N(0, \Sigma)$, where $\theta = [\theta_1, \theta_2]$ and*

$$\Sigma = \begin{bmatrix} \sigma_1^2 & \rho\sigma_1\sigma_2 \\ \rho\sigma_1\sigma_2 & \sigma_2^2 \end{bmatrix}.$$

Let $h(\theta) = \theta_1/\theta_2$ so that

$$h'(\theta)^\top = \left[\frac{1}{\theta_2}, \frac{-\theta_1}{\theta_2^2} \right].$$

Then $\hat{\theta}_1/\hat{\theta}_2$ is asymptotically normal with mean θ_1/θ_2 and variance

$$\left(\frac{1}{n} \right) h'(\theta)^\top \Sigma h'(\theta) = \frac{1}{n\theta_2^2} \left\{ \sigma_1^2 - 2\rho\sigma_1\sigma_2 \left(\frac{\theta_1}{\theta_2} \right) + \sigma_2^2 \left(\frac{\theta_1}{\theta_2} \right)^2 \right\}.$$

10.7 Wald, Score, and Likelihood Ratio Tests

There are three important classes of asymptotic tests: Wald tests, score tests (also called Rao-score tests, or Lagrange multiplier tests), and likelihood ratio tests. Here we give a very brief overview when applied to likelihood equations.

Suppose we have data vectors on n individuals, $x_1, \ldots, x_n$. Assume that the associated random variables, $X_1, \ldots, X_n$ are independent. Because of independence, we can write the likelihood as $f(x; \theta) = \prod_{i=1}^{n} f_i(\theta)$, where $f_i(\theta) \equiv f_i(x_i; \theta)$ is the probability density function or probability mass function, and θ is a vector. For the asymptotics of this section, we use the loglikelihood (i.e., the log of the likelihood) in order to work with sums of independent terms and use the central limit theorem. Let $\ell(x; \theta) = \sum_{i=1}^{n} \log(f_i(\theta))$ be the loglikelihood. The derivative of the loglikelihood is called the *score statistic*, which is a $p \times 1$ vector,

$$U(\mathbf{x};\theta) = \sum_{i=1}^{n} \frac{\partial \log\{f_i(\mathbf{x}_i;\theta)\}}{\partial\theta} = \sum_{i=1}^{n} U_i(\mathbf{x}_i;\theta). \qquad (10.3)$$

A useful estimate of θ is the value of θ that maximizes the likelihood. Let this maximum likelihood estimate be $\hat{\theta}$. Under usual regularity conditions (Note N3), the score statistic can be used as an estimating equation and $\hat{\theta}$ is the value that solves $U(\mathbf{x};\hat{\theta}) = 0$.

The regularity conditions (Note N3) will be required for most of the results of this section. We mention some common examples that are excluded by the regularity conditions. First, the conditions exclude the case where the maximum (or supremum) occurs at the edge of the parameter space.

Example 10.6 *Consider a two-sample study comparing a new treatment to a control arm and the responses are binary representing a treatment success. A common model is a logistic regression with an intercept and a treatment effect parameter representing the log odds of ratio of success on treatment compared to control. Suppose all the responses in the control arm are 0 and some of the binary responses in the treatment arm are 1, then there is no finite parameter value that maximizes the likelihood, and the supremum is infinity.*

A closely related case that does not meet the regularity conditions is when the null parameter value is at the edge of its parameter space, such as testing the null hypothesis that a random effect variance is 0 (see, e.g., Self and Liang, 1987, Case 5). The regularity conditions also exclude cases where the support of the data (the possible values) depend on the parameter. For example, the conditions would be violated for the model where $X_1, \ldots, X_n$ come from a uniform distribution on $[0,\theta]$. Another exemption is when the dimension of θ grows with the sample size. For example, paired data that allow a nuisance parameter for the average effect for each pair. Another example is fitting a proportional odds model (see Equation 9.4) to continuous data, which will have the number of nuisance parameters growing with the sample size.

For asymptotic inferences based on the score statistic, it is useful to have notation and terminology for the expectation of its derivative. Let

$$\Omega(\theta) \equiv \sum_{i=1}^{n} \Omega_i(\theta)$$

$$= \sum_{i=1}^{n} \mathrm{E}\left[-\frac{\partial U_i(\mathbf{X}_i;\theta)}{\partial\theta^\top} \right] = \sum_{i=1}^{n} \mathrm{E}\left[U_i(\mathbf{X}_i;\theta)U_i(\mathbf{X}_i;\theta)^\top \right], \qquad (10.4)$$

where $\Omega \equiv \Omega(\theta)$ is the Fisher information for the total sample, and $\Omega_i \equiv \Omega_i(\theta)$ is the Fisher information for the ith individual. For n independent and identically distributed samples, $\Omega_1 = \cdots = \Omega_n$ so that $\Omega = n\Omega_1$. In that case, typically Ω_1 is called *the* Fisher information matrix (see, e.g., Boos and Stefanski, 2013), but to avoid confusion we will keep the modifiers "total" for Ω and "individual" for Ω_1.

For independent and identically distributed samples, we typically use the estimates, $\hat{\Omega}(\mathbf{x}) = n\hat{\Omega}_1(\mathbf{x})$, where the $p \times p$ matrix

$$\hat{\Omega}_1(\mathbf{x}) \equiv \frac{1}{n}\sum_{i=1}^{n}\left[-\frac{\partial U_i(\mathbf{x}_i;\theta)}{\partial\theta^\top} \right]_{\theta=\hat{\theta}},$$

is called the *observed information* (for an individual). Another estimator which we mention for historical interest is the *expected information* (for the ith individual),[1] where we replace θ from the individual information definition to get $\Omega_i(\hat{\theta})$. Efron and Hinkley (1978) showed that the observed information is generally better to use than the expected information.

Assume that $X_1, \ldots, X_n$ are iid (independent and identically distributed), so that $f_i(\mathbf{x}_i; \theta) = f(\mathbf{x}_i; \theta)$ for all i. Then under the usual regularity conditions (Note N3), we have that

$$\hat{\theta} \text{ is } AN\left(\theta, \hat{\Omega}^{-1}\right), \tag{10.5}$$

where $\hat{\Omega}^{-1} = n^{-1}\hat{\Omega}_1^{-1}$ (see Note N4 for a reference). Then *Wald tests* are approximate inferences using Expression 10.5 in an analogous manner to the exact inferences in Section 10.4.

As an example of a Wald test, parametrize θ so that $\theta = [\beta, \xi]$, where β is an $r \times 1$ parameter vector of interest, and ξ is a nuisance parameter vector. Let the null hypothesis be $H_0 : \beta = \beta_0$ and let the alternative be $H_1 : \beta \neq \beta_0$. Let

$$\hat{\Omega}^{-1} = \begin{bmatrix} V_\beta(\hat{\theta}) & V_{\beta\xi}(\hat{\theta}) \\ V_{\beta\xi}(\hat{\theta})^\top & V_\xi(\hat{\theta}) \end{bmatrix}. \tag{10.6}$$

Then a $100(1 - \alpha)\%$ confidence region for β is $\{\beta_0 : Q_W(\hat{\beta}; \beta_0) \leq F_{\chi_r^2}^{-1}(1 - \alpha)\}$, where

$$Q_W(\hat{\beta}; \beta_0) = (\hat{\beta} - \beta_0)^\top V_\beta(\hat{\theta})^{-1}(\hat{\beta} - \beta_0)$$

and $F_{\chi_r^2}^{-1}(1 - \alpha)$ is the $1 - \alpha$ quantile of a chi-squared random variable with r degrees of freedom.

With the *score test*, we start with the asymptotic normality of the score statistic under the null hypothesis $H_0 : \beta = \beta_0$. Let $\tilde{\theta}(\beta_0) = [\beta_0, \tilde{\xi}(\beta_0)]$ be the value of θ that maximizes the likelihood under the null hypothesis that $\beta = \beta_0$. Let $\tilde{U}(\beta_0) = [U(\mathbf{x}, \tilde{\theta}(\beta_0))]_{1:r}$ (i.e., the first r elements of the $p \times 1$ score statistic after replacing θ with $\tilde{\theta}(\beta_0)$). Let $\tilde{V}_\beta(\beta_0) = V_\beta(\tilde{\theta}(\beta_0))$. Then under the null hypothesis and the usual regularity conditions (Note N3),

$$\tilde{U}(\beta_0) \text{ is } AN\left(0, \tilde{V}_\beta(\beta_0)^{-1}\right).$$

(see Boos and Stefanski, 2013, p. 288). A $100(1 - \alpha)\%$ confidence region for β is

$$\left\{\beta_0 : Q_S(\beta_0) \leq F_{\chi_r^2}^{-1}(1 - \alpha)\right\},$$

where

$$Q_S(\beta_0) = \tilde{U}(\beta_0)^\top \tilde{V}_\beta(\beta_0)\tilde{U}(\beta_0).$$

The third important way to make asymptotic inferences is with a *likelihood ratio test*. The likelihood ratio statistic is

$$Q_{LR}(\hat{\theta}; \tilde{\theta}(\beta_0)) = 2\left\{\ell(\mathbf{x}; \hat{\theta}) - \ell(\mathbf{x}; \tilde{\theta}(\beta_0))\right\}.$$

---

[1] The *expected information* terminology is somewhat confusing, because the expected information is an estimator using the $\Omega_i(\cdot)$ function which is defined by an expectation (Equation 10.4), but it is not an expectation itself since it is a function of $\hat{\theta}$ which depends on the data.

Although this statistic is not a quadratic form such as Q_W and Q_S, it also is asymptotically distributed as a chi-square with r degrees of freedom under the usual regularity conditions (Note N3). Thus, $Q_W(\hat{\beta}; \beta_0)$, $Q_S(\tilde{\theta}(\beta_0))$, and $Q_{LR}(\hat{\theta}; \tilde{\theta}(\beta_0))$ are all asymptotically equivalent.

These three asymptotic methods can be used when there is no nuisance parameter (i.e., $\theta = \beta$), and the expressions simplify in the obvious way (e.g., $\hat{\Omega}^{-1} = V_\beta(\hat{\beta})$).

All three asymptotic methods are routinely used in applied statistics. We list some of the advantages and disadvantages of each.

Wald test advantages:

- Since $\hat{\theta}$ maximizes the likelihood, it is often an efficient estimator.
- The Wald test only requires estimation of θ and it gives natural confidence intervals on it.

Wald test disadvantages:

- The Wald tests are not invariant under reparametrization. For example, tests on σ may be different from tests on σ^2.
- In some cases when β is far away from its null value β_0, the power can be very low (Væth, 1985).

Score test advantages:

- The score test only depends on $\tilde{\theta}(\beta_0)$. If there are no nuisance parameters, then $\tilde{\theta}(\beta_0) = \beta_0$, and even when there are nuisance parameters, $\tilde{\theta}(\beta_0)$ may be much easier to calculate than $\hat{\theta}$.
- The score test is invariant under reparametrization.
- The score test often has better coverage and Type I error rates than Wald test and likelihood ratio test (see simulations).

Score test disadvantages:

- It is not always straightforward to create confidence intervals for β from the score test, for example, when it is hard to calculate $\tilde{\theta}(\beta_0)$ for some values of β_0.
- In some cases when β is far away from its null value, β_0, the power of the score test can be very low. For example, in the one-sample Cauchy case with sample size fixed, the score test power can approach 0 as the parameter approaches infinity, while the Wald test approaches 1 (but for any fixed θ both methods are asymptotically equivalent) (see Lehmann, 1999, p. 532).

Likelihood ratio test (LRT) advantages:

- The LRT is invariant under reparametrization.
- The LRT generally has closer to nominal Type I error rates than the Wald test.

LRT disadvantage:

- The LRT requires calculation of both $\hat{\theta}$ and $\tilde{\theta}(\beta_0)$.

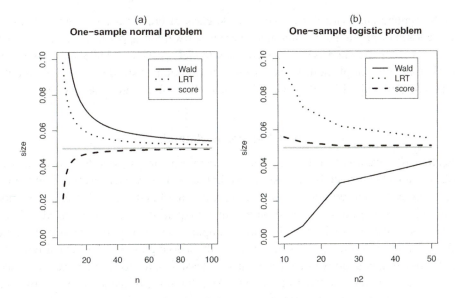

Figure 10.1 Comparison of the Type I error rate of the asymptotic Wald, score, and likelihood ratio test. Part (a) gives application to the one-sample normal problem and the Type I error rate is calculated theoretically. Part (b) gives the application to a one-sample logistic regression and the Type I error rates is calculated (at $n = 10, 15, 25$, and 50) by simulation. The gray line is the nominal rate of 0.05.

We explore the Type I error rates of the three methods. First, consider the one-sample normal case where a one-sample t-test would be used, and compare the three methods as an illustration. In Figure 10.1(a) we compare the size (i.e., Type I error rate) for testing $H_0 : \mu = \mu_0$ versus $H_1 : \mu \neq \mu_0$ from the single sample normal model (i.e., $X_1, \ldots, X_n$ are independent with $X_i \sim N(\mu, \sigma^2)$ with σ unknown) using the Wald, score, and likelihood ratio tests (see, e.g., Boos and Stefanski, 2013, p. 134). We see that the Wald test is quite anti-conservative, the LRT is closer to the nominal 0.05 but still anti-conservative, while the score test is conservative and closest to the nominal.

Now consider an example where $\mathbf{X}_i = [Y_i, Z_i]$ and $Y_i \sim Bernouli(0.5)$ and independently $Z_i \sim N(0, 1)$. Further, $\mathbf{X}_1, \ldots, \mathbf{X}_n$ are independent of each other. Fit a logistic regression model, where if the model does not converge (i.e., there is no finite maximum likelihood estimate, see Example 10.6), then fail to reject the null for all three methods,[2] but otherwise test by comparing Q_W, Q_S, and Q_{LR} to a chi-square with one degree of freedom and reject at the 5% level. We simulate 10^5 replications for each sample size and give the simulated rejection rates in Figure 10.1(b). Here the LRT is the most anti-conservative, the Wald test is very conservative, and the score test has the closest to the nominal error rate.

The methods based on asymptotic normality are derived using approximations using the first few terms of Taylor series expansions. Taylor series expansions can include more terms, and there are other higher-order asymptotic methods such as Edgeworth expansions and saddle-point approximations. An issue with higher order asymptotic methods is that they

---

[2] In practice, if the model does not converge, other methods would be used. See end of Section 14.3.2, page 258.

cannot be applied automatically to such a wide class of problems and must be tailored to specific classes of problems. Despite this, there is software to apply these methods to some very general problems, such as generalized linear models (see the end of Section 14.3.1, page 256). Alternatively, simple adjustments using t-distributions instead of normal distributions (such as a Satterthwaite-like adjustment: Example 10.3) often lead to better coverage.

10.8 Sandwich Methods

For Section 10.7 we assumed that the data generating model is (or at least is close to) the probability model described by the likelihood. In this section, we use central limit theorems to make inferences and do not require that we know the likelihood to describe the model. We can use "working models" based on likelihood equations (e.g., generalized estimating equations, see Section 14.5), but the large sample inferences will approach asymptotic validity even when the data are not generated by the working models. We can also use other general estimating equations like those related to least squares to fit the parameters, which do not need to be tied to likelihood equations. We relax some of the assumptions of Section 10.7. For example, we do not necessarily need identically distributed individual responses. This more general approach is called *M-estimation* or the estimating equation approach. We use some of the same notation as Section 10.7, but the symbols are more general here.

The idea is to start with an estimating equation,

$$U(\beta) = \sum_{i=1}^{K} U_i(\beta) = 0, \tag{10.7}$$

where U_i is a function of the data used to estimate β. For example, consider a linear model where the expected response for the ith individual is model as $E(Y_i) = \mathbf{X}_i\beta$, where β is a $p \times 1$ vector of parameters and $\mathbf{X}_i$ is a $1 \times p$ vector of covariates. To minimize the sum of squared errors, we take the derivative of $\sum(Y_i - \mathbf{X}_i\beta)^2$ with respect to β and solve for 0. Dividing both sides by 2, gives Equation 10.7 with $U_i(\beta) = \mathbf{X}_i^\top (Y_i - \mathbf{X}_i\beta)$. Alternatively, U_i could be a score statistic from a working likelihood model (we give some regularity conditions on U_i below). The data could represent K individual responses, or K clusters of individual responses. Let $\hat{\beta}$ be the value of β that gives $U(\beta) = 0$. We use the same notation as previously, because if U_i is the derivative of a loglikelihood function, then $U(\beta)$ is the score equation and $\hat{\beta}$ will be the maximum likelihood estimate under the standard assumptions of Section 10.7. However, in this section we do not require such a specific model. We instead assume that there exists some β_0 such that $E(U(\beta_0)) = 0$. We assume that the correlation between individual terms goes to 0 as the sample size gets large, i.e., $cov(U_i(\beta_0), U_j(\beta_0)) \to 0$ for $i \neq j$ as $K \to \infty$. We assume that $\hat{\beta}$ is a consistent estimator of β_0, i.e., $\hat{\beta} - \beta_0 \to 0$ as $K \to \infty$. As in Section 10.7, we assume that we can approximate each term of the estimating equation with the first term of a Taylor series as

$$U_i(\beta) \approx U_i(\hat{\beta}) - \hat{\Omega}_i(\beta - \hat{\beta}),$$

except in this section, $\hat{\Omega}_i$ is a consistent estimator of $\Omega_i(\hat{\beta}) = \frac{-\partial U_i(\beta)}{\partial \beta}\big|_{\beta=\hat{\beta}}$ that represents the derivative of only β, and we allow $\Omega_i \neq \Omega_j$. Let $\hat{\Omega} = \sum_{i=1}^{K} \hat{\Omega}_i$ and assume that $\hat{\Omega}^{-1}$ exists. Using the inverse, and summing over the i, since $\sum U_i(\hat{\beta}) = 0$, we get

$$\hat{\beta} - \beta \approx \hat{\Omega}^{-1}\left(\sum_{i=1}^{K} U_i(\beta)\right).$$

We assume that $\hat{\Omega}^{-1}$ is well behaved (i.e., it is approximately constant in the neighborhood of $\hat{\beta}$), so that the variance of $\hat{\beta}$ can be estimated by

$$V_s = \hat{\Omega}^{-1}\left(\sum_{i=1}^{K} U_i(\hat{\beta})U_i(\hat{\beta})^{\top}\right)\hat{\Omega}^{-1}. \tag{10.8}$$

This is known as a *sandwich estimator of variance*, since the middle part is sandwiched between two identical pieces. For inferences, if K is large we treat $\hat{\beta}$ as $AN(\beta, V_s)$ and we can use Wald-type inferences.

As with the likelihood-based methods the Wald inferences tend to be anti-conservative and adjustments for that usually need to be made (see, e.g., Fay and Graubard, 2001; Kauermann and Carroll, 2001). These adjustments can be especially important when some elements of the β vector are essentially estimated from very few individuals.

Example 10.7 *Fay et al. (1997) performed a meta-analysis of dietary fats and their effect on mammary tumors in rodents. The rodents were fed diets with different types of dietary fats, then given carcinogens to see which diets lead to higher mammary tumor rates. The data for the main analysis used $K = 124$ sets of experiments on rats, and the working model was a conditional logistic regression with main effects for calorie restriction and percent of calories from four types of dietary fats (saturated, monounsaturated, ω-6 polyunsaturated, and ω-3 polyunsaturated). The working model assumes that the parameters are the same across all clusters (i.e., sets of experiments), but using a sandwich estimator on the clusters allows us to assume the true parameters randomly fluctuate around some mean (i.e., they are random effects). The asymptotics of Wald-type test used with the sandwich estimator works well when there is an "effectively" large number of clusters. The issue is that for most of the studies of the meta-analysis, there were only trace amounts of ω-3 polyunsaturated fats (the kind found predominately in fish oils), so the parameter related to that type of fat was estimated from essentially very few studies. So even though $K = 124$, the effective number of clusters for the ω-3 polyunsaturated fat parameter was much less. The method of Fay and Graubard (2001) adjusts for that by using a different t-distribution for each parameter and estimates an effective sample size for each parameter using a Satterthwaite-type estimate.*

We can also get generalized score tests created using the asymptotic normality of $U(\tilde{\beta})$ (see Boos and Stefanski, 2013, chapter 8). These generalized score tests are often conservative, but the simple adjustment of multiplying the quadratic form by $K/(K-1)$ can give closer to nominal size in some cases (Guo et al., 2005).

In summary, if the likelihood regularity conditions of Note N3 are met, then the middle part of the sandwich estimator is approximately $\hat{\Omega}$ (see Equation 10.18 with no nuisance parameters so that $\beta = \theta$), and $V_s \approx \hat{\Omega}^{-1}$. But when we relax the usual regularity conditions then

$$\Omega_i(\beta) \equiv \mathrm{E}\left\{-\frac{\partial U_i(\beta)}{\partial\beta^{\top}}\right\} \neq \mathrm{E}\left\{U_i(\beta)U_i(\beta)^{\top}\right\},$$

and we cannot use $\hat{\Omega}^{-1}$ to estimate the variance of β and need to use a sandwich estimator.

10.9 Permutation Tests

10.9.1 Overview and Definition

Permutation tests scramble aspects of the data to calculate the *p*-values. Thus, this is a very general class of tests and is a very useful tool for the applied statistician. Permutation tests are especially useful for testing a strong null hypothesis, meaning that the null is that the treatment or other covariate of interest has no effect on the response. These strong null hypotheses are easy to state, but since they are not tied to specific probability models it is less obvious how to measure effects and confidence intervals on those effects. Because of this we start with a definition of the *p*-value from a general permutation test, and later we discuss the assumption necessary for its validity.

The idea is to permute the data, meaning shuffle a component of the data set. For example, in a two-treatment study we could consider all possible ways that the two treatments could have been assigned to the individuals in the study. Let m be the number of permutations and let π_j be the jth permutation. Let $\mathbf{x}$ be the data vector and $\mathbf{x}(\pi_1), \ldots, \mathbf{x}(\pi_m)$ be the set of m permutations of the data (see Example 10.8 for an example with details). Let $T(\cdot)$ be some function of the data that returns a scalar real value such that larger values suggest the effect that you want to detect. The permutation *p*-value represents how extreme the function from the observed data is compared to the functions computed from the scrambled data sets, and is defined as

$$p(\mathbf{x}) = \frac{\sum_{j=1}^{m} I\left\{T(\mathbf{x}(\pi_j)) \ge T(\mathbf{x})\right\}}{m}. \tag{10.9}$$

For, m we discuss approximations for that *p*-value in Subsections 10.9.2, 10.9.4, and 10.9.5. A sufficient condition for the validity of a permutation *p*-value is that

$$\Pr[T(\mathbf{X}(\pi_i)) \ge T(\mathbf{X})] = \Pr[T(\mathbf{X}(\pi_j)) \ge T(\mathbf{X})], \tag{10.10}$$

for all $i \ne j$ under the null hypothesis. This condition is easier to justify than one might think, because we do not need to be able to calculate the probabilities, we just need to know they are the same. There are many ways to define null probability models for the data such that Equation 10.10 holds. We begin with an "example" that is quite general and useful.

Example 10.8 An $n!$ permutation test: *Suppose we have n individuals, where the ith one has a vector of responses, $\mathbf{y}_i$, and covariates, $\mathbf{z}_i$. Let the data $\mathbf{x} = [\mathbf{y}, \mathbf{z}]$, be an $n \times (q + r)$ matrix, where the data for the ith individual is the ith row, $\mathbf{x}_i = [\mathbf{y}_i, \mathbf{z}_i]$, and $\mathbf{y}_i$ is a $q \times 1$ vector and $\mathbf{z}_i$ is an $r \times 1$ vector. Let $\pi_0 = [1, 2, \ldots, n]$ and let $\pi_1, \ldots, \pi_m$ be the complete set of all $m = n!$ unique permutations of π_0, so that $\pi_0 \in \{\pi_1, \ldots, \pi_m\}$. Let $\pi_j = [\pi_{j1}, \ldots, \pi_{jn}]$. Then we write the jth permutation of the data set as $\mathbf{x}(\pi_j)$, where we permute the responses only, and where $\mathbf{x}(\pi_j) = [\mathbf{x}_1(\pi_j), \ldots, \mathbf{x}_n(\pi_j)]$, with $\mathbf{x}_i(\pi_j) = [\mathbf{y}_{\pi_{ji}}, \mathbf{z}_i]$. We get equivalent p-values if instead of permuting the responses we permute the covariate vectors in all $m = n!$ ways (Problem P4).*

Consider a null hypothesis probability model and let $\mathbf{X}$ be the random vector associated with $\mathbf{x}$. We consider two probability models for which Equation 10.10 holds:

(1) The randomness comes through the covariates, so $\mathbf{X} = [\mathbf{y}, \mathbf{Z}]$. For example, if $\mathbf{z}_i$ is the random allocation to treatment arm in a randomized trial. Under the null hypothesis, as

long as the randomization is not adaptive, the distribution of $\mathbf{Z}_i$ is independent of $\mathbf{y}_i$ (for example, the response for the ith individual would not change if $\mathbf{z}_i$ changed).

(2) The randomness comes through the responses, so $\mathbf{X} = [\mathbf{Y}, \mathbf{z}]$. The random vectors associated with the $\mathbf{y}_i$, say $\mathbf{Y}_i$ for $i = 1, \ldots, n$, are all independent and distributed $\mathbf{Y}_i \sim F$, where F does not depend on $\mathbf{z}_i$, and F can be completely general (e.g., infinite dimensional).

In the example, we have not given the form of the test statistic function, T, since it can be any function where large values represent "far away" from the null hypothesis. We can think of T as a way of choosing a direction in the multidimensional space of the data to define extremeness. Many common tests are equivalent to a permutation test in the form of Example 10.8. For example, some common two-sample tests can be written in that form, such as the Wilcoxon–Mann–Whitney test, and Fisher's exact test (see Section 10.9.2). Here is a more specific $n!$ permutation test that is more complicated.

Example 10.9 Test of r covariates: *Suppose that for the ith subject in a study, y_i is a binary indicator of the presence ($y_1 = 1$) or absence ($y_i = 0$) of a specific disease, and $\mathbf{z}_i$ is an $r \times 1$ vector of measurements from r different cytokines (chemicals that send signals between cells) measured in the blood of the ith subject. Suppose we wanted to check if any of the levels of the cytokines are elevated or reduced when the disease is present. If $r = 1$ we could let $T(\mathbf{x}) = (\bar{z}_1 - \bar{z}_0)^2$, where*

$$\bar{z}_a = \frac{\sum_{i=1}^{n} I(y_i = a)z_i}{\sum_{i=1}^{n} I(y_i = a)},$$

for $a = 0, 1$. Then the permutation test would test the null hypothesis that there is no association between the cytokine and the disease. But if $r > 1$ we could perform a hypothesis test that combines the tests on all r cytokines in one permutation test to test the null that there is no association between the responses and any of the r cytokines. To do this, let

$$T(\mathbf{x}) = \max_{j \in \{1, \ldots, r\}} (\bar{z}_{1j} - \bar{z}_{0j})^2,$$

where

$$\bar{z}_{aj} = \frac{\sum_{i=1}^{n} I(y_i = a)z_{ij}}{\sum_{i=1}^{n} I(y_i = a)},$$

for $a = 0, 1$. This test can be more powerful than doing a simple Bonferroni adjustment (see Chapter 13) on the set of r individual tests on each cytokine, because it accounts for the correlation that is likely among the cytokines, since the levels of some cytokines increase or decrease the levels in others.

Another class of permutation tests is for paired data. For example, consider a before and after experiment comparing a response before an intervention and after an intervention on n pairs.

Example 10.10 Permutation of n pairs: *Suppose we have n pairs of responses, $\mathbf{x} = [\mathbf{x}_1, \ldots, \mathbf{x}_n]$ with the ith pair $\mathbf{x}_i = [y_{i1}, y_{i2}]$. Let the set of permutations be all the $m = 2^n$ possible orderings of the paired responses. Here we can represent the jth permutation vector, $\pi_j = [\pi_{j1}, \ldots, \pi_n]$ by an $n \times 1$ vector of 0s and 1s. Then the ith response pair under the* *j*th permutation is

$$\mathbf{x}_i(\pi_j) = \begin{cases} [y_{i1}, y_{i2}] & \text{if } \pi_{ji} = 0 \\ [y_{i2}, y_{i1}] & \text{if } \pi_{ji} = 1. \end{cases}$$

For this example, π_j is not a permutation of the indices, but it is a vector used to permute the pairwise indices. The permutation validity condition (Equation 10.10) is met if, under the null, the distribution is the same for each of the pairs of responses, i.e., $Y_{i1} \sim F_i$ and $Y_{i2} \sim F_i$ for all i.

It is often convenient to think of the different values of the m permutations as random variables coming from a discrete distribution. Let $T_1, \ldots, T_m$ be the different values of the m permutations, with $T_j = T(\mathbf{x}(\pi_j))$. Then let W be a simple random draw from the set $\{T_1, \ldots, T_m\}$, so that the probability that $W = T_j$ is $1/m$ for all j. We call the distribution of W the *permutation distribution*, and the permutation *p*-value is $p(\mathbf{x}) = \Pr[W \geq T(\mathbf{x})]$.

Finally, note that the permutation validity condition (Equation 10.10) may be easy to satisfy under strong null hypotheses where the distribution of the responses does not depend on the covariates at all, but it is harder to justify under alternative hypotheses. That means that confidence intervals using the permutation tests are not straightforward, and generally more assumptions need to be made. For example, in Section 6.5.3, more assumptions are needed for confidence intervals associated with Spearman's correlation than just testing whether the correlation is different from 0, which can be done easily with a permutation test.

10.9.2 Equivalent Forms

We have used very general forms of permutation tests, but there can be some simplifications. For many permutation tests there are many equivalent forms of the test (e.g., redefining the function T so that the *p*-values do not change). Let an equivalence rule be a way to define equivalent permutation tests. We give one equivalence rule that applies to all permutation tests (ER 1) and one that applies to $n!$ permutation tests (ER 2).

Equivalence Rule 1 *From the definition of the permutation p-value, we see that two test statistic functions, T and T^*, will give equivalent p-values if*

$$I\left\{T(\mathbf{x}(\pi_j)) \geq T(\mathbf{x})\right\} = I\left\{T^*(\mathbf{x}(\pi_j)) \geq T^*(\mathbf{x})\right\} \text{ for all } j = 1, \ldots, m.$$

For example, letting $T^*(\mathbf{x}) = g(T(\mathbf{x}))$ for any strictly monotonic function, g, means that the permutation p-value based on T will be equivalent to the one based on T^*. The function g may even depend on any statistic of the data that is invariant to permutations.

For example, consider the complete permutation of n responses from (Example 10.8) for a two-sample study with $z_i = 0$ (for standard treatment) or $z_i = 1$ (for new treatment), with sample sizes of n_0 and $n_1 = n - n_0$ in the two groups. Then having T be the difference in means between the two groups,

$$T(\mathbf{x}) = \frac{\sum_{i=1}^n z_i y_i}{n_1} - \frac{\sum_{i=1}^n (1 - z_i) y_i}{n_0} \tag{10.11}$$

is equivalent to $T^*(\mathbf{x}) = \sum_{i=1}^n y_i z_i$ (Problem P3).

Recall that for $n!$ permutation tests we can permute on the responses or equivalently permute on the covariates. Simplifications occur if there are many ties in either the responses (i.e., all responses are binary) or the covariates (a two-treatment study where the covariates represent treatment only).

Equivalence Rule 2 *If the permuted variable is a scalar and there are ties in it, you do not need to calculate all $n!$ permutations, you only need to do the permutations that lead to unique values of the ordered permuted vector.*

For example, suppose we are permuting on the covariate vector, the covariate vector is a scalar, and we have only two possible values of the covariate, n_0 individuals with $z_i = 0$ and n_1 with $z_i = 1$. Then there are only $\binom{n}{n_1} = \frac{n!}{n_0! n_1!}$ unique permutations of the covariates $\mathbf{z} = [z_1, \ldots, z_n]$, since for each of the $n!$ permutations we can get $n_0! n_1!$ identical covariate permutations by permuting within the values where $z_{\pi_{ji}} = 0$ and for each of those permuting within the values where $z_{\pi_{ji}} = 1$. Thus, we can use $n_0! n_1!$ times fewer permutations, but we have to pick the unique ones. The savings can be substantial; if $n_0 = n_1 = 5$ then $n! = 3,628,800$ but $\binom{n}{n_1} = 252$.

10.9.3 Direction of Effects

It is difficult to get confidence intervals from permutation tests, because the permutation validity condition (Equation 10.10) often requires more assumptions than just one scalar parameter equaling a null value (see Section 9.5.5). Nevertheless, we can choose $T(\mathbf{x})$ to highlight a certain direction of effect that we are interested in.

Consider the case where we are interested in a parameter, θ, with an estimator $\hat{\theta}(\mathbf{x})$. Suppose that when the permutation validity condition (Equation 10.10) holds that this implies $\theta = \theta_0$. If we want to have good power to reject with large values of θ, then let $T(\mathbf{x}) = \hat{\theta}(\mathbf{x}) - \theta_0$, since by Equation 10.9 if $\hat{\theta}(\mathbf{x})$ is large, then $T(\mathbf{x})$ is large in the permutation distribution, and the p-value is small. Let the associated permutation p-value be $p_U(\mathbf{x})$. To detect small values of θ, then let $T(\mathbf{x}) = \theta_0 - \hat{\theta}(\mathbf{x})$ and let the associated p-value be $p_L(\mathbf{x})$.

If we want to detect effects in either direction, then we can calculate a two-sided p-value in one of two ways. Similar to the way central p-values are usually formulated, we can define the central-like two-sided permutation p-value as $p_c(\mathbf{x}) = \min(1, 2p_U(\mathbf{x}), 2p_L(\mathbf{x}))$. We call $p_c(\mathbf{x})$ "central-like" instead of "central," because the p-value may not be strictly central in the sense of Section 2.3.2. The problem is that even though the permutation validity condition implies $\theta = \theta_0$, this does not mean that $\theta = \theta_0$ implies the permutation validity condition, so one-sided tests on θ are not straightforward. Another way to create two-sided p-values is by using $T(\mathbf{x}) = |\hat{\theta}(\mathbf{x}) - \theta_0|$ in Equation 10.9, which we call the absolute value p-value, $p_a(\mathbf{x})$. For small sample sizes and when there are ties, the two two-sided p-values can be different. Often historically $p_a(\mathbf{x})$ may be more common, but $p_c(\mathbf{x})$ may be better for creating compatible confidence intervals (and creating confidence intervals may require more assumed structure in the probability model than needed for validity of the permutation test). For large sample sizes, the differences between the two two-sided p-values are usually not an issue. For details related to the Mann–Whitney parameter and compatible confidence intervals with the exact Wilcoxon–Mann–Whitney test see Fay and Malinovsky (2018).

10.9.4 Permutational Central Limit Theorem

Suppose that you have an $n!$ permutation test, where we can find an equivalent form, such that

$$T(\mathbf{x}) = \sum_{i=1}^{n} a_i \mathbf{b}_i,$$

where a_i represents some type of scalar response and $\mathbf{b}_i$ represents some type of covariate vector. For example, a_i could represent the rank of the ith individual's response out of the n responses and $\mathbf{b}_i$ could represent a transformation of $\mathbf{z}_i$. We require that the formation of $\mathbf{a} = [a_1, \ldots, a_n]$ does not depend on the covariates, and the formation of $\mathbf{b} = [\mathbf{b}_1, \ldots, \mathbf{b}_n]$ does not depend on the responses. We permute either all $n!$ unique permutations of the a_i or of the $\mathbf{b}_i$. Then we call the associated permutation test a *linear* permutation test.

For large n and under some regularity conditions (Note N5), we can approximate the permutation distribution with a multivariate normal one. Before giving the permutational central limit theorem, we first give the mean and variance of the permutation distribution. Let W_n be a random variable representing a simple random sample from the permutation set $\{T_1, \ldots, T_m\}$, where $T_j = T(\mathbf{x}(\pi_j))$. Then the actual mean and variance (not estimators) of W_n are

$$\mathrm{E}(W_n) = n\bar{a}\bar{\mathbf{b}} \tag{10.12}$$

and

$$\mathrm{Var}(W_n) = \frac{1}{n-1} \left(\sum_{i=1}^{n} (a_i - \bar{a})^2 \right) \left(\sum_{j=1}^{n} (\mathbf{b}_j \mathbf{b}_j^T - \bar{\mathbf{b}}\bar{\mathbf{b}}^\top) \right), \tag{10.13}$$

where $\bar{a} = n^{-1} \sum a_i$ and $\bar{\mathbf{b}} = n^{-1} \sum \mathbf{b}_i$.

Theorem 10.7 Permutational central limit theorem: *Under regularity conditions given in Note N5, W_n is asymptotically normal with mean $\mathrm{E}(W_n)$ and $\mathrm{Var}(W_n)$ given by Equations 10.12 and 10.13.*

10.9.5 Monte Carlo Implementation

Complete enumeration of the permutation distribution for calculation of the permutation p-value is rarely practical (even with $n = 20$ we have $20! \approx 2.4 \times 10^{18}$). But we can take a pseudorandom sample from the permutation distribution by computer simulation. Let M be the number of Monte Carlo samples from the permutation distribution, then we can estimate the exact permutation p-value (Equation 10.9) with

$$p_{mc}(\mathbf{x}) = \frac{1 + \sum_{j=1}^{M} I\left\{W_j \geq T(\mathbf{x})\right\}}{M+1},$$

where W_j is a simple pseudorandom sample from the permutation distribution. If we did not add 1 to the numerator and denominator, that p-value estimator is no longer valid, as can be seen because it would be possible to get $p_{mc}(\mathbf{x}) = 0$. If we are interested in testing a hypothesis at level α, then under the null hypothesis we can get the probability of rejecting the null to be exactly equal to α if we choose M such that $(M+1)\alpha$ is an integer. So for example, if $\alpha = 0.05$, then use $M = 999$ instead of $M = 1,000$ (see Section 5 of Fay and Follmann, 2002, for details).

Deciding on how large M should be is based on computer time and how accurate it is desired to get the p-value. If the random number generator were perfect, then $\sum_{i=1}^{M} W_i \sim Binomial(M, p(\mathbf{x}))$, where $p(\mathbf{x})$ is the p-value calculated with complete enumeration. Thus, we can use exact central confidence intervals on a single binomial (Section 4.2) to gauge the precision of the Monte Carlo estimate of $p(\mathbf{x})$. We can also base the selection of M on the ratio of power using $p_{mc}(\mathbf{x})$ compared to power using $p(\mathbf{x})$. Davidson and Hinkley (1997, p. 155) give an approximation for this ratio that only depends on the significance level α and M. This approximate ratio when $\alpha = 0.05$ gives 0.83 (for $M = 99$), 0.95 (for $M = 999$) and 0.98 (for $M = 9,999$). So there is only about a 5% relative reduction in power from using $p_{mc}(\mathbf{x})$ with $M = 999$ instead of using $p(\mathbf{x})$.

10.10 Bootstrap

10.10.1 Overview

The bootstrap is a variety of methods based on the idea of resampling from the data with replacement. The bootstrap is very good for automatically getting reasonable inferences for a wide variety of problems. In our brief description, we focus on the nonparametric bootstrap, and do not discuss parametric bootstrap methods.

10.10.2 Nonparametric Bootstrap

To describe the basic idea of the bootstrap, consider n independent observations. Let $\mathbf{X} = \{X_1, \ldots, X_n\}$ represent the random variables of the data with $\mathbf{X}_i \sim F$. Suppose we are interested in a parameter $\theta = t(F)$ that is a functional of F. For example, for scalar X_i, then θ could be the mean, median, variance or many other parameters that can be defined using F. The bootstrap is consistent for most, but not all such parameters, see Note N6. For a bivariate $\mathbf{X}_i$, then θ could be the correlation between the two components of $\mathbf{X}_i$. Suppose our estimator of θ is $\hat{\theta} = t(\hat{F})$, where $\hat{F} \equiv \hat{F}_{\mathbf{x}}$ is the empirical distribution function of $\mathbf{x}$ (e.g., for continuous $\mathbf{X}_i$, then $\hat{F}$ is the discrete distribution with probability of $1/n$ at each of $x_1, \ldots, x_n$). We wish to estimate the distribution of $R(\mathbf{X}, F)$, a function of $\mathbf{X}$ and F, which, for example, we could define as

$$R(\mathbf{X}, F) = \hat{\theta} - \theta = t(\hat{F}) - t(F). \tag{10.14}$$

A bootstrap approximation to the distribution of $R(\mathbf{X}, F)$ uses the distribution of

$$R(\mathbf{X}^*, \hat{F}) = \hat{\theta}^* - \hat{\theta} \equiv t(\hat{F}^*) - t(\hat{F}), \tag{10.15}$$

where $\mathbf{X}^* = \{\mathbf{X}_1^*, \ldots, \mathbf{X}_n^*\}$ is a bootstrap sample, where for each i we have independently that $\mathbf{X}_i^* \sim \hat{F}$, and $\hat{F}^* \equiv \hat{F}_{\mathbf{X}^*}$ is the empirical distribution of $\mathbf{X}^*$. In other words, we want to estimate the distribution of $R(\mathbf{X}, F)$ where F is the fixed data generating distribution and $\mathbf{X}$ is a set of n draws from F, so we approximate it with the distribution of $R(\mathbf{X}^*, \hat{F})$ where $\hat{F}$ is the fixed bootstrap generating distribution and $\mathbf{X}^*$ is a set of n draws from $\hat{F}$.

A common way to operationalize the distribution of $R(\mathbf{X}^*, \hat{F})$ is to sample the data with replacement to create B bootstrap data sets. Let the first bootstrap data set be $\mathbf{X}^{*1} = \{\mathbf{X}_1^{*1}, \ldots, \mathbf{X}_n^{*1}\}$, where for $i = 1, \ldots, n$ the value $\mathbf{X}_i^{*1}$ is a sample with replacement from $\{\mathbf{x}_1, \ldots, \mathbf{x}_n\}$. Since it is sampling with replacement, every $\mathbf{X}_i^{*1} \in \mathbf{x}$, but not every $\mathbf{x}_i \in \mathbf{X}^{*1}$ since some individual data vectors are repeated and some are missing. Now, repeat this bootstrap resampling $B - 1$ more times to create B bootstrap data sets, $\mathbf{X}^{*1}, \ldots, \mathbf{X}^{*B}$. Then we treat $R^{*1}, \ldots, R^{*B}$ as a random sample from the distribution of $R(\mathbf{X}^*, \hat{F})$, where $R^{*i} = R(\mathbf{X}^{*i}, \hat{F})$. So for example, a bootstrap estimator of the variance of $R(\mathbf{X}, F)$ is the sample variance of the bootstrap samples,

$$V^* = \frac{1}{B-1} \sum_{i=1}^{B} \left(R^{*i} - \bar{R}^* \right)^2, \tag{10.16}$$

where a bar denotes an average, so that $\bar{R}^* = n^{-1} \sum R^{*i}$. When $R(\mathbf{X}^*, \hat{F})$ is defined by Equation 10.15, because $\hat{\theta}$ is fixed for all R^{*i}, we have

$$V^* = \frac{1}{B-1} \sum_{i=1}^{B} \left(\hat{\theta}^{*i} - \bar{\hat{\theta}}^* \right)^2,$$

where $\hat{\theta}^{*i} = t(\hat{F}(\mathbf{X}^{*i}))$ is the bootstrap estimator of θ from the ith bootstrap data set, and $\bar{\hat{\theta}}^* = B^{-1} \sum \hat{\theta}^{*i}$.

Other functions for R are possible. In a simpler example, $R(\mathbf{X}, F) = \hat{\theta} = t(\hat{F}(\mathbf{X}))$, which is a function of $\mathbf{X}$ only. In that case, a bootstrap variance is V^* given by Equation 10.16. Sometimes it is useful to work with an approximate pivot, such as

$$R \equiv R(\mathbf{X}, F) = \frac{\hat{\theta} - \theta}{\hat{\sigma}},$$

where $\hat{\sigma}$ is the standard deviation of $\hat{\theta}$, and we use R to represent the random variable as well as the function. If the distribution of $\hat{\theta}$ is approximately normal, then R is approximately standard normal or perhaps a central t-distribution. The bootstrap theory says that the distribution of R is approximately the same as the distribution of

$$R^* \equiv R(\mathbf{X}^*, \hat{F}) = \frac{\hat{\theta}^* - \hat{\theta}}{\hat{\sigma}^*}, \tag{10.17}$$

where the $*$ superscript denotes bootstrap sample. Let q_a^* be the ath quantile of R^* of Equation 10.17, so that for continuous random variables

$$\Pr[q_{\alpha/2}^* \le R^* \le q_{1-\alpha/2}^*] = 1 - \alpha.$$

Then a bootstrap-*t* confidence interval is motivated by the main idea that the distribution of R^* is approximately equal to the distribution of R, so we have (for continuous R^*),

$$
\begin{aligned}
1 - \alpha &= \Pr[q^*_{\alpha/2} \le R^* \le q^*_{1-\alpha/2}] \\
&\approx \Pr[q^*_{\alpha/2} \le R \le q^*_{1-\alpha/2}] \\
&= \Pr\left[q^*_{\alpha/2} \le \frac{\hat{\theta} - \theta}{\hat{\sigma}} \le q^*_{1-\alpha/2}\right] \\
&= \Pr\left[\hat{\theta} - q^*_{1-\alpha/2}\hat{\sigma} \le \theta \le \hat{\theta} - q^*_{\alpha/2}\hat{\sigma}\right].
\end{aligned}
$$

The $100(1 - \alpha)$ bootstrap-*t* confidence interval is $\left(\hat{\theta} - q^*_{1-\alpha/2}\hat{\sigma}, \hat{\theta} - q^*_{\alpha/2}\hat{\sigma}\right)$. Although often the bootstrap-*t* interval has very good coverage compared to other bootstrap confidence intervals (see Section 10.10.4 for details), one problem with the bootstrap-*t* confidence interval is that it is not transformation respecting. Thus, for example, if you are really interested in inferences on $\log(\theta)$, you would get different confidence intervals if you take the log of the bootstrap-*t* interval of θ than if you took the bootstrap-*t* intervals of $\log(\theta)$ itself. We discuss some transformation respecting bootstrap confidence intervals in Section 10.10.3.

10.10.3 Transformation Respecting Bootstrap Confidence Intervals

Suppose we want to get a $100(1 - \alpha)\%$ central confidence interval for $\theta = t(F)$. In this section, we let $\hat{\theta}(\mathbf{X})$ be any estimator of θ; it does not have to be $t(\hat{F})$. A very simple method is the (nonparametric) percentile interval,

$$
\left(q\{\alpha/2, \hat{\theta}^*\}, q\{1 - \alpha/2, \hat{\theta}^*\}\right),
$$

where $\hat{\theta}^* = \hat{\theta}(\mathbf{X}^*)$ is a bootstrap random variable and $q(a, W)$ is the a quantile of a random variable, W. We can estimate the percentile interval using B nonparametric bootstrap samples. Specifically, for the $100(1 - \alpha)\%$ central interval use the kth and $(B + 1 - k)$th ordered values from the bootstrap sample, where

$$
k = \lfloor (B + 1)\alpha/2 \rfloor,
$$

and $\lfloor a \rfloor$ is the largest integer $\le a$. For example, the 95% percentile interval when $B = 999$ is the 25th largest and 975th largest of the $\hat{\theta}(\mathbf{X}^{*j})$. The percentile method is transformation respecting. It has good coverage (see Note N7) if there exists some monotonic transformation, g (we do not need to know the specific form of g), such that the distribution of $g(\hat{\theta}) - g(\theta)$ is symmetric about 0 (see, e.g., Boos and Stefanski, 2013, Section 11.5.2). Thus, in order to have good coverage, $\hat{\theta}$ should be median unbiased (i.e., its median should equal θ).

Since we cannot generally assume median unbiasedness of $\hat{\theta}$, a better confidence interval is the BC_a, bias-corrected and accelerated bootstrap interval (see, e.g., Efron and Tibshirani, 1993, Section 14.3). The acceleration allows the variance of $\hat{\theta}$ to depend on the mean. The BC_a interval assumes that there exists a monotonic transformation, g such that

$$
\frac{g(\hat{\theta}) - g(\theta)}{\tau(\theta)} \sim N(-b, 1)
$$

where $\tau(\theta) = c(1 + a\theta)$ is the variance of $g(\hat{\theta})$, c is a constant and a is the acceleration. As with the percentile interval, we do not need to know the transformation function, g, we just need to assume that it exists. For the BC_a interval, we take B bootstrap samples and estimate $\hat{\theta}$ from each one, creating $\hat{\theta}^{*1}, \ldots, \hat{\theta}^{*B}$. The BC_a interval is the resulting percentile interval except the levels are adjusted by a function, s. In other words, the BC_a interval is

$$\left(q\{ s(\alpha/2), \hat{\theta}^* \}, q\{ s(1 - \alpha/2), \hat{\theta}^* \} \right),$$

where s is the bias-corrected and accelerated adjustment function of the quantile, γ,

$$s(\gamma) = \Phi \left(\hat{b} + \frac{\hat{b} + \Phi^{-1}(\gamma)}{1 - \hat{a}\left(\hat{b} + \Phi^{-1}(\gamma) \right)} \right),$$

where

$$\hat{b} = \Phi^{-1} \left(\frac{\sum_{i=1}^{B} I\left(\hat{\theta}^{*i} < \hat{\theta} \right)}{B} \right)$$

is the estimated bias, and the acceleration is estimated by

$$\hat{a} = \frac{\sum_{i=1}^{n} D_i^3}{6\left\{ \sum_{i=1}^{n} D_i^2 \right\}^{2/3}},$$

where D_i is some estimate of the change in $\hat{\theta}$ due to the ith observation. A nonparametric value for D_i uses the jackknife (Efron and Tibshirani, 1993, p. 186)

$$D_i = \hat{\theta}_{(\cdot)} - \hat{\theta}_{(i)},$$

where $\hat{\theta}_{(i)} = \hat{\theta}(\mathbf{x}_{(i)})$, $\mathbf{x}_{(i)}$ is the data without the ith observation, and $\hat{\theta}_{(\cdot)} = n^{-1} \sum \hat{\theta}_{(i)}$. Alternatively, if $\hat{\theta}$ has the form $t(\hat{F})$, we can use the empirical influence function,

$$D_i = \lim_{\epsilon \to 0} \frac{t\left((1 - \epsilon)\hat{F} + \epsilon \delta_i \right) - t(\hat{F})}{\epsilon},$$

where δ_i is a point mass at $\mathbf{x}_i$ (Efron and Tibshirani, 1993, equations 22.29 and 21.14). There are many other estimators of the acceleration, a, for parametric situations (see Shao and Tu, 1995, pp. 136–140). If $\hat{b} = 0$ and $\hat{a} = 0$, then $s(\gamma) = \gamma$ and no adjustment is made.

Finally, note that there is an approximate BC_a interval called the ABC interval that uses numerical derivatives instead of bootstrap sampling. We do not give the form of the ABC interval (see, e.g., Efron and Tibshirani, 1993, p. 328).

10.10.4 Small Sample Coverage and Calibration

Shao and Tu (1995, Section 4.4) and Davidson and Hinkley (1997, Section 5.7) review and give references to some simulations comparing the length and coverage of the different bootstrap confidence intervals in various situations. Overall, the percentile and BC_a interval can have substantial undercoverage (for example about 80% coverage for a 90% confidence interval), while the bootstrap-t interval tends to be closer to nominal coverage and have wider intervals. Unfortunately, the advantage of the bootstrap-t confidence interval does not hold over all situations; for confidence intervals on the correlation coefficient from the bivariate normal model, the BC_a has conservative coverage closer to nominal than the more conservative bootstrap-t interval.

The overall conclusion is that all the bootstrap methods are good for getting automatic confidence intervals in complicated situations where it is difficult to use other methods; however, the coverage of those intervals may be poor. If proper coverage is important for the problem, then bootstrap calibration could be used. With calibration, we estimate a level using bootstrap methods so that when that level is used in the original bootstrap confidence interval, the coverage is close to the nominal target level. This calibration would be done separately for the lower and the upper interval.

For example, suppose $L(\mathbf{x}, 1 - \alpha/2)$ is the one-sided $100(1 - \alpha/2)$th lower limit for θ by the ABC bootstrap method. Then if the confidence interval was exact, $\Pr[L(\mathbf{X}, 1 - \alpha/2) \leq \theta] = 1 - \alpha/2$. If the limit systematically under- or over-covers θ, then we want to calibrate it by finding the a such that $\Pr[L(\mathbf{X}, 1 - a/2) \leq \theta] = 1 - \alpha/2$. We can estimate a, using bootstrap methods. Let $\mathbf{X}^{*1}, \ldots, \mathbf{X}^{*B}$ be a bootstrap sample, then we estimate that probability with

$$\Pr[L(\mathbf{X}, 1 - a/2) \leq \theta] \approx \frac{\sum_{i=1}^{B} I\left\{L(\mathbf{X}^{*i}, 1 - a/2) \leq \hat{\theta}\right\}}{B}.$$

We can estimate a by trying many values of a and using linear interpolation to estimate the one that solves $\Pr[L(\mathbf{X}, 1 - a/2) \leq \theta] = 1 - \alpha/2$. DiCiccio and Efron (1996, p. 203) discuss an algorithm for that calculation.

10.11 Melding: Combining Confidence Intervals in the Two-Sample Case

Consider a two-sample problem such as a randomized trial comparing two treatments. Suppose we have a method for calculating $100(1 - \alpha)\%$ central confidence intervals for parameters related to each of the two groups, say θ_1 and θ_2. One simple thing we can do is see if the confidence intervals overlap. For example, if the lower bound of the 95% central confidence interval for θ_2 is greater than the upper bound of the 95% central confidence interval for θ_1, then we can say that θ_2 is significantly larger than θ_1 at the 5% level. (Actually, it is even significant at the 4.9375% level, see Problem P5). If we can assume that the estimators of θ_1 and θ_2 are normal with the same known variance, then we can be much more efficient. In that case, if the lower bound of the 83.4% central confidence interval for θ_2 is greater than the upper bound of the 83.4% central confidence interval for θ_1, then we can say that θ_2 is significantly larger than θ_1 at the 2.5% significance level (Problem P6). In fact, with those normal assumptions we can be even more efficient. The melding confidence interval method automatically gives the most efficient confidence interval for that case.

Melding is a very general way of combining two nested, valid, central confidence intervals from independent sources that gives inferences that are automatically efficient and are designed to be valid for any sample size. The independent sources can come from the same study such as two arms from a randomized trial. A nested confidence interval procedure is one where the $(1 - \alpha)$ confidence interval from that procedure using any data set contains all values within the $(1 - \alpha^*)$ confidence interval using that same data set whenever $(1 - \alpha) > (1 - \alpha^*)$.

Before defining the melded confidence interval, let us define the notation and the problem. Let $\mathbf{X}_1$ be the random data vector for Group 1 and similarly define $\mathbf{X}_2$ for Group 2. Assume $\mathbf{X}_1$ and $\mathbf{X}_2$ are independent of each other. We are interested in comparing two scalar parameters from the two groups, say θ_1 and θ_2. We use some comparison function, $g(\theta_1, \theta_2)$, that is defined such that it is decreasing in θ_1 and increasing in θ_2. For example, $g(\theta_1, \theta_2) = \theta_2 - \theta_1$. Note $g(\cdot, \cdot)$, only needs to meet the monotonicity constraints within the allowable values of the parameter. So for parameters that take on only positive values, we may additionally use $g(\theta_1, \theta_2) = \theta_2 / \theta_1$ and for $\theta_a \in (0, 1)$ we may additionally use the odds ratio comparison function, $g(\theta_1, \theta_2) = \theta_2(1 - \theta_1) / \{\theta_1(1 - \theta_2)\}$. Let the comparator parameter be $\beta = g(\theta_1, \theta_2)$, and we use the subscripts d, r, and or to denote the difference, ratio or odds ratio comparison. The comparison function is useful for defining effects and getting confidence intervals on those effects. For example, instead of testing $H_0 : \theta_2 \leq \theta_1$ versus $H_1 : \theta_2 > \theta_1$, we express the hypotheses in terms of β. For example, using the difference we can equivalently test $H_0 : \beta_d \leq 0$ versus $H_1 : \beta_d > 0$ or using the ratio we can equivalently test $H_0 : \beta_r \leq 1$ versus $H_1 : \beta_r > 1$. The advantage of writing in terms of β is we can get accompanying confidence intervals.

Suppose we have a valid, nested, central confidence interval procedure for each of two parameters, θ_1 and θ_2, and that procedure allows for constructing confidence intervals associated with varying α levels. Let the $100(1 - \alpha)\%$ central confidence procedure for θ_a applied to the data for Group a, say $\mathbf{x}_a$, be $\{L_a(\mathbf{x}_a, 1 - \alpha/2), U_a(\mathbf{x}_a, 1 - \alpha/2)\}$. We use this notation because of the centrality of the interval, so that both $\{L_a(\mathbf{x}_a, 1 - \alpha/2), \infty\}$ and $\{-\infty, U_a(\mathbf{x}_a, 1 - \alpha/2)\}$ are one-sided $100(1 - \alpha/2)\%$ confidence intervals. The melded $100(1 - \alpha)\%$ central confidence interval for β is created using two steps. First, for each of the confidence interval functions, we create a lower and upper *confidence distribution random variable* (CDRV). Let

$$T_{La} = L_a(\mathbf{x}_a, A_a) \text{ and } T_{Ua} = U_a(\mathbf{x}_a, B_a)$$

be the lower and upper CDRVs respectively for Group a, where A_1, B_1, A_2, and B_2 are independent uniform $(0, 1)$ random variables that have nothing to do with the data. For continuous random variables, $\mathbf{X}_a$, then the distribution of T_{La} and the distribution of T_{Ua} are the same and called a *confidence distribution* of θ_a. We can think of the confidence distribution like a frequentist posterior distribution. Whereas the confidence interval is defined only for a single value of α, the confidence distribution can be used to get intervals for any α. Since it is frequentist, the randomness of the CDRV only comes from the uniform random variables, since the data are fixed in the confidence limit functions. We add in extra randomness as a tool. One use of this tool is to use the CDRVs to create p-values. To test $H_0 : \theta_2 \leq \theta_1$ a p-value is $p = \Pr[T_{L2} \leq T_{U1}]$ and analogously to test $H_0 : \theta_2 \geq \theta_1$ a p-value is $p = \Pr[T_{U2} \geq T_{L1}]$.

The quantiles of the CDRVs give back the group confidence intervals. Specifically, we can rewrite the $100(1 - \alpha)\%$ confidence interval for Group a,

$$\{L_a(\mathbf{x}_a, 1 - \alpha/2), U_a(\mathbf{x}_a, 1 - \alpha/2)\} \quad \text{as} \quad \{q(\alpha/2, T_{La}), q(1 - \alpha/2, T_{Ua})\},$$

where $q(r, Y)$ is the rth quantile of the random variable Y. But a main reason why we have defined the CDRVs is to create the $100(1 - \alpha)\%$ melded central confidence interval on $\beta = g(\theta_1, \theta_2)$, which is

$$[q\{\alpha/2, \ g(T_{U1}, T_{L2})\}, \ q\{1 - \alpha/2, \ g(T_{L1}, T_{U2})\}].$$

Example 10.11 *For the two-sample problem with normal responses where θ_1 and θ_2 are the means, and the variances are unknown and not equal, the melded confidence interval on the difference in means gives the Behrens–Fisher confidence interval.*

Example 10.12 *Consider the two-sample problem with binary responses where θ_1 and θ_2 are the true proportions and (L_a, U_a) are the exact central binomial interval for Group a. The melded confidence interval method on β_d, β_r, or β_{or} using the exact central binomial confidence intervals for each group gives central confidence intervals that are compatible with the central Fisher's exact test.*

10.12 Within-Cluster Resampling

Suppose we have n independent clusters of data, and within the ith cluster we observe m_i exchangeable replicates (i.e., the ordering of the replicates within a cluster is irrelevant). Within-cluster resampling shows how to perform inferences by repeatedly sampling one replicate from each cluster, and then properly combining the data. The key requirement for implementing within-cluster resampling is that we know a method for making proper inferences if we had only one replicate per cluster.

For example, suppose the clusters were litters of mice, and there were m_i pups in the ith litter. First consider the case where there is only one replicate per cluster and all $m_i = 1$. Let the data be $\mathbf{X} = [\mathbf{X}_{11}, \ldots \mathbf{X}_{n1}]$, where $\mathbf{X}_{i1}$ is the data vector for the ith cluster (the second subscript denotes the first replicate per cluster). Suppose we wanted inference on some parameter, β, and if there was only one observation per cluster we have an estimator $\hat{\beta}(\mathbf{X})$ of β that is asymptotically normal with variance estimator $\hat{\Sigma}(\mathbf{X})$. Then if n is large, we can treat $\hat{\beta}(\mathbf{X})$ as $AN(\beta, \hat{\Sigma}(\mathbf{X}))$ and use methods as in Section 10.4.

Now suppose that some of the litters had $m_i > 1$. Because of the correlation of the data within a litter, we cannot treat all the data from each pup as independent. Let the data for the ith cluster be represented by $\mathbf{X}_i = [\mathbf{X}_{i1}, \ldots, \mathbf{X}_{im_i}]$, where $\mathbf{X}_{ij}$ is the data vector for the jth replicate of the ith cluster. A within-cluster resampled data set is one where we randomly sample a data vector from each cluster, and each data vector within a cluster is equally likely to be sampled. Let $\mathbf{X}^{*1} = [\mathbf{X}_{1J_{11}}, \ldots, \mathbf{X}_{nJ_{n1}}]$ be the first within-cluster resampled data set, where J_{i1} is a sample from $\{1, \ldots, m_i\}$ with each element equally likely. Take M within cluster data sets, $\mathbf{X}^{*1}, \ldots, \mathbf{X}^{*M}$, then for within-cluster resampling we treat the average of the estimates, $\bar{\beta} = M^{-1} \sum_{k=1}^{M} \hat{\beta}(\mathbf{X}^{*k})$, as asymptotically normal with mean β and variance,

$$V_{wcr} = \left(\frac{\sum_{k=1}^{M} \hat{\Sigma}(\mathbf{X}^{*k})}{M} \right) - \left(\frac{\sum_{k=1}^{M} \{\hat{\beta}(\mathbf{X}^{*k}) - \bar{\beta}\} \{\hat{\beta}(\mathbf{X}^{*k}) - \bar{\beta}\}^{\top}}{M - 1} \right),$$

where some divide the second term by M instead of $M - 1$ (see Hoffman et al., 2001).

Hoffman et al. (2001) introduced within-cluster resampling, emphasizing its advantage over generalized estimating equations (GEE) (see Section 14.5) when the m_i is related to the effect. This might happen for example on studies of tooth decay where each person is a cluster, and each tooth is a replicate. In that case the number of teeth is likely related to the outcome, and the GEE methods are not valid, while within-cluster resampling is valid. Follmann et al. (2003) studied more applications of the method including how to use it to combine *p*-values. Follmann and Fay (2010) applied within-cluster resampling to exact permutation tests.

10.13 Other Methods

There are several more very general methods for frequentist inference that are useful to applied statisticians. We mention a few here for completeness.

When there are nuisance parameters it is sometimes useful to condition on sufficient statistics for the nuisance parameter, and then work with the conditional likelihood. That is the way Fisher's exact test can be derived for the two-sample binomial problem. We condition on the total number of failures in both samples combined, and the likelihood simplifies to the extended hypergeometric distribution that only depends on the odds ratio (see Section 7.3). We can generalize this to logistic regression, and condition on sufficient statistics for the nuisance parameters to perform conditional logistic regression. Although these conditional methods are very useful, it is rare in consulting that the applied statistician will need to work out the conditional likelihood from scratch.

Another way to handle nuisance parameters is to use a profile likelihood. Suppose a likelihood has parameters $\theta = [\beta, \xi]$, where β is the parameter of interest, and the nuisance parameter, ξ, may even be infinite dimensional (for example, the baseline hazard function in a proportional hazards model). Suppose the full likelihood is $L(\theta) = L(\beta, \xi)$. Let $\hat{\xi}(\beta)$ be the maximum likelihood of ξ if β is known. Then the profile likelihood maximizes, $L_{prof}(\beta)$ with respect to β, where

$$L_{prof}(\beta) = L(\beta, \hat{\xi}(\beta)).$$

Often the profile likelihood may be treated like a usual likelihood and tests like score, Wald, or likelihood ratio test may be done on it (Murphy and van der Vaart, 2000).

A beautiful method developed by Owen and others is the empirical likelihood (see, e.g., Owen, 2001). This gives a nice nonparametric way to get a confidence interval for a mean, for example. It is essentially a way to get a likelihood ratio test on the mean without specifying the form of the distribution, and it can be shown to be distributed chi-square with one degree of freedom when the true mean equals the point null mean. There are many other applications of empirical likelihood.

10.14 Summary

This chapter briefly introduces a number of general methods that can be applied in a wide variety of settings.

We reviewed a likelihood approach when evaluating a parametric model with a single parameter. There are three steps: formulate a likelihood, then check that the likelihood ratio is monotonic for some statistic, and then calculate a *p*-value based on the null distribution of that statistic. Examples are shown.

A pivot can be used to address a more complicated parametric model, where θ is a vector which can be reparametrized into β, a scalar of interest and ξ, a vector of nuisance parameters. A pivotal quantity can be used to create tests about β. A function of the data and θ is a pivot if its distribution does not depend on θ and the distribution of the data is completely described by θ. A pivot is often not obtainable, but there are a few special cases, and a one-sample t-test is a classic example.

It is straightforward to test statistics that are normally distributed. We can sometimes approximate the distribution of a statistic by the normal distribution as the sample size grows large. Further, sometimes, a transformation of the estimate may be preferred to yield approximate normality with a more reasonable sample size; the delta method can be used to find the mean and variance of the approximately normal transformed estimate.

There are three important classes of asymptotic tests: Wald tests, score tests, and likelihood ratio tests. We provide general advantages and disadvantages of the three approaches with respect to ease of calculation, invariance to reparameterization, and Type I error rates. Nonetheless, the relative performance of the three methods will depend on the specific scenario at hand.

In those settings where observations are clustered, a more general method, the estimating equation approach, can be used. The variance can be estimated with a sandwich estimator (known as such because the middle part is sandwiched between two identical pieces). If sample sizes are large, we can treat $\hat{\beta}$ as approximately normal with mean β and variance based on the sandwich variance, and then use Wald-type or score-type inferences.

A permutation test for calculation of p-values is a highly general tool and very helpful to the applied statistician. The idea is to permute the data; for example, in a comparison of two treatments, the treatment assignments are scrambled in all possible ways that the patients could have been allocated. To determine the p-value, one would define a function of the data that reflects a treatment effect, and then compute the proportion of scrambled allocations that return a value of this function that is more extreme than the functions value of the actual observed allocation. This method allows calculation of p-values for treatment effect metrics that have complex distributions. In some settings the permutation distribution can be approximated by a multivariate normal one. Complete enumeration of permutations is impractical for larger sample sizes, so it is very common to compute a p-value from Monte Carlo samples from the permutation distribution.

Another useful tool for getting reasonable inference for a wide range of problems is the bootstrap which is based on resampling from the data with replacement. A common approach is to sample the data with replacement in B bootstrap data sets to learn about the distribution of some function in the original data set. For example, we can compute a bootstrap estimator of the variance of some function of the original data set (e.g., the median, the inverse of the mean, etc.). Say we want to estimate the variance of the median from the original sample. This is computed as the sample variance of the medians across the bootstrap data sets. This approach is consistent for many different functions of the data set and is asymptotic, but there are calibration methods for small samples.

Melding is a general approach for computing a confidence interval of some comparison function of parameters from two groups, if we have a method for computing a $100(1 - \alpha)\%$ central confidence interval for the relevant parameter within each group. For example, if we can compute confidence intervals for any value of α for θ_1 from the first group, and likewise for θ_2 from the second group, melding can produce a 95% confidence interval for the difference of the two parameters. When melding two valid confidence intervals, the resulting melded interval is conjectured to be valid as well and is preferred for small samples. While there are some limitations of what comparison functions are allowed (e.g., ratios are only allowed for positive parameters), this approach has very wide applicability.

Within-cluster resampling can be employed when there are independent clusters of data, as long we know a method to estimate the parameter of interest for a setting where there is only a single replicate per cluster. We randomly sample one observation per cluster, and compute an estimate for the parameter of interest, using the single replicate method; this process is repeated many times. We then treat the mean of these estimates as asymptotically normal with a special variance formula for the within-cluster resampling setting.

10.15 Extensions and Bibliographic Notes

Rather than give primary references, here we give mostly references to good textbooks that in turn will give long lists of references. A very comprehensive book on the theory of hypothesis tests from the standpoint of finding the optimal test given the hypotheses (e.g., a full description of the Neyman–Pearson lemma, and much much more) is Lehmann and Romano (2005). The first chapters of Graybill (1976) and Vonesh and Chinchilli (1997) give nice summaries of matrix and multivariate normal results. Lehmann (1999) is a nice book on asymptotics that is similar to the level of this book. Boos and Stefanski (2013) gives a much fuller treatment of the methods described in this chapter, such as asymptotic likelihood theory (including sandwich estimators), bootstrap and permutation methods. Efron and Tibshirani (1993) and Davidson and Hinkley (1997) are both nice accessible books on the bootstrap.

Statistical inference using confidence distributions is a frequentist way of rigorously using some ideas from the fiducial distribution of Fisher. For a review on confidence distributions see Xie and Singh (2013) and Schweder and Hjort (2016). Fay et al. (2015) gives details on melded confidence intervals.

10.16 R Functions and Packages

Since this chapter is on general methods, we will not discuss specific R functions. Nevertheless, the boot R package does many of the bootstrap methods, and the abcnonHtest function of the asht R package gives the nonparametric ABC method in the usual hypothesis testing format (e.g., does one-sided and two-sided testing).

10.17 Notes

N1 **Neyman–pearson lemma:** Suppose a probability model is completely described by a single scalar parameter θ, and we wish to test $H_0 : \theta = \theta_0$ versus $H_1 : \theta = \theta_1$. Let $f_0(x)$ is the probability density function for continuous data or the probability mass function for discrete data under the null hypothesis and let $f_1(x)$ be the analogous function under the alternative hypothesis. For discrete data $f_a(x) = \text{Pr}_{\theta_a} [X = x]$ for $a = 0, 1$. Let $\delta(\cdot)$ be a (possibly randomized) level α decision function, so that $E_0(\delta(X)) \leq \alpha$, where

E_a is the expectation when $\theta = \theta_a$, $a = 0, 1$. Let $\mathcal{D}_{NP}$ be the set of $\delta(\cdot)$ such that $E_0(\delta(X)) = \alpha$ and

$$\delta(x) = \begin{cases} 1 & \text{if } f_1(x) > kf_0(x) \\ 0 & \text{if } f_1(x) < kf_0(x), \end{cases}$$

for some constant k. (Note, $\mathcal{D}_{NP}$ may include randomized decision rules that take on a value $0 < \gamma < 1$ when $f_1(x) = kf_0(x)$.) Let the set of most powerful $\delta(\cdot)$ be $\mathcal{D}_{MP}$. Then

(a) there exists some $\delta \in \mathcal{D}_{NP}$,

(b) if $\delta \in \mathcal{D}_{NP}$ this implies $\delta \in \mathcal{D}_{MP}$, and

(c) if $\delta \in \mathcal{D}_{MP}$ this implies either $\delta \in \mathcal{D}_{NP}$ or that $E_0(\delta(X)) < \alpha$ and $E_1(\delta(X)) = 1$.

This statement of the lemma is based on Lehmann and Romano (2005, p. 60), and a proof is given there.

N2 **Liapounov central limit theorem:** Let $X_i \sim F_i$ with $\mathrm{E}(X_i) = \mu_i$, $\mathrm{Var}(X_i) = \sigma_i^2$, and suppose that X_i all have finite third moments. Let $\bar{\mu}$ and $\bar{\sigma}^2$ be the average of the means and variances. Then

$$\frac{\sqrt{n}\,(\bar{X} - \bar{\mu})}{\bar{\sigma}} \to N(0, 1)$$

if

$$\frac{\left\{\mathrm{E}\left(\sum_{i=1}^{n} \|X_i - \mu_i\|^3\right)\right\}^2}{\left(\sum_{i=1}^{n} \sigma_i^2\right)^3} \to 0.$$

N3 **Regularity conditions for likelihood-based asymptotics:** $X_1, \ldots, X_n$ are independent with the same density function $f(x; \theta)$ and distribution $F(x; \theta)$. Let $\mathcal{X}$ be the support (i.e., the possible values) of X_i (from Boos and Stefanski, 2013, p. 286).

(a) If $\theta_1 \neq \theta_2$, then $F(x; \theta_1) \neq F(x; \theta_2)$ for some x.

(b) The support, $\mathcal{X}$, does not depend on θ.

(c) For each $\theta \in \Theta$, the first three partial derivatives of $\log f(x; \theta)$ with respect to θ exist for all $x \in \mathcal{X}$.

(d) For each $\theta^* \in \Theta$, there exists a function $g(x)$ (possibly depending on θ^*), such that in a neighborhood of θ^* and for all $j, k, \ell \in \{1, \ldots, p\}$,

$$\left\| \frac{\partial^3}{\partial\theta_j \partial\theta_k \partial\theta_\ell} \log f(x; \theta) \right\| \leq g(x),$$

for all x and where $\int g(x) dF(x; \theta^*) < \infty$.

(e) For each $\theta \in \Theta$, $E\left[\frac{\partial}{\partial\theta} \log f(X_i; \theta)\right] \equiv \mathrm{E}[U(\mathbf{X}_i; \theta)] = 0$, for $i = 1, \ldots, n$,

$$\Omega_i(\theta) \equiv \mathrm{E}\left\{-\frac{\partial^2}{\partial\theta\partial\theta^\top} \log f(X_i; \theta)\right\} = \mathrm{E}\left\{-\frac{\partial U(\mathbf{X}_i; \theta)}{\partial\theta^\top}\right\}$$

is nonsingular (i.e., is invertible), and since the data are iid, $\Omega_1(\theta) = \cdots = \Omega_n(\theta)$. Additionally,

$$\Omega_i(\theta) = \mathrm{E}\left\{U(\mathbf{X}_i; \theta)U(\mathbf{X}_i; \theta)^\top\right\}. \tag{10.18}$$

N4 **Fisher information and asymptotic variance:** Typically, Expression 10.5 comes from something like theorem 6.7 of Boos and Stefanski (2013) which shows $\hat{\theta}$ is $AN(\theta, n^{-1}\Omega_1(\theta)^{-1})$, and then replacing the Fisher information from one individual, $\Omega_1(\theta)$, with a consistent estimator, specifically, the average of the individual observed informations gives: $\Omega_1(\theta)^{-1} \approx \left(n^{-1}\sum \hat{\Omega}_i\right)^{-1} = n\hat{\Omega}^{-1}$.

N5 **Regularity conditions on permutational central limit theorem:** First, assume that b_i is a scalar, for $i = 1, \ldots, n$. We introduce notation to explicitly denote dependence on the total sample size, so that b_i becomes b_{ni}, and we let $\mathbf{a}_n = \{a_{n1}, a_{n2}, \ldots, a_{nn}\}$ and $\mathbf{b}_n = \{b_{n1}, b_{n2}, \ldots, b_{nn}\}$. Here are the conditions:

(a) The sequence, $\mathbf{a}_1, \mathbf{a}_2, \ldots$ must satisfy the Noether condition, i.e.,

$$\lim_{n \to \infty} \left[\frac{\max_{1 \le i \le n}\{a_{ni} - \bar{a}_n\}^2}{\sum_{i=1}^{n}\{a_{ni} - \bar{a}_n\}^2} \right] = 0,$$

where

$$\bar{a}_n = \frac{1}{n}\sum_{i=1}^{n} a_{ni}.$$

(b) The sequence, $\mathbf{b}_1, \mathbf{b}_2, \ldots$ must also satisfy the Noether condition.
(c) Both sequences together must satisfy the Lindeberg condition, i.e., for every $\epsilon > 0$,

$$\lim_{n \to \infty} \left[\frac{1}{n}\sum_{i=1}^{n}\sum_{j=1}^{n} \delta_{nij}^2 I\left\{|\delta_{nij}| > \epsilon\right\} \right] = 0,$$

where

$$\delta_{nij} = \frac{\{a_{ni} - \bar{a}_n\}\{b_{nj} - \bar{b}_n\}}{\left[n^{-1}\sum_{h=1}^{n}\{a_{nh} - \bar{a}_n\}^2 \sum_{k=1}^{n}\{b_{nk} - \bar{b}_n\}^2\right]^{\frac{1}{2}}},$$

for $i, j = 1, \ldots, n$.

For the case when $\mathbf{b}_{ni}$ is a vector, we use the fact that to show normality of a vector, it is sufficient to show that every arbitrary linear combination of its elements are normal (see, e.g., Seber, 1984, p. 19). Therefore, we give the conditions for the asymptotic multivariate normality of $\sum a_i \mathbf{b}_{ni}$ (when $\mathbf{b}_{ni}$ is a vector) in two steps. First, define $\mathbf{c}_n = [c_{n1}, \ldots, c_{nn}]$, where $c_{ni} = \sum_{j=1}^{p} w_j b_{nij}$ and w_j is an arbitrary constant, and b_{nij} is the jth element of the ith $p \times 1$ vector $\mathbf{b}_{ni}$. Second, show that $\mathbf{a}_n$ and $\mathbf{c}_n$ meet the same three conditions as described above for $\mathbf{a}_n$ and the $n \times 1$ vector $\mathbf{b}_n$. See, for example, Sen (1985).

N6 **Bootstrap failure example:** The bootstrap does not give a consistent estimate in all situations. For example, if $Y_1, \ldots, Y_n$ are independent from a uniform distribution with support $(0, \theta)$. We cannot consistently estimate θ by a nonparametric bootstrap (Bickel and Freedman, 1981). For regularity conditions on the bootstrap, see Bickel and Freedman (1981) or Beran (1997).

N7 **Accuracy of bootstrap intervals:** Let $C(\mathbf{X}, 1 - \alpha)$ be a $100(1 - \alpha)\%$ confidence interval for θ based on data with n observations. We say the confidence interval is first order accurate if

$$\Pr[\theta \in C(\mathbf{X}, 1 - \alpha)] = 1 - \alpha + O(n^{-1/2})$$

and second order accurate if

$$\Pr[\theta \in C(\mathbf{X}, 1 - \alpha)] = 1 - \alpha + O(n^{-1}).$$

The bootstrap-t and BC_a confidence intervals are second-order accurate, while the percentile confidence interval is generally only first-order accurate (there are exceptions, for example, if the confidence interval is two-sided and $\hat{\theta}$ is asymptotically normal then the percentile interval is second order accurate (see Boos and Stefanski, 2013, p. 430)).

10.18 Problems

P1 **Randomized test:** Consider the application of the Neyman–Pearson lemma to discrete data. Suppose X represents a discrete random variable with support having k distinct values, so that $X \in \{\mathcal{X}_1, \ldots, \mathcal{X}_k\}$. Suppose we are interested in testing a simple null versus a simple alternative hypothesis: $H_0 : \theta = \theta_0$ versus $H_1 : \theta = \theta_1$. Let $f_0(x)$ and $f_1(x)$ be the probability mass functions under the null and alternative. Let $Q(x) = f_1(x)/f_0(x)$ if $f_0(x) > 0$ and define $Q(x) = \infty$ if $f_0(x) = 0$. Let the decision rule for testing H_0 vs. H_1 be

$$\delta(x) = \begin{cases} 1 & \text{if } Q(x) > k \\ \gamma & \text{if } Q(x) = k \\ 0 & \text{if } Q(x) < k, \end{cases}$$

where k is chosen so that $\Pr_{\theta_0}[Q(x) > k] \le \alpha$ and $\Pr_{\theta_0}[Q(x) \ge k] \ge \alpha$, and γ is chosen so that $E_0(\delta(X)) = \alpha$, where E_0 is the expectation under the null hypothesis. If $\Pr_{\theta_0}[Q(x) = k] = 0$ then set $\gamma = 1$, and δ gives the most powerful test. If $\Pr_{\theta_0}[Q(x) = k] > 0$ then set

$$\gamma = \frac{\alpha - S_0(k)}{S_0(k-) - S_0(k)},$$

where $S_0(q) = \Pr_{\theta_0}[Q(X) > q]$ and $S_0(q-) = \Pr_{\theta_0}[Q(X) \ge q]$, so that δ gives the most powerful test. Show that when $\Pr_{\theta_0}[Q(x) = k] > 0$ then $E_0(\delta(X)) = \alpha$.

P2 Follow the three step process (find likelihood, check monotonicity, find p-value) for a sample of n independent and identically distributed Poisson random variables with mean θ to find a one-sided p-values for testing $H_0 : \theta \ge \theta_0$. In the process show that the sum of the n Poisson random variables is Poisson with mean $n\theta$. Check the p-value against poisson.test in base R.

P3 Show that in the two-sample permutation test in the form of Example 10.8) with T equal to the difference in means (Equation 10.11) is equivalent to the one with $T^*(\mathbf{x}) = \sum y_i z_i$.

P4 Show that the permutation p-value for a test of the form Example 10.8 will be equivalent to the one that keeps the responses fixed and permutes the covariate vectors. Hint 1: since the numbering of the individuals is arbitrary (i.e., the indices for $\mathbf{x}_1, \ldots, \mathbf{x}_n$ are arbitrary), T must be invariant to the original numbering of the individuals. Hint 2: show that for each $\pi_j \in \Pi$, there is a unique $\pi_j^* \in \Pi$ such that $T([\mathbf{y}(\pi_j), \mathbf{z}]) = T([\mathbf{y}, \mathbf{z}(\pi_j^*)])$.

P5 **Nonoverlapping confidence intervals:** Let (L_1, U_1) and (L_2, U_2) be two $100(1 - \alpha)\%$ central confidence intervals for parameters θ_1 and θ_2, respectively. Suppose the data used to calculate each confidence interval are independent (e.g., two independent samples from a two-sample study). Show that $\Pr[U_1 < L_2] \leq \alpha - \alpha^2/4$ if $\theta_1 = \theta_2$.

P6 **Nonoverlapping Normal confidence intervals:** Suppose $\hat{\theta}_1 \sim N(\theta_1, \sigma^2)$ and independently $\hat{\theta}_2 \sim N(\theta_2, \sigma^2)$. Let (L_1, U_1) and (L_2, U_2) be two $100(1-\gamma)\%$ central confidence intervals for θ_1 and θ_2. Show that the value of γ such that $\Pr[U_1 < L_2] = \alpha/2$ is $\gamma = 1 - 2\{1 - \Phi(\frac{\sqrt{2}}{2}\Phi^{-1}[1 - \alpha/2])\}$. (see Schenker and Gentleman, 2001, for details).

11

k-Sample Studies and Trend Tests

11.1 Overview

This chapter covers a wide variety of different tests. We consider situations where the ith individual's data is represented by the pair, (y_i, z_i), and y_i is the response and z_i is the group membership variable. For convenience let $z_i \in \{1, 2, \ldots, k\}$, representing the k groups. Throughout this chapter, we assume that within each group we have independent responses, and $Y_i|z_i = j \sim F_j$, where F_j is the distribution of the responses in Group j. We allow the responses to be binary, nominal, ordinal, or numeric, and the groups may have one of three types of relationships. Table 11.1 gives the type of data that are covered in this chapter with the accompanying section.

An important distinction is between cases where there is an inherent ordering of the groups or not, and this is reflected in the testing. By ordering of the groups, we mean that either the responses from Group j tend to be larger than those from i for all $i < j$, or conversely, the responses from Group j tend to be *smaller* than those from i for all $i < j$. For example, suppose the groups are three treatments in a randomized trial: placebo ($z_i = 1$), treatment A ($z_i = 2$), and treatment B ($z_i = 3$). If treatment A and treatment B are totally different drugs, then we will usually not want to assume an ordering of the responses prior to running the study. However, if treatment A and treatment B are the same drug, but treatment A has half the dose as treatment B (and the placebo has a dose of 0), then it is often (but not always) reasonable to assume that the responses will be ordered, with the effect of treatment A being between the effect of placebo and that of treatment B.

	Responses			
Groups	Binary	Nominal	Ordinal	Numeric
No order in groups (all group comparisons equally important)	11.2.1 11.7	11.2.1	11.2.4	11.2.2, 11.2.3 11.5
No order in groups (one group is reference)	11.7		11.7	11.6 11.7
Ordered Groups	11.3.1 11.7		11.3.2 11.7	11.7

Table 11.1 *Section where different types of data are covered in the chapter*

Whether or not there is an ordering of the groups, the overall null hypothesis is $H_0 : F_1 = F_2 = \cdots = F_k$, so that we are testing the null that there is independence between the responses and the group variable. Under the independence null, the group variable does not give any information about the responses. When we hypothesize an ordering of the groups, we call such an independence test a *trend* test (see Section 11.3), and when there is no such ordering of the groups we call it a *k-sample* test (see Section 11.2). The main difference between these two types of independence tests is that the trend tests account for the ordering and hence can be more powerful under alternatives where the suspected ordering occurs.

Throughout this book we have emphasized accompanying tests with estimates and confidence intervals. For a test of independence (*k*-sample tests or trend tests), it is less straightforward to choose a single estimand that describes either independence or lack of independence, because there are many ways for dependence to manifest itself. For scientific inferences, it is often important to follow a statistically significant independence test (e.g., a *k*-sample test) with follow-up tests (e.g., several pairwise comparisons between the groups) in order to learn more about the nature of the dependence. We treat the overall independence test and the follow-up tests as one family of tests, and because of this we address the problem of multiple comparisons. Chapter 13 deals generally with multiple comparisons, and in this current chapter we focus on the multiple comparison problem in the context of these types of families of tests. These families are special in that there are some logical relationships among the null hypotheses. In Section 11.4 we define the familywise Type I error rate (FWER). Then we study methods that control the FWER; in Section 11.5 the family of tests is all pairwise comparisons, and in Section 11.6 the family is pairwise comparisons of a reference group to each of the other groups. The classic example of the latter family is the set of tests of each new treatment compared to a common control. Finally, in Section 11.7 we briefly outline a very general approach that can address many of the multiple comparison problems for this type of data (see also Section 13.5).

11.2 Unordered Groups: *k*-Sample Tests

11.2.1 Unordered Categorical Responses

In this section we describe independence tests on contingency tables where there is no ordering of either the rows (e.g., the groups) or the columns (e.g., the outcomes). For the binary response case there are only two columns, and hence ordering does not matter in the two-sided tests.

The formal way to represent random variables where the responses are one of c possible nominal values is to let $\mathbf{y}_i$ be a $c \times 1$ vector with indicators for the c categories of responses. Let δ_j be a $c \times 1$ vector of 0s, except a 1 in the jth row. Then $\mathbf{y}_i \in \{\delta_1, \ldots, \delta_c\}$. These type of data are typically represented as a $k \times c$ contingency table, with the ajth entry equal to

$$x_{aj} = \sum_{i:z_i=a} I(\mathbf{y}_i = \delta_j).$$

Then the random variable associated with the ath row of the table is $X_a = [X_{a1}, \ldots, X_{ac}]$, and X_a is multinomial with parameters n_a and $\mathbf{f}_a = [f_{a1}, \ldots, f_{ac}]$, where

$$f_{aj} = \Pr[\mathbf{Y}_i = \delta_j | z_i = a].$$

The unconditional probability (or marginal probability) is $\mathbf{f}_. = [f_{.1}, \ldots, f_{.c}]$ with

$$f_{.j} \equiv \Pr[\mathbf{Y}_i = \delta_j] = \sum_{a=1}^{k} w_a f_{aj},$$

where $w_a = n_a/n$, so that $\sum_{a=1}^{k} w_a = 1$. Under independence, $f_{aj} = f_{.j}$ for all $a \in \{1, \ldots, k\}$ and $j \in \{1, \ldots, c\}$.

One way to measure distance between two distributions is to use the Kullback–Leibler distance function, and this distance between $\mathbf{f}_a$ and $\mathbf{f}_.$ is

$$KL(\mathbf{f}_a, \mathbf{f}_.) = \sum_{j=1}^{c} f_{aj} \log\left(\frac{f_{aj}}{f_{.j}}\right).$$

The weighted average of the Kullback–Leibler distances of the empirical distributions is proportional to the likelihood ratio test statistic for independence,

$$T_{LR} = 2n \sum_{a=1}^{k} \sum_{j=1}^{c} w_a \hat{f}_{aj} \log\left(\frac{\hat{f}_{aj}}{\hat{f}_{.j}}\right),$$

where $\hat{f}_{aj} = x_{aj}/n_a$ and

$$\hat{f}_{.j} = \frac{\sum_{a=1}^{k} x_{aj}}{n} = \sum_{a=1}^{k} w_a \hat{f}_{aj},$$

and $0 \log(0)$ is defined as 0. Under the null hypothesis of independence T_{LR} is asymptotically distributed chi-square with $v = (k-1)(c-1)$ degrees of freedom. Also under independence, T_{LR} is asymptotically equivalent to Pearson's chi-squared statistic,

$$T_{chisq} = \sum_{a=1}^{k} \sum_{j=1}^{c} \frac{\left(x_{aj} - n_a \hat{f}_{.j}\right)^2}{n_a \hat{f}_{.j}} = n \sum_{a=1}^{k} \sum_{j=1}^{c} \frac{w_a \left(\hat{f}_{aj} - \hat{f}_{.j}\right)^2}{\hat{f}_{.j}}.$$

So the p-value of the likelihood ratio test is $p_{LR} = 1 - F_{\chi_v^2}(T_{LR})$ and of Pearson's chi-squared test is $p_{chisq} = 1 - F_{\chi_v^2}(T_{chisq})$.

Another way to measure how close the observed data is to independence is to define data less likely under the independence assumption to be "further" from independence. This can be done by conditioning on both marginal totals of the contingency tables, so that the probability distribution of any table under the null hypothesis is a multivariate version of the hypergeometric distribution. Let $\mathbf{x}$ be the observed matrix of counts in the contingency table and $\mathbf{X}$ be the associated random variable, then

$$g_x(\mathbf{x}) = \Pr[\mathbf{X} = \mathbf{x}] = \frac{\prod_{a=1}^{k} n_a! \prod_{j=1}^{c} m_j!}{n! \prod_{a=1}^{k} \prod_{j=1}^{c} x_{aj}!}, \tag{11.1}$$

where $m_j = \sum_{a=1}^{k} x_{aj}$. Then for two tables $\mathbf{x}$ and $\mathbf{x}^*$ with the same row and column totals (i.e., same values for $n_1, \ldots, n_k$ and $m_1, \ldots, m_c$), we may say that $\mathbf{x}$ is closer to independence than $\mathbf{x}^*$ if $g_x(\mathbf{x}) > g_x(\mathbf{x}^*)$. Freeman and Halton (1951) proposed an exact test based on this ordering, and since it is a generalization of the Fisher–Irwin test, it is

Eye color	Hair color			
	Black	Brunette	Red	Blond
Brown	6	11	2	0
Blue	2	8	1	9
Hazel	1	5	1	1
Green	0	2	1	1

Table 11.2 *Modification of data from Snee (1974). Modification is to divide the original counts by 10 and take the integer portion*

usually called Fisher's exact test (but is sometimes called the Freeman–Halton test). Let $T_{FH}(\mathbf{y}, \mathbf{z}) = -g_x(\mathbf{x}) \equiv -g(\mathbf{y}, \mathbf{z})$ be the test statistic as a function of the responses and the group variables, so that smaller values are closer to independence. Then the exact p-value is obtained from a permutation test. Let $\mathbf{z}_1, \ldots, \mathbf{z}_N$ be all the $N = \frac{n!}{\prod_{a=1}^{k} n_a!}$ unique permutations of $\mathbf{z}$. The p-value is

$$P_{eFH} = \sum_{i : T_{FH}(\mathbf{y}, \mathbf{z}_i) \geq T_{FH}(\mathbf{y}, \mathbf{z})} g(\mathbf{y}, \mathbf{z}_i). \tag{11.2}$$

We can get exact versions of the likelihood ratio tests or the chi-squared test by substituting T_{LR} or T_{chisq} for T_{FH} in Equation 11.2.

To compare the three exact tests and the two asymptotic versions, we use the eye and hair color data from Snee (1974) (see the HairEyeColor data set in the datasets R package), except we modified it in order not to have all the tests be extremely significant (Table 11.2). Here are the p-values from the different tests (using StatXact-11):

	Likelihood ratio	Chi-squared	Freeman–Halton (Fisher's exact)
Asymptotic	0.0285	0.0796	
Exact	0.0459	0.0745	0.0276

The asymptotic approximations appear to do fairly well even though there are some table cells with 0 counts. Readers should not overgeneralize the ordering of the p-values, since each of the three statistics emphasizes a different one-dimensional direction away from independence. Historically Pearson's chi-squared test was the first statistic and is very common. For a rigorous and interpretable description of a confidence interval for a metric of dependence (i.e., parameters whose estimates are simple functions of T_{chisq}), see Diaconis and Efron (1985).

It is helpful to use Poisson regression, that assumes the counts are independent, to explain the dependence in contingency tables. This gives reasonable answers even when one or more of the marginal totals may be fixed (Baker, 1994). The statistic T_{chisq}/ν has an interpretation as an estimate of the linear overdisperion from the Poisson model (see Note N1). The asymptotic likelihood ratio test comparing the Poisson model of the counts with main effects for the row and column effects to the saturated model is called an analysis of deviance test (see Problem P1).

We have no strong preference between the three tests, but we prefer exact tests to the asymptotic ones because the exact tests are valid while the asymptotic ones are only valid asymptotically.

11.2.2 Numeric Responses: One-Way Analysis of Variance

Suppose that Y_i are numeric responses. Let n_a, μ_a, and σ_a^2 be respectively, the sample size, the mean, and the variance for Group a. When there are two groups, we are often interested in $\mu_2 - \mu_1$, but to summarize the effect with $k > 2$ groups we need another effect parameter. We propose using a weighted sum of squared differences in the group means,

$$\tau^2 = \sum_{a=1}^{k} w_a (\mu_a - \mu)^2,$$

where $w_a = n_a/n$ and $\mu = \sum_{a=1}^{k} w_a \mu_a$ is the overall mean. Another parameter of interest is the proportion reduction of variance due to the group effect,

$$\rho^2 = \frac{\mathrm{Var}(Y_i) - \mathrm{Var}(Y_i|z_i)}{\mathrm{Var}(Y_i)} = \frac{\tau^2}{\tau^2 + \sigma^2},$$

where $\sigma^2 = \sum_{a=1}^{k} w_a \sigma_a^2$. This is a general parameter related to R-squared for models (see, e.g., Agresti, 2013, p. 56). Thus, the parameter τ^2 may be interpreted as the $\mathrm{Var}(Y_i)$ if all $\sigma_a^2 = 0$ (i.e., within a group, all responses are the same). Parameters τ^2 and σ^2 are also often referred to as between-group variance and within-group variance, respectively.

We first consider the case where $\sigma_1^2 = \cdots = \sigma_k^2 = \sigma^2$ and the distributions are normal. In this case, we get the traditional one-way analysis of variance (see Note N2 about terminology). Here we focus on not just getting the p-value but getting estimates and confidence intervals on τ^2 or ρ^2, so our presentation is different from the usual one for one-way analysis of variance. Let $\hat{\mu}_a = n_a^{-1} \sum_{i:z_i=a} y_i$ and $\hat{\mu} = \sum_{a=1}^{k} w_a \hat{\mu}_a$. The pooled variance estimate is

$$\hat{\sigma}^2 = \frac{\sum_{a=1}^{k} (n_a - 1)\hat{\sigma}_a^2}{\sum_{a=1}^{k} (n_a - 1)} = \frac{\sum_{a=1}^{k} \sum_{i=1}^{n} I(z_i = a)(y_i - \hat{\mu}_a)^2}{n - k},$$

where $\hat{\sigma}_a^2 = (n_a - 1)^{-1} \sum_{i:z_i=a} (Y_i - \hat{\mu}_a)^2$ is the usual unbiased sample variance. The estimate $\hat{\sigma}^2$ is called the *mean squared error*. The treatment mean square is

$$MSTR = \frac{n \sum_{a=1}^{k} w_a (\hat{\mu}_a - \hat{\mu})^2}{k - 1}.$$

The F statistic is $T_F = MSTR/\hat{\sigma}^2$. Under the null hypothesis that $\mu_1 = \cdots = \mu_k$ (and still under the equal variance and normality assumptions), T_F is distributed F with degrees of freedom $v_1 = k - 1$ and $v_2 = n - k$. So the p-value for testing that null hypothesis is $p = 1 - F_F(T_F, v_1, v_2)$, where $F_F(t, v_1, v_2)$ is the cumulative distribution of an F random variable with v_1 and v_2 degrees of freedom, evaluated at t.

Under the alternative hypothesis, T_F is distributed noncentral F with v_1 and v_2 degrees of freedom and noncentrality parameter, $n\tau^2/\sigma^2$, so that

$$\mathrm{E}(T_F) = \frac{v_2\left(v_1 + \frac{\tau^2}{\sigma^2/n}\right)}{v_1(v_2 - 2)}.$$

Setting T_F equal to its mean and solving for τ^2, and then substituting $\hat{\sigma}^2$ for σ^2, we get an estimator of τ^2,

$$\hat{\tau}^2 = \frac{v_1\hat{\sigma}^2}{n}\left(\frac{(v_2 - 2)T_F}{v_2} - 1\right),$$

and because τ^2 must be nonnegative, we use $\hat{\tau}^{*2} = \max(0, \hat{\tau}^2)$. We can get an exact confidence interval for $\xi = \tau^2/\sigma^2$ by using the noncentral F distribution. Specifically, letting t_F be the observed value of T_F, if $p < \alpha$, a lower one-sided $100(1 - \alpha)\%$ confidence limit for ξ as the value ξ_L such that $F_F(t_F, v_1, v_2, n\xi_L) = \alpha$, while if $p \geq \alpha$, then set $\xi_L = 0$. Analogously, we can get an upper one-sided $100(1 - \alpha)\%$ confidence limit for ξ as the value ξ_U such that $F_F(t_F, v_1, v_2, n\xi_U) = \alpha$. We can convert the estimate and confidence interval into one for $\rho^2 = \xi/(\xi + 1)$ using $\hat{\rho}^2 = \hat{\tau}^{*2}/(\hat{\tau}^{*2} + \hat{\sigma}^2)$ and

$$(\rho_L^2, \rho_U^2) = \left(\frac{\xi_L}{\xi_L + 1}, \frac{\xi_U}{\xi_U + 1}\right).$$

We get an approximate confidence interval for τ^2 using $(\hat{\sigma}^2\xi_L, \hat{\sigma}^2\xi_U)$.

If we relax the assumption of equal variances, there are no simple exact solutions, but we can get approximate p-values and confidence intervals by generalizing Welch's t-test to the k-sample case. Following Brown and Forsythe (1974a) we replace the pooled variance estimator $\hat{\sigma}^2$ with

$$\tilde{\sigma}^2 = (k - 1)^{-1}\sum_{a=1}^{k}(1 - w_a)\hat{\sigma}_a^2$$

and replace the denominator degrees of freedom using Satterthwaite's approximate degrees of freedom,

$$\tilde{v}_2 = \left(\sum_{a=1}^{k}\frac{c_a^2}{n_a - 1}\right)^{-1},$$

where

$$c_a = \frac{(1 - w_a)\hat{\sigma}_a^2}{\sum_{b=1}^{k}(1 - w_b)\hat{\sigma}_b^2}.$$

Under normality, Brown and Forsythe (1974b) performed simulations under different variances and show that this adjustment has Type I error roughly close to the nominal (e.g., Type I error rates of 6.5% for nominal 5% tests with six groups with samples sizes of $n = [4, 6, 6, 8, 10, 12]$ and standard deviations of $\sigma = [1, 1, 2, 2, 3, 3]$).

The Brown–Forsythe adjustment may not have Type I error rates close to nominal if the data are far from normal. It is mathematically difficult to be precise about how the Type I error rates change as distributions get more different from normal. Nevertheless, using

Laguerre polynomials, Tan (1982) explored this issue and made the following conclusion when the variances are equal: for the test based on T_F "one need not worry about non-normality unless the departure is in the extreme." We expect similar robustness for moderate departures from normality for the Brown–Forsythe adjustment.

11.2.3 Numeric Responses: Studentized Range Test

Another k-sample test is Duncan's studentized range test; this procedure is also referred to as Tukey's studentized range test when the samples sizes are the same in all k groups. The parameter of interest is the range of the true means, $\psi = (\max_{a \in \{1,2,\dots,k\}} \mu_a) - (\min_{b \in \{1,2,\dots,k\}} \mu_b)$, where μ_a is the mean for Group a. This method allows us to determine a confidence interval on this metric.

First, assume all responses are normal. If $Y_i | z_i = a \sim N(\mu_a, \sigma^2)$ for each group, then the sample means within each group are independently, $\hat{\mu}_a \sim N(\mu_a, \sigma^2/n_a)$. As in Section 11.2.2, let $\hat{\sigma}^2$ and $\hat{\sigma}_a^2$ be the pooled variance estimate and Group a sample variance. Then under the normality assumption, $\hat{\sigma}^2$ is independent of the vector of sample means, and $(n-k)\hat{\sigma}^2/\sigma^2$ is distributed chi-squared with $n-k$ degrees of freedom. Let

$$T_{ab} = \frac{\hat{\mu}_a - \hat{\mu}_b}{\hat{\sigma}\sqrt{\frac{1}{2}\left(\frac{1}{n_a} + \frac{1}{n_b}\right)}}. \tag{11.3}$$

Similar to the derivation of the t-test, under the null when $\mu_1 = \mu_2 = \cdots \mu_k = \mu$ then the distribution of T_{ab} does not depend on μ or σ. We reject the null that $\mu_1 = \mu_2 = \cdots = \mu_k$ when

$$T_{max} \equiv \max_{a,b} |T_{ab}|$$

is large. In practice, to determine the p-value, we either simulate (since the null distribution of T_{ab} does not depend on μ and σ, we can use $\mu = 0$ and $\sigma = 1$), or we approximate the distribution using the studentized range distribution. The *studentized range distribution* with parameters k and ν is defined as the distribution of $\max_{a,b} |Z_a - Z_b| / \sqrt{\chi^2/\nu}$, where $Z_1, \dots, Z_k$ are independently distributed standard normal, and independently χ^2 is distributed chi-squared with ν degrees of freedom. When $n_1 = n_2 = \cdots = n_k$ then T_{max} follows the studentized range distribution with parameters k and $n-k$. Then the two-sided p-value is $p = 1 - F_{tukey}(T_{max}; k, n-k)$ and the $100(1-\alpha)\%$ confidence interval on ψ is

$$\left(\hat{\mu}_h - \hat{\mu}_\ell\right) \pm \frac{\hat{\sigma}}{\sqrt{n_1}} F_{tukey}^{-1}(1 - \alpha; k, n-k), \tag{11.4}$$

where $\hat{\mu}_h$ and $\hat{\mu}_\ell$ are the highest and lowest sample means, and we let the cumulative distribution function (cdf) and inverse cdf of the studentized range distribution with parameters k and ν be $F_{tukey}(\cdot; k, \nu)$ and $F_{tukey}^{-1}(\cdot; k, \nu)$, respectively (in R the functions are ptukey and qtukey). When the group sample sizes are not equal, p is a conservative p-value (see Hochberg and Tamhane, 1987, p. 118), and the associated approximate confidence interval is calculated by replacing n_1 in Equation 11.4 with $\tilde{n}_{h\ell}$, where $\tilde{n}_{h\ell} = (0.5(1/n_h + 1/n_\ell))^{-1}$ (see Hochberg and Tamhane, 1987, Section 3.2.1).

By assuming normality on the responses, the sample means are also normal and are independent of the pooled variance. When we do not assume normality, the central limit theorem implies that the sample means will be close to normal for large samples and finite variance distributions. Further, as n gets large, then the chi-squared distribution approaches a normal distribution as well, so the central limit theorem acts on the pooled variance also. Thus, for large samples, if the pooled variance is approximately independent of the means, then the studentized range test and confidence intervals may have reasonable Type I error rates and coverage rates.

11.2.4 Ordinal Responses: Kruskal–Wallis Test

The traditional nonparametric version of the one-way analysis of variance (ANOVA) test is the Kruskal–Wallis test. As with any test for ordinal data, this method can be used if the data are inherently ordinal, or if the data are numeric but we opt to analyze by ranks to be robust from deviations from normality. The null hypothesis in this case is $H_0 : F_1 = F_2 = \cdots = F_k$, where we make no assumptions on the distributions. The Kruskal–Wallis test is a permutation test version of the one-way ANOVA where the responses are replaced by their respective midranks (i.e., ranks automatically adjusted for ties by averaging the ranks of tied values). We present it in terms of empirical distributions and h_{MW}, the Mann–Whitney functional (see Equation 9.3, page 130), so that it can be more readily interpreted in terms of a parameter that is a function of the $F_1, F_2, \ldots, F_k$. Define the Kruskal–Wallis parameter as

$$\tau_{KW}^2 = \sum_{a=1}^{k} w_a \left\{ h_{MW} \left(F_a, F \right) - \frac{1}{2} \right\}^2, \tag{11.5}$$

where $F = \sum_{a=1}^{k} w_a F_a$ is the distribution of a randomly sampled response out of the total n responses. Then the Kruskal–Wallis test is a permutation test where the test statistic is

$$T_{KW} = \sum_{a=1}^{k} w_a \left\{ h_{MW} \left(\hat{F}_a, \hat{F} \right) - \frac{1}{2} \right\}^2, \tag{11.6}$$

where $\hat{F}$ is the empirical distribution of all n responses, and $\hat{F}_a$ is the empirical distribution of only the n_a responses from group a. In terms of ranks,

$$T_{KW} = \frac{1}{n^2} \sum_{a=1}^{k} w_a \left(R_a - \frac{n+1}{2} \right)^2, \tag{11.7}$$

where R_a is the mean of the midranks in Group a and $(n + 1)/2$ is the mean of the midranks in the sample of all n responses (see Problem P2). An asymptotic approximation for the permutation p-value based on T_{KW} is

$$p = 1 - F_{\chi_{k-1}^2} \left(t_{KW} / V \right),$$

where t_{KW} is the observed value of T_{KW}, $V = (n + 1)t/12$ and t is the same tie adjustment factor used in the Wilcoxon–Mann–Whitney (WMW) test (see Equation 9.22, page 144). See Lehmann (1975, Appendix) for the asymptotic justification.

To our knowledge, confidence intervals for τ_{KW}^2 have not been developed but this deficit is not too important since for many practical applications the focus will often be on pairwise tests performed after the Kruskal–Wallis test is significant (see Section 11.4).

11.3 Ordered Groups: Trend Tests

In this section we discuss two well-known tests when the groups are inherently ordered, the Cochran–Armitage test for binary responses, and the Jonckheere–Terpstra test for ordinal responses. We do not present any test for numeric responses, since this kind of analysis will typically be done as part of a regression analysis.

11.3.1 Binary Responses: Cochran–Armitage Trend Test

Suppose the responses are binary and the groups are ordered. The group ordering could be inherent and a reasonable assumption, or the group ordering could reflect a structure that we want to have high power to detect. In the latter case, the tests are still valid (although less powerful) when the group ordering assumption fails, because validity depends only on the veracity of the null hypothesis assumptions.

Define the score for Group a as s_a, and index the groups such that we hypothesize $s_1 < s_2 < \cdots < s_k$. For example, suppose the groups represent arms of a randomized experiment that are given different doses of some treatment. Then the score, s_a, could represent the dose or log(dose) for Group a. If there is no obvious score for the ordered groups, we can let $s_a = a$, representing equal spaced effects between the groups. Let $\pi_a = \Pr[Y_i = 1|z_i = a]$. Suppose we expect that higher scores lead to higher probability of a success (i.e., $Y_i = 1$). Following the development in Tarone and Gart (1980), consider a model $\pi_a = H(\beta_0 + \beta_1 s_a)$, where H is a monotonic and twice differentiable function (e.g., the inverse logit function). Then a permutation version of the score test (see Section 10.7) is equivalent to the permutation based on the test statistic, $t_{CA} = \sum_{i=1}^{n} y_i x_i$, where x_i is the score associated with the ith individual (i.e., if $z_i = a$, then $x_i = s_a$). As with many permutation tests, we can create one- or two-sided p-values related to the test statistic (see Section 10.9.3). Any of these versions are called a Cochran–Armitage test. Because of this relationship to the score test, the Cochran–Armitage test has good power locally (i.e., in alternatives close to the null hypothesis of $\pi_1 = \pi_2 = \cdots = \pi_k$). Further, since H does not need to be completely specified, the test is robust.

If a parameter related to the test is needed, then an easier alternative approach is to pick a typical H function and make tests and confidence intervals from that model. For example, using the inverse logit function gives a logistic regression, and inferences about the β_1 parameter can be interpreted within the context of that model (see Section 14.3). Alternatively, we can relate t_{CA} directly to a parameter by rewriting t_{CA} in terms of the proportions in each group. Let $\hat{\pi}_a$ be the proportion of successes in Group a, and let

$w_a = n_a/n$. Then $t_{CA} = n \sum_{a=1}^{k} w_a \hat{\pi}_a s_a$. So the natural parameter is a linear combination of π_a values: $\sum_{a=1}^{k} w_a \pi_a s_a$. We do not pursue developing confidence intervals on that parameter, since the modeling approach is already well developed. The advantage of the Cochran–Armitage test is its simplicity, and since it can be implemented as a permutation test, it is relatively easily to calculate exact p-values.

11.3.2 Ordinal Responses: Jonckheere–Terpstra Test

The Jonckheere–Terpstra test addresses the case where the groups are ordered, and the responses are ordinal; it uses combinations of Mann–Whitney parameters as the parameter of interest. Suppose the ordering implies that when $a < b$ then responses in Group b tend to be equal or larger than those in Group a, implying that $h_{MW}(F_a, F_b) \geq 1/2$, where we are using the same notation as Section 11.2.4. The Jonckheere–Terpstra test statistic just takes a weighted average of the sample Mann–Whitney statistics in all the cases where $a < b$. It is proportional to

$$T_{JT} = \sum_{b=2}^{k} \sum_{a=1}^{b-1} w_a w_b h_{MW} \left(\hat{F}_a, \hat{F}_b \right). \tag{11.8}$$

Letting the null hypothesis be $H_0 : F_1 = F_2 = \cdots = F_k$, we can create a permutation test by permuting the group membership variable to create a permutation distribution associated with T_{JT}. To describe the permutation test, let $T_{JT}(\mathbf{y}, \mathbf{z})$ be the observed value of T_{JT} given by Equation 11.8, and let $T_{JT}(\mathbf{y}, \mathbf{Z})$ be the random variable associated with the permutation distribution, where $\mathbf{Z}$ is a random sample of all of the $\frac{n!}{\prod_{a=1}^{k} n_a!}$ unique permutations of $\mathbf{z}$. The p-value can be defined as either $p_U = \Pr[T_{JT}(\mathbf{y}, \mathbf{Z}) \geq T_{JT}(\mathbf{y}, \mathbf{z})]$ (upper tail p-value), or $p_L = \Pr[T_{JT}(\mathbf{y}, \mathbf{Z}) \leq T_{JT}(\mathbf{y}, \mathbf{z})]$ (lower tail p-value), or as $p_{TS} = \min(1, 2\min(p_U, p_L))$ (two-sided p-value) (see Section 10.9.3 for another way of defining a two-sided p-value). This can be implemented by normal approximation, Monte Carlo, or complete enumeration (see Section 6.2 of Hollander et al. (2014) and Section 11.10). It is generally not helpful to estimate θ_{JT}, the parameter defined by replacing the empirical distributions with the true values in T_{JT}. Thus, we do not pursue getting a confidence interval on θ_{JT} since that would require some additional assumptions, as is done with the confidence intervals on $h_{MW}(F_1, F_2)$ (see Section 9.5.5).

11.4 Follow-Up Tests after Independence Test and Familywise Error Rates

After an independence test rejects the null hypothesis of independence, we often wish to test several follow-up hypotheses to learn more about the nature of the dependence. Consider a randomized trial with five treatment groups. After rejecting independence, we may want to test all $\binom{5}{2} = 10$ pairwise comparisons (see Section 11.5). Alternatively, if the five treatments were four new experimental treatments and one control treatment, we may want to follow-up with four subsequent hypotheses comparing each of the four new treatments to the control (see Section 11.6).

As with any case where we test many hypotheses, we often want to adjust our inferences to account for the number and types of hypotheses being tested. Chapter 13 explores multiple hypothesis testing in detail. Here, we introduce the concepts of familywise error rate (FWER), and weakly or strongly controlling that FWER within the context of this chapter. Let $H_0^{(1)} : F_1 = F_2 = \cdots = F_k$ be the null hypothesis for the k-sample test, and let $H_0^{(2)}, \ldots, H_0^{(m)}$ be a series of null hypotheses on subsets of the groups. For example, if we test the k-sample test and all pairwise comparisons, then $m = 1 + \binom{k}{2}$. Let $\mathbf{d} = [d_1, \ldots, d_m]$ be the vector of decisions, with $d_j = 1$ if we reject $H_0^{(j)}$ and $d_j = 0$ otherwise. Let the true state of nature be denoted by $\mathbf{a} = [a_1, \ldots, a_m]$ where $a_j = 0$ if $H_0^{(j)}$ is true, and $a_j = 1$ otherwise (if the alternative is true whenever the null is false [i.e., the union of the null and alternative is the set of all possible models], then a_j is an indicator of the truth of the jth alternative hypothesis). No errors are made if $\mathbf{d} = \mathbf{a}$. Let D_j be the random variable associated with d_j. Let the number of Type I errors (i.e., rejections when the null hypothesis is true) be $R_0 = \sum_{j=1}^{s} D_j(1 - a_j)$. Let the *FWER* $= \Pr[R_0 > 0]$. A multiple testing procedure controls the FWER at α if $\Pr[R_0 > 0] \le \alpha$. If $\Pr[R_0 > 0] \le \alpha$ holds only over the global null model (i.e., the model that is the intersection of all the null hypotheses: $H = \cap_{j \in \{1,2,\ldots,m\}} H_0^{(j)}$), then the procedure *weakly* controls the FWER. For the k-sample problem, the global null hypotheses is $H = H_0^{(1)} : F_1 = \cdots = F_k$. If $\Pr[R_0 > 0] \le \alpha$ holds over all possible null models (specifically, any model that is an intersection of any subset of the null hypotheses), then the procedure *strongly* controls the FWER.

11.5 All Pairwise Comparisons

11.5.1 Tukey–Welsch Step-Down Procedure

Before describing the Tukey–Welsch step-down procedure, we consider the simpler Fisher's least significant difference (LSD) procedure. Fisher's LSD procedure is a two-stage procedure where at stage 1 we perform a k-sample test, and only perform subsequent tests in stage 2 if the k-sample test is significant, using the same significance level for both stages of comparisons. Fisher first proposed this procedure using the one-way ANOVA test, followed by pairwise t-tests, but we expand the definition to include other versions (e.g., Krusal–Wallis at step 1, followed by Wilcoxon–Mann–Whitney tests). Although trivially, Fisher's LSD weakly controls the FWER for any $k \ge 3$, it strongly controls the FWER when $k = 3$, and not when $k > 3$. For proof of the strong control when $k = 3$, see Note N3.

Because of the strong FWER control when $k = 3$, we can define the adjusted p-value for any of the three pairwise tests as the maximum of the p-value from the k-sample test and the unadjusted p-value for the pairwise test. If we reject any hypothesis when these adjusted p-values are less than or equal to α, then the FWER is strongly controled at α. In practice, this means that whenever we compare exactly three groups, we have full license to test all the pairwise comparisons at α, as long as we only do this testing if the global test is significant at α.

Example 11.1   Fake data: *Consider an experiment on 15 individuals that are randomized equally to three groups A, B, and C. The responses are: Group A: 3.4, 4.2, 3.2, 5.6, 4.3; Group B: 3.2, 4.5, 4.7, 4.6, 3.7; and Group C: 7.5, 6.4, 5.4, 3.8, 7.1. Assume the data are sufficiently close to normal that we can use one-way ANOVA followed by t-tests. A one-way ANOVA test using the Brown–Forsythe adjustment gives a p-value of $p_{ANOVA} = 0.034$ and Welch's t-test on all pairs gives: $p_{AB} = 1.0$ for A vs B; $p_{BC} = 0.043$ for B vs C; and $p_{AC} = 0.048$ for A vs C. Because p_{ANOVA} is less than p_{AB}, p_{BC}, and p_{AC}, none of the pairwise p-values needs to be adjusted for multiple comparisons. So at the FWER of $\alpha = 0.05$, we reject not only $H_0 : \mu_A = \mu_B = \mu_C$, but additionally reject $H_0 : \mu_B = \mu_C$ and $H_0 : \mu_A = \mu_C$, where μ_A, μ_B, and μ_C are the means from the three groups.*

When $k > 3$ Fisher's LSD procedure only weakly controls the FWER. There is no simple step-down procedure that strongly controls the FWER for $k > 3$, but the Tukey–Welsch procedure (described below) strongly controls the FWER for all $k \geq 3$.

The Tukey–Welsch procedure starts with testing all k groups, then tests all subsets of $k - 1$ groups, and continues in this manner until it tests all pairwise tests, adjusting the required significance level at each stage. The type of test applied for each hypothesis could be any of the three k-sample tests previously described (with slight modifications). Exact versions of the tests strongly control the FWER, and asymptotic versions only asymptotically strongly control the FWER. Hochberg and Tamhane (1987) gives details and references, describing the strong control of the FWER for the Tukey–Welsch procedure when using one-way ANOVA (see p. 111), studentized range tests (see p. 117–118), or asymptotic control for Kruskal–Wallis tests (p. 248). The procedure is known by several names. For example, in SAS these types of procedures are called Ryan–Einot–Gabriel–Welsch procedures (Westfall et al., 1999).

The Tukey–Welsch procedure is different from Fisher's LSD in two major ways. First, instead of having only two stages, we have up to $k - 1$ stages. Second, each stage uses a possibly different α–level. We label the stages by the number of distributions in the null hypotheses at each stage, and proceed from stage k, to stage $(k - 1)$, stepping all the way down to stage 2. Any test performed within stage j uses a significance level of α_j for $j = k$, $k - 1, \ldots, 2$, and where $\alpha_k = \alpha_{k-1} = \alpha$ and for $j < k - 1$,

$$\alpha_j = 1 - (1 - \alpha)^{j/k}.$$

At the kth stage we start by performing the k-sample test of the null hypothesis that all k groups have the same distribution. If we fail to reject that null hypothesis, we stop. If we reject the hypothesis at stage k, we proceed to stage $(k - 1)$, where we test all $\binom{k}{k-1} = k - 1$ null hypotheses that have $k - 1$ distributions equal. If we fail to reject any one of those hypotheses, we do not test any null hypotheses at lower stages that contain the nonrejected one. For example, suppose $k = 4$, and at stage 3 we fail to reject the null that $F_1 = F_2 = F_4$. That means that we automatically would fail to reject (and hence do not need to test) the following three hypotheses in stage 2: $F_1 = F_2$ and $F_1 = F_4$ and $F_2 = F_4$. We continue in this manner until there are no more hypotheses to test. For the one-way ANOVA and studentized range test version, we use the same pooled variance estimate from all k groups for all the hypotheses.

We can define adjusted p-values for this procedure, as the smallest α for each hypothesis that would lead to the rejection of that hypothesis (Note N4).

Although the Tukey–Welsch procedure tends to have slightly larger power than other procedures (see Section 11.5.3), the number of individual tests that need to be performed increases at a very rapid rate. Let $\eta(k)$ be the number of tests done when there are k groups. Then $\eta(4) = 11$, $\eta(8) = 247$, $\eta(16) = 65,519$, and $\eta(32) \approx 4.3 \times 10^9$. Thus, this procedure is not tractable for large k.

11.5.2 Tukey–Kramer Procedure

The Tukey–Kramer procedure (also called Tukey's honest significant difference) is very similar to the studentized range test and confidence interval of Section 11.2.3, except instead of testing that $\psi = 0$ (recall ψ is the maximum mean minus the minimum mean) and getting confidence intervals on ψ, we perform the test and confidence intervals on each pairwise difference. Specifically, to test the null that $\mu_b - \mu_a = 0$ the p-value is $p = 1 - F_{tukey}(T_{ab}; k, n - k)$ and the $100(1 - \alpha)\%$ confidence interval on $\mu_b - \mu_a$ is

$$\left(\hat{\mu}_b - \hat{\mu}_a\right) \pm \frac{\hat{\sigma}}{\sqrt{\tilde{n}_{ab}}} F_{tukey}^{-1}(1 - \alpha; k, n - k), \tag{11.9}$$

where all the notation is the same as with the k-sample studentized range test (see Section 11.2.3). When the responses are normal, the p-values and confidence intervals are exact when $n_1 = n_2 = \cdots = n_k$ and tend to be conservative otherwise (see Hochberg and Tamhane, 1987, p. 92–93). Note that because the p-value from the global test of the null hypothesis of $\mu_1 = \mu_2 = \cdots = \mu_k$ is just the minimum p-value of the pairwise tests (see Section 11.2.3), there is no need to first reject the global test before testing the pairwise hypotheses.

11.5.3 Comparison of Procedures

To compare different procedures, we perform a simulation. In the simulations, we have k different groups, with means $\mu = [\mu_1, \ldots, \mu_k]$ and sample sizes $\mathbf{n} = [n_1, \ldots, n_k]$. We simulate the power to reject $\mu_1 = \mu_2$, in several situations. We compare four methods that all approximately strongly control the FWER (exactly for the normal [F = N] with equal sample sizes, and approximately for Poisson [F = P] or unequal sample sizes), and for comparison we include the uncorrected two-sample t-test that does not strongly control the FWER. First, we perform a t-test (assuming equal variances) on each pairwise comparison and make the Holm adjustment for multiple comparison (see Chapter 13). The other three methods that approximately strongly control the FWER were mentioned above: the Tukey–Kramer test or the Tukey–Welsch procedure using either the studentized range test, or the one-way ANOVA.

The results are in Table 11.3. In general, the Tukey–Welsch method is more powerful. The Holm and Tukey–Karmer methods are slightly less powerful but have advantages of simplicity (Holm) and ease of calculating the compatible confidence interval. We see in the last scenario, where the errors are Poisson, that the Tukey–Welsch tests are slightly anti-conservative.

F	$[\mu_1, \ldots, \mu_k]$ $[n_1, \ldots, n_k]$	Unadj. t-test	Holm adj. t-test	Tukey– Kramer	T–W stud. range	Tukey– Welsch ANOVA
N	**[1, 0, 0]** $[10, 10, 10]$	54.8	37.9	40.7	46.9	48.6
N	**[1, 0, 0, 0, 0, 0]** $[8, 12, 10, 8, 12, 10]$	52.8	14.7	23.1	27.0	25.5
N	**[1, 0, 50, 50, 50, 50]** $[8, 12, 10, 8, 12, 10]$	52.7	23.2	23.7	39.7	39.7
N	**[0, 0, 50, 50, 50, 50]** $[8, 12, 10, 8, 12, 10]$	4.8	0.8	0.6	1.8	1.8
P	**[1, 2, 2]** $[10, 10, 10]$	39.5	23.0	23.3	27.4	28.2
P	**[4, 2, 1.5, 1.5, 1.5, 1.5]** $[8, 12, 10, 8, 12, 10]$	61.6	17.7	57.0	66.8	67.7
P	**[4, 4, 1.5, 1.5, 1.5, 1.5]** $[8, 12, 10, 8, 12, 10]$	4.9	0.6	3.1	6.6	6.7

Table 11.3 *Simulated power (%) to reject* $\mu_1 = \mu_2$ *at the* 0.05 *level using different methods. Used* 10,000 *simulations*

11.6 Many-to-One Comparisons

In this section, instead of being interested in all pairwise comparisons, we consider only comparing $k - 1$ of the groups to a single reference group (e.g., each new treatment is compared to control). We study the case of numeric responses where the means are approximately normal, and (like the usual derivation of a t-test) there is a common variance estimate that is independently distributed gamma (i.e., a scaled chi-squared).

Assume the responses are normally distributed with a mean shift for each group. When comparing a reference group to all other groups, Dunnett's test (Dunnett, 1955) is similar to the Tukey–Kramer procedure, except that instead of getting confidence intervals for all pairwise difference in means, we only get them for the $k - 1$ differences between the reference group and all the others. Because of this change, we cannot use the studentized range distribution to calibrate the confidence interval widths, but use a similar idea. Let the kth group be the reference group. Let $T_a = T_{ak}/\sqrt{2}$, where T_{ak} is given in Equation 11.3. Then under the null hypothesis, the vector $[T_1, \ldots, T_{k-1}]$ has a multivariate t distribution with $n - k$ degrees of freedom, and with correlation Λ, where the abth element of Λ is $\sqrt{\frac{n_a n_b}{(n_a + n_k)(n_b + n_k)}}$ (Bretz et al., 2011, p. 73). Then the $100(1 - \alpha)\%$ simultaneous confidence interval for $\mu_a - \mu_k$ is

$$\hat{\mu}_a - \hat{\mu}_k \pm c\hat{\sigma}\sqrt{\frac{1}{n_a} + \frac{1}{n_k}},$$

where c is the solution to

$$1 - \alpha = \int_{-c}^{c} \cdots \int_{-c}^{c} f_{multT}([x_1, \ldots, x_{k-1}]; n - k, \Lambda) dx_1 \cdots dx_{k-1}, \qquad (11.10)$$

where $f_{multT}(\cdot; n-k, \Lambda)$ is the density of the multivariate t distribution with $n-k$ degrees of freedom and correlation Λ. The two-sided p-value for testing $\mu_a = \mu_k$ is $p_a = 1 - \Pr[|T_a| \leq |t_a|]$, where t_a is the observed value of T_a, i.e.,

$$p_a = 1 - \int_{-t_a}^{t_a} \cdots \int_{-t_a}^{t_a} f_{multT}([x_1, \ldots, x_{k-1}]; n - k, \Lambda) dx_1 \cdots dx_{k-1}.$$

For one-sided tests of null hypotheses $\mu_a \geq \mu_k$, the analysis is similar. Then the $100(1-\alpha)\%$ one-sided upper simultaneous confidence interval for $\mu_a - \mu_k$ is

$$\hat{\mu}_a - \hat{\mu}_k + c_U \hat{\sigma} \sqrt{\frac{1}{n_a} + \frac{1}{n_k}},$$

where c_U is the c that solves

$$1 - \alpha = \int_{-\infty}^{c} \cdots \int_{-\infty}^{c} f_{multT}([x_1, \ldots, x_{k-1}]; n - k, \Lambda) dx_1 \cdots dx_{k-1}.$$

The associated one-sided p-value is

$$p_a = 1 - \int_{-\infty}^{t_a} \cdots \int_{-\infty}^{t_a} f_{multT}([x_1, \ldots, x_{k-1}]; n - k, \Lambda) dx_1 \cdots dx_{k-1}.$$

The one-sided tests and confidence intervals in the other direction are analogous.

11.7 Max-*t*-Type Procedures

A max-*t*-type procedure is a generalization of the studentized range procedure and can handle many of the postindependence test situations already studied, plus some more. That method is described in Section 13.5 and Bretz et al. (2011). The general idea is that we have a $k \times 1$ vector of parameter estimates for each of the k samples, say $\hat{\theta}$. By premultiplying that vector by an $h \times k$ contrast matrix, say $\mathbf{C}$, then we can create either all pairwise comparisons, the many-to-one comparisons or some other set of comparisons. For example, if $k = 4$ and Group 1 is the reference group, then a many-to-one contrast matrix is

$$\mathbf{C} = \begin{bmatrix} -1 & 1 & 0 & 0 \\ -1 & 0 & 1 & 0 \\ -1 & 0 & 0 & 1 \end{bmatrix}. \qquad (11.11)$$

Let $\mathbf{c}_j^\top$ be the jth row of $\mathbf{C}$, and the h null hypotheses are given by $H_0 : \mathbf{c}_j^\top \theta = 0$ for $j = 1, \ldots, h$, where superscript $\top$ denotes transpose. Then if we can assume asymptotic normality of $\hat{\theta}$ and we have an estimate of the covariance matrix for $\hat{\theta}$, then we can use the multivariate normal or perhaps the multivariate t-distribution to get simultaneous confidence intervals and adjusted p-values. For example, when $\hat{\theta}$ represents the k means and $\mathbf{C}$ is defined as a many-to-one set of contrasts as in Equation 11.11, we can recreate the tests of

Section 11.6. By changing the contrasts to all pairwise contrasts, we get another option for testing all pairwise comparisons. For constrasts for testing trends or ordered groups, see Section 4.3 of Bretz et al. (2011).

For ordinal responses, we can define $\hat{\theta}$ to represent average ranks of the groups. Konietschke et al. (2012) describe the details of such a procedure. They use a $\hat{\theta}_a = h_{MW}(\hat{F}_a, \hat{F})$ for $a = 1, \ldots, k$. This can be defined either using $\hat{F} = k^{-1} \sum_{a=1}^{k} \hat{F}_a$ or using $\hat{F} = \sum_{a=1}^{k} w_a \hat{F}_a$, where $w_a = n_a/n$. For this situation we tend to get better coverage using a multivariate delta method and a Fisher-transformation. Konietschke et al. (2012) point out that this way of framing the problem, as opposed to using pairwise comparisons based on $h_{MW}(\hat{F}_a, \hat{F}_b)$, avoids some problems with intransitivity (see Section 9.5.2 or Thangavelu and Brunner (2007)).

For binary responses, we can define $\hat{\theta}$ to represent log odds for the k groups from a logistic regression model (see Bretz et al., 2011, Section 4.7).

11.8 Summary

This chapter addresses testing when there are more than two groups; we test if there is independence between the response and group variable. When there is no ordering in groups, we refer to this as a k-sample test, and when there is ordering, this is a trend test.

When the response is binary or nominal, and the groups are unordered, we can use either of the following simple asymptotic procedures: the chi-squared test (the oldest method which is in very common use) or the likelihood ratio test. Furthermore, exact versions of both methods exist. Another exact procedure is the Freeman–Halton test which is a generalization of Fisher's exact test, and is sometimes referred to as Fisher's exact test even when there are more than two possible outcomes. We have no strong preferences between likelihood ratio, chi-squared or Fisher's exact test, but we do prefer the exact procedures to asymptotic ones. As shown in an example, even the three exact procedures produce different p-values, this is because each emphasizes a different deviation from independence.

Unordered group comparisons of numeric responses have traditionally used one-way ANOVA. The classic one-way ANOVA is analogous to a t-test, it assumes normality and equal variance. The assumption of equal variance can be relaxed by generalizing Welch's t-test to the k-sample test. Defining a treatment effect parameter is less straightforward than for the two-sample comparison, but we propose two measures. One parameter, τ^2, is the weighted sum of squared differences in the means, and the other, ρ^2, is the proportion reduction of variance due to the group effect. We provide confidence intervals for τ^2 or a couple of functions of both parameters.

The studentized range test is another procedure for numeric data for unordered group comparisons. The corresponding parameter of interest is the difference between the largest and smallest mean. This method should be powerful for detecting treatment differences that would be especially large with respect to this parameter. It is designed for equal sample sizes, although there are some adjustments for unequal sample sizes.

Ordinal responses for unordered groups can be analyzed by the Kruskall–Wallis test. This is a permutation test version of the one-way ANOVA where responses are replaced by their respective midranks.

The Cochran–Armitage trend test is used for binary responses when the groups are ordered. This requires specifying a score for each group; for example, in the case of varying drug doses, the score could be the dose or log(dose). The procedure is a permutation test based on the sum of score values among the successes. The Jonckheere–Terpstra test handles ordinal responses with ordered groups. It is a generalization of the Wilcoxon–Mann–Whitney test.

To provide useful interpretation of results, we often follow a significant global test of independence with additional tests (such as differences between pairs of treatments) to learn about the nature of the group differences. These multistep processes need to account for the multiplicity. R_0 denotes the number of Type I errors made among all conducted tests. The familywise Type I error rate is $FWER = \Pr[R_0 > 0]$. The multiple testing process controls FWER at α if $\Pr[R_0 > 0] \leq \alpha$. A procedure that controls the FWER only under the assumption of global independence (i.e., all k distributions are equal) is considered to have weak control, whereas as a procedure that controls the FWER generally (i.e., without needing to assume global independence) is considered to have strong control.

The Fisher's LSD is a two-stage procedure when a significant k-sample test is followed by all pairwise two-sample comparisons, without any adjustment to α. Interestingly, in the special case where k = 3 and the k-sample test is significant with respect to α, then we can test each of the three pairwise comparisons without any adjustment (i.e., test at α) without any concern about inflating the FWER. When $k > 3$, then Fisher's LSD only weakly controls the FWER. Another method, Tukey–Welsch, approximately strongly controls the FWERs for any k; instead of two stages there are up to $k - 1$ stages, and the significance level becomes more stringent at later stages. A third approach, Tukey–Kramer, provides a special adjustment to evaluate all pairwise comparisons. Based on our simulations, in general, the Tukey–Welsch method is among the more powerful of these procedures for testing pairwise differences. However, as discussed in Section 11.5.3, there are other motivations for choosing other procedures.

We then consider the setting where there is a single reference group, such as a clinical trial where multiple unordered treatments are studied alongside a common control group. When the response variable is numeric, one can use Dunnett's test. This provides a $100(1 - \alpha)\%$ simultaneous confidence interval for the true mean of each treatment minus the true mean in the reference group.

Finally, we end with a discussion of max-t-type procedures which provides a very general framework. This approach can be used to handle many of the follow-up tests situations already presented, plus some more.

11.9 Extensions and Bibliographic Notes

A seminal book on contingency tables and their analysis is Bishop et al. (1975). Brown and Forsythe (1974a) discuss Satterthwaite's approximate degrees of freedom for one- and two-way ANOVA. Hochberg and Tamhane (1987) give a theoretical overview of many of the classical methods of multiple comparisons, including the studentized range test and the Tukey–Welsch, Dunnett, and Tukey–Kramer procedures. Bretz et al. (2011) is a more up-to-date book about multiple comparisons that not only discusses important multiple comparison concepts (i.e., closure principle, partition principle, max-t tests), but also gives detailed descriptions of applications using their multcomp package. See Chapter 13 for more references on multiple comparisons.

11.10 R Functions and Packages

In the asht R package the function anovaOneWay performs the one-way ANOVA tests and confidence intervals of Section 11.2.2. The p-values assuming equal variances may also be calculated using anova(lm(y~as.factor(z))). Also in asht the tukeyWelsch function does the Tukey–Welsch test. The Tukey–Kramer procedure is done with the TukeyHSD function. The Jonckheere–Terpstra upper tail p-value is calculated with the pJCK function in the NSM3 R package. To calculate the lower tail p-value, multiply the responses by -1.

The multcomp R package is a comprehensive package for multiple comparisons (Bretz et al., 2011). The glht function in that package can perform many of the tests. The Dunnett test uses glht(aov(y z),linfct=mcp(z="Dunnett")) with simulataneous p-values using summary and confidence intervals using confint. All pairwise comparisons use the same code, except "Dunnett" is replaced by "Tukey"." The nparcomp R package performs the max-t nonparametric methods described in Konietschke et al. (2012); see also Konietschke et al. (2015).

11.11 Notes

N1 **Quasi-Poisson model:** Let $X_{aj}|\gamma_{aj} \sim Poisson(\gamma_{aj})$, where γ_{aj} is a random effect distributed gamma with mean $n_a f_{aj}$ and variance $n_a f_{aj}(\sigma^2 - 1)$ for $\sigma^2 > 1$. Then unconditionally, X_{aj} is negative binomial with mean $n_a f_{aj}$ and variance $\sigma^2 n_a f_{aj}$. If you fit a model with main effects for the rows and columns of the contingency table, then that model has $p = 1 + (c - 1) + (k - 1)$ parameters, leaving $ck - p$ degrees of freedom. Thus $T_{chisq}/(ck - p)$ is a reasonable estimator of σ^2. The quasi-likelihood only assumes $E(X_{aj}) = n_a f_{aj}$ and $Var(X_{aj}) = \sigma^2 n_a f_{aj}$, and also uses $\hat{\sigma}^2 = T_{chisq}/(ck - p)$. This is the quasipoisson family in glm in R (see McCullagh and Nelder, 1989, Sections 6.2.3 and 6.2.4).

N2 **ANOVA terminology:** Analysis of variance (ANOVA) is a very old and general way for analyzing many different types of data. The presentation of the results typically involves a table that decomposes the sum of the squared error into different rows. One-way ANOVA refers to having only one group variable, and two-way ANOVA has two group variables. Although ANOVA can be applied to many problems, often it is easier to use linear models and random effects models to attack the same problems. For a discussion on these points see Gelman (2005).

N3 **Fisher's LSD strongly controls the FWER when $k = 3$:** Let the overall null hypothesis be $H_0^{(1)} : F_1 = F_2 = F_3$, and let the pairwise null hypotheses be $H_0^{(2)} : F_1 = F_2$, $H_0^{(3)} : F_1 = F_3$ and $H_0^{(4)} : F_2 = F_3$. As in Section 11.4, let $R_0 = \sum_{j=1}^{s} D_j(1 - a_j)$ be the number of rejections when the null hypothesis is true, where **D** and **a** are indicators of rejection or indicators of the truth of the alternative hypothesis (technically, **a** are indicators of the falsity of the null), respectively. The familywise Type I error rate is $FWER = Pr[R_0 > 0]$. We want to show that $FWER \le \alpha$ for Fisher's LSD procedure, where Fisher's LSD procedure means that if $D_1 = 0$, then $D_2 = D_3 = D_4 = 0$, and all tests are α−level (i.e., $\Pr[D_j = 1 | a_j = 0] \leq \alpha$ for $j = 1, 2, 3, 4$). We do the proof by partitioning the possible probability models into five cases and showing that $FWER \leq \alpha$ for all five cases. Here are the five cases:

Case number	Description of state	True state vector
Case 1	$F_1 = F_2 = F_3$	$\mathbf{a}_1 = [0, 0, 0, 0]$
Case 2	$F_1 = F_2 \neq F_3$	$\mathbf{a}_2 = [1, 0, 1, 1]$
Case 3	$F_1 = F_3 \neq F_2$	$\mathbf{a}_3 = [1, 1, 0, 1]$
Case 4	$F_1 \neq F_2 = F_3$	$\mathbf{a}_4 = [1, 1, 1, 0]$
Case 5	$F_1 \neq F_2 \neq F_3$	$\mathbf{a}_5 = [1, 1, 1, 1]$

Here is the proof (all summations are from $j = 1$ to 4):

$$Case\ 1: \quad \Pr\left[\sum D_j(1 - a_j) > 0 | \mathbf{a}_1\right] = \Pr\left[\sum D_j > 0 | \mathbf{a}_1\right]$$

$$= \Pr[D_1 = 1 | a_1 = 0] \leq \alpha$$

$$Case\ 2: \quad \Pr\left[\sum D_j(1 - a_j) > 0 | \mathbf{a}_2\right] = \Pr[D_2 > 0 | \mathbf{a}_2] = \Pr[D_2 = 1 | a_2 = 0] \leq \alpha$$

$$Case\ 3: \quad \Pr\left[\sum D_j(1 - a_j) > 0 | \mathbf{a}_3\right] = \Pr[D_3 > 0 | \mathbf{a}_3] = \Pr[D_3 = 1 | a_3 = 0] \leq \alpha$$

$$Case\ 4: \quad \Pr\left[\sum D_j(1 - a_j) > 0 | \mathbf{a}_4\right] = \Pr[D_4 > 0 | \mathbf{a}_4] = \Pr[D_4 = 1 | a_4 = 0] \leq \alpha$$

$$Case\ 5: \quad \Pr\left[\sum D_j(1 - a_j) > 0 | \mathbf{a}_5\right] = \Pr[0 > 0 | \mathbf{a}_5] = 0 \leq \alpha.$$

To see why Fisher's LSD procedure does not strongly control the FWER for $k = 4$, and other related examples, see Proschan and Brittain (2020).

N4 **Tukey–Welsch adjusted p-values:** Let p_{jh} be the unadjusted p-value associated with $H_0^{(jh)}$, the hth null hypothesis at stage j. Let Ω_{jh} be the set of index pairs of null hypotheses that contain the null hypothesis associated with p_{jh}. In other words, for all $(ab) \in \Omega_{jh}$ then $H_0^{(jh)} \subseteq H_0^{(ab)}$. For example, suppose $k = 4$ and the null hypothesis at stage 4 is $H_0^{(4, 1)} : F_1 = F_2 = F_3 = F_4$ and those at stage 3 are $H_0^{(3, 1)} : F_1 = F_2 = F_3$, $H_0^{(3, 2)} : F_1 = F_2 = F_4$, $H_0^{(3, 3)} : F_1 = F_3 = F_4$, and $H_0^{(3, 4)} : F_2 = F_3 = F_4$. Suppose $H_0^{(2, 3)} : F_1 = F_4$, then

$$\Omega_{2, 3} = \{(2, 3), (3, 2), (3, 3), (4, 1)\}.$$

The adjusted p-value associated with $H_0^{(jh)}$ is $\hat{p}_{jh}$, the smallest value of α that would lead to rejection of $H_0^{(jh)}$

$$p_{jh}^* = \max_{(ab) \in \Omega_{jh}} \tilde{p}_{ab},$$

where

$$\tilde{p}_{ab} = \begin{cases} p_{ab} & \text{if } a = k, k-1 \\ 1 - (1 - p_{ab})^{k/a} & \text{if } a < k-1 . \end{cases}$$

11.12 Problems

P1 Write an R program to input a contingency table and fit the elements of the table with a Poisson model with main effects rows and columns using glm. Write another program to calculate T_{LR}. Compare the residual deviance given in the output from glm from the Poisson model with T_{LR}. Try several tables. Extra credit: show that the two statistics are always equal (see Agresti, 2013, Section 4.5.2).

P2 Prove Equation 11.7. Hints: show that the midrank for the ith observation is $nh_{MW}(\delta_i, \hat{F}) + 1/2$, where $\delta_i(x) = I(y_i \leq x)$. You may use the fact that $\sum_{a=1}^{k} w_a h_{MW}(F_a, G) = h_{MW}(\bar{F}, G)$, where $\bar{F} = \sum_{a=1}^{k} w_a F_a$, for any nonnegative constants w_a such that $\sum_{a=1}^{k} w_a = 1$ and any distributions $F_1, \ldots, F_k$ and G (see Fay and Shih, 1998).

12

Clustering and Stratification

12.1 Overview

In this chapter, we deal mostly with two-sample tests when observations are no longer independent, but there is some correlation structure between the observations. Let the ith observation be $x_i = [y_i, z_i, s_i]$, where y_i is the response, z_i defines the treatment or other effect of interest, and s_i denotes the grouping variable (e.g., strata or clustering variable). Suppose there are k possible groups for s_i. In this chapter, we use the term *clustering* when all observations within a group have the same value of z, and we use *stratification* when within each stratum there are observations with each of the possible values of z.[1] For example, consider a randomized trial of two possible treatments among k villages. If villages are randomized to treatment and every participant in a village gets the same treatment, this is a cluster randomized trial. If, instead, we apply a village-specific randomization scheme where, within each village, some participants are assigned to one of the two possible treatments and some are assigned to the other treatment, this is a two-sample study, stratified by village.

12.2 Clustering

12.2.1 Overview

In this section we consider the case where we are interested in the effect of one covariate (z), and all observations within a cluster have the same value of the z covariate.

Example 12.1 *Webster et al. (1983) did experiments on birth defects in mice from alcohol, we consider only a small part of the experiment. Pregnant mice are exposed to two doses of alcohol at gestational Day 7. There are two groups, where the separation between doses was either four hours (n = 15 litters) or six hours (n = 10 litters). There are about 130 live fetuses, with an average of 5.5 per litter in the first group and 4.8 per litter in the second. Consider the response observation the weight of the fetus at birth.*

Example 12.2 *Consider a hypothetical study of the effect of vitamin supplements on birth weights of children born in different villages. In some villages prenatal vitamins are distributed to pregnant mothers and in others the prenatal vitamins are not distributed. Over*

---

[1] In other places in the book, clustering and stratification are more general terms and are not restricted in this way. For example, in other chapters, members of different treatment groups may be in the same cluster.

the course of three years the weight at birth of all children born alive in the participating villages are recorded as the response.

In the first example, the clusters are litters, and in the second example the clusters are villages. In either example, if we simply did a t-test comparing the fetus (or newborn) weights in one group to the weights in the other, that would not be correct because it ignores the likely correlation of the weights within the cluster (e.g., a litter or a village). There are two ways to handle that correlation. We could use an *individual-level analysis* and estimate the intervention effect using the individual observations (e.g., fetus weights) and account for the within-cluster correlation when performing inferences (e.g., getting confidence intervals). Alternatively, we could treat each cluster as an observation, using a *cluster-level summary* of the individual observations (e.g., average fetus weight per cluster), and then perform inferences treating those cluster-level responses as independent observations. Hayes and Moulton (2009) compare these two approaches in the context of cluster randomized trials. We give a brief overview.

There are several issues to consider when deciding between the two general strategies. One is simplicity, and for this, the cluster-level summary analysis wins out over the individual-level analysis. Once we summarize the cluster, we can treat the cluster summary as the observation and can often assume that those cluster summaries are independent. Then we can use all the tools we already have for analyzing two samples (or three or more samples) with independent observations.

Another issue is efficiency. A potential pitfall of using cluster-level summaries is that we may lose some efficiency by treating each cluster-level summary as having equal importance. Consider a simple example with normally distributed errors where clusters are randomized to either treatment or control, and we are interested in estimating the mean of individuals on treatment (μ_1) or control (μ_0). We summarize each cluster with its mean. The most efficient estimate of μ_1 (or μ_0) is the weighted average of the cluster means of the treated clusters (or control clusters) with the weight for the jth cluster equal to

$$w_j = \frac{m_j}{m_j + \sigma_w^2/\sigma_b^2}, \tag{12.1}$$

where m_j is the number of individuals in cluster j and σ_w^2 and σ_b^2 are the within- and between-cluster variances. The details are in Note N1. When $m_1 = m_2 = \cdots = m_J$ for all J clusters, then regardless of the variance ratio, σ_w^2/σ_b^2, we can efficiently estimate the control and treatment means by just taking the average of the cluster-level means over the appropriate clusters. If the within-cluster variance is much smaller than the between-cluster variance, then $w_j \approx 1$ and each cluster gets about equal weight, while if the opposite is true (i.e., there is no substantial cluster effect), then w_j is approximately proportional to m_j and each cluster gets weighted based on the number of individuals in that cluster. Section 12.2.4 discusses a more sophisticated analysis that can be in between those two choices when σ_w^2/σ_b^2 is neither very large nor very small.

Another factor for deciding on an analysis is whether the number of observations per cluster is related to the response. For the mice litter example (Example 12.1) it is highly likely that the weight of a fetus is related to the number of live births in its litter. But for the vitamin example (Example 12.2), it is a more reasonable assumption that the weight of any newborn is independent of how many live births occurred in its village. This consideration is especially important for individual-level analyses. Some individual-level analyses (for example, using generalized estimating equations to account for within-cluster correlation) will give biased estimators if the number of observations per cluster is related to the response.

In this chapter there will be more choices for analysis than we have space to discuss, so we only discuss some common choices. For example, an applicable method not discussed in this chapter is within-cluster resampling (see Section 10.12). In many cases, it is useful to embed the problem in a model, so we discuss some properties of different models in this chapter that will not be formally introduced until Chapter 14.

12.2.2 Binary Responses

Let $y_i = 1$ if the event of interest occurred for the ith individual and $y_i = 0$ otherwise. Let $r_j = \sum_{i:s_i=j} y_i$ be the number events and m_j be the total number of (binary) responses in the jth cluster. Let $\theta_j = E(R_j)/m_j$, the expected probability of an event per individual. Let z_j^* be the value of z_i for all individuals in cluster j. We allow that the θ_j may be random variables themselves, with $E(\theta_j|z_j^* = h) = \beta_h$.

We first consider individual-level analyses, and suppose that m_j is independent of θ_j. In this case, it is often useful to use an overdispersed generalized linear model (see Section 14.3) for binary responses, also called a quasi-binomial model in glm in R. In the quasi-binomial model, the cluster effect is ignored in the model of the mean but accounted for in the model of the variance. Assume $R_j \sim Binomial(m_j, \theta_j)$, where θ_j is a random variable with $E(\theta_j|z_j^* = h) = \beta_h$ and $var(\theta_j|z_j^* = h) = \phi\beta_h(1 - \beta_h)$. Since the distribution is left unspecified except by its mean and variance, this is a quasi-likelihood model. Another model is the beta-binomial model where the θ_j are distributed as a beta distribution. In general, the quasi-likelihood is easier to fit than the full beta-binomial model, and inferences on it are made using asymptotic normality (see McCullagh and Nelder, 1989, for details on quasi-likelihood models). Other modeling approaches (mixed effects model, and the generalized estimating equation [GEE] model) are more complicated and are introduced in Section 14.5. The GEE model requires the assumption that m_j is independent of θ_j, while the mixed effects model does not require that assumption.

Suppose z_i has only two levels, and we are interested in testing the null hypothesis that $\Delta = \beta_2 - \beta_1 = 0$. We can perform a cluster-level analysis where we summarize each cluster by $\hat{\theta}_j = r_j/m_j$ and perform Welch's t-test on, for example, the $\hat{\theta}_j$ values, comparing clusters with all $z_i = 1$ to those with all $z_i = 2$. In addition to reasons of simplicity discussed earlier, another reason to consider this approach is to yield the proper Type I error rate when we cannot assume that m_j is independent of θ_j.

We simulate two types of examples. First, consider the case where m_j is independent of θ_j. If you do an inappropriate analysis that ignores the clustering by fitting the typical logistic model (see Section 12.3.3) with a main effect for only the z_i covariate, you may have a dramatically inflated Type I error rate. We give the results of some simulations (Problem P1 gives details). We simulate 50 clusters in each group where the simulated Type I error rate for a nominally 5% two-sample test is 27.3% when using standard logistic regression.

In contrast, an analysis that accounts for the clustering, the quasi-binomial regression, is much closer to the nominal, giving 5.8%. The Type I error rate is even better using Welch's t-test on the $\hat{\theta}_j$ with simulated Type I error rate of 4.9%. Using a mixed effects model using a logit link, with a random normal effect for each cluster (using **glmer** function of the **lme4 R** package, version 1.1-21), we find that about 3.6% of the simulated data sets do not converge to a solution, so we set those p-values to 1, and after doing that we reject 5.9% of all simulated data sets.

Now consider cases where m_j is related to θ_j. Problem P2 gives the details of a simulation of one such case, and in that case the simulated Type I error rates are 51.1% (for the standard logistic regression), 13.1% (for the quasi-binomial model), 5.0% (for Welch's t-test), and 4.8% (for the mixed effects model after setting the p-value to 1 for 30.5% of the simulated data sets that did not converge). We see that when m_j is related to θ_j, then either using the cluster-level analysis (Welch's t-test) or the mixed effects model retains the nominal Type I error rate. However, the 30.5% nonconvergence rate for the mixed effects model is unacceptable in practice, and for a real data set different algorithms might be tried in order to get converged results of the mixed effects model.

Zhang et al. (2011) present several simulations using mixed effects models for these types of problems comparing different software. They found that the coverage and bias may be worse for nonnull effects (i.e., when $\Delta \neq 0$). In that case the SAS procedure NLMIXED appeared to perform better than existing R packages at the time which could have poor coverage, but because of the speed of development in R, there may have been improvements on those R packages since that time.

We give more details on interpretation of parameters in mixed effects models and GEE models in Sections 14.5.2 and 14.5.3, respectively.

In the binary outcome case, the choice on whether to use a cluster-level analysis versus an individual-level analysis that takes clustering into account, such as the quasi-binomial model, may boil down to the variability of sample sizes per cluster. If sample sizes across clusters are very similar, both types of analyses would probably lead to similar conclusions. In this case the cluster-level analysis has the advantage of simplicity. The individual-level analysis has the advantage of leading to a confidence interval of the odds ratio, whereas the cluster-level analysis might lead to a confidence interval on the difference in proportions with values outside -1 and 1. On other hand, if the sample sizes across clusters vary considerably, then care must be taken if the sample sizes might be related to the group membership or outcome. In that case, the quasi-binomial and GEE analyses can give inflated Type I error rates, while the mixed effects model does allow for a more appropriate way to take sample size into account in an individual-level model. When the sample sizes are related to group membership, then the cluster-level analysis gives appropriate Type I error rates, although there is a possibility of a loss of efficiency compared to the mixed effects model.

12.2.3 Numeric Responses

For testing numeric responses using an individual-level analysis, such as the mixed effects models and GEE models, many of the same issues as discussed in Section 12.2.2 apply.

For example, if the mean per cluster depends on the sample size, then the mixed effects model is better than the GEE model. For details, see Section 14.5.

In the simulation in this section, we focus on the case when the sample size for each cluster is independent of the mean for that cluster. Under this lack of dependency, we focus on efficiency. Consider a linear model with random effect for each cluster. Let

$$Y_i = \beta_k I(z_i = k) + \delta_j I(s_i = j) + \epsilon_i, \qquad (12.2)$$

where $\delta_j \sim N(0, \sigma_b^2)$ and $\epsilon_i \sim N(0, \sigma_w^2)$ are the random cluster effects and random (individual) errors, and all δ_j and ϵ_i are independent. The simulated model has 20 clusters (10 with $z = 1$ and 10 with $z = 2$) with sample sizes drawn with replacement from $\{10, 20, 40\}$. Let $\sigma_w = 1$ and σ_b be either $0.01, 0.1, 1$ or 10. We use $\Delta = \beta_2 - \beta_1 = 0$.

We compare three methods for analyzing the simulated data using 10,000 replications. First, we inappropriately ignore the clustering and treat each individual response as if it were independent and perform a Welch's t-test on the individual responses (with the associated estimates and confidence intervals on Δ). As expected, the 95% confidence interval for Δ has poor coverage if σ_b is too large: the simulated coverage is 95% when $\sigma_b = 0.01$, 92% when $\sigma_b = 0.1$, 37% when $\sigma_b = 1$, and 26% when $\sigma_b = 10$. Second, we take the mean of each cluster and perform a Welch's t-test treating the cluster means as observations. In this case, the 95% confidence intervals have reasonable coverage: all cases the simulated coverage is 95% (rounded to the nearest percentage point). Third, we use a mixed model with random effects for the cluster using the lme4 R package (version 1.1-21). We estimate the degrees of freedom using a Satterthwaite degrees of freedom estimator with the lmerTest R package (version 3.1-0, Kuznetsova et al., 2017). This results in reasonable coverage for 95% confidence intervals: simulated coverage 96% ($\sigma_b = 0.01, 0.1$) and 95% ($\sigma_b = 1, 10$). We repeated the simulations with five clusters for each treatment and got similar results, in particular the coverage was equal or greater than 95% for both the mixed effects model and the Welch's t-test using the cluster means even with such a small sample size.

We compare the mean squared error (MSE) of the estimate and the confidence interval lengths of all three methods graphically in Figure 12.1; the linear mixed model is the reference method, and the other two methods are presented relative to this reference. If $\sigma_b \ll \sigma_w$ then analyzing "ignoring cluster" (i.e., ignoring the cluster and performing a t-test on individuals) will be the most efficient (because the MSE ratio is less than 1). On the other hand, if σ_b is larger ($\sigma_b \geq \sigma_w$ in this simulation scenario), then analyzing "by cluster" (i.e., treating the means of clusters as observations) appears to have similar efficiency as the linear mixed effects model. But notice that the linear mixed effects model has close to the best MSE regardless of the value of σ_b, so appears to be the robust choice. Although this simulation assumes normal errors, because of the central limit theorem (CLT) (which tells us that averages under "regular" conditions with enough observations tend to be approximately normal, see Section 10.5) the conclusion may be approximately true for many other situations. If we use a t-test or Wilcoxon–Mann–Whitney test on cluster-level means, this can be an even more robust choice to assumptions than the linear mixed effects model. Although as the simulation shows, those tests may be inefficient when the within-cluster variance is much larger than the between cluster variance and the sample sizes vary between clusters.

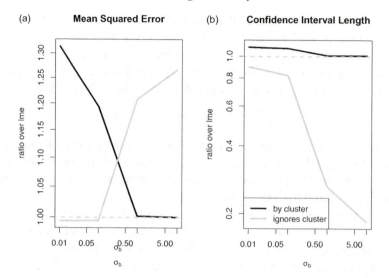

Figure 12.1 Part (a) is the ratio of the MSE of the estimate of interest over the MSE of the estimate from the linear mixed effects model; "by cluster" uses the mean per cluster as an observation, and "ignores cluster" uses the individual as an observation without taking cluster into account. Part (b) is the ratio of confidence interval lengths.

12.2.4 Ordinal Responses

For ordinal responses, a semiparametric model such as a proportional odds model or a proportional hazards can be used, where the cluster effect is modeled as a main effect. We performed a simulation using the same model as Equation 12.2, except we used a logistic distribution instead of a normal distribution, and we rounded the responses to the nearest integer. This will give a proportional odds model (Problem P3). We simulated only $\sigma_b = 0.01, 0.1$, and 1, since there were convergence issues with $\sigma_b = 10$. We tried two models, a proportional odds model ignoring clustering with only an effect for treatment (z_i), and a mixed effects model with a random effect for cluster using the ordinal R package (version 2019.12-10). We simulate the effect under the null when $\beta_1 = \beta_2$, and calculate confidence intervals on $\Delta = \beta_2 - \beta_1$ using asymptotic normality and standard error estimates (with Hess = TRUE). We use 10,000 replications.

As expected, in terms of coverage, the mixed effects model does much better than the independence model that ignores clustering when σ_b is larger: for the proportional odds model that ignores the clustering the simulated coverage of the 95% confidence interval on Δ is 95% ($\sigma_b = 0.01$), 92% ($\sigma_b = 0.1$), and 39% ($\sigma_b = 1$), while for the random effects model the simulated coverage is 95% ($\sigma_b = 0.01$), 93% ($\sigma_b = 0.1$), and 89% ($\sigma_b = 1$). Although the coverage of the mixed effects model is lower than 95% when σ_b is large, it is much better than ignoring the cluster effects. For this model we have made no attempt to comprehensively study all the available R packages (nor have we studied all other software packages). There may be better software packages. Often when implementing newer methodology or complicated models (like mixed effects models), there are many algorithms or software packages available and it is good to either read some review of the software, check the internal validation of the software, or run some simulations modeled after your data to see that the coverage is reasonable before choosing one algorithm. Because software improvements are common, we do not make a definitive software recommendation.

For some projects, it may be important to have a valid p-value. If so, a permutation approach may be used, where we permute the z_i values associated with each cluster (keeping the clustering structure so that within each permutation data set all individuals within a cluster have the same z_i value). For the test statistic of the permutation test, we could use $\hat{\Delta}$ from the mixed effects model, and see if the observed value is extreme in the permutation distribution (see Section 10.9). For rank permutation tests on clustered data, see Fay and Shih (1998). For asymptotic rank tests and the issues of defining parameters related to ranks for clustered and stratified data (especially factorial designs), see Brunner et al. (2017).

12.3 Stratification

12.3.1 Overview

As in the section on clustering (Section 12.2), the data for the ith individual is $x_i = [y_i, z_i, s_i]$, where y_i is the response, z_i is the covariate of interest, and $s_i = j$ represents the jth stratum. Unlike the clustering section, in this section, within each level of the strata variable we have different values of z_i. To keep it simple, we let z_i have only two values (e.g., $z_i = 1$ for control and $z_i = 2$ for treatment, or $z_i = 1$ for unexposed and $z_i = 2$ for exposed). As usual, we let upper case values represent random variables.

There are numerous contexts for stratification (i.e., doing a stratified analysis). Stratified analyses can be employed to adjust for potentially biasing imbalances in observational data. Stratified analyses can also be used for randomized trial data, when randomization is performed within each stratum (such as the village in the earlier example), or to ensure better balance when randomization is not done within a stratum, and to gain efficiency. Meta-analysis uses stratified analyses, where each study is a stratum.

We first ask: do we need to adjust in some way for the strata effect? The answer to this is context dependent, as we discuss in Section 12.3.2 for the simple case of binary response and only two strata. This is followed by a brief discussion on stratified analysis choices for binary responses (Section 12.3.3) and other responses (Section 12.3.4).

We then focus on stratified analyses in special settings. First up is meta-analysis, where we consider another question: does the treatment effect remain constant over all the strata (fixed effect) or does it vary between strata (random effect)? We discuss these two models in Section 12.3.5 about meta-analyses. We consider the case where within each stratum we have an estimate of the treatment effect and its variance, and we want to combine those estimates across all strata to get an overall mean treatment effect. In Section 12.3.6 we discuss direct standardization of stratified rates. For the other sections, we briefly discuss analysis choices for binary responses (Section 12.3.3) and other responses (Section 12.3.4).

12.3.2 Simpson's Paradox: Is Stratification Needed?

In this section, y_i, z_i, and s_i are all binary. Let $\theta_{hj} = \Pr[Y_i = 1 | Z_i = h, S_i = j]$ and $\theta_h = \Pr[Y_i = 1 | Z_i = h]$. In terms of parameters, Simpson's paradox is present when there are only two strata ($k = 2$) and

$$\theta_{2j} > \theta_{1j} \quad \text{for both } j = 1 \text{ and } j = 2, \text{ but}$$
$$\theta_2 < \theta_1.$$

We also call it Simpson's paradox when both inequalities are reversed or the parameters are replaced by estimates, such as sample proportions. Suppose $Z_i = 2$ is exposed and $Z_i = 1$ is unexposed, then it seems strange that the exposed have a higher probability of the event in every stratum, but if we ignore the strata variable, the unexposed have a higher probability of the event. Simpson's paradox rarely occurs with designed experiments, where the proportion of individuals with $Z = 1$ within each stratum can be designed to be similar, and it is a potential issue more often with observational studies. A classic real-world example is death penalty verdicts by race in Florida between 1976 and 1987: overall, whites were more likely to get a death penalty verdict than blacks, but if the victim was black *OR* if the victim was white, then blacks were more likely to get a death penalty verdict than whites. This paradox occurred because the death penalty was much more likely to be handed down for those cases with white victims than for those with black victims, and since defendants and victims tend to be of the same race, the marginal death penalty verdict rate was higher for white defendants (see, e.g., Agresti, 2013, p. 48).

The paradox is explained by writing $\Pr[Y_i = 1 | Z_i = h] = \theta_h$ for $h = 1, 2$ as weighted means of the stratum-specific probabilities. Simpson's paradox:

$$\theta_{2j} > \theta_{1j} \quad \text{for } j = 1, 2$$

and

$$\theta_2 \equiv \frac{\sum (1 - \lambda_j) \eta_j \theta_{2j}}{\sum (1 - \lambda_j) \eta_j} < \frac{\sum \lambda_j \eta_j \theta_{1j}}{\sum \lambda_j \eta_j} \equiv \theta_1, \tag{12.3}$$

where all summations are over $j = 1, 2$, and η_j and λ_j are the expected sample size and proportion with $z_i = 1$, respectively, in stratum j. From Expression 12.3 it is clear that the weights in the weighted means may be different for the two groups. Further, if the expected proportion with $Z_i = 1$ (e.g., the proportion unexposed) within each stratum are the same (i.e., $\lambda_1 = \lambda_2$) then the weights are the same for both weighted means and Simpson's paradox cannot occur.[2]

As discussed in Pearl (2009b, chapter 6), when we are trying to make causal inferences about Z_i on Y_i, then we need extra-statistical information about the study design (i.e., more information than just knowledge about the distribution of $X_i = [Y_i, Z_i, S_i]$) to decide on whether it is better to pay attention to the conditional parameters (conditional on the stratum: θ_{1j} and θ_{2j}) or the unconditional or marginal parameters (θ_1 and θ_2). In Table 12.1 we give an example of Simpson's paradox, where we let $\beta_{ra} = \theta_{a2}/\theta_{a1}$ for $a = 1, 2$. We then give two study design stories to go with the fake data: Example 12.3 would

---

[2] We get a similar exclusion of the paradox when the paradox is defined by estimates (e.g., $\hat{\lambda}_1 = \hat{\lambda}_2$ are the observed proportions with $z_i = 1$).

$S = 1$

	$Y = 0$	$Y = 1$	
$Z = 1$	560	140	700
$Z = 2$	270	30	300
	830	170	1000

$\hat{\theta}_{11} = 20\%$ $\hat{\beta}_{r1} = 0.50$
$\hat{\theta}_{21} = 10\%$

$S = 2$

	$Y = 0$	$Y = 1$	
$Z = 1$	20	80	100
$Z = 2$	540	360	900
	560	440	1000

$\hat{\theta}_{12} = 80\%$ $\hat{\beta}_{r2} = 0.50$
$\hat{\theta}_{22} = 40\%$

Total

	$Y = 0$	$Y = 1$	
$Z = 1$	580	220	800
$Z = 2$	810	390	1200
	1390	610	2000

$\hat{\theta}_1 = 27.5\%$ $\hat{\beta}_r = 1.182$
$\hat{\theta}_2 = 32.5\%$

Table 12.1 *Fake data demonstrating Simpson's paradox*

make causal inferences based on the stratum-specific comparisons, and Example 12.4 would make causal inferences based on the marginal parameters.

Example 12.3   *Consider a retrospective study of 2,000 people infected with a very serious disease. Consider two treatments, a standard treatment ($Z = 1$) and an experimental treatment ($Z = 2$). Suppose that generally people either die ($Y = 1$) or get better ($Y = 0$) by 30 days after presenting with symptoms. We select 1,000 people who present to the hospital with mild symptoms ($S = 1$) and 1,000 people who present to the hospital with severe symptoms ($S = 2$). The experimental treatment tended to be used more on people that presented with severe symptoms. The scientific question is: which treatment should people get in the future? Suppose we can assume that the only information used in assigning treatment was severity at presentation (S) so that within each stratum, we can assume treatment assignment was random. Under these assumptions, we look at the stratum-specific rate ratios, $\hat{\beta}_{r1} = 0.50$ and $\hat{\beta}_{r2} = 0.50$, which show that the experimental treatment has failure rates half the size of the failure rates of the standard treatment. We calculate 95% confidence intervals using the melding method (*binomMeld.test in the exact2x2 R package), and they appear significantly different from 1 (the 95% confidence interval for β_{r1} is $(0.33, 0.73)$ and the one for β_{r2} is $(0.44, 0.58)$). Thus, the experimental treatment is recommended in the future, regardless of symptom severity at presentation. The stratification by symptom severity is needed for proper inferences.*

Example 12.4   *Consider a second story to go with the data in Table 12.1. The story is similar to Example 12.3, we are still studying a severe disease, with the same response (Y is equal to death [$Y = 1$] or survival [$Y = 0$] at 30 days since presenting), and the same covariate ($Z = 1$ is standard treatment, and $Z = 2$ is an experimental treatment). But in this second scenario, suppose that all 2,000 people present to the hospital with mild symptoms, and are randomized to either standard or experimental treatment. Here let S be the symptoms 24 hours after treatment initiation, with $S = 1$ being mild and $S = 2$ being severe.*

It appears that $900/1,200 = 75\%$ *of the individuals that got the experimental treatment had severe symptoms by 24 hours, but only* $100/800 = 12.5\%$ *of individuals that got the standard treatment had severe symptoms by 24 hours. Thus, it appears that the experimental treatment is worsening symptoms more quickly. To decide on the treatment for the future, we should look at the 30-day failure rates (27.5% for the standard treatment, but 32.5% for the new treatment). The marginal ratio of failure rates is greater than* 1, $\beta_r = 1.18$ *(95% CI by melding:* 1.03, 1.37*), and the standard treatment is recommended in the future to those presenting with mild symptoms. In this scenario, stratifying by severity of symptoms would give you the wrong answer. In general, stratifying by post-randomization variables in a randomized setting should be strictly avoided, except for specialized causal analyses (see Sections 15.4.3 and 15.4.4).*

This section has illustrated the extreme case where doing a stratified analysis will provide a completely different answer from an analysis that ignores the stratification. Furthermore, it illustrates that the decision to adjust or not is context specific, that even if the effect of stratification is dramatic, it may be wrong to do a stratified analysis. This concept is addressed more formally and precisely in Chapter 15.

When the context calls for a stratified analysis (e.g., Example 12.3), we would do this even if the setting is much less extreme than that seen in Simpson's paradox. A stratified analysis can mean either that we estimate a separate effect for each stratum, or that we get an overall effect after adjusting for the strata in some way. As an example of the latter, suppose the direction of the effect within strata is the same as the direction of the overall sample, but the magnitudes differ. In this case, the stratified analysis of an overall effect may not change the direction of the point estimate of the treatment effect but may change its magnitude. This is the more common scenario.

For randomized studies, often there is baseline stratification factor that is hypothesized to potentially have an effect on the treatment. One option is to ignore the stratification factor at randomization and perform a stratified analysis to deal with potential imbalance (especially in a small study) and to gain efficiency. Another option is to do a stratified randomization, where we randomize within each stratum, and in that case although imbalances within strata will be less likely, we often also perform a stratified analysis. We next discuss general methods for conducting a stratified analysis.

12.3.3 Binary Responses

The same modeling methods that are discussed in Section 12.2.2 may also be used with stratified data (quasi-likelihood methods, generalized estimating equations, and mixed effects models). The difference with stratified data is that since within each stratum we have both levels of z, we can estimate a fixed effect for each stratum. Consider a logistic regression model for the stratified binary response data and we are interested in a common (fixed) treatment effect. As previously, for the ith individual the response is y_i, the treatment is z_i, and stratum is s_i. Let $\pi(z,s) = \Pr[Y_i = 1 | z_i = z, s_i = s]$. The logistic model with main effects for strata and treatment is

$$\log\left(\frac{\pi(z_i, s_i)}{1 - \pi(z_i, s_i)}\right) = \gamma_j I(s_i = j) + \beta I(z_i = 2),$$

where $z_i \in \{1, 2\}$ and $s_i \in \{1, \ldots, k\}$. Then the odds ratio (odds of Treatment 2 over the odds of Treatment 1) is $\psi = \exp(\beta)$.

So if the sample sizes are large enough within each stratum, then the maximum likelihood estimate (MLE) of β can be nearly unbiased, but for small sample sizes within a stratum there is bias (see Section 14.5.2). Alternatively, we can condition on the total number of $Y_i = 1$ within each stratum, then we do not need to estimate the γ_j and can get a conditional estimator of β. These conditional inferences can be asymptotic or exact (see Section 14.10 for references), and the exact version is a way to extend Fisher's exact test to account for stratification. See Chapter 14 for more details on general modeling methods, where stratification is a special case.

One simple method for stratification with binary responses is the Mantel–Haenszel method, which is useful because it is consistent if either the sample size within each stratum goes to infinity or the sample size per stratum does not change but the number of strata goes to infinity.

Assume that the stratum-specific odds ratios are all equal (i.e., $\psi_1 = \psi_2 = \cdots = \psi_k \equiv \psi$), but we may have $\theta_{hj} \neq \theta_{hj'}$ for some h and $j \neq j'$. Then the *Mantel–Haenszel estimator* of ψ is

$$\tilde{\psi} = \frac{\sum_{j=1}^{k} w_j \hat{\theta}_{2j} \left(1 - \hat{\theta}_{1j}\right)}{\sum_{j=1}^{k} w_j \hat{\theta}_{1j} \left(1 - \hat{\theta}_{2j}\right)},$$

where $w_j = \left(1/m_{1j} + 1/m_{2j}\right)^{-1}$ and m_{hj} is the number of individuals with $z_i = h$ and $s_i = j$. Robins et al. (1986) proposed a variance estimator for $\tilde{\psi}$ that is consistent when either the number of strata k is fixed and the sample sizes per stratum go to infinity, or the sample sizes per stratum are bounded but the number of strata goes to infinity (see Note N2). Then using the asymptotic normality of $\tilde{\psi}$ (under either asymptotic model) and the delta method, Robins et al. (1986) proposed a $100(1 - \alpha)\%$ CI for $\log(\psi)$,

$$\log\left(\tilde{\psi}\right) \pm \Phi^{-1}(1 - \alpha/2)\sqrt{\tilde{V}},$$

where $\tilde{V}$ is an estimator of $Var\left\{\log\left(\tilde{\psi}\right)\right\}$ (see Note N2).

12.3.4 Ordinal and Numeric Responses

For ordinal and numeric responses, we only give a brief discussion, because stratification is often handled within a more complex model (see Chapter 14).

First, consider ordinal and numeric responses. For generalizing the Wilcoxon–Mann–Whitney test to multiple strata, it is conceptually straightforward to get a p-value; a permutation approach is used, where we permute treatment labels only within each stratum. The details follow. First, we rank all responses within each stratum, define a test statistic that is a function of those ranks, permute the z_i values within each stratum and recalculate the test statistic, and finally, see how extreme the observed value of the test statistic is within the distribution of the test statistics calculated for each permutation (see Section 10.9 for details on a general permutation test). When the sample size for the jth stratum is m_j and we want an efficient test, we can define the test statistic used in the permutation procedure as $T(\mathbf{x}) = \sum_{j=1}^{k} \frac{W_j}{m_j+1}$, where W_j is the sum of the ranks with $z_i = 1$ within stratum j (Lehmann, 1975, pp. 135, 145). This procedure is known as the *van Elteren test*. See Kawaguchi and Koch (2015) for an accompanying effect estimate.

An alternative is to use a proportional odds (i.e., ordinal logistic) model which works well if there are not too many response categories. If the sample sizes are large within each stratum, then we can include a main effect for each stratum, otherwise we can model those main strata effects as random effects (we can use the ordinal R package as discussed in Section 12.2.4).

For normally distributed responses, a common method is to incorporate the strata effects as main effects in a linear model. As with other types of responses we can use a linear (fixed effects) model if there are large numbers of observations per stratum or use a random effects model if there are a large number of strata but small numbers in each stratum. We can also use generalized estimating equations. All these models are discussed in Chapter 14.

For Poisson distributed responses, direct standardization can be used (see Section 12.3.6), or the problem may be modeled using either generalized linear models with a main effect for strata, quasi-Poisson models, mixed effects models, or generalized estimating equation models (see Chapter 14).

12.3.5 Meta-Analysis

Now consider the case where we summarize each stratum by a treatment effect and its variance and perform all analyses using only that information. This is common in meta-analyses, where we have limited information about many studies (e.g., we do not have all the individual responses within each study, but just have summaries of treatment effects). In the meta-analysis case, each study is a stratum. We could also apply these methods to a stratified randomized study.

Suppose within the jth stratum we estimate the treatment effect with $\hat{\beta}_j$ and estimate the variance of $\hat{\beta}_j$ with $v_j = \widehat{var}(\hat{\beta}_j)$. Suppose that $\hat{\beta}_j$ is approximately normally distributed with mean β_j and variance v_j (in other words, we treat v_j as if it were the true variance of $\hat{\beta}_j$).

Example 12.5 *Suppose within each stratum we estimate the difference in means of the responses in the treatment group (individuals with $z_i = 2$) and control group ($z_i = 1$). Then $\hat{\beta}_j = \bar{y}_{2j} - \bar{y}_{1j}$, where $\bar{y}_{hj}$ is the mean of the responses with $z_i = h$ and $s_i = j$. We estimate the variance with the sum of the square of the standard error of those means, $v_j = s_{2j}^2/m_{2j} + s_{1j}^2/m_{1j}$, where s_{hj}^2 and m_{hj} are respectively, the sample variance of the responses and the number of responses, with $z_i = h$ and $s_i = j$.*

Example 12.6 *Suppose within each stratum we estimate the log odds ratio of mortality comparing the treatment group to the control group. Let x_{hj} be the number of events (e.g., deaths) among the m_{hj} individuals with $z_i = h$ and $s_i = j$. Then the mortality rate for treatment Group h in stratum j is $r_{hj} = x_{hj}/m_{hj}$, and the sample log odds ratio is*

$$\hat{\beta}_j = \log\left(\frac{r_{2j}(1 - r_{1j})}{r_{1j}(1 - r_{2j})}\right)$$

and using the delta method (Section 10.6) we estimate the variance of $\hat{\beta}_j$ with

$$v_j = \frac{1}{m_{2j}r_{2j}(1 - r_{2j})} + \frac{1}{m_{1j}r_{1j}(1 - r_{1j})}.$$

We consider two models of the overall treatment effect. In the fixed effects model, all the strata have the same treatment effect, $\beta_1 = \cdots = \beta_k \equiv \beta$, and we estimate β with the weighted average of the stratum-specific effect, where the weights are proportional to the inverse of the variances,

$$\hat{\beta}(0) = \frac{\sum \frac{\hat{\beta}_j}{v_j}}{\sum \frac{1}{v_j}},$$

and the summations are from 1 to k. A more general model is the random effects model where the treatment effects vary from stratum to stratum. We assume normality so that $\beta_j \sim N(\beta, \tau^2)$. Under the random effects model $\hat{\beta}_j \sim N(\beta, v_j + \tau^2)$ and assuming τ^2 is known, the efficient weighted average (where the weights are proportional to the inverse variances) is

$$\hat{\beta}(\tau^2) = \frac{\sum \frac{\hat{\beta}_j}{v_j + \tau^2}}{\sum \frac{1}{v_j + \tau^2}}.$$

Then we can show that (Problem P5)

$$\hat{\beta}(\tau^2) \sim N\left(\beta, \frac{1}{\sum \frac{1}{v_j + \tau^2}}\right). \tag{12.4}$$

In practice we estimate τ^2 and of course the v_j values are estimates. For now, we ignore the latter problem which will be small if the sample sizes within all strata are large enough. DerSimonian and Laird (1986) developed a simple method of moments estimator for τ^2 and DerSimonian and Kacker (2007) discuss a slightly better estimator first proposed by Paule and Mandel (1982) (Note N3). Although the estimate of τ^2 is often treated as known and the normal distribution is used for inferences, better confidence interval coverage is achieved by plugging in the estimator of τ^2 but using the t-distribution to account for the fact τ^2 was estimated from only k strata. An approximate $100(1 - \alpha)\%$ confidence interval for β under the random effects model is

$$\hat{\beta}(\hat{\tau}^2) \pm se(\hat{\beta}(\hat{\tau}^2)) F_{t,df}^{-1}(1 - \alpha/2), \tag{12.5}$$

where $\hat{\tau}^2$ is the estimate of τ^2, $se(\hat{\beta}(\hat{\tau}^2)) = 1/\sqrt{\sum(v_j + \hat{\tau}^2)^{-1}}$, and $F_{t,df}^{-1}(a)$ is the ath quantile of a t-distribution with df degrees of freedom. The fixed effects confidence interval uses $\hat{\tau}^2 = 0$ and $df = \infty$ (to give a standard normal quantile). In Figure 12.2, we show that we get much better coverage if we use a t-distribution with $k - 1$ degrees of freedom, rather than relying on a normal distribution (see Brittain et al., 2012, Supplement, Section 3, for motivation on using $k - 1$ degrees of freedom). These are all approximate methods even if the assumption of v_j being known is true, but, more recently, a valid (i.e., exact) formulation under that assumption was developed by Michael et al. (2019).

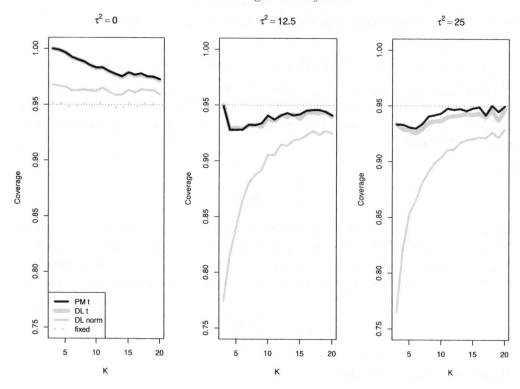

Figure 12.2 Simulated coverage for different values of τ^2, where the v_j range from 1 to 25. "DL norm" is the DerSimonian–Laird estimator of τ^2 and using normal-based confidence intervals; "DL t" is the DerSimonian–Laird estimator of τ^2 and using t-based confidence intervals with $df = k - 1$, where k is the number of strata; "PM t" is the Paule–Mandel estimator of τ^2 using t-based confidence intervals with $df = k - 1$; and "fixed" is the fixed effects normal confidence intervals. In the panels with $\tau^2 > 0$ the fixed effects confidence intervals have coverage less than 75% and are not shown. For details of the simulation see Note N4.

Because the fixed effects model is a special case of the random effects model (when $\tau^2 = 0$), the random effects model requires fewer assumptions. However, if there are very few strata (e.g., $k = 2, 3,$ or 4), the price we pay for the more robust random effects model can be very large. To see the extent of the price, consider the width of the confidence intervals. In the confidence interval of Equation 12.5, we ignore the changes due to changing $\tau^2 = 0$ in the fixed effects model to using $\tau^2 = \hat{\tau}^2$ in the random effects model (i.e., we ignore the change from using $\hat{\beta}(0)$ and $se(\hat{\beta}(0))$ to using $\hat{\beta}(\hat{\tau}^2)$ and $se(\hat{\beta}(\hat{\tau}^2))$), and only compare the changes in the standard quantile factor going from a normal to a t-distribution. For the 95% confidence interval, using the t-distribution rather the normal increases the confidence interval width by a factor of 6.5 ($k = 2$), 2.2 ($k = 3$), 1.6 ($k = 4$), or 1.2 ($k = 10$). So the price of assuming a more robust model can be quite severe with a very small number of strata. However, as shown in our simulation results, if the effects across strata are truly highly variable, then the fixed effects model would produce confidence intervals with poor coverage, and thus is unacceptable.

12.3.6 Direct Standardization

Often with observational studies, we stratify based on a factor, s_i, that is associated with both the response, y_i, and the effect of interest, z_i. A classic example is comparing cancer rates over time, where the strata are age groups. In this setting, we may want to standardize our rates according to some external stratum weights, such as the age distribution in a particular year, as in this example. Many types of cancer become much more prevalent as people get older, and additionally a population may be aging over time because of medical or other societal advances that prolong life. In this example, $z_i = 1$ or 2 represent two time periods, $s_i = j$ represents the jth age group ($j = 1, \ldots, k$), and $y_i = 0$ or 1 indicates whether the ith individual had a cancer incidence in the time period given by z_i when they were in age group s_i. Let the data when $z_i = h$ be called time period h, for $h = 1, 2$. Let $\theta_{hj} = \Pr[Y_i = 1 \mid z_i = h, s_i = j]$. Let x_{hj} be the number of incident (i.e., new) cases and n_{hj} the number at risk among individuals with $z_i = h$ and $s_i = j$. In other words, $x_{hj} = \sum I(y_i = 1)$ and $n_{hj} = \sum 1$, where the summations are over all individuals with $z_i = h$ and $s_i = j$.[3] We assume $X_{hj} \sim Poisson(n_{hj}\theta_{hj})$, so the expected crude rate in the year coded as $z_i = h$ is

$$\mu_h^* = \frac{\sum_{j=1}^{k} n_{hj}\theta_{hj}}{\sum_{j=1}^{k} n_{hj}}.$$

Differences between μ_1^* and μ_2^*, could be due to either the differences in the age specific rates, θ_{hj}, or the differences in the populations, n_{hj}, or both. A fairer comparison is to use directly standardized rates that weigh the rates using the same age-specific factors for each time period,

$$\mu_h = \frac{\sum_{j=1}^{k} N_j\theta_{hj}}{\sum_{\ell=1}^{k} N_\ell},$$

where N_j is the population size in age stratum, j, from some standard population (e.g., US population at the year 2000). Then the directly standardized rate, μ_h, reduces to the expected crude rate (i.e., total incidence divided by total population) for time period h, if the age specific rates (the θ_{hj}) come from time period h and additionally the population has the same age distribution as the US population in 2000. This kind of direct standardization is similar to the way dollars are inflation adjusted so that we can compare prices during different time periods.

Under the Poisson assumption on the X_{hj}, we estimate μ_h with

$$\hat{\mu}_h = \frac{\sum_{j=1}^{k} N_j\hat{\theta}_{hj}}{\sum_{\ell=1}^{k} N_\ell} = \sum_{j=1}^{k} w_{hj}X_{hj},$$

where we use $\hat{\theta}_{hj} = X_{hj}/n_{hj}$ so that $w_{hj} = N_j/(n_{hj}\sum N_\ell)$. Let $v_h \equiv \mathrm{Var}(\hat{\mu}_h) = \sum_j w_{hj}^2 n_{hj}\theta_{hj}$, and let $\hat{v}_h$ be v_h with $\hat{\theta}_{hj}$ replacing each θ_{hj} value.

---

[3] For cancer surveillance, the x_{hj} and n_{hj} are actually slightly more complicated because over any year some in the population are born into (or move into) and die out of (or move out of) the catchment area that defines the population. Thus, n_{hj} represents an estimate of the average population over the time period. But the Poisson assumption is still a good approximation for this situation. For a discussion of that assumption in demographic data see Brillinger (1986).

If every X_{hj} is large, then we could use a normal approximation assuming that $\hat{\mu}_h$ is approximately normal with mean μ_h and variance approximately $\hat{v}_h$. A method that is more likely to provide at least nominal coverage, especially for small X_{hj} uses the gamma distribution (Fay and Feuer, 1997). We call this the *gamma confidence interval,* and although originally derived without reference to confidence distribution random variables (CD-RVs), we motivate them using CD-RVs (see Section 10.11). Specifically, the $100(1 - \alpha)\%$ gamma confidence interval is

$$\{q(\alpha/2, G_L), q(1 - \alpha/2, G_U)\},$$

where $q(a, G)$ is the ath quantile of a random variable G, and G_L and G_U are the lower and upper CD-RVs, with G_L equal to a gamma random variable with mean $\hat{\mu}_h$ and variance $\hat{v}_h$, and G_U equal to a gamma random variable with mean $\hat{\mu}_h + w_{h,max}$ and variance $\hat{v}_h + w_{h,max}^2$ and $w_{h,max} = \max_{j \in \{1,...,k\}} w_{hj}$. The addition of the $w_{h,max}$ factor is a conservative adjustment because of the discreteness of the X_{hj}, because it is the maximum possible change in $\hat{\mu}_h$ from adding 1 to only one of the X_{hj} values. Other smaller values for the added factor have been tried, but they can lead to under coverage of confidence intervals in certain situations (Ng et al., 2008). When the population in time period h is proportional at each age group to the standard population, then $w_{h1} = \cdots = w_{hk} = w$, the the gamma confidence interval reduces to the exact central one. In other cases, it tends to be conservative, while with small θ_{hj} and large w_j other approximations can have coverage substantially less than the nominal (e.g., 95% intervals having less than 80% coverage, see Fay and Feuer (1997)). For intervals closer to nominal coverage on average, we can use a mid-p version of the gamma intervals (Fay and Kim, 2017).

If we are interested in the ratio μ_2/μ_1, then an approximate $100(1 - \alpha)\%$ conservative confidence interval uses the ratio of two gamma confidence distribution random variables,

$$\{q(\alpha/2, G_{2L}/G_{1U}), q(1 - \alpha/2, G_{2U}/G_{1L})\}, \tag{12.6}$$

where G_{hL} and G_{hU} are the lower and upper confidence distributions for μ_h, $h = 1, 2$. Fay (1999a) proposed a confidence interval similar to this except $w_{h,max}$ associated with G_{hU} is replaced with

$$w_{h,max}^* = \max_{j:x_{3-h,j}>0} w_{hj}$$

(e.g., $w_{1,max}$ is the largest w_{1j} only among the j with $x_{2j} > 0$). We call this the *F-interval* because of the ratio of two independent gamma distributions is distributed as an F-distribution (Problem P6). If both populations are proportional to the standard population, then the F-interval is equivalent to an exact central interval after conditioning on the total number of incident cases within each stratum.

12.4 Summary

This chapter deals with two-sample comparisons of treatment (or some other exposure of interest) where the observations are not independent. In this chapter, we use the term clustering to denote the case where everyone in the same cluster (e.g., village, classroom, etc.) has the same treatment. We use the term stratification to denote the case where we are interested in the result within stratum (e.g., village, classroom, etc.), and where there is a mix of treatments within each stratum.

There are two general approaches for analyzing clustered data. The first approach simply takes a cluster-level summary measure, such as a mean, as the unit of analysis; then the use of an appropriate two-sample test procedure can test the treatment effect. The second approach compares the treatments by analyzing the individual responses, while taking the within-cluster correlation into account. While we sometimes study a third approach (analyses that completely ignore the cluster) for illustration, these should be strictly avoided since the Type I error rate can be wildly inflated.

We compare the performance of the different strategies under the null hypothesis in two simulations, both having binary outcomes. In the first simulation, the sample size per cluster was unrelated to the outcome. In that case, a Welch's t-test which analyzes the proportions in each cluster essentially matches the nominal Type I error rate. A quasi-binomial model, an individual level analysis which accounts for the clustering, has slight inflation of the Type I error rate; the same is true for the mixed model (with a logit link); however, this method had a high nonconvergence rate. Importantly, in a second simulation when the sample size per cluster is correlated with the outcome, the quasi-binomial method's Type I error rate is unacceptably high, whereas the Type I error rate of Welch's t-test and the mixed model was essentially unaffected. We briefly discuss considerations for the analysis choice in the binary case; we expect there will be little difference if sample sizes are essentially the same across the clusters.

We then compare performance for methods when the outcome is numeric. In this scenario, the linear mixed model (i.e., an individual-level analysis which accounts for the within-cluster correlation) appears to clearly beat the cluster-level analysis, even when the simulation was run with just five clusters per treatment. Furthermore, generalized estimating equation models should not be used in the setting where the outcome is associated with cluster sample size (e.g., a litter is a cluster and the outcome is body weight). In contrast, when the outcome is ordinal, mixed models were associated with less than nominal coverage when the variance between clusters is large; we discuss permutation tests as a possible remedy, if exact results are needed.

We begin the discussion of stratification by introducing Simpson's paradox where, puzzlingly, the treatment has a positive effect within each stratum, and yet a negative effect overall (or vice versa). However, we illustrate that even in this extreme case we may not want to do a stratified analysis, and that this determination depends on extra-statistical considerations. That said, we may want to adjust for strata when the effect is much less extreme than the Simpson's paradox setting, but, again, the decision to do so depends on considerations beyond statistical evaluation. We also may do a stratified analysis to get more precision in a randomized study, or to reflect the randomization if the randomization is performed within each stratum. When the outcome is binary, we can do a stratified logistic analysis, or the simpler Mantel–Haenszel test. When the outcome is numeric, a linear model with main strata effects can be employed. When the outcome is ordinal, a van Elteren test (essentially a stratified Wilcoxon rank sum test) or a stratified proportional odds model can be used.

We address meta-analysis as a special case of stratification where we only have summary data for each stratum; in this context, each study is a stratum. We discuss the distinction between fixed and random effects models. With fixed effects, the assumption is that the true treatment effect is the same across studies; whereas, under random effects, we allow the treatment effects to vary across studies. We show that the coverage for a t-distribution-based random effect confidence interval is much closer to 95% than that of a normal-based confidence interval; however, neither approach provides guaranteed coverage. We note a more recently proposed procedure has guaranteed coverage for random effects models. A random effect model can have a much wider confidence interval than the corresponding fixed effect model for the same data set, especially if the number of studies (i.e., strata) is small. However, if the effects truly vary across studies, then the fixed effect model could have substantial inflation in its Type I error rate, and thus would be unacceptable. Unfortunately, when there are few studies, there is little basis for assessing the variability of study effects.

The final topic considers a special case where it is desirable to compute estimates using stratum weights from an external population. The case we discuss is estimation and comparison of cancer rates in two different years. To provide direct comparability, we could standardize both rates to match the age distribution in the year 2000, for example. Confidence interval methods based on the gamma distribution provide better coverage than other methods.

12.5 Extensions and Bibliographic Notes

Hayes and Moulton (2009) is a very practical book on cluster randomized trials, and their perspective shaped Section 12.2. References on modeling are deferred until Chapter 14. Chapter 6 of Pearl (2009b) gives a very thorough discussion of Simpson's paradox from a causal perspective, including historical controversies. For methods for directly standardized rates, see Fay and Kim (2017) and the references therein.

12.6 R Functions and Packages

To perform the meta-analyses of Section 12.3.5 see the metaNorm function in the asht package. Michael et al. (2019) developed the rma.exact package for their exact meta-analysis method. The wspoissonTest function in the asht R package does the gamma confidence interval, its mid-p version, and several others that have been proposed. The ordinal R package does proportional odds models allowing for random effects.

12.7 Notes

N1 **Estimating means with random effects:** Suppose the data for the ith observation is $x_i = [y_i, z_i, s_i]$, where y_i is the response, z_i is treatment and equals 0 or 1, and $s_i = j$ when the ith observation is in cluster j. Consider the model,

$$Y_i = (1 - z_i)\mu_0 + z_i\mu_1 + I(s_i = j)\delta_j + \epsilon_i,$$

where $\delta_j \sim N(0, \sigma_b^2)$ and $\epsilon_i \sim N(0, \sigma_w^2)$ are the independent errors due to cluster and individual respectively, and we are interested in the mean values μ_0 (of individuals with $z_i = 0$) and μ_1 (of individuals with $z_i = 1$). So σ_w^2 represents the variability within clusters and σ_b^2 represents the variability between clusters. Suppose there are m_j observations with $s_i = j$. Let $\bar{Y}_j = \frac{1}{m_j}\sum_{i:s_i=j} Y_i$ be the mean response for cluster j. Suppose cluster j has only individuals with $z_i = 0$ then

$$\bar{Y}_j \sim N\left(\mu_0, \sigma_b^2 + \frac{\sigma_w^2}{m_j}\right),$$

and the most efficient estimate of μ_0 is a weighted average of the $\bar{Y}_j$ with $z_i = 0$ with weights proportional to the inverse of the variance, $1/(\sigma_b^2 + \sigma_w^2/m_j)$ (e.g., w_j in Equation 12.1, because multiplying every weight by a constant like $1/\sigma_b^2$ will not change the weighted average). Similarly, for μ_1.

N2 **Variance estimator of the Mantel–Haenszel estimator:** Here is $\tilde{V}$, the estimator of $\text{Var}(\log(\tilde{\psi}))$ of Robins et al. (1986) (in their notation it was V_{US}). First, let $R_j = w_j \hat{\theta}_{2j}(1 - \hat{\theta}_{1j})$ and $T_j = w_j \hat{\theta}_{1j}(1 - \hat{\theta}_{2j})$. Then $\tilde{\psi} = \sum R_j / \sum T_j$, where the summations (and subsequent ones) are over $j = 1$ to k. Let $R = \sum R_j$ and, $T = \sum T_j$, $P_j = w_j(m_{2j}\hat{\theta}_{2j} + m_{1j} - m_{1j}\hat{\theta}_{1j})$, and $Q_j = w_j(m_{1j}\hat{\theta}_{1j} + m_{2j} - m_{2j}\hat{\theta}_{2j})$.

$$\tilde{V} = \frac{\sum P_j R_j}{2R^2} + \frac{\sum (P_j T_j + Q_j R_j)}{2RT} + \frac{\sum Q_j T_j}{2T^2}.$$

The associated estimator of $\text{Var}(\tilde{\psi})$ is $\tilde{V}\tilde{\psi}^2$, which is consistent under either asymptotic model.

N3 **DerSimonian and Laird estimator:** The DerSimonian and Laird estimator of τ^2 is

$$\hat{\tau}_{DL}^2 = \max\left(0, \frac{\left[\sum w_j(0)(\hat{\beta}_j - \hat{\beta}(0))^2\right] - (k - 1)}{\sum w_j(0) - \frac{\sum [w_j(0)]^2}{\sum w_j(0)}}\right),$$

where summations are from $j = 1$ to k, and $w_j(\tau^2) = 1/(v_j + \tau^2)$ and other notation is defined in Section 12.3.5. The Paule–Mandel estimator is the value of τ^2 that solves,

$$\left[\sum w_j(\tau^2)(\hat{\beta}_j - \hat{\beta}(\tau^2))^2\right] - (k - 1) = 0,$$

which may be solved iteratively.

N4 **Figure 12.2 simulation details:** We modeled our simulation exactly after the first set of simulations described in Section 3 of Michael et al. (2019), so the $\sqrt{v_j}$ are even spaced between 1 and 5 (inclusive) for each value of $K = 3, \ldots, 20$ (so the v_j are unevenly [quadratically] spaced between 1 and 25). We used 10000 replications.

12.8 Problems

P1 We did a simulation with 10^4 replications, each simulating 100 beta-binomial observations with $m_j - 1 \sim Poisson(10)$ and beta parameters $(1, 3)$ and exactly half of the beta-binomial variates had $z = 0$ and the other half had $z = 1$. Simulate the rejection probability at significance level $\alpha = 0.05$ testing for an effect of z using either the standard logistic model (e.g., a test that the main effect for z is 0), the quasi-binomial model, and Welch's t-test on the $\hat{\theta}_j$. Plot the beta-probability density function for those parameters. Try your own simulations with beta parameters $(0.1, 0.3)$ and $(10, 30)$. What are the simulated Type I error rates? How are they related to the beta distribution?

P2 Repeat the previous simulation, except make $m_j - 1 \sim Poisson(10^{1+\theta_j})$, where $\theta_j \sim Beta(1, 3)$ are the beta values used in creating the beta-binomial responses. Simulate Type I error rates as in the previous problem.

P3 Show that a simple proportional odds model can be simulated by using logistic errors and grouping the results into ordered categories (for example, by rounding). Hint: write the proportional odds model as $F(x)/(1 - F(x)) = \theta G(x)/(1 - G(x))$, then show that for this model there exists some strictly increasing transformation $b(x)$, such that $b(X) \sim F^*$, where F^* is a standard logistic and $G(x) = F^*(b(x) - \beta)$, where $\theta = \exp(\beta)$ (see Fay and Malinovsky, 2018, Supplement).

P4 For the death penalty example discussed in Section 12.3.2, is the more relevant analysis the one that uses the conditional effects, conditioned by stratum (θ_{1j} and θ_{2j}) or the marginal effects (θ_1 and θ_2)? What is your reasoning?

P5 Show Expression 12.4 assuming the $\hat{\beta}_j$ are independently normal with mean β and variance $\tau^2 + v_j$, for $j = 1, \ldots, \ell$.

P6 Show that the $100(1 - \alpha)\%$ confidence interval of Equation 12.6 may be written as

$$\left\{ q\left(\alpha/2, \frac{\hat{\mu}_2}{\hat{\mu}_1 + w_{1,max}} F_L\right), q\left(1 - \alpha/2, \frac{\hat{\mu}_2 + w_{2,max}}{\hat{\mu}_1} F_U\right) \right\},$$

where F_L is an F random variable with degrees of freedom (d_2, d_1^*) and F_U is an F random variable with degrees of freedom (d_2^*, d_1), where $d_h = 2\hat{\mu}_h^2/\hat{v}_h$ and $d_h^* = 2(\hat{\mu}_h + w_{h,max})^2/(\hat{v}_h + w_{h,max}^2)$ (Fay, 1999a). Bonus: what is the form of the associated p-value for the central two-sided test with null hypothesis $H_0 : \mu_2/\mu_1 = 1$? Answer: see Fay et al. (2006, Appendix).

13

Multiplicity in Testing

13.1 Overview

This chapter deals with interpretation of results when many hypothesis tests from a single study are performed. The fundamental problem is that if we test very many hypotheses, the probability of making at least one Type I error can be very high. So emphasizing a single rejection of a null hypothesis can be misleading if the results are not put in the proper context of how many tests where done. Thus, it is important to report how many hypotheses were performed, or even better, to adjust the p-values so that they have an interpretation that makes sense in light of the number of hypotheses that were tested. These multiple comparison adjustments are the topic of this chapter.

While multiplicity comes in many forms, the classic example is testing m independent hypotheses, each tested at level α. The probability of rejecting at least one true null hypothesis (the familywise error rate, FWER) when all null hypotheses are true is $1 - (1 - \alpha)^m$. Even if there is considerable correlation among the tests, the FWER may still be high. Suppose instead of testing m new treatments each with its own control – which would be m independent tests, we test m new treatments against the same control. Consider the simple case where each comparison uses a difference in means, and the responses are all normal with the same known variance. Then each test statistic comparing one of the m treatments to control is distributed standard normal, and the correlation between any two of those test statistics is $\rho = 0.5$ (Problem P1). Here are some values of the FWER assuming all null hypotheses are true under those two scenarios:

	$m = 1$	$m = 2$	$m = 5$	$m = 10$	$m = 20$	$m = 100$
FWER (independent)	0.0500	0.0975	0.2262	0.4013	0.6415	0.9941
FWER ($\rho = 0.5$)	0.0500	0.0907	0.1829	0.2869	0.4188	0.7777

The first step in multiple hypothesis testing is defining the family of hypotheses. Section 13.2 addresses when to group hypothesis tests together into a family. Often the family of hypothesis tests is bigger than it appears. For example, we may graphically examine the data to pick hypotheses (see Example 13.4). If we only formally perform tests on a subset of covariates after looking at graphs of all covariates, that does not mean that the family of hypotheses is restricted to the subset of covariates on which formal tests were done. In that case we have used the data to select the hypotheses, and that procedure could lead to very large FWERs. The rest of the chapter assumes that the family of m hypotheses has been picked.

Controlling the FWER is sometimes too stringent. For example, in genetic studies where we are testing thousands of hypotheses, then controlling the FWER using a traditional significance level of 0.05 may make it virtually impossible to find any significant effects. An alternative property is the false discovery rate (FDR), the rate of rejections that are false out of the total number of rejections. Section 13.3 formally defines both the FWER and the FDR, as well as the FWER adjusted p-value and the FDR adjusted p-value.

Section 13.4 gives some simple procedures for controlling the FWER or FDR using only the m p-values associated with the m hypotheses in the family. Section 13.5 describes the maxT multiple comparison procedure based on multivariate t-distributions that gives a widely applicable method that accounts for the correlation among the m tests. The method may be applied whenever the test statistics for the m hypotheses are multivariate normal or multivariate t. Further, the method returns adjusted p-values as well as simultaneous confidence intervals related to the m parameters associated with the m hypotheses. A $100(1 - \alpha)\%$ set of valid *simultaneous confidence intervals* are m confidence intervals such that the probability that any one of the m intervals does not cover the true parameter value is bounded by α. For maxT procedures, the adjusted p-values are compatible with the simultaneous confidence intervals. Section 13.6 describes another approach for accounting for the correlation among the m tests, using resampling-based methods such as permuting or bootstrapping. Section 13.7 shows a very useful graphical method for depicting sequential multiple testing procedures. Section 13.8 briefly describes how logical constraints between hypotheses can be used to increase the power of multiple testing procedures. Finally, Section 13.9 gives a very brief introduction to the theoretical idea of closure that motivates some of the methods.

13.2 Defining the Family of Hypothesis Tests

It is hard to give simple rules on how to define a family of hypothesis tests, because the definition of the family is integrally related to the context of the scientific problem and the context of the communication about it. However, to illuminate the issues we give some examples in which the family definition is more straightforward.

The first two examples define families from groups of hypotheses that at the design stage are equally scientifically important and significance in any one of them would change the direction of future scientific research on the topic.

Example 13.1 *Consider a study comparing subjects with a chronic parasitic infection to subjects without such an infection. Suppose that on each subject a blood sample is evaluated, and m $=$ 20 cytokines (proteins that signal cells) are measured. Suppose there is little knowledge about how any of the 20 cytokine levels are related to this parasitic infection. If we test for differences between the 2 types of subjects on each of the 20 cytokines, then it is reasonable to define those 20 tests as a family.*

Example 13.2 *Consider the same scenario as Example 13.1, except now suppose that there was one cytokine, say interleukin-5 (IL-5), that is strongly suspected (through many prior studies) to be elevated in subjects with the infection. Of the other 19 cytokines, there is no prior information to suspect elevated (or decreased) levels in infected subjects. In this* *case, it is natural to group the tests into 2 families, the test on IL-5 and the 19 tests on the other cytokines. The test on IL-5 would be confirming earlier studies and there would be no need to correct for the fact that 19 other cytokines are being tested, while there is no a priori reason to suspect that any one of the other 19 cytokines would have significantly different levels between the infected and the noninfected.*

How one statistically handles the family would depend on the situation. For Example 13.1 the default would be to correct for the multiple comparisons in the family, and if that was not done it should be justified. For Example 13.2 there are several possibilities for multiplicity adjustment for the two families (the IL-5 test and the 19 other tests). The first possibility is to define the test on IL-5 as the primary hypothesis and make no adjustment, and then make a multiplicity adjustment to bound the FWER on the family of the 19 other tests. A second possibility is treat the IL-5 test as primary but make no adjustment for the other 19 tests because they are just exploratory and will need confirmation in later studies. A third possibility is to combine the two families into one big family where the FWER is controlled at level α, but we partition the overall α into $\alpha/2$ for the test on IL-5 and $\alpha/2$ for the test on the other 19 tests. There are other possibilities. For deciding on the adjustment approach, context is key. Sometimes protecting against criticisms of lack of reproducibility is important and more adjustments would be made, while at other times the added complexity of explaining the multiplicity adjustment would obscure the message of the study which may just be exploratory and looking for which of the many tests deserves follow-up study.

Now consider a case where there are many possible outcomes related to an intervention effect. To avoid adjustment for multiple comparisons, it is often useful to prespecify one outcome as the primary one. If there is a group of hypotheses that help elucidate different aspects of the primary effect, then it is usually not necessary to treat them as a multiple comparison family. (An exception is the case where there is an intent to make some special claim about a secondary endpoint, which sometime occurs in the regulatory setting.) Here is an example:

Example 13.3 *Skou et al. (2015) reported a randomized trial comparing patients randomized to either 12 weeks of nonsurgical treatment or total knee replacement followed by 12 weeks of nonsurgical treatment. There are many ways to measure the results of the two interventions. Skou et al. (2015) followed standard practice in clinical trials and picked one prespecified "primary endpoint," change from baseline to one year using $KOOS_4$ ("mean score on four Knee Injury and Osteoarthritis Outcome Score [KOOS] subscales covering pain, symptoms, activities of daily living, and quality of life"). There were at least 20 additional tests on secondary outcomes that compared the 2 groups (Skou et al., 2015, Table 2 gives confidence intervals on 10 between-group measures, each one with and without an adjustment for covariates). The 10 measures are: 5 subscales from the KOOS, 2 timed activity tests, 2 measures of a self-report quality of life questionnaire and weight change. Because each of the secondary outcomes measures a different aspect of the intervention, there is no need to define the secondary outcomes as a family or to correct for multiple comparisons. The important points here are that (1) there was only one primary endpoint that was the focus of the study, and (2) the secondary endpoints were used as supplementary descriptors of the differences between the groups and the descriptors that were significant were not unduly highlighted and the others ignored.*

An important closely related issue is selection of hypotheses. Often care is not taken to record all the hypotheses that have been tested or could have been tested. There may be some informal ad hoc process to select hypotheses that are more likely to be significant.

Example 13.4 *Graphical selection: A researcher may make many scatter plots of different pairs of covariates, and only test for association between pairs of covariates that look substantially correlated. Although a formal test of correlation was not done for each pair, the graphical inspection will bias toward choosing pairs that are more likely to show significant correlation. In this case, the family is all pairs that were graphed, regardless of whether or not a formal follow-up test was done.*

Example 13.5 *Statistical methods selection: A researcher may perform many statistical hypotheses on the same variables. For example, consider looking for an age effect on a response. We could test for a correlation between age and response or perform a t-test or a Wilcoxon–Mann–Whitney (WMW) test comparing responses between children and adults. Any of these three approaches might be reasonable, but it is not valid to perform all three tests and pick the one with the most significant result and not report that the other two were done. It is acceptable to report all three tests using unadjusted p-values if appropriate caution is advised, or to define the three tests as a family and give just one multiple-comparison adjusted p-value. If all three tests are significant, or all are nonsignificant, then it may not be necessary to report all three unadjusted p-values (depending on space constraints), since the scientific inference would be similar whether or not those p-values are known. However, if the scientific implications where not similar (e.g., the three unadjusted p-values using $\alpha = 0.05$ are $p = 0.055$, $p = 0.60$, and $p = 0.80$), one should still be careful not to cherry-pick the result for presentation to avoid biasing the conclusion.*

Simmons et al. (2011) convincingly show that statistical methods selection can be a serious problem. They ran a purposely misleading study, where many covariates were collected, the data were repeatedly analyzed after controlling for different covariates, the analyses were repeated as more observations were collected, and only the most "interesting" result was reported. The problem need not be due to a desire to fraudulently achieve a significant result, it could be due to naïvely trying to get a "good fit" to a model. We discuss this issue with respect to model building in Section 14.7.

Although we will not give inflexible and binding advice for all situations regarding how multiple comparison families should be defined, in general, there are two predominant concerns: scientific reproducibility and clear communication. Scientific reproducibility pushes researchers into complete disclosure of every step of their data collection, design, and analysis. If many analyses were performed, then hiding some can seriously compromise other users' ability to replicate the study. On the other hand, clear communication requires highlighting the important analyses, and too many details can muddle the message. Two ways out of this predicament are: (1) predetermine the primary outcome for your study, so that no multiple comparison adjustments need to be made for the main message; or (2) adjust the inferences to account for multiple comparisons, so that if you perform many tests, you properly account for that in your presentation by, for example, using adjusted p-values. Optimally, we would prespecify the approach to multiple comparison adjustment, to avoid any bias in the choice of the adjustment method.

The rest of the chapter assumes that the family of hypotheses have been chosen.

13.3 Properties of Multiple Testing Procedures

For this chapter, let a family of null hypotheses be $H_1, \ldots, H_m$ and let the associated alternative hypotheses be $K_1, \ldots, K_m$. Let $\mathbf{d} = [d_1, \ldots, d_m]$ be the decision vector, where $d_j = 1$ means reject H_j and $d_j = 0$ is fail to reject. Let $\mathbf{a} = [a_1, \ldots, a_m]$ be the unknown vector of true states, where $a_j = 0$ means that H_j is true and $a_j = 1$ means that H_j is false (typically, but not necessarily, implying that the alternative K_j is true). Let $r_1 = \sum_{j=1}^m d_j a_j$ be the number of correct rejections and $r_0 = \sum_{j=1}^m d_j (1 - a_j)$ be the number of incorrect rejections. Let D_j and R_i be the random variables associated with d_j ($j = 1, \ldots, m$) and r_b ($b = 0, 1$). Let $R_0 + R_1 = R$.

Here are three important definitions:

FWER = Familywise error rate is $\Pr[R_0 > 0]$.

FDR = False discovery rate is $E[R_0/R]$, with $0/0$ defined as 0.

PCER = Per comparison error rate is $E[R_0]/m$.

Typically, we want to bound the error rates at some value α, for any true model. For the FWER, sometimes we speak of *weakly* controlling the FWER, which means that $\Pr[R_0 > 0 | H] \leq \alpha$, where $H = \cap_{j=1}^m H_j$ is the global null hypothesis (i.e., all m null hypotheses are true). To differentiate, we sometimes refer to *strongly* controlling the FWER, when $\Pr[R_0 > 0] \leq \alpha$ regardless of the true model. If there is no modifier, the FWER control is assumed to be strong.

Example 13.6 *Consider a family of hypotheses coming from a k-sample study. Assume the responses are normal with the same variance regardless of group, and responses from Group a have a mean of μ_a. Let the m null hypotheses be $H_1 : \mu_1 = \cdots = \mu_k$ plus all pairwise comparisons, $H_2 : \mu_1 = \mu_2$, $H_3 : \mu_1 = \mu_3, \ldots, H_m : \mu_{k-1} = \mu_k$, where $m = 1 + \binom{k}{2}$. In this case, $H = \cap_{j=1}^m H_j = H_1$. Consider Fisher's least significant difference (LSD) procedure, where we test H_1 using one-way ANOVA, and only if that is significant go on to test all pairwise comparisons with t-tests at level α. Trivially, this procedure weakly controls the FWER, with strong control of FWER only when $k = 3$ (see Section 11.5.1).*

Under the global null hypothesis, $FDR = FWER$, but more generally (i.e., when some of the null hypotheses are false), then $PCER \leq FDR \leq FWER$ (Problem P3). So if we develop a multiple comparison procedure to bound the FDR rather than the FWER, it can have more power.

There are other properties of multiple comparison procedures that are used less often. For example, $FWER(k) = \Pr[R_0 \geq k]$ is the probability of making k or more Type I errors. So the usual FWER is FWER(1). See Benjamini (2010) for a list of more properties with references.

Suppose that $d_j(\alpha)$ is the decision rule for the jth hypothesis in the family ($j = 1, \ldots, m$) by a multiple comparison procedure that strongly controls the FWER at α. Then the associated adjusted p-value for H_j is roughly the smallest α that rejects H_j. More precisely,

$$p_j^* = \min \left\{ a : d_j(\alpha) = 1 \text{ for all } \alpha \geq a \right\}. \tag{13.1}$$

We can define the FDR adjusted p-value analogously, by replacing $d_j(\alpha)$ in Equation 13.1 with the jth decision rule associated with a procedure that controls the FDR at α.

13.4 Procedures Using *p*-Values Only

Suppose we have an (unadjusted) *p*-value associated with each of the m hypothesis tests in the family. Let $p_{(1)} \le p_{(2)} \le \cdots \le p_{(m)}$ be the set of ordered *p*-values and let $H_{(j)}$ denote the null hypothesis associated with $p_{(j)}$.

The simplest procedure is the *Bonferroni procedure*, where we reject $H_{(j)}$ if $p_{(j)} \le \alpha/m$, so that the associated adjusted *p*-value is $p^*_{(j)} = \min\left(1, mp_{(j)}\right)$. The Bonferroni procedure strongly controls the FWER regardless of the correlation among the m unadjusted *p*-values (or their associated test statistics).

The Bonferroni procedure is popular because of its simplicity, but a uniformly more powerful procedure that also strongly controls the FWER is the *Holm procedure* that rejects $H_{(j)}$ if

$$p_{(i)} \le \frac{\alpha}{m - i + 1} \text{ for all } i \le j.$$

The Holm adjusted *p*-value is defined iteratively (starting with $j = 1$ and letting $p^*_{(0)} = 0$) as

$$p^*_{(j)} = \min\left\{1, \max\left((m - j + 1)p_{(j)}, p^*_{(j-1)}\right)\right\}.$$

The Holm procedure is also known as the *sequentially rejective* Bonferroni procedure because of the following way of describing it. We start with the Bonferroni adjustment for the smallest *p*-value. If we reject $H_{(1)}$, then there are $m - 1$ null hypotheses left that might be true, so we reject $H_{(2)}$ if $p_2 \le \alpha/(m - 1)$, using the Bonferroni adjustment on $p_{(2)}$ out of those $m - 1$ null hypotheses. And so on.

The Holm procedure for any hypothesis uses information about the tests of other hypotheses. In the literature, the Holm procedure is known as a *step-down procedure*, because we step through the different hypotheses from the most significant tests on down to the least significant hypotheses (see Table 13.1). A *step-up procedure* proceeds in the opposite direction, and either type is called a stepwise procedure. The Bonferroni is a single-step procedure because it does not need to use information about other hypotheses to make its adjustment. It is easier to find the associated simultaneous confidence intervals with single-step procedures. For example, the $100(1 - \alpha)\%$ simultaneous confidence intervals associated with the Bonferroni adjustment are simply the set of m unadjusted $100(1 - \alpha/m)\%$ confidence intervals. Constructing simultaneous confidence intervals associated with stepwise procedures such as the Holm procedure is difficult (Strassburger and Bretz, 2008).

When all *p*-values are independent or the test statistics have certain kinds of positive dependence (Note N1), Hochberg developed a procedure which strongly controls the FWER, and is easier to reject than Holm's procedure. The Hochberg procedure is a step-up version of the Holm one: reject $H_{(j)}$ if

$$p_{(i)} \le \frac{\alpha}{m - i + 1} \quad \text{for some } i \ge j.$$

The associated Hochberg adjusted *p*-value is given in Problem P4. Still another procedure which is uniformly more powerful than Hochberg's procedure and also strongly controls the FWER under the same conditions is due to Hommel (1988).

Ordered	Ordered	Step-down		Step-up	
p-value	test stat.	test	decision	test	decision
$P_{(1)}$	$T_{(8)}$	1	1	-	1
$P_{(2)}$	$T_{(7)}$	1	1	-	1
$P_{(3)}$	$T_{(6)}$	1	1	-	1
$P_{(4)}$	$T_{(5)}$	1	1	-	1
$P_{(5)}$	$T_{(4)}$	0	0	-	1
$P_{(6)}$	$T_{(3)}$	-	0	1	1
$P_{(7)}$	$T_{(2)}$	-	0	0	0
$P_{(8)}$	$T_{(1)}$	-	0	0	0

Symbols: 1 = reject null, 0 = fail to reject null,
and "-" = test not done

Table 13.1 *Example comparing a step-down procedure to a step-up procedure with m = 8, where we reject for large values of the test statistic. Step-down starts from the most significant and stops testing at the first nonrejection, while step-up starts from the least significant and stops testing at the first rejection. Generally, the step-up methods are more powerful, but need more assumptions to control the Type I error rate*

Benjamini and Hochberg (1995) proposed a step-up procedure that bounds the FDR at α when all *p*-values are independent or when the test statistics are multivariate normal with nonnegative correlation (see Note N2 for less restrictive conditions), reject $H_{(j)}$ if

$$p_{(i)} \leq \frac{i\alpha}{m} \text{ for some } i \geq j.$$

See Problem P4 for the adjusted *p*-value.

Benjamini and Yekutieli (2001) gave a modified procedure that always controls the FDR at α, reject $H_{(j)}$ if

$$p_{(i)} \leq \frac{i\alpha}{m\left(\sum_{h=1}^{m} \frac{1}{h}\right)} \text{ for some } i \geq j.$$

13.5 Max-*t*-Type (Generalized Studentized Range) Inferences

13.5.1 Linear Models

For readers unfamiliar with linear models, first review Section 14.2. Suppose we have a linear model. Let the response vector be $\mathbf{Y} = [Y_1, \ldots, Y_n]$, the $n \times k$ matrix of covariates with rank k be $\mathbf{X}$, the $k \times 1$ parameter vector be β, and the $n \times 1$ error vector be ϵ. The standard linear model is

$$\mathbf{Y} = \mathbf{X}\beta + \epsilon,$$

where ϵ is independent of $\mathbf{X}$ and $\epsilon \sim N(0, \sigma^2 \mathbf{I}_n)$, where $\mathbf{I}_n$ is an $n \times n$ identity matrix. The maximum likelihood estimate of β is

$$\hat{\beta} = \left(\mathbf{X}^\top \mathbf{X}\right)^{-1} \mathbf{X}^\top \mathbf{Y},$$

and the model-based estimate of $\mathbf{V} = \text{Var}(\hat{\beta})$ is

$$\hat{\mathbf{V}} = \hat{\sigma}^2 (\mathbf{X}^\top \mathbf{X})^{-1},$$

where $\hat{\sigma}^2 = (n-k)^{-1} \sum_{i=1}^{n} (Y_i - \hat{Y}_i)^2$ and $\hat{\mathbf{Y}} = \mathbf{X}\hat{\beta}$. Suppose we are interested in testing m contrasts of the parameters. Let the $m \times k$ matrix $\mathbf{C}$ be the contrast matrix and let $\mathbf{C}\beta = \theta = [\theta_1, \ldots, \theta_m]$ be the $m \times 1$ vector of contrasts. Let $\theta_0 = [\theta_{01}, \ldots, \theta_{0m}]$ be the null hypothesis value of θ. Then $H_1, \ldots, H_m$ are the null hypotheses, with $H_j : \theta_j = \theta_{0j}$. Here are two examples of these mutliple-comparisons hypotheses from linear models:

Example 13.7   *Control versus m treatments: We return to the example of Equation 11.11. Consider an experiment, where individuals are randomized to either the control group or one of m different new treatments. Let y_i be the response for the ith individual and $z_i = j$, where $j = 1$ represents the control group, or $j = 2, \ldots, m+1$ represents each of the m new treatments. Let $\mathbf{y}$ be the $n \times 1$ vector of responses and $\mathbf{X}$ be the matrix of covariates. Let the jth column of $\mathbf{X}$ be a vector of indicators, with ith element $I(z_i = j)$. So $\text{E}(Y_i | z_i = j) = \beta_j$ for $j = 1, \ldots, m+1$. Then if $m = 3$ we want to test whether the mean of each new treatment is different from the control we use $\mathbf{C}$ given in Equation 11.11, so that $\theta_{j-1} = \beta_j - \beta_1$ and $\theta_0 = [0, 0, 0]$. In other words, the null hypotheses are $H_1 : \theta_1 = 0$, $H_2 : \theta_2 = 0$, and $H_3 : \theta_3 = 0$, which can be equivalently written as $H_1 : \beta_2 = \beta_1$, $H_2 : \beta_3 = \beta_1$, and $H_3 : \beta_4 = \beta_1$.*

Example 13.8   *Testing for linear effects on m covariates: Suppose we are exploring the relationship of a response, y_i, to a vector of covariates, $\mathbf{x}_i^\top = [x_{i1}, \ldots, x_{im}]$ on n individuals. Let $\mathbf{X}$ be an $n \times (m+1)$ matrix, with the first column being an $n \times 1$ vector of 1s, and the other columns representing the m covariates. In other words, the ith row of $\mathbf{X}$ is $[1, x_{i1}, \ldots, x_{im}]$. The linear model is $Y_i = \beta_0 + \sum_{j=1}^{m} \beta_j x_{ij} + \epsilon_i$, where $\epsilon_i \sim N(0, \sigma^2)$ and is independent of x_{ij} for all j. So β_0 represents the intercept (if all the covariates are standardized to have mean 0, then the intercept is the overall mean), and β_j represents the additional linear effect of the jth covariate on the response given that the other covariates are in the model. To test $H_j : \beta_j = 0$ for $j = 1, \ldots, m$ let $\mathbf{C}$ be the $m + 1$ identity matrix without the first row and θ_0 be an $m \times 1$ vector of 0s.*

Because of the properties of the multivariate normal distributions (Section 10.4),

$$\hat{\theta} = \mathbf{C}\hat{\beta} \sim N(\theta, \Sigma),$$

where $\Sigma = \mathbf{C}\mathbf{V}\mathbf{C}^\top$ and it is estimated with $\hat{\Sigma} = \mathbf{C}\hat{\mathbf{V}}\mathbf{C}^\top$. Let $T_j(\theta_j) = (\hat{\theta}_j - \theta_j)/\sqrt{\hat{\Sigma}_{jj}}$, where $\hat{\Sigma}_{jj}$ is the jjth element of $\hat{\Sigma}$. Let $\mathbf{D}$ be an $m \times m$ diagonal matrix with jjth element equal to Σ_{jj}, so that $\Lambda = \mathbf{D}^{-1/2} \Sigma \mathbf{D}^{-1/2}$ is the correlation matrix of $\hat{\theta}$. Then Λ does not depend on σ (Problem P5), so we can equivalently write it as

$$\Lambda = \hat{\mathbf{D}}^{-1/2} \hat{\Sigma} \hat{\mathbf{D}}^{-1/2}, \tag{13.2}$$

where $\hat{\mathbf{D}}$ and $\hat{\Sigma}$ are $\mathbf{D}$ and Σ with σ^2 replaced by $\hat{\sigma}^2$ (but for Λ the $\hat{\sigma}^2$ values cancel out anyway). Thus, the variance-covariance matrix of $\mathbf{T}(\theta) = \hat{\mathbf{D}}^{-1/2}(\hat{\theta} - \theta) = [T_1(\theta_1), \ldots, T_m(\theta_m)]^\top$ is Λ, which is known, and $\mathbf{T}(\theta)$ is multivariate t with $\nu = n - k$ degrees of freedom and correlation Λ. So, under the null hypotheses $\mathbf{T}(\theta_0)$ has a known distribution.

For the two-sided hypotheses, we reject H_j when $|t_j(\theta_{0j})|$, the observed value of $|T_j(\theta_{0j})|$, is large for any $j = 1, \ldots, m$. If we only care about the PCER, then the two-sided p-value for H_j is the probability of observing something equal or more extreme than $|t_j(\theta_{0j})|$:

$$p_j = 1 - \int_{-|t_j(\theta_{0j})|}^{|t_j(\theta_{0j})|} f_{t,n-k}(x)dx$$

$$= 2\left\{1 - F_{t,n-k}(|t_j(\theta_{0j})|)\right\},$$

where $F_{t,n-k}(x)$ and $f_{t,n-k}(x)$ are the cumulative distribution function and probability density function (pdf) of the central t-distribution with $n - k$ degrees of freedom. For strongly controlling the FWER, we use the max-t test where we reject when $|t_j(\theta_{0j})|$ is large compared to $\max_{i \in \{1,\ldots,m\}} |T_i(\theta_{0i})|$, so the adjusted p-value is

$$p_j^* = 1 - \int_{-|t_j(\theta_{0j})|}^{|t_j(\theta_{0j})|} \cdots \int_{-|t_j(\theta_{0j})|}^{|t_j(\theta_{0j})|} f_T([x_1, \ldots, x_m]; n - k, \Lambda)dx_1 \, dx_2 \cdots dx_m,$$

where $f_T(\cdot; n - k, \Lambda)$ is the pdf of the multivariate t-distribution with $n - k$ degrees of freedom and correlation Λ. The associated $100(1 - \alpha)\%$ simultaneous confidence interval for θ_j (for any $j = 1, \ldots, m$) is

$$\hat{\theta}_j \pm c\sqrt{\hat{\Sigma}_{jj}},$$

where c solves

$$1 - \alpha = \int_{-c}^{c} \cdots \int_{-c}^{c} f_T([x_1, \ldots, x_m]; n - k, \Lambda)dx_1 \, dx_2 \cdots dx_m.$$

The one-sided confidence intervals are analogous (Note N3).

In Section 11.7, we discussed the application of the max-t test to k-sample tests, including all pairwise comparisons and comparing each new treatment to a control. To explore the max-t procedure in more generality we simulated data from highly correlated covariates, and uncorrelated covariates. We find that for this type of data, inferences are highly dependent on the correlation among the covariates (columns of $\mathbf{X}$).

Example 13.8 *(continued) For notational ease, we write the linear model without the i indices: $Y = \beta_0 + \sum_{j=1}^{m} \beta_j X_j + \epsilon$, and we call the covariates, $X_1, \ldots, X_m$. Suppose that $\beta_1 = 1$ and $\beta_j = 0$ for $j = 0, 2, 3, \ldots, m$. Suppose the covariates are multivariate normal with mean 0 (i.e., an $m \times 1$ vector of 0s) and covariance matrix Σ, with all diagonal elements equal to 1 and all off diagonal elements equal to ρ. We simulate data with $n = 200$, $m = 8$, and $\rho = 0.9$. The unadjusted p-value for H_1 is $p_1 = 0.0052$ and the max-t adjusted p-value is $p_1^* = 0.04$. For $j = 2, \ldots, 8$ the range of the p_j is $(0.077, 0.997)$ and the range of the p_j^* is $(0.45, 1.00)$. So once the first covariate, X_1, is in the model, none of the other covariates $(X_2, \ldots, X_m)$ appear to have any additional significant effect (at the 0.05 level), either before or after adjustment. The incorrect interpretation of these results is that none of the other covariates $(X_2, \ldots, X_m)$ has any information about the response. In fact, since $X_2, \ldots, X_m$ are highly correlated with X_1 they are each correlated with Y with correlation $0.9/\sqrt{2} = 0.636$. If we form seven different linear models with an intercept and only X_j for*

$j = 2, \ldots, 8$ *the p-values for testing the slope associated with* X_j *are all less than* 10^{-10}, *so their adjusted p-values are all highly significant (the Bonferroni correction with either seven or eight in the family is still highly significant).*

If we simulate new data from the same model except $\rho = 0$, *then all the covariates are independent of each other, and* $X_2, \ldots, X_8$ *are independent of Y. Now both* $p_1 < 10^{-10}$ *and the max-t adjusted p-value,* $p_1^* < 10^{-10}$, *because none of the other covariates add any useful information to the model. For* $j = 2, \ldots, 8$ *the range of the* p_j *is* $(0.114, 0.767)$ *and the range of the* p_j^* *is* $(0.623, 1.00)$. *Now when we form seven different linear models with an intercept and only* X_j *for* $j = 2, \ldots, 8$ *the p-values for testing the slope are in the range* $(0.041, 0.908)$, *which is closer to what we would expect from uniformly distributed p-values. When we do the Holm correction the apparent significant effect of* $p_6 = 0.041$ *is no longer significant,* $p_6^* = 0.284$. *(Problem P6).*

13.5.2 Asymptotic Generalizations

The max-t method is much more general than just a linear model. Consider any case where $\hat{\theta}$ is asymptotically normal with mean θ and covariance Σ, which is consistently estimated by $\hat{\Sigma}$, and where $\hat{\theta}$ and $\hat{\Sigma}$ are asymptotically independent. Then we can get asymptotic strong control of the FWER if we use the same methods as the previous section by either using the multivariate t-distribution (using Λ in Equation 13.2 as correlation and some estimate of degrees of freedom) or the multivariate normal distribution (using Λ in Equation 13.2 as covariance). Bretz et al. (2011) has many parametric examples: logistic regression, parametric survival regression, and mixed effects models. Also, the nonparametric framework of Konietschke et al. (2012) for k-sample tests described in Section 11.7 is a max-t style method.

13.6 Permutation- or Bootstrap-Based Methods

Now we consider resampling methods for performing multiple comparisons. These can be based on a permutation null distribution or a bootstrap null distribution. Here we describe a resampling method using a permutation null distribution. Resampling methods can be applied to a wide range of applications, require few distributional assumptions, and automatically adjust for dependencies between different hypotheses.

Consider a set of m null hypotheses, $H_1, \ldots, H_m$, where associated with the jth hypothesis is a test statistic, t_j, and its associated unadjusted p-value, p_j. Suppose that large values of t_j suggest rejection of H_j. For example, t_j could be an F statistic and p_j could be its associated p-value calculated using an F distribution. Without loss of generality, let $t_1 \geq t_2 \geq \cdots \geq t_m$ so that $p_1 \leq p_2 \leq \cdots \leq p_m$. In other words, the most significant test is testing H_1 and the least significant test is testing H_m. Suppose the data may be represented by an $n \times 1$ response vector, $\mathbf{y}$, and an $n \times m$ matrix of covariates, $\mathbf{x}$. Suppose H_j represents the null hypothesis that $\mathbf{y}$ and the jth column of $\mathbf{x}$ are independent. Now suppose that we do not necessarily believe that the assumptions that created the unadjusted p-value (e.g., the t_j may not actually follow an F distribution under the null hypothesis, it is just a poor approximation). In this case, we can use these test statistics or the unadjusted p-value as test statistics in a permutation test.

Suppose we can test H_j by a permutation test using t_j as a test statistic that is a function of $\mathbf{y}$ and the jth column of $\mathbf{x}$, say $\mathbf{x}_j$. Simulate B Monte Carlo permutation data sets by permuting $\mathbf{y}$, and for each permutation data set calculate the m test statistics associated with the m hypotheses. Let $\mathbf{T}^*$ be the $B \times m$ matrix of those test statistic values, where T_{ij}^* represents the ith permuted data set and the jth test statistic. Let $\mathbf{P}^*$ be the $B \times m$ matrix of unadjusted p-values.

A multiple comparison adjusted p-value for testing H_1 is

$$p_{a1} = \frac{1 + \sum_{i=1}^{B} I\left(\max_{j \in \{1,\ldots,m\}} T_{ij}^* \geq t_1\right)}{B + 1}. \tag{13.3}$$

In other words, we compare the largest test statistic, t_1, to the largest test statistic out of the m test statistics for each permutation data set. For testing H_2, we remove all the test statistics related to H_1 from consideration and proceed in a similar manner. Specifically, we remove the first column from $\mathbf{T}^*$ (or $\mathbf{P}^*$) from consideration and calculate the adjusted p-value as

$$p_{a2} = \max\left(p_{a1}, \frac{1 + \sum_{i=1}^{B} I\left(\max_{j \in \{2,\ldots,m\}} T_{ij}^* \geq t_2\right)}{B + 1}\right). \tag{13.4}$$

We continue in the same manner to test $H_3, \ldots, H_m$. Taking the maximum iteratively enforces a type of monotonicity such that $p_{a1} \leq p_{a2} \leq \cdots \leq p_{am}$.

Other methods that enforce monotonicity are the *free step-down resampling method* (see Westfall and Young, 1993, Algorithm 2.8, p. 66) or the similar step-down minP procedure (Dudoit and Van Der Laan, 2008, Procedure 5.6).

Westfall and Troendle (2008) state the assumption needed to control the FWER using the adjusted p-values. Let $M = \{1, \ldots, m\}$ be the set of indices for all m null hypotheses and let $J \subset M$ be some subset of M. Let $H_J = \cap_{j \in J} H_j$. Then the following assumption is needed to control the FWER:

Subset pivotality: The distribution of $\max_{j \in J} T_{ij}^* | H_J$ and $\max_{j \in J} T_{ij}^* | H_M$ are identical for all $J \subset M$.

It we replace the permutation data sets with bootstrap data sets, then we replace the subset pivotality assumption with an asymptotic version, where the distribution of $\max_{j \in J} T_{ij}^* | H_J$ and $\max_{j \in J} T_{ij}^* | H_M$ are asymptotically equivalent, and T_{ij}^* now represents the jth test statistic for the ith bootstrap data set.

If there is a strong positive correlation among the unadjusted p-values, then the maxT method (either the resampling implementation of this section, or the implementation of Section 13.5) will generally have better power than the simple Holm adjustment (Goeman and Solari, 2014).

The minP method is similar to the maxT method. One way to define the minP method is to replace the maximum of the T_{ij}^* in Equations 13.3 and 13.4, with the minimum of P_{ij}^*. By that definition, the two methods are equivalent when the test statistics are identically distributed under the null hypotheses (see Dudoit and Van Der Laan, 2008, p. 212). Another way to define the minP method is to use minimum of p-values, where P^* is replaced by $\tilde{P}^*$ where $\tilde{P}_{ij}^* = B^{-1} \sum_{h=1}^{B} I(P_{hj}^* \leq P_{ij}^*)$ (Goeman and Solari, 2014). The idea of the minP method is to put all the m tests on a similar scale. In general, the minP will have many more ties, and hence may need much larger values of B (Goeman and Solari, 2014).

13.7 Sequential Testing of Different Hypotheses

Bretz et al. (2009) show how to formulate sequential hypothesis tests that strongly control the FWER at level α. The sequential procedure is represented graphically, where each hypothesis is a node, with arcs showing how the α-level is transferred after any rejection. Before getting into the details, we show a graph associated with a simple sequential procedure:

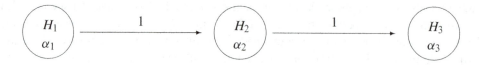

Here is a description of the procedure. Test H_1 at level α_1; if H_1 is rejected then test H_2 at level $\alpha_2^* = \alpha_1 + \alpha_2$, otherwise test H_2 at $\alpha_2^* = \alpha_2$. Test H_2 at level α_2^*; if H_2 is rejected then test H_3 at level $\alpha_3^* = \alpha_2^* + \alpha_3$, otherwise test H_3 at level $\alpha_3^* = \alpha_3$. The basic idea of the graph is that once the jth null hypothesis is rejected, its significance level is passed on to other hypotheses in proportion to the arrows coming from that hypothesis (see below for a graph with multiple arrows coming out of hypotheses). In this procedure if $\alpha_1 = \alpha$ and $\alpha_2 = \alpha_3 = 0$ this is known as a *fixed sequence method* or more informally as a *gatekeeping* approach and testing a hypothesis at level 0 means that the hypothesis is not tested. If $0 < \alpha_1 < \alpha$ and $\alpha_2 > 0$ and perhaps $\alpha_3 > 0$, this is known as the *fallback method*. For the latter method, even if H_1 is not rejected, there is still a chance to test other hypotheses, although at a lower significance level.

More complicated procedures are possible, where the order of the testing is not preset. For example, the Holm procedure with $m = 3$ is as follows:

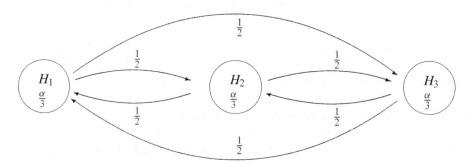

We can define a wide class of multiple testing procedures using graphs such as this. The procedure is defined by the vector of initial significance levels $[\alpha_1, \ldots, \alpha_m]$, where $0 \leq \alpha_j \leq \alpha$ and $\sum_{j=1}^m \alpha_j = \alpha$, and an $m \times m$ transition matrix, $\mathbf{G}$, with ijth elements g_{ij}, such that $0 \leq g_{ij} \leq 1$, $g_{ii} = 0$, and $\sum_{j=1}^m g_{ij} \leq 1$ for all $i, j = 1, 2, \ldots, m$. The ith row of $\mathbf{G}$ gives the weight to the arrows going from H_i to the other hypotheses, and g_{ij} is the proportion of the significance level that goes from H_i to H_j when H_i is rejected. Here we reproduce the algorithm of Bretz et al. (2009) in terms of adjusted p-values. Let $p_1, \ldots, p_m$ be the unadjusted p-values for the m hypotheses and let $w_j = \alpha_j / \alpha$. Let $M = \{1, \ldots, m\}$.

Step 0 Set $J = M$ and $p_{max} = 0$.

Step 1 Let $k = \arg \min_{j \in J} \frac{p_j}{w_j}$.

Step 2 Calculate $p_k^* = \max \left\{ \frac{p_k}{w_k}, p_{max} \right\}$ and then set $p_{max} = p_k^*$.

Step 3 Set J equal to J with k removed.

Step 4 If J is the empty set then stop, otherwise continue.

Step 5 Update the graph:

- For $i \in J$ set w_i equal to $w_i + w_k g_{ki}$. For $i \notin J$ set $w_i = 0$.
- For $h, i \in J$ set g_{hi} equal to $(g_{hi} + g_{hj} g_{ki})/(1 - g_{hk} g_{ki})$; otherwise set $g_{hi} = 0$.

Step 6 Return to Step 1.

At the end of the algorithm, you have the adjusted p-values, $p_1^*, \ldots, p_m^*$, which defines the multiple comparison procedure. That procedure rejects H_j when $p_j^* \leq \alpha$ for $j = 1, \ldots, m$ and it strongly controls the FWER at α. See Bretz et al. (2009) for the description of the associated simultaneous confidence intervals.

These multiple comparison graphs can be generalized so that the hypotheses may be grouped into subfamilies, such that all hypotheses in a subfamily must be rejected before continuing with other hypotheses. These are represented graphically using infinitesimal transition weights (see Bretz et al., 2009, Section 3.3).

13.8 Logical Constraints

Sometimes there are logical constraints between hypotheses. For example, in Chapter 11 Note N3 we showed that for a k-sample study when $k = 3$ Fisher's LSD procedure strongly controls the FWER. This is because of logical constraints. Recall that Fisher's LSD procedure first tests the global null hypothesis that all k distributions are equal, if that is not rejected no more tests are done, otherwise all pairwise comparisons are tested. When $k = 3$, if the global null hypothesis is false, then no more than one of the pairwise comparisons can be true. That is why no multiple correction is needed between those three pairwise comparisons when the global null is rejected.

Example 13.9 *Here is an example of a sequential testing that accounts for logical constraints. Consider a study of adults with seasonal allergic rhinitis (stuffy nose). The study evaluates three regimens: a placebo (Group 0), a single drug treatment (Group 1), and a combination drug treatment (Group 2). Let μ_0, μ_1, and μ_2 be the respective means. The plan is to first test the combination drug treatment against the placebo, $H_1 : \mu_0 = \mu_2$. If this is significant at level α, then test both $H_2 : \mu_0 = \mu_1$ (the single drug against placebo) and $H_3 : \mu_1 = \mu_2$ (the combination against the single drug treatment), each tested at level α. This procedure strongly controls the FWER at α. The issue is logical constraints.*

If H_1 is false, then one and only one of H_2 and H_3 can be true, so that both can be tested at level α (Problem P8).

For some sets of hypotheses there is a logical relationship between them. For example, consider a three-sample study where the means for the three samples are μ_1, μ_2, and μ_3. Suppose the null hypotheses to be tested are $H_1 : \mu_1 = \mu_2 = \mu_3$, $H_2 : \mu_1 = \mu_2$, $H_3 : \mu_1 = \mu_3$, and $H_4 : \mu_2 = \mu_3$. There are logical relationships between the hypotheses. For example, if H_2 and H_3 are false, this implies that H_1 and H_4 must be false. So more power can be had in sequential hypothesis testing when we only count the future tests in the sequence that have a positive probability of being true given the previous rejections are not in error (see, e.g., Westfall, 1997).

13.9 Closure Method

This section discusses the closure method, one important theoretical idea that is often used in the development of new multiple comparison methods. We include this section not to help with any specific application, but because the closure method is frequently mentioned in the literature on multiple comparisons, and many new methods are still being developed in this area.

The closure method is a way of creating a procedure that strongly controls the FWER. Let $J \subset \{1, \ldots, m\}$ be a subset of the indices of the m null hypotheses, $H_1, \ldots, H_m$. Let $H_J = \cap_{j \in J} H_j$ be the probability model where all null hypotheses with indices in J are true. Define δ_J as a valid decision rule for testing H_J for any J. In other words, δ_J is a function of the data and significance level α that rejects H_J when $\delta_J = 1$ and $\Pr[\delta_J = 1 | H_J] \le \alpha$. The closure method rejects any H_i if and only if $\delta_J = 1$ for all $J \subset \{1, \ldots, m\}$ with $i \in J$. The closure method strongly controls the FWER at level α (Marcus et al., 1976).

The closure method is difficult to apply if m is large, since the number of subsets grows rapidly with m. There are some shortcut procedures that can be made if certain assumptions hold. For example, the subset pivotality assumption mentioned in Section 13.6 together with the maxT method for testing hypotheses, ensures that the maxT method will give the same results as the full closure method, so all subsets of the closure do not need to be tested.

13.10 Summary

Multiple testing comes in many forms, common examples occur in a study that compares each pair of multiple treatment groups or in a study that separately tests several different outcomes. If we test many hypotheses, the chance of making at least one Type I error can be very high. While there is judgment about whether every setting with multiple testing needs multiple comparison adjustments, we discuss a range of possible approaches.

Defining families of hypothesis tests for a given study context is important for determining which sets of hypotheses need to be considered and potentially controlled jointly. While defining these families requires some judgment, the examples provided in Section 13.2 are intended to illuminate this process.

Decisions about if or how to adjust for multiplicity can be nuanced, but, in the interest of scientific reproducibility and clear communication, our primary recommendation is to avoid reporting results in a misleading manner. For example, if a study is relatively exploratory and the number of multiple tests is not very large, it may be adequate to report the number of hypotheses tested and to urge caution in interpretation. (Although, the reported number of tests would need to be a fair account, if one used preliminary informal examination of the data to exclude some variables from testing, then this would need to reported as well.) At the other end of the spectrum, if a study is intended to provide a definitive answer or involves a very large number of tests, then it would often be appropriate to adjust for multiple comparisons, and to prespecify the adjustment method prior to seeing the data. That said, a legitimate and common approach for even a definitive study is to choose to focus on a single prespecified primary analysis and make the outcome of this test the main message of the study results. This example illustrates the family of tests concept: the primary endpoint is one family, and the secondary endpoints are jointly another family, and one might decide that there is no need to control the overall error for the secondary family.

There are three definitions that represent important properties of any multiple testing procedure. Family wise error rate (FWER) is the probability that at least one null hypothesis is rejected. False discovery rate (FDR) is the expected proportion of rejected hypotheses that are null. Per comparison error rate (PCER) is the expected proportion of tests where a null hypothesis is rejected. For any multiple testing procedure, $PCER \leq FDR \leq FWER$.

Commonly used procedures based on the multiple p-values are presented in Section 13.4; these methods allow for a rejection decision for each test. The simplest procedure is the Bonferroni adjustment where each of m tests is evaluated at α/m to provide $FWER \leq \alpha$, i.e., this strongly controls the FWER. The Holm procedure also strongly controls the FWER, and is uniformly more powerful than Bonferroni, but is slightly more complicated to explain and implement. The Hochberg is more powerful still, however, it only controls the FWER when all p-values are independent or under some conditions if they are not. All of these methods can also provide adjusted p-values.

The maxT approach is very general. In the linear model setting, it can address multiplicity of groups and multiplicity of covariates. It can also be applied to a variety of models such as logistic regression, parametric survival, and mixed effects. Further, when there is uncertainty about the applicability of multivariate t and multivariate normal distributions, one can use permutation- or bootstrap-based methods for inference. The maxT approach can also be more powerful than the Holm adjustment, such as when there is strong positive correlation between unadjusted p-values.

Another very useful, and potentially powerful, approach is sequential (or hierarchical or gatekeeper) testing of hypotheses. Some multiple hypotheses are naturally sequential. For example, say we are testing a high dose drug, low dose drug, and placebo, and are nearly certain that the high dose drug will have a treatment effect at least as large as the low dose. Then, we could first test high dose versus placebo at 0.05, and if that is significant, then test the low dose versus placebo at 0.05, and if that is significant, then test high versus low dose at 0.05. That is, because of the inherent ordering of the test questions, this procedure potentially allows us to test all three comparisons at 0.05, with strong control of the FWER. The downside is we cannot test further if an earlier test is not successful, but if the setting is such that the further questions are of little interest, this is a cheap price to pay. An analogous approach can be used to order tests of endpoints. These types of procedures are gaining in popularity, especially in settings where formal declarations of statistical significance are important. For example, a new drug might get approved mostly based on the group comparison on the primary endpoint; however, if there were a prespecified plan to test the next most important endpoint given a rejection of the test on the primary endpoint, it might be possible for the drug to have a formal claim on providing benefit with respect to this secondary outcome, where a competitor has no such claim. The great advantage of this approach is that it can be done without making the test on the primary endpoint more stringent, almost like getting something for nothing. Furthermore, these procedures can be set up in a more complicated manner than these simple examples, that is, α might be split between various hypotheses, and passed on to later hypotheses when significant.

13.11 Extensions and Bibliographic Notes

Bretz et al. (2011) gives a good overview of the issues of multiple comparisons that is not overwhelming. It describes many of the procedures and focuses on the maxT procedure. More theoretically oriented texts are Hochberg and Tamhane (1987) (classical procedures), Westfall and Young (1993) (resampling-based procedures), Dudoit and Van Der Laan (2008) (resampling-based procedures focusing on genetic applications).

Tamhane and Gou (2017) gives updates on more complicated but more powerful multiple comparison procedures based on p-values.

13.12 R Functions and Packages

The p.adjust function in the stats R package gives adjusted p-values for the methods in Section 13.4.

The multcomp R package has functions for implementing the maxT method described in Section 13.5. That package also has options for accounting for logical constraints as in Section 13.8 (see Bretz et al., 2011, pp. 98–99). The nparcomp R package has a function for nonparametric multiple comparisons mentioned in Section 13.5.2.

The multtest R package has functions for implementing the maxT and minP methods described in Section 13.6 using either permuting or bootstrapping.

13.13 Notes

N1 **Dependence assumptions such that Hochberg procedure controls FWER:** Sarkar and Chang (1997) give several examples of dependence between test statistics such that the Hochberg procedure controls the FWER: equicorrelated multivariate normal with nonnegative correlation, absolute-valued equicorrelated multivariate normal, absolute-valued central multivariate t, central multivariate F. More general cases exist, such as any multivariate normal with nonnegative correlation, see Tamhane and Gou (2017) for details with references.

N2 **Dependence assumptions such that Benjamini–Hochberg procedure controls FDR:** For the general form of dependence on the test statistics needed and theorem that shows the Benjamini–Hochberg procedure strongly controls the FDR see Theorem 1.2 of Benjamini and Yekutieli (2001).

N3 **One-sided simultaneous CIs associated with max-t tests (see page 243):** The one-sided $100(1 - \alpha)\%$ upper simultaneous confidence limit for θ_j (for any $j = 1, \ldots, m$) is

$$\hat{\theta}_j + c\sqrt{\hat{\Sigma}_{jj}},$$

where here c solves

$$1 - \alpha = \int_{-\infty}^{c} \cdots \int_{-\infty}^{c} f_T([x_1, \ldots, x_m]; n - k, \Lambda) dx_1 \, dx_2 \cdots dx_m.$$

The analogous lower limit is $\hat{\theta}_j - c\sqrt{\hat{\Sigma}_{jj}}$, using the same c.

13.14 Problems

P1 Consider a study with $m + 1$ arms, one control and m treatments, with n individuals in each arm. Let $z_i = j$ when the ith individual is in treatment j, with $j = 0$ for control and $j = 1, \ldots, m$ for treatments. Let the responses be normal: $Y_i | z_i = j \sim N(\mu_j, \sigma^2)$. Suppose σ^2 is known. Let $T_j = (\hat{\mu}_j - \hat{\mu}_0) / \sqrt{2\sigma^2/n}$, where $\hat{\mu}_j$ is the sample mean for individuals with $z_i = j$. Show that under the null that $\mu_j = \mu_0$ that T_j is standard normal. Then show that the correlation between T_j and T_k is 0.5.

P2 Calculate the FWER for testing m hypotheses when all m null hypotheses are true and when using m test statistics that are each normally distributed under their respective null hypothesis and are equally correlated with correlation ρ. Try several values of m and ρ. Hint: use the pmvnorm function in the mvtnorm R package.

P3 Show that $PCER \leq FDR \leq FWER$. Hint: $\Pr[R_0 > 0] = E\{I(R_0 > 0)\}$. See Bretz et al. (2011, p. 14).

P4 *Adjusted p-values:* Show that the Hochberg adjusted p-value is defined iteratively. Start with $j = m$ and letting $p^*_{(m+1)} = 1$, and then decrease j by 1 for each iteration:

$$p^*_{(j)} = \min \left\{ 1, \min \left((m - j + 1) p_{(j)}, p^*_{(j+1)} \right) \right\}.$$

Next, show that the Benjamini–Hochberg adjusted p-value for $H_{(j)}$ is similarly defined iteratively as

$$p^*_{(j)} = \min \left\{ 1, \min \left((m/j) p_{(j)}, p^*_{(j+1)} \right) \right\}.$$

P5 Show that Equation 13.2 is true. Hint: write Σ as $\sigma^2 C(X^\top X)^{-1} C^\top$, then show that $\mathbf{R}$ does not depend on σ^2.

P6 Reproduce the analyses from Example 13.8. The data were simulated using R 3.4.1 with library(mvtnorm); set.seed(1); R=diag(k);R[R==0]=rho; X=rmvnorm(n,sigma=R); Y=X[,1] + rnorm(n); d=data.frame(X,Y). The max-t analyses used glht from the multcomp R package. Try different seeds, sample sizes, and correlations.

P7 Program the algorithm of page 247. Set **G** and $[\alpha_1, \alpha_2, \alpha_3]$ according to the graph of the Holm procedure on page 246. Try calculating the Holm adjusted p-values for some unadjusted p-value using your program. Check that they match the usual Holm adjusted p-values (e.g., using the R function p.adjust).

P8 Show that the procedure of Example 13.9 strongly controls the FWER. Hint: list all possible types of true models (e.g., $\mu_0 = \mu_1 = \mu_2$, $\mu_0 \neq \mu_1 = \mu_2$, etc.), and show that the FWER $\leq \alpha$ for each type.

14

Testing from Models

14.1 Overview

In this chapter we give a quick overview of several common useful models. Here we consider data where the ith individual has a scalar response y_i and a vector of covariates, $\mathbf{x}_i = [x_{i1}, \ldots, x_{ip}]$. One very broad class of models is where we try to model the mean of the response given the covariates, in other words we model $E(Y_i|\mathbf{x}_i)$. Most of the models we consider use a linear combination of the covariates (or functions of the covariates, such as the square of one of them) within the model. The linear combination of covariates is for ease of interpretation of the result as well as for computational ease. There is a whole range of more complicated nonlinear models that we do not discuss. The range of possible values of y_i primarily motivates which type of models of the mean are considered. If $y_i \in (-\infty, \infty)$, then the linear model may be appropriate (see Section 14.2). When y_i is positive, then the responses may be log transformed (or some other transformation may be used) and a linear model is used on the transformed responses. If y_i is binary (i.e., $y_i \in [0, 1]$), then a logistic model or other binary response generalized linear model (GLM) is appropriate (Section 14.3). If y_i is count data (i.e., $y_i \in [0, 1, 2, \ldots)$), then a Poisson model or other count GLM may be used.

Another broad class of models that still uses a linear combination of covariates is a semiparametric linear transformation model, which models the ranks of the responses (Section 14.4). These models are important for ordinal data. Some special cases are the proportional odds model and the proportional hazards model. In Section 14.5 we study models for clustered or stratified data. In Section 14.6 we study a few methods for model selection, and in Section 14.7 we briefly mention some inherent problems with testing from models. In Section 14.8 we mention one approach for hypothesis testing after model building.

14.2 Linear Models

14.2.1 Standard Assumptions

The linear model represents the ith individual's response as

$$Y_i = \beta_1 x_{i1} + \beta_2 x_{i2} + \cdots + \beta_p x_{ip} + \epsilon_i,$$

where $\beta = [\beta_1, \ldots, \beta_p]$ is the parameter vector and ϵ_i is the random error term. It is convenient to write the model for all n individuals in matrix notation as

$$\mathbf{Y} = \mathbf{X}\beta + \epsilon,$$

where $\mathbf{X}$ is an $n \times p$ matrix and ϵ and $\mathbf{Y}$ are $n \times 1$ vectors. Assume that $\epsilon \sim N(0, \sigma^2 \mathbf{I})$, where $\mathbf{I}$ is the identity matrix. This means that the errors are independent (since the correlation is zero and the errors are multivariate normal), and the errors all have the same variance (i.e., they are *homoscedastic*). We discuss those assumptions more in Section 14.2.2. We estimate β using the least squares estimate (which is the maximum likelihood estimate under normality), $\hat{\beta} = (\mathbf{X}^\top \mathbf{X})^{-1} \mathbf{X}^\top \mathbf{y}$.

A basic test from the linear model is

$$H_0 : \mathbf{C}\beta = \mathbf{c}$$
$$H_1 : \mathbf{C}\beta \neq \mathbf{c},$$

where $\mathbf{C}$ is a $q \times p$ matrix of linear contrasts, and $\mathbf{c}$ is a $q \times 1$ vector of null values (typically a vector of 0s). This test can be tested from a quadratic form statistic, Q, which is like the square of the distance from $\mathbf{C}\hat{\beta}$ to $\mathbf{c}$ weighted by the inverse of the estimated variance matrix of $\mathbf{C}\hat{\beta}$. Here are the details. Let $\hat{V} = \hat{\sigma}^2 (\mathbf{X}^\top \mathbf{X})^{-1}$ be the estimate of the variance-covariance matrix of $\hat{\beta}$, and let the estimate of σ^2 be $\hat{\sigma}^2 = (n - p)^{-1} \sum_{i=1}^{n} (y_i - \hat{y}_i)^2$, where $\hat{y} = \mathbf{X}\hat{\beta}$ is the vector of predicted values of $\mathbf{Y}$. Then the quadratic form is

$$Q = \frac{1}{q} (\mathbf{C}\hat{\beta} - \mathbf{c})^\top (\mathbf{C}\hat{V}\mathbf{C}^\top)^{-1} (\mathbf{C}\hat{\beta} - \mathbf{c}). \tag{14.1}$$

Under the null hypothesis and the normality assumption on the errors, then $Q \sim F_{q,n-p}$ (i.e., Q is distributed as an F distribution with q and $n - p$ degrees of freedom). This is a generalized likelihood ratio test (see, e.g., Graybill, 1976, Theorem 6.3.1).

In the special, but common, case where $\mathbf{C}$ is a $1 \times p$ vector of all 0s except a 1 in the jth row and $\mathbf{c} = 0$, then $\mathbf{C}\beta = \beta_j$, $Q = \hat{\beta}_j^2 / \widehat{var}(\hat{\beta}_j)$, so that $T = \text{sign}(\hat{\beta}_j)\sqrt{Q} \sim t_{n-p}$, where t_{n-p} is a t-distribution with $n - p$ degrees of freedom. Often software will present a table with a p-value for each test of the null that $\beta_j = 0$ for $j = 1, \ldots, p$, and present the associated T statistics and confidence intervals (Problem P1). Because such a table is testing p different hypotheses, interpreting the table is a case of multiple comparisons (Chapter 13). For example, the probability of finding a significant parameter in a model with $p = 10$ parameters when in truth all parameters are 0 can be close to 40% (Problem P2). In other examples, testing one parameter will be the primary hypothesis and the other parameters are nuisance parameters (see Section 14.7).

Typically, the first term in the linear model is an intercept term, so that $x_{i1} = 1$ for all i. A categorical covariate with b categories is represented in a linear model with an intercept term with $b - 1$ covariates. For example, suppose there is a study with n individuals randomized to b treatment arms, represented by $t = [t_1, \ldots, t_n]$, where $t_i = j$ when the ith individual is in the jth treatment arm. A simple model using an intercept and t translates to the linear model with the $n \times b$ matrix $\mathbf{X}$, where the first column is a vector of 1s, and the j column (for $j = 2, \ldots, b$) is a vector with ith element equal to $I(t_i = j)$. Then to test the null hypothesis that no treatment has an effect on the response, we let $\mathbf{c} = \mathbf{0}_{b-1}$ and

$\mathbf{C} = [\mathbf{0}_{b-1}\mathbf{I}_{b-1}]$, where $\mathbf{0}_{b-1}$ is a $b-1$ vector of 0s, and $\mathbf{I}_{b-1}$ is a $(b-1) \times (b-1)$ identity matrix. Software can easily handle the details of creating the $\mathbf{C}$ matrix for you (Problem P3).

14.2.2 Checking Assumptions and Transformations

One major assumption is that the linear relationship to the dependent variables holds. With only one variable that assumption can be checked graphically by plotting the response by the covariate. For many variables, other methods such as partial residual plots can be used.

We can check normality and homoscedasticity (i.e., equal variance assumption) together as follows. Let the ith residual be $r_i = y_i - \mathbf{x}_i^\top \hat{\beta}$. If we plot y_i by r_i, we want the errors to be approximately normally distributed about 0 and we do not want the spread of the residuals to change with y_i. If the residuals get more spread out for larger y_i, then this implies the homoscedasticity assumption fails. To get more stable variances, we might first transform the responses, refit a new linear model on the transformed responses, and then recheck the residuals against the transformed responses to see if they appear approximately normally distributed and do not change spread as the transformed responses change. An important class of transformations is the Box–Cox transformation (Note N1). Another fix for heteroscedasticity (i.e., different variances for the error terms) is to use sandwich estimators of variance, which can also deal with some kinds of dependence such as autocorrelation (Zeileis, 2004).

If the normality assumption (including homoscedasticity) holds on the responses, then we do not need any assumption about the covariates (the columns of $\mathbf{X}$), for example, some covariates might be binary and other covariates might be very skewed, but still linearly related to the responses. However, if the errors are clearly departing from independent homoscedastic normality, this is more likely to be a problem (in terms of the Type I error rates of the tests) if a small percentage of the individuals have a large influence on the result, which depends on the distribution of $\mathbf{X}$ (Note N2). Thus, it may sometimes make sense to make a log transformation on a covariate if that covariate is all positive with one or two very large values, since that may reduce the influence of the outliers in that covariate, although the distribution of the residuals should be checked before and after the log transformation to assess which fitted model seems more likely to reflect the true model. Any type of transformation (on the responses or on a covariate) of course changes the model and changes the interpretation, but those transformations may be desirable in order to lead to more useful inferences.

14.3 Generalized Linear Models

14.3.1 General Framework

Another way to write the linear model is in terms of conditional expectations,

$$\mathrm{E}\left(Y_i | \mathbf{x}_i\right) = \beta_1 x_{i1} + \beta_2 x_{i2} + \cdots + \beta_p x_{ip} = \mathbf{x}_i^\top \beta = \mu_i,$$

and to assume $Y_i \sim N(\mu_i, \sigma^2)$. This form of the model more easily generalizes to different types of responses.

We can generalize the linear model by allowing different distributions for Y_i and allowing transformations of the conditional mean (not on the original responses). First, consider the transformations on the conditional mean. Let $E(Y_i|\mathbf{x}_i) = \mu_i$, and define a generalized linear model (GLM) as

$$g(\mu_i) = \mathbf{x}_i^\top \beta,$$

where $g(\cdot)$ is the link function that links the (conditional) mean to the linear predictor, $\mathbf{x}_i^\top \beta \equiv \eta_i$. Alternatively, we can write the GLM as

$$\mu_i = g^{-1}(\eta_i) = g^{-1}(\mathbf{x}_i^\top \beta),$$

where g^{-1} is the inverse of g (so that $g^{-1}(g(\mu_i)) = \mu_i$). The function $g^{-1}(\cdot)$ takes as its argument the linear predictor value, which can be any value in $(-\infty, \infty)$, and (usually) transforms it so that $g^{-1}(\mathbf{x}_i^\top \beta)$ is always within the range of allowable values of μ_i. For example, for binary responses then $\mu_i \in (0, 1)$ and a common link function is the logit link, $g(u) = \log(u/(1-u))$, so that $g^{-1}(\eta) = 1/(1 + \exp(-\eta))$, and $g^{-1}(\eta) \in (0, 1)$ for all $\eta \in (-\infty, \infty)$. As another example, for count responses then $\mu_i \in (0, \infty)$ and a common link function is the log link, $g(u) = \log(u)$ giving $g^{-1}(\eta) = \exp(\eta) \in (0, \infty)$ for all $\eta \in (-\infty, \infty)$.

The other generalization of the linear model is that the conditional distribution of Y_i given $\mathbf{x}_i$ can be other distributions besides the normal. For example, when Y_i is binary then the distribution is Bernoulli with mean μ_i. For count data the simplest distribution is the Poisson. An important point about these nonnormal distributions is that the variances typically depend on the mean. For example, independent binary responses have $E(Y_i|\mathbf{x}_i) = \mu_i$ and $\text{Var}(Y_i|\mathbf{x}_i) = \mu_i(1 - \mu_i)$, and for independent Poisson responses $E(Y_i|\mathbf{x}_i) = \mu_i = \text{Var}(Y_i|\mathbf{x}_i)$. In many situations the GLM is a more natural way to handle heterogeneity of variances than transformation of the responses and using a linear model (Section 14.2.2). For example, with count responses with 0s, the log transformation on the responses cannot be used straightforwardly, since there is no finite definition for $\log(0)$. To use the transformation, we typically add a little bit (say $\delta = 1/2$) to the 0s so that the log transformation has a finite value. In contrast, the Poisson model may be used without needing to choose an arbitrary value of δ to add to the 0s.

Inferences using the GLM typically use asymptotic normality (see Section 10.7). Brazzale et al. (2007) use higher-order asymptotics in the hoa R package that may give better inferences with small samples.

14.3.2 Binary and Binomial Responses

For binary responses the most common link is the logit link that leads to logistic regression. Each parameter β_j represents the log of the odds ratio of adding one unit to $\mathbf{x}_{ij}$, holding all other covariates constant (Problem P4).

Example 14.1 *Consider a logistic regression that models death in 2016. Suppose we have a sample of n individuals, and we predict whether or not they died in 2016, and model this prediction based on covariates measured at baseline on December 31, 2015: age, sex, indicators of heart disease and cancer. A very simple model is a logistic model where $y_i = 1$*

if the ith individual died in 2016 and $y_i = 0$ otherwise. Let the covariates be $x_{i1} = 1$ (an intercept), x_{i2} =age (in years), $x_{i3} = I(sex_i = Female)$, $x_{i4} = I(heart Disease_i = Yes)$, $x_{i5} = I(sex_i = Female)I(heart Disease_i = Yes)$, and $x_{i6} = I(cancer_i = Yes)$. This model says that the log(odds) of dying are related to baseline age, sex, heart disease, and cancer, and that the effect of heart disease may be different based on sex. After we fit the model, an estimate of the odds ratio of a person diagnosed with cancer at baseline of dying compared to a person of the same age, sex, and heart disease status at baseline is $\exp(\hat{\beta}_6)$. An estimate of the odds ratio of a woman with heart disease at baseline dying compared to a man without heart disease at baseline (but both of the same age and baseline cancer status) is $\exp(\beta_3 + \beta_5)$. An estimate of the odds ratio of dying for a person age 50 compared to a person of age 70 with all other covariates the same is $\exp(-20\hat{\beta}_2)$.

Sometimes binary response data are more conveniently written as binomial responses, since the sum of m independent Bernoulli responses with mean μ is binomial with parameters m and μ. Suppose the data for Example 14.1 are written as a two-column table where each row represents a unique $\mathbf{x}_i$ vector and there are columns for deaths and total number of people that have that set of covariates. Write the data from that table as the covariate vector for the jth row, $\mathbf{x}_j^*$, the number of deaths (y_j^*) and people (n_j) from that row. Software for logistic regression often allows the binary logistic model to be written using the binomial form, but it is essentially the same model as the associated one using binary responses: the residual deviance and degrees of freedom may change, but the estimates and inferences are the same (Problem P5).

Sometimes data is grouped in binomial form, not because of the unique covariate vector, but because of some different cause of clustering.

Example 14.2 *Suppose we are interested in modeling birth defects in a mouse experiment. The binary responses are each pup and whether it has a defect or not. The covariates may relate to the experimental interventions done to the mother. Typically, we treat the data as binomial for each mother, instead of each unique vector of covariates, because we expect pups from the same mother to be more alike (i.e., more correlated) than from different mothers.*

One way to handle the within-mother clustering of Example 14.2 is to have a random effect for each mother. These random effects models are briefly mentioned in Section 14.5. A simple alternative is to allow overdispersion in the binomial model: let the variance for the binomial response for jth mother (number of pups with defects Y_j^* out of n_j pups born) be $\text{Var}(Y_j^*|\mathbf{x}_j) = \sigma^2 n_j \mu_j (1 - \mu_j)$, where for the binomial model $\sigma^2 = 1$. When $\sigma^2 > 1$ then the model says that the pups from the same mother will be more alike than from different mothers with the same values of the covariates (see, e.g., McCullagh and Nelder, 1989, Section 4.5), and in glm in R this is called a *quasi-binomial family*. Usually underdispersion ($\sigma^2 < 1$) is unlikely to be true, and if $\hat{\sigma}^2 < 1$, then the standard binomial model (i.e., $\sigma^2 = 1$) is used.

When fitting some GLMs, there may be convergence issues on the boundaries, such as the following example.

Example 7.4 *(continued) Let us return to the Ebola example of where there were 20 subjects randomized to standard of care (SOC) and 24 subjects randomized to ZMapp+SOC,* *and only 4 subjects died and they were all in the SOC group. Suppose we fit a logistic model (GLM on a binary variable with a logit link) with only an intercept ($x_{i1} = 1$ for all i) and a treatment group indicator ($x_{i2} = 1$ if the ith subject is in the ZMapp+SOC group and $x_{i2} = 0$ otherwise). Since zero deaths happen in ZMapp+SOC group, there is no value of β_2 that maximizes the likelihood, since the value $\hat{\beta}_2$ that fits best is $-\infty$ so that the response prediction when $x_{i2} = 1$ would be exactly 0.*

It can be less obvious that there are convergence issues if there are many covariates. A way to recognize a lack of convergence is to be aware of extreme parameter estimates with extreme standard deviation estimates (Problem P6). This issue is also called *separation* since, for example, one covariate could have the failures and nonfailures separated, with all of the smaller values of the covariate being failures and all of the larger values being nonfailures (or vice versa). Heinze (2006) compares two methods to handle this issue, exact logistic regression (Mehta and Patel, 1995) and penalized maximum likelihood (Firth, 1993).

14.3.3 Count Responses

Suppose the responses are counts. Often these counts are in terms of rates or concentrations.

Example 14.3   Rate of a recurring event: *Suppose we follow the ith individual for t_i amount of time and let y_i be the number of events that occur for that individual during that time. If the events follow a Poisson process with intensity λ_i (i.e., no two events happen at the same time and the time between events are independent exponential random variables with mean $1/\lambda_i$), then $Y_i \sim Poisson(t_i\lambda_i)$.*

Example 14.4   Concentration: *Suppose we are measuring the concentration of malaria parasites in the circulating blood. Suppose the ith individual has a concentration of λ_i and we take a blood sample of size t_i and count the number of parasites in that sample as y_i. If the parasites are randomly mixed, then $Y_i \sim Poisson(t_i\lambda_i)$. Actual assays for this are more complicated (Note N3).*

Suppose the response for the ith individual is $Y_i \sim Poisson(\mu_i)$, where $\mu_i = t_i\lambda_i$, the value t_i is known (see examples), and the covariate vector is $\mathbf{x}_i$. Suppose we want to model the rate, λ_i, as a function of the covariates. We use a GLM with a log link:

$$\log\{E(Y_i|\mathbf{x}_i)\} = \log(\mu_i) = \log(t_i) + \mathbf{x}_i^\top \beta,$$

where the term $\log(t_i)$ is known as an "offset"; it is like an additional covariate in the linear model where the parameter is known to be 1. The offset is used so that inferences are on the rate not the count, since we can rewrite the model as

$$\log(\mu_i) - \log(t_i) = \log(\lambda_i) = \mathbf{x}_i^\top \beta.$$

Then relative risks or rate ratios are expressed as $\exp(\beta)$.

Example 14.5   *Consider a randomized trial where the ith subject is randomized to either a new vaccine ($x_{i2} = 1$) or control ($x_{i2} = 0$) and let $x_{i1} = 1$ for all i be the intercept term. Suppose the ith subject has several baseline variables that are incorporated into the rest of the covariate vector, $x_{i3}, \ldots, x_{ip}$. Let t_i be the time that the ith subject is followed and at risk of getting malaria and let y_i be the number of malaria infections during the follow-up* *period (malaria may be treated and the subjects may get reinfected). Then fit a Poisson GLM with a log link and an offset of* $\log(t_i)$ *for the ith subject:* $\log(\mu_i) = \log(t_i) + x_{i1}\beta_1 + x_{i2}\beta_2 + \cdots x_{ip}\beta_p$. *Then the rate of malaria infections on subjects in the vaccine arm compared to those in the control arm is estimated by* $\exp(\hat{\beta}_2)$ *and the estimate of the vaccine efficacy is* $1 - \exp(\hat{\beta}_2)$.

Typically it is safer to not assume the Poisson distribution but to allow overdispersion from the Poisson model so that $\text{Var}(Y_i|\mathbf{x}_i) = \sigma^2\mu_i$, and we estimate σ^2 with a scaled Pearson chi-squared statistic,

$$\hat{\sigma}^2 = \frac{1}{n-p} \sum_{i=1}^{n} \frac{\left(y_i - \hat{\mu}_i\right)^2}{\hat{\mu}_i}.$$

This is called a *linear overdispersion model* (McCullagh and Nelder, 1989, p. 199) or a *quasi-Poisson model* (in glm in R). A quadratic overdispersion model is derived using a gamma random effect. Let γ_i be distributed gamma with mean 1 and variance τ^2, so that the model with the random effect has $\text{E}(Y_i|\mathbf{x}_i, \gamma_i) = \gamma_i\mu_i$, and unconditionally (without conditioning on γ_i) $\text{E}(Y_i|\mathbf{x}_i) = \mu_i$ and $\text{Var}(Y_i|\mathbf{x}_i) = \mu_i + \tau^2\mu_i^2$. This is the same as a negative binomial model on Y_i. The quadratic overdispersion (negative binomial) model is slightly harder to calculate than the linear overdispersion. Zeileis et al. (2008) discuss the negative binomial model and other extensions to the Poisson model such as zero-inflated Poisson models that allow $\Pr(Y_i = 0)$ to be larger than expected from the Poisson model.

While these methods are often useful for count data, it is also important to recognize that sometimes count data are not well modelled by the Poisson model, or even the more generalized versions of them discussed above. Inappropriate use of these models could result in inflated Type I error rates or low power. In such scenarios, a semiparametric model such as proportional odds or proportional hazards regression might be a better choice. For a two-sample problem, the proportional odds model is very closely related to a Wilcoxon rank sum test when there are no covariates (see Section 9.2, Fay and Malinovsky (2018)). The semiparametric models, furthermore, easily accommodate covariate adjustment, and they are introduced in the next section.

14.4 Ordinal Responses or Ranked Responses

In this section we show how to generalize the two-sample proportional odds model and proportional hazards model to handle a set of k covariates.

14.4.1 Ranking of Continuous Responses

Consider the linear transformation model on a continuous response,

$$h(y_i) = \mathbf{x}_i^\top \beta + \epsilon_i,$$

where $\epsilon_i \sim F$ and F is a known distribution, $h(\cdot)$ is an unknown increasing transformation, and β is a $p \times 1$ parameter vector of interest. The model is semiparametric because $h(\cdot)$ is infinite dimensional, and β has a finite dimension. Because $h(\cdot)$ is unknown, the model only uses the ranks of the responses, since any monotonic transformation would give the same β

parameters (but with of course a different $h(\cdot)$ function). When F is the logistic distribution, this is the proportional odds model and when F is the extreme value distribution this is the proportional hazards model (more often used in survival analysis, see Chapter 16). An equivalent formulation of the model is

$$g\left(\Pr\left[Y_i \leq t | \mathbf{x}_i\right]\right) = h(t) - \mathbf{x}_i^\top \beta \text{ for any } t,$$

where $g(q) = F^{-1}(q)$. For the logistic distribution $F^{-1}(q) = \text{logit}(q) = \log(q/(1-q))$, and for the extreme value distribution $F^{-1}(q) = \log\{-\log(1-q)\}$. These are common link functions (the logit link and the complementary log-log link).

With continuous data the proportional hazards model can be written as

$$S_i(t) = S_0(t)^{\exp(-\mathbf{x}_i^\top \beta)}, \tag{14.2}$$

where $S_i(t) = \Pr[Y_i > t | \mathbf{x}_i]$ is called the survival function and $S_0(t) = \exp\{-e^{h(t)}\}$ is the baseline survival function (i.e., the survival function when $\mathbf{x}_i = \mathbf{0}_{p \times 1}$). The proportional hazards model can also be written as

$$\lambda_i(t) = \lambda_0(t) \exp(-\mathbf{x}_i^\top \beta), \tag{14.3}$$

where λ_i and λ_0 are the hazard function given $\mathbf{x}_i$ and the baseline hazard function, respectively (Problem P7). Specifically,

$$\lambda_i(t) = \lim_{\delta \to 0^+} \frac{\Pr\left[t \leq Y_i \leq t + \delta | Y_i \geq t, \mathbf{x}_i\right]}{\delta}.$$

If Y_i is a continuous time-to-event response, we interpret the hazard function at t as the instantaneous rate of having the event given that the event has not happened before time t.

The proportional hazards model has a close relationship to the Mann–Whitney parameter (Equation 9.3, p. 130), through the win ratio.

Example 14.6 Win ratio: *Consider a clinical trial comparing a new treatment to control where the endpoint is time until death. For the ith individual, let y_i be the time from start of the study until death, and let $x_{i2} = 1$ if they received treatment and $x_{i2} = 0$ if they received control, and let $x_{i1} = 1$ for all i. Then under the (continuous) proportional hazards model the hazard ratio for treatment is $\theta = \exp(-\beta_2)$. Let W_0 and W_1 be a random response on control or treatment, respectively. Then $\theta^{-1} = \Pr[W_1 > W_0]/\Pr[W_0 > W_1]$ is called the* win *ratio, which is the probability that a random individual on treatment would die after a random individual on control (a win for treatment) divided by the probability that a random individual on treatment would die before a random individual on control (a loss for the treatment). The hazard ratio, θ, is sometimes called the* loss *ratio. Because the win ratio is defined using the Mann–Whitney parameter (without the tie correction), it does not depend on the proportional hazards assumption. However, if the study is stopped before all individuals have died (typical for most studies), the proportional hazards assumption is necessary for the loss-ratio interpretation of the hazard ratio (Oakes, 2016).*

Typically, the proportional hazards model is easier to fit than the proportional odds model because we can use a partial likelihood such that λ_0 does not need to be estimated in order for β to be estimated (see, e.g., Kalbfleisch and Prentice, 2002). For the proportional odds model, methods that estimate $h(\cdot)$ can be used such as profile likelihoods (Murphy et al., 1997). Simpler estimators that work for any linear transformation model use estimating equations based on $I(Y_i \geq Y_j)$ for all pairs i, j (see Cheng et al., 1995; Thas et al., 2012).

The proportional odds model has the advantage over the proportional hazards model that the former model is palindromically invariant, meaning the inferences will change in the appropriate way if we reverse the order of the responses (i.e., if we transform both the responses and the covariates by multiplying by -1, then β does not change). The proportional hazards model is not palindromically invariant. The proportional hazards model has advantages for time-to-event data where some observations are right censored and the study is stopped before all individuals have had the event (see Example 14.6 and Chapter 16).

14.4.2 Ordinal Responses

Suppose that the response is ordered categorical with k categories, and write that response for the ith individual as a scalar Y_i with possible values in $\{1, \ldots, k\}$. As with the rest of the chapter let $\mathbf{x}_i$ be a $p \times 1$ vector of covariates. To relate Y_i to the previous section, suppose there is some unobserved continuous response Y_i^* such that

$$Y_i = j \text{ if } \xi_{j-1} < Y_i^* \leq \xi_j.$$

In other words, Y_i is a grouped version of Y_i^*. Suppose Y_i^* follows a linear transformation model of the previous section, then

$$g\left(\Pr\left[Y_i^* \leq \xi_j | \mathbf{x}_i\right]\right) = h(\xi_j) - \mathbf{x}_i^\top \beta \text{ for any } t,$$
$$\Rightarrow \quad g\left(\Pr\left[Y_i \leq j | \mathbf{x}_i\right]\right) = \gamma_j - \mathbf{x}_i^\top \beta \text{ for } j = 1, \ldots, k,$$

where $\gamma_j = h(\xi_j)$, and when g is the logit function we get the proportional odds model and when g is the complementary log-log link we get the proportional hazards model. We see that β from the linear transformation model is the same as β from the grouped responses, Y_i. For grouped responses, we only need to estimate $h(\cdot)$ at k points. In this case, the likelihood is

$$l(\mathbf{y}, \mathbf{X}, \beta, \gamma) = \prod_{i=1}^{n} \prod_{j=1}^{k} \left\{ g^{-1}\left(\gamma_j - \mathbf{x}_i^\top \beta\right) - g^{-1}\left(\gamma_{j-1} - \mathbf{x}_i^\top \beta\right) \right\}^{I(y_i=j)},$$

where $\gamma_0 = -\infty$. We can use asymptotic likelihood methods (Section 10.7) for inferences.

14.5 Clustered Data

14.5.1 Overview

In this section, clustered data are data where individuals may be grouped (or clustered) into typically similar units. For example, individual students are clustered into schools, individual pups are clustered into litters, and individual teeth are clustered into mouths.[1] Longitudinal

---

[1] In Chapter 12 we differentiated between when all individuals in a group had the same covariates (in that case we called the grouping "clustering") and when individuals in a group had different covariates (in that case we called the grouping "stratification"). In this Section we make no such differentiation and call both types of data, clustered data.

data are a type of clustered data where individuals are measured repeatedly over time, and for that type of data individual responses in time are clustered within each individual.

We discuss two main types of models for clustered data, mixed effects models (which can also be analyzed by conditional likelihood models) and marginal models (which are often analyzed with generalized estimating equations). Loosely speaking, the mixed effect model combines fixed effects that are our primary interest and random effects that introduce random cluster-level error into our inferences. The marginal models do not explicitly model the cluster-level error on the means but do account for it using within-cluster correlation (typically exemplified by observations within a cluster being more similar than observations between clusters).

We will emphasize a few main points in this section. First, if your interest is in an effect that acts across clusters, and you have a nuisance parameter for each cluster and a small number of observations per cluster (e.g., two eyes clustered into each individual), then, as we show later in Example 14.7a, there may be bias in your estimates even if you have very many clusters. Second, a generalized linear model version of the mixed effects model (GLMMIX model) and the associated marginal model are estimating different estimands, unless your link function is the identity link. For example, with the logit link for binary data the fixed effects parameters from mixed effects model and the apparently identical parameters from the marginal model are estimating different quantities. Finally, we emphasize that if the number of observations per cluster is related to the mean, then care must be taken using marginal models, because what you are estimating depends on how many observations per cluster you measure. For example, if the response is the number of cavities per tooth in adults and teeth are clustered by person, then the expected number of cavities per tooth may be related to how many teeth a person has left in their mouth (since poor dental health can result in loss of teeth as well as more cavities in the remaining teeth), and adults with more teeth will get more weight in a marginal model.

14.5.2 Mixed Effects and Conditional Models

These models are very general, but to highlight the main points we focus on some specific examples. Consider a generalized linear model applied to a multicenter clinical trial. Suppose the design of the trial is that there are n_c centers, and in each center there are m individuals randomized $1:1$ to a treatment and control, where m is even and in each center we ensure $m/2$ in each treatment group. Notationally, it is easier to use double indices for center and person-within center, so for the ith center the m individual responses are $\mathbf{y}_i = [y_{i1}, \ldots, y_{im}]$, and the vector of covariates (perhaps including a 1 for the intercept term) is $\mathbf{x}_{ij}$, and within $\mathbf{x}_{ij}$ there is a treatment indicator z_{ij}, where $z_{ij} = 0$ if randomized to control and $z_{ij} = 1$ if randomized to treatment. Let Y_{ij} be the random variable associated with y_{ij} and let $\mu_{ij} = E(Y_{ij}|\mathbf{x}_{ij})$. Consider the generalized linear model (Problem P8),

$$g(\mu_{ij}) = \gamma_i + z_{ij}\beta, \tag{14.4}$$

where γ_i is a shift effect for the ith center and β is the treatment effect that is the same across all centers. We are interested in β and the shift effects are nuisance parameters.

Unfortunately, the usual maximum likelihood method may be biased if there are a large number of parameters compared to the sample size.

Example 14.7a *Suppose we have a logistic regression following Equation 14.4, so that Y_{ij} are binary values and g is the logit link. We simulate 10,000 data sets where $n_c = 100$, $m = 2$, $\beta = 1$, and the $\gamma_1, \ldots, \gamma_m$ are independent normal with $\gamma_i \sim N(-1, 1)$ for all i. For each simulated data we fit a logistic model where we treat the center effects $(\gamma_1, \ldots, \gamma_m)$ as fixed effects and use* glm *in* R. *The mean of the 10,000 simulated estimates of β is 2.05, about twice the true value. So there is clearly some bias. In this situation, there are n = 200 total individuals and 101 parameters. If we repeat the simulation with 10 times more centers, the bias remains. Let $n_c = 1,000$ and $m = 2$ so the total sample size is n = 2,000 but there are 1,001 parameters. Since the calculations are much slower with 1,001 parameters, we only simulate 1,000 replicates, but the mean of the estimates of β is 2.01. In both cases there is about one parameter for every two observations.*

To reduce the bias, we need more individuals per center. We repeat the simulation with $n_c = 100$ but with $m = 4, 6, 10$, or 20 so that the total number of individuals grows from 400 to 2,000 with still 101 parameters. The means of the simulated estimates of β for the different m values are 1.35 (m = 4), 1.21 (m = 6), 1.12 (m = 10), and 1.05 (m = 20).

When the number of parameters increases with the sample size, even as the total sample size goes to infinity the maximum likelihood estimate of a fixed parameter (i.e., one that does not change with sample size) may asymptotically have substantial bias. This is called the Neyman–Scott problem after Neyman and Scott (1948).

One way to handle the Neyman–Scott problem is to condition on sufficient statistics for the nuisance parameters (i.e., the parameters that increase with sample size). For Example 14.7a this amounts to performing conditional logistic regression (details not covered, see Section 14.10 for references). If we repeat the simulation of Example 14.7a with $n_c = 100$ and $m = 2$ and use conditional logistic regression, then the mean estimate of β from 10,000 replicates is 1.03, much closer to the true value of 1 than the mean of 2.05 for the unconditional logistic regression with fixed effects for center.

Another approach is to treat the nuisance parameters as random effects, and treat the other parameters as fixed effects, so that the combined model is called a *mixed effects* model. We now describe the generalized linear mixed effects (GLMMIX) model. Similar to above, let Y_{ij} and $\mathbf{x}_{ij}$ be, respectively, the response and fixed effects covariate vector for the jth individual within the ith cluster. Let $\mathbf{c}_{ij}$ be the random effects covariate vector associated with Y_{ij}. The GLMMIX model, can be written as

$$g(\mu_{ij}) = \mathbf{x}_{ij}^{\top}\beta + \mathbf{c}_{ij}^{\top}\mathbf{b}_i, \tag{14.5}$$

where $\mu_{ij} = \mathrm{E}\left(Y_{ij}|\mathbf{x}_{ij}, \mathbf{c}_{ij}, \mathbf{b}_i\right)$, β is a vector of fixed effects parameters, and $\mathbf{b}_i$ is a vector of random effects parameters associated with the ith cluster. The model assumes that the random effects parameters $\mathbf{b}_i$ are independent and distributed with some known distributional form described by a few (usually) unknown parameters. For convenience, we typically assume $\mathbf{b}_i$ are multivariate normal with mean $\mathbf{0}$ (a vector of all 0s) and covariance **G**. The form of **G** depends on the specific application, but if the dimension of b_i is low it can be left completely unspecified.

Example 14.7b   *(continuation of Example 14.7a) Here is a GLMMIX model applied to the data design with binary responses from n_c centers, with m responses in each center, half to treatment (when $z_{ij} = 1$) and half to control ($z_{ij} = 0$):*

$$g(\mu_{ij}) = \beta_1 + z_{ij}^\top \beta_2 + b_i, \tag{14.6}$$

*where we assume $b_i \sim N(0, \sigma_b^2)$. In this model, there are only three parameters that need to be estimated (β_1, β_2, and σ_b). If we let $\beta_1 = -1$, $\beta_2 = 1$, and $\sigma_b = 1$, then this is essentially the same model as Example 14.7a, except previously we treated the parameters $\gamma_1, \ldots, \gamma_{n_c}$ as fixed effects, now we treat those same effects as a fixed intercept parameter, β_1, plus the random effects, $b_1, \ldots, b_{n_c}$. Using the same data generating model as Example 14.7a with $n_c = 100$ and $m = 2$, we again simulate 10,000 replicates and use the **glmer** function in the **lme4** R package to estimate the treatment effect, here denoted with β_2. The mean of the estimates is 0.97. We see there is much less bias than using the fixed effects model.*

14.5.3 Marginal Models

In the previous subsection, we explicitly modeled the effects for each cluster in the means (either using a fixed effect for each cluster such as $\gamma_1, \ldots, \gamma_m$ in Example 14.7a, or using random effects such as $\mathbf{b}_1, \ldots, \mathbf{b}_{n_c}$ in Example 14.7b). In this subsection, we account for the cluster effect not through the means, but through the covariance structure. It is important to realize that if the link function is not the identity link, this means that the effects of interest that act across all clusters will not be equivalent between the two models. We explain this difference by supposing we have a GLMMIX data generating model, but we model it with a marginal model. But first, we explain the marginal model.

To explain how a marginal model can account for the within-cluster correlation we describe a very general approach, called *generalized estimating equations* (GEE). Suppose we have n_c clusters of data that are independent. For example, each cluster could be a set of responses from a litter of mice. Let $\mathbf{Y}_i = [Y_{i1}, \ldots, Y_{im_i}]$ be an $m_i \times 1$ vector of responses for the ith cluster and let $\mathbf{x}_i$ be its $m_i \times r$ matrix of covariates. Let $\theta = [\beta^*, \xi]$ be the parameter vector. Here we use β^* to emphasize that the marginal parameters are different from the parameters from the conditional model or the fixed effects parameters from the mixed model. The parameters do not necessarily represent the complete description of a probability model. The β^* parameter vector describes the mean of the responses, and the ξ parameters (together with β^*) describe the variance. As with generalized linear models, let

$$\mathrm{E}(\mathbf{Y}_i | \mathbf{x}_i) = g^{-1}(\mathbf{x}_i^\top \beta^*) \equiv \mu_i^*, \tag{14.7}$$

where g^{-1} is an inverse link function (see Section 14.3).[2] Let $D_i(\beta^*) = \partial \mu_i^* / (\partial \beta^*)$ and let $V_i(\beta^*, \xi)$ be a working variance estimate of Y_i. The term *working variance estimate* means that the asymptotic accuracy of the method does not depend on correctly specifying

---

[2] We assume the link functions and their inverses act elementwise. For example, for a vector $\mathbf{u} = [u_1, \ldots, u_m]$ then $g(\mathbf{u}) = [g(u_1), \ldots, g(u_m)]$.

the variance model, it is just a working approximation. The GEE method partitions the working variance estimate into parts representing a scalar dispersion parameter and the correlation matrix, which is then pre- and postmultiplied by a diagonal matrix of elements proportional to the standard deviations (Note N4). This allows us to try different working correlation models, such as (i) an independence model where all observations within a cluster are uncorrelated, (ii) an exchangeable correlation structure where $\text{Corr}(Y_{ij}, Y_{ik}) = \rho$ for all $j \neq k$, or (iii) a first-order autoregressive process for time series data where the correlation is related to how far apart observations are in time.

Inferences are made using sandwich estimators (Section 10.8), where the estimating equation is

$$U_i(\beta^*) = D_i^T(\beta^*) V_i^{-1}(\beta^*, \xi) \left\{ Y_i - \mu_i^*(\beta^*) \right\} = \mathbf{0}.$$

There are regularity conditions such as: a consistent estimator of the derivative of $U_i(\beta^*)$ is $\hat{\Omega}_i = D_i^T(\hat{\beta}^*) V_i^{-1}(\hat{\beta}^*) D_i(\hat{\beta}^*)$, the mean is correctly specified (i.e., $\text{E}(Y_i) = \mu_i^*(\beta^*)$ for some β^*), and the sample size n_i does not depend on β^*. Under those conditions then we can treat $\hat{\beta}^*$ as $AN(\beta^*, V_s)$ and V_s is given in Equation 10.8.

Now we explain the difference between β from Section 14.5.2 and β^*. Suppose the data are generated by a GLMMIX model,

$$g(\mu_i) = \mathbf{x}_i^\top \beta + \mathbf{c}_i^\top \mathbf{b}_i, \tag{14.8}$$

where $\mu_i = [\mu_{i1}, \ldots, \mu_{im_i}]$, $\mu_{ij} = \text{E}(Y_{ij} | \mathbf{x}_{ij}, \mathbf{c}_{ij}, \mathbf{b}_i) \equiv \text{E}(Y_{ij} | \mathbf{b}_i)$, g is the link function, $\mathbf{b}_i$ is an $r \times 1$ random effect parameter vector, and $\mathbf{c}_i$ is an $m_i \times r$ matrix of known covariate values or constants. Assume the $\mathbf{b}_i$ are independent and multivariate normal with mean 0 and covariance $\mathbf{G}$. Taking g^{-1} of both sides gives

$$\text{E}(\mathbf{Y}_i | \mathbf{b}_i) = g^{-1}\left(\mathbf{x}_i^\top \beta + \mathbf{c}_i^\top \mathbf{b}_i\right). \tag{14.9}$$

Now suppose that we use a marginal model on data from this data generating mechanism, so we model it as

$$\text{E}(\mathbf{Y}_i) = g^{-1}\left(\mathbf{x}_i^\top \beta^*\right). \tag{14.10}$$

If the link function is the identity (as in a linear model), then g^{-1} is also the identity function and taking expectation over the random effects from equation 14.9 just gives $\mathbf{x}_i^\top \beta$, so in that case $\beta = \beta^*$. But if the link function is the logit link or the log link, then taking the expectation over the random effects will give $\beta \neq \beta^*$ (Note N5).

Example 14.7c *(continuation of Example 14.7a). Here is a GEE model applied to the binary responses from n_c centers, with m responses in each center, half to treatment (when $z_{ij} = 1$) and half to control ($z_{ij} = 0$). We model the mean responses using*

$$g\left(\text{E}(Y_{ij} | \mathbf{x}_{ij})\right) = \beta_1^* + z_{ij}^\top \beta_2^*, \tag{14.11}$$

where g is the logit link and β_1^* and β_2^* are analogous to (but different from) β_1 and β_2 from Example 14.7b. We use a GEE with working independence correlation. Using the same 10,000 simulated data sets as previously (i.e., with treatment effect, $\beta_2 = 1$, and standard deviation of the random center effects, $\sigma_b = 1$), the mean of the estimated β_2^* is 0.84. The true value is $\beta_2^* \approx \beta_2 / \sqrt{1 + 0.346\sigma_b^2} = 0.862$ (see Note N5).

The important point is that the marginal parameter, β^*, is not equal to the mixed effects (or conditional) parameter β. Sometimes the marginal parameter is called the *population averaged* parameter, and the conditional parameter is called the *subject-specific* parameter (Zeger et al., 1988).

Because the marginal model is taking the average over the population, we need to be careful about the nature of the number within each cluster. If you do not want the number of individual observations to affect the mean parameters, you can use within-cluster resampling (see Section 10.12).

Example 14.8 *Suppose you want to model whether a high sugar diet affects the amount of cavities in teeth. If your data is $y_{ij} = 1$ if the jth tooth in the ith subject has a cavity or filling, and $y_{ij} = 0$ otherwise, then it is very likely that m_i the number of teeth for the ith person is related to the presence or absence of cavities in those teeth, because poor dental health is related to both losing teeth and the presence of cavities in existing teeth. If you use a marginal model, then individuals with a large number of teeth will have more weight than individuals with few teeth.*

14.6 Covariate Selection

There are two main choices in model selection. First, we must choose the form of the model. For example, for count data we might choose between a Poisson regression or a linear model on the log-transformed responses. Another model form choice is the link in a generalized linear model. For example, if the responses are binary, we may choose a GLM from the binomial family with either a logit link (i.e., a logistic model) or a complementary log-log link. Once you choose the form of the model, the second choice is which covariates to include in your model. In this section, we focus on the second choice.

First consider a series of linear models, all trying to predict a response using a different set of covariates. If you take a linear model and add another covariate, the fit of the model will be better in the sense that the maximum likelihood estimate of variance, $\hat{\sigma}^2$ (i.e., the average of the squared residuals), will decrease. Alternatively, we can judge the fit on the R^2 value (higher is better) or the loglikelihood (higher is better). The problem with these measures of goodness-of-fit is that apparently "better" fit will happen even if the added covariate in truth is not related to the response; even adding a parameter associated with an independent random covariate to a model will slightly improve those goodness-of-fit measures. To correct for this, Akaike developed the AIC (Akaike's information criterion), which is

$$AIC = -2\log(L(\hat{\theta})) - 2k,$$

where $\log(L(\hat{\theta}))$ is the loglikelihood evaluated at $\hat{\theta}$, the maximum likelihood estimate of the parameter vector, and k is the dimension of θ. When interpreting the AIC, lower is better. When you add a parameter that is just modeling random noise, under some regularity conditions the asymptotic expected value of the AIC will not increase. As a practical matter, you can only compare the AIC between models that all use the same individuals, so if some parameters are missing values and the software automatically deletes individuals without those values, then the AIC values will not be comparable.

The AIC can be used to choose among a large set of models. Consider a model with up to p possible parameters. We could fit each of the p single parameter models, then all $\binom{p}{2}$ models with two parameters, etc., until all models have been tried. Then we could pick the model with the lowest AIC. However, as p increases, the number of models quickly increases making the algorithm intractable. Another procedure is to use stepwise regression, where from the current model we try either adding each additional parameter not already in the model (forward stepwise) or subtracting each parameter in the model (backward stepwise). The stepwise procedure proceeds until no further decrease in the AIC is achieved.

Another very general way of selecting variables is through a model selection method with a tuning parameter (the tuning parameter allows more or less variables to be included into the model as it changes), combined with a method for evaluating the model at each value of the tuning parameter. A conceptually simple method for evaluation is cross-validation, and a very useful model selection method is the lasso.

Tibshirani (1996) introduced the *lasso*, which stands for least absolute shrinkage and selection operator. We consider the generalized linear model (GLM) generalization of the lasso called L_1−regularized GLM, where the response, y_i, may be continuous (as in linear regression), binary (as in logistic regression), or counts (as in Poisson regression). The predictor variables, $\mathbf{x}_i = [x_{i1}, \ldots, x_{ip}]$, are standardized so that $(1/n) \sum_{i=1}^{n} x_{ij} = 0$ and $(1/n) \sum_{i=1}^{n} x_{ij}^2 = 1$. The model for the expected responses is $E(Y_i | \mathbf{x}_i) = \beta_0 + \sum_{j=1}^{p} x_{ij} \beta_j$. Let $\ell(\beta_0, \beta)$ be the loglikelihood, which is a function of β_0 and $\beta = [\beta_1, \ldots, \beta_p]$ and perhaps σ^2 in the case of linear models. We separate β_0, the intercept, because it is treated differently from the rest of the parameters. In the linear model, the loglikelihood is a function of the sum of squared errors. The L_1−regularized GLM gives the parameters β_0 and β that solve

$$\min_{(\beta_0, \beta)} \left\{ -\ell(\beta_0, \beta) + \lambda \sum_{j=1}^{p} |\beta_j| \right\},$$

where λ is the tuning parameter and the summation does not include β_0. The L_1−regularized GLM shrinks the parameter estimates toward 0, and also selects parameters for the model (for some values of λ many parameter estimates may be exactly 0, effectively defining them as not selected).

One problem in model building is overfitting, that is, developing a model that fits the current data set really well, but does not generalize to an independent set of data. Overfitting comes from using the same data set to fit the model as we use to evaluate how well the model fits. A solution to avoid overfitting is cross-validation, where we repeatedly partition the data set into a large training data set and the rest of the data, which we call the *validation data set*. For each partitioning, we evaluate the fit on the validation data set using the model created with the training data set.

Here are more details. In cross-validation we randomly partition the data set into r approximately equal sized data sets. For example, suppose we have data on $n = 91$ individuals, and the data on the ith individual is the response y_i and the $1 \times p$ covariate vector $\mathbf{x}_i = [x_{i1}, \ldots, x_{ip}]$. Let $r = 9$ and randomly partition the data set into 9 groups of approximately 10 individuals each, with the 9 groups defined by sets of indices: $G_1 = \{1, 2, \ldots, 10\}$, $G_2 = \{11, 12, \ldots, 20\}, \ldots, G_9 = \{81, 82, \ldots, 90, 91\}$. Then we fit the model r times, each time leaving out one of the groups of individuals. Let the model parameter estimate from leaving out the jth group be $\hat{\theta}^{(-j)}$. Let the predicted response for the ith individual using a parameter estimate $\hat{\theta}$ be $\hat{y}_i(\hat{\theta})$. Calculate the mean squared error of the residuals using the fit for the individuals in each group from the model created without the individuals in that group,

$$MSE = \frac{1}{n} \sum_{j=1}^{r} \sum_{i \in G_j} \left(y_i - \hat{y}_i \big(\hat{\theta}^{(-j)} \big) \right)^2 .$$

Typical values of r are 5 or 10, and when $r = n$, this is known as leave-one-out cross-validation. For details about bias and variance of cross-validation estimates see Hastie et al. (2009, chapter 7).

We can combine a selection method such as the lasso with cross-validation. Let $MSE(\lambda)$ be the cross-validated mean squared error when the tuning parameter is fixed at λ, then choose the model associated with the value of λ that minimizes $MSE(\lambda)$.

14.7 Some Problems When Testing from Models

When we make inferences from models, there are issues of selection (which model we will use for inferences), and issues of multiple testing (sometimes we test many hypotheses from a single model). This section will necessarily be short and will highlight some of the problems with testing from models. As this is a very active area of research, the solutions we point to may no longer be the best ones at the time you are reading this. But the reader should be aware of the problems.

In the earlier parts of this chapter, we have assumed that the full model (both the form and the choice of parameters) is known, and the inferential problem is to estimate the parameters and test hypotheses on the parameters. Many times in applications, the model is not known. Thus, often model selection is part of the inference problem. A multiplicity problem can arise because hypothesis tests are model dependent.

Example 14.9 *Suppose we want to know if we can predict a response, y, from a covariate of interest, x_1, in the presence of a nuisance covariate, x_2. Consider two linear models:*

$$\mathrm{E}(Y_i) = \beta_0 + x_{i1}\beta_1,$$

and

$$\mathrm{E}(Y_i) = \beta_0^* + x_{i1}\beta_1^* + x_{i2}\beta_2^*.$$

Although our primary interest is in the relationship between x_{i1} and y_i, the way we measure that relationship is different in the two models. In the second model, β_1^* represents the change in $\mathrm{E}(Y_i)$ due to x_{i1} after adjusting for the effect of x_{i2}, while the first model makes no adjustment for x_{i2}. In other words, $\beta_1 \neq \beta_1^*$. So if our inferential procedure is a two step procedure: (1) pick one of the two models based on the data, (2) test whether the parameter associated with x_{i1} is 0 or not, then it is not straightforward to understand the properties of the two-step procedure, because sometimes we test $H_0 : \beta_1 = 0$ and other times we test $H_0^* : \beta_1^* = 0$.

Note that using AIC or cross-validation will not eliminate multiplicity issues with hypothesis testing. We review this issue with a simulation study of a linear model with one response and 10 covariates. We first consider the case where we do not have a particular covariate effect that is of primary interest, and the analysis is exploratory and any subset of the covariates could relate to the response.

Example 14.10 *Consider the following simulation design. Let $n = 100$ be the total number of subjects. Let the responses be distributed standard normal, $Y_i \sim N(0,1)$ for $i = 1, \dots, n$ and all are independent. In addition, suppose we have 10 covariates that are all standard normal and independent of each other and the response; in other words, $X_{ij} \sim N(0,1)$ for $i = 1, \dots, n$ and $j = 1, \dots, 10$ and all random variables (covariates and responses) are independent. Suppose we are interested in finding any significant covariate that is related to the response. We test the (familywise) Type I error rate by simulating the rate of rejection of the most significant covariate. We consider three modeling and testing strategies, all rejecting at the $\alpha = 0.05$ level, and use simulation to determine the true rejection rate:*

(1) *Identify the most significant single variable by fitting a single covariate model for each of the 10 covariates. For the jth covariate, fit the linear model which models the response for the ith individual as $E(Y_i) = \beta_0 + \beta_1 X_{ij}$. Test the null that $\beta_1 = 0$ for each model and pick the model with the lowest p-value and estimate its rejection rate.*

(2) *Identify the most significant single variable in the presence of the other covariates in a large model. Put all 10 covariates into the model as main effects, $E(Y_i) = \beta_0 + \sum_{j=1} \beta_j X_{ij}$. Perform 10 tests of the 10 null hypotheses that $\beta_j = 0$ for $j = 1, \dots, 10$. Estimate the rejection rate of the most significant covariate.*

(3) *Identify the most significant single variable in the presence of the other covariates after selecting the "best fit" model. Here we use stepwise regression using the AIC to determine the best fit. We use the R function* **step** *to do stepwise regression, where the function steps through different models by picking the best next model by adding (direction = "forward") or subtracting (direction = "backward") the covariate that decreases the AIC the most. We use the default in* **step** *(R version 3.5.1) which is direction = "both" and is a combination of forward and backward selection. Then estimate the rejection rate of the most significant covariate from the final selected model.*

Because all the covariates are independent of each other and the response, Strategy 1 is like doing 10 independent tests at $\alpha = 0.05$, so we would expect to reject at least 1 of the 10 tests with probability $1 - (0.95)^{10} = 0.4013$ (see page 235). Our simulated rejection rates (using 10^4 replicates) for the most significant covariate from the three strategies are similar: 0.4000 for Strategy 1, 0.3835 for Strategy 2, and 0.4450 for Strategy 3. So even if we only select one model by a stepwise strategy for example, we still have a multiple testing problem when we focus on the most significant covariate.

Suppose we have one covariate for our focus, such as a treatment indicator in a randomized study, and we want to adjust for other nuisance covariates. Does trying many models affect the Type I error rate of the covariate of interest? This depends on how you decide among the models, and on the correlations among the nuisance covariates and the response. First consider the case where there is no correlation among the covariates.

Example 14.11 *We repeat the simulation design of Example 14.10, but now our interest is in only the first covariate, and the other nine covariates are just adjusting for nuisance effects. We study the Type 1 error rate for rejecting the null with $\alpha = 0.05$ that the first covariate is 0. The three strategies are now:*

(1*) *Fit the single covariate model for the covariate of interest, say X_{i1}, i.e., $\mathrm{E}(Y_i) = \beta_0 + \beta_1 X_{i1}$. Test the null that $\beta_1 = 0$ estimate its rejection rate.*

(2*) *Fit the model in 1*, but additionally fit the nine other models that add one of the other covariates, i.e., $\mathrm{E}(Y_i) = \beta_0 + \beta_1 X_{i1} + \beta_2 X_{ij}$ for $j = 2, \ldots, 10$. Pick the model based on the most significant effect (i.e., lowest p-value) of the 10 tests of $H_0 : \beta_1 = 0$. Then estimate the rejection rate based on that pick.*

(3*) *Perform a stepwise regression as in Strategy 3, except always include the first covariate, $\mathbf{X}_1$. Then estimate the rejection rate from the final selected model for the test $H_0 : \beta_1 = 0$, where β_1 is the parameter associated with $\mathbf{X}_1$.*

As expected from the theory, we reject from Strategy 1* about 5% of the time, with simulated rejection rate 0.0480. Strategy 2* is guaranteed to reject more often, but the simulated Type I error rate is slightly inflated, 0.0654. Strategy 3* has simulated Type I error rate of 0.0571, and there is not much rate inflation because the fit is picked based on the AIC not on the p-value for $H_0 : \beta_1 = 0$.

The example is a little artificial in that none of the covariates are related to each other. Consider a different and even more artificial example.

Example 14.12 *We use the same design as Example 14.11, but now all covariates and the responses are correlated with $\rho = 0.5$ except the covariate of interest and the response which have correlation 0. So there is independence between the response and the covariate of interest. For this simulation design, we get simulated rejection rates of 0.0506 for Strategy 1*, 0.9998 for Strategy 2*, and 1.000 for Strategy 3*. Since there is independence between the response and $\mathbf{X}_1$, the p-values for testing $H_0 : \beta_1 = 0$ when only the intercept and $\mathbf{X}_1$ are in the model are uniformly distributed and give rejection rates of approximately $\alpha = 0.05$. But why do Strategies 2* and 3* reject nearly every time? The issue is that if you build the linear model, $\mathrm{E}(Y_i) = \beta_0 + X_{i1}\beta_1 + X_{i2}\beta_2$ with the trivariate data drawn independently from a multivariate normal,*

$$\begin{bmatrix} Y_i \\ X_{i1} \\ X_{i2} \end{bmatrix} \sim MN \left(\begin{bmatrix} 0 \\ 0 \\ 0 \end{bmatrix}, \begin{bmatrix} 1 & 0 & 0.5 \\ 0 & 1 & 0.5 \\ 0.5 & 0.5 & 1 \end{bmatrix} \right),$$

then $\beta_1 \neq 0$ even though Y_i and X_{i1} are independent. The dependence among the other variables ensures that $\beta_1 \neq 0$. Note that the inflation of the rejection rate in this setting is not only about selection of the covariates, as some substantial inflation would still occur even if there only one additional prespecified covariate, and thus nothing to select.

Deciding on what covariates to add to the model when causal inference is desired is a complicated and tricky issue (see Chapter 15 and, e.g., Pearl et al., 2016). But to partially understand the issue of Example 14.12, we consider another simpler example where the covariate of interest is uncorrelated with the response but correlated with a covariate.

Example 14.13 *Consider the following fake data. Suppose we are collecting rates of Type 2 diabetes in adult men. Generally, it is known that Type 2 diabetes is related to weight, but we want to know if it is additionally related to height. Consider the following cross-sectional data as a contingency table:*

	Height < 6ft		Height ≥ 6ft	
	Diab.	No diab.	Diab.	No diab.
Weight < 180 *lbs*	68	1072	2	78
Weight ≥ 180 *lbs*	102	458	28	192

If we ignore weight, then men with Height < *6 feet as well as men with Height* ≥ *6 feet have an observed 10% rate of Type 2 diabetes, and a logistic regression estimates the odds ratio for height to be 1.00, with confidence interval and two-sided p-value using Wald methods (95% CI: 0.66, 1.51, p = 1). In other words, height and diabetes appear independent. If we perform a logistic regression with main effects for height and weight, then both effects would be significant, where shorter height increases the odds of diabetes by a factor of 1.61 (95% CI: 1.05, 2.46; p = 0.028) and lower weight decreases the odds of diabetes by a factor of 0.277 (95% CI: 0.202, 0.380; p < 0.001). Should we include weight as a factor when our interest is the effect of height on the incidence of diabetes? We answer that question on page 296 after studying causal methods.*

14.8 Inference after Model Building

In this section, we briefly mention a few methods for making inferences after going through a model selection process. To control the familywise Type I error rate, a simple approach is a two-step approach. In step 1, divide up your data set into a training and a validation data set. Build the model using the training data set. Then in step 2, do your formal hypothesis testing based on the selected model from step 1.

You can use Holm's procedure if there are many hypotheses, or depending on the model, a maxT method can be used to correct for multiple comparisons. Bühlmann et al. (2014) mentions this and other options with suggestions of R packages.

The work on this problem is new and quite extensive, so we limit ourselves to some bibliographic notes in Section 14.10.

14.9 Summary

We begin the chapter with the workhorse of models: the linear model, where an individual's response equals a linear combination of covariates plus a normally distributed error term. The model assumes that the observations are all independent and that the variance of the model errors is the same across individuals (i.e., homoscedasticity). We estimate the parameters using least squares estimates which is equivalent to the maximum likelihood estimate under normality. Tests on a linear contrast of the parameters assume normality, but the independence and homoscedasticity assumptions are more critical than normality, as only large departures from normality are likely to lead to more than negligible inflated Type I error rates. Plots of responses versus residuals ideally show that the residuals are approximately normally distributed about 0, and their spread does not change as a function of the outcome. If spread increases with larger outcomes, then a transformation on the outcome variable (e.g., a log transformation) could better correspond with the assumptions. While the normality of the outcome variable is assumed, the covariates need not be normally distributed, for example, an indicator variable for gender is perfectly fine. However, if the residuals appear to be very skewed, and a small subset of the sample has a large influence on the result, a transformation on this covariate may lead to more robust conclusions.

The generalized linear model allows different distributions for the response variable, such as binary or count data. It uses a link function that links the conditional mean of the response to the linear combination of covariates. Binary responses typically use the logit link, which leads to logistic regression, where each parameter represents the log of the odds ratio of adding one unit to each covariate, holding all other covariates constant. Subtle convergence problems can arise with logistic regression or when fitting other GLMs. We show how this problem can be identified and suggest some solutions in Section 14.3.2.

Poisson GLM models with a log link can be used for count data. It is typically safer to allow an overdispersion from the Poisson model to allow a variance that differs from the mean. The negative binomial model is also often used for count data; however this is more susceptible to convergence problems. One should also be aware that sometimes distributions of count responses are not consistent with these models, and that lack of fit can lead to inflated Type I error rates. In such a scenario, an alternative is to use methods for ordinal or ranked data.

Two-sample proportional odds and proportional hazards models can be generalized to handle covariates, to analyze ordinal or ranked outcome data. Consider a model where some unknown transformation of ranked responses equals a linear combination of covariates plus an error term. When the error is distributed as a logistic, the model is a proportional odds model. When the error is distributed as an extreme value distribution, this leads to a proportional hazards model. Proportional hazards models can be easier to fit than proportional odds, and censored outcomes can be incorporated. The advantage of the proportional odds model is that it is palindromically invariant, meaning the inferences will change in the appropriate way if we reverse the order of responses.

Sometimes observations are not independent, such as where individuals are clustered into classrooms, litters, and so on; in these cases, the correlation structure of the data needs to be incorporated into the model. Longitudinal data is another important category of clustered data, where outcomes over time are clustered within an individual. There are two main types of models for clustered data: mixed effects models and marginal models (e.g., generalized estimating equations).

The mixed effects model treats nuisance parameters, such as an effect for each individual in the study, as random effects and the parameters of interest as fixed effects. We typically assume the random effects parameters are multivariate normal with mean 0 and covariance $\mathbf{G}$, where the form of $\mathbf{G}$ depends on the specific application. In contrast, marginal models account for the cluster effects not through the cluster means, but through the covariance structure. Often the parameters will not be equivalent in the two models, and we illustrate this difference by applying both approaches to the same example. One can view the mixed effects model as holding the random effects constant, and the marginal model as averaging over random effects.

There are principled strategies for choosing covariates to include in the model. Unlike an index such as R^2 which will improve whenever a variable is added to the model, the AIC attempts to distinguish between an additional variable which adds only random noise versus one that truly improves the model's fit. This can be used to anchor a stepwise procedure that adds or deletes variables guided by change in the AIC. Another more general approach is to combine a variable selection method, such as lasso, with cross-validation to identify a model with important variables, while not overfitting to the data set at hand.

We briefly alert readers to problems that can occur when testing from models and show a number of simulation results indicating areas of concern. For example, if one uses a model selection method, the resulting model will affect the test of the primary question of interest, and this two-stage process can inflate the Type I error rate. A rigorous approach is to divide data into two subsets, and develop the model on the training set, and then do formal testing on the test set, using the model built on the training set.

14.10 Extensions and Bibliographic Notes

The classic and still useful book on generalized linear models is McCullagh and Nelder (1989).

For a review of Akaike's Information Criterion, see, e.g., Burnham and Anderson (2002), pp. 60–64. For AIC and other model selection methods such as cross-validation see Hastie et al. (2009, chapter 7). Since the introduction of the lasso, there has been much subsequent work, see Park and Hastie (2007), Hastie et al. (2009), and Tibshirani (2011).

An exciting and potentially very useful tool in model selection are knockoffs (Barber and Candès, 2015; Candès et al., 2018).

We did not cover conditional logistic regression, but it is covered in many textbooks (see, e.g., Agresti, 2013, Section 7.3). Conditional logistic regression involves some complicated algorithms which can also be used with Cox proportional hazards (Gail et al., 1981). Inferences for the conditional logistic regression rely on either asymptotics such as Fisher's information or a sandwich estimator of variance (see Fay et al., 1998), or on exact methods (Mehta and Patel, 1995).

For likelihood ratio tests in nonstandard conditions, such as testing whether a random effects variance is zero, see Self and Liang (1987).

14.11 R Functions and Packages

Generalized linear models are fit by glm in the stats package. The sandwich package does sandwich estimators of variance for linear models with bias corrrections (Zeileis, 2004). The penalized likelihood method of Firth (1993) to handle separation in logistic regression is in the logistf package.

The pscl package handles negative binomial regression and other models for count data including hurdle models and zero-inflated Poisson models. The glmnet package does fast calculation of the lasso and several generalizations of it (Friedman et al., 2010). See also glmpath for lasso with GLMs and proportional hazards models (Park and Hastie, 2007).

The continuous proportional hazards models may be solved using the pim package with the logit link (De Neve et al., 2019). The proportional hazards can be handled using the coxph function in the survival package. For conditional logistic regression using traditional asymptotics see the clogit function of the survival package or using sandwich estimators see the clogistCalc function of the saws package.

14.12 Notes

N1 **Box–Cox transformation:** Box and Cox (1964) introduced a class of transformations where the transformed responses are y_i^λ, and if $\lambda = 0$ then you use the log transformation. If you have repeated observations at each unique $\mathbf{x}_i$, then λ may be estimated by $\hat{\lambda} = 1 - \hat{\beta}_\lambda$, where $\hat{\beta}_\lambda$ is the slope of the least squares line fit using the log of the sample standard deviation of each set of replicates as the response and the log of the mean of each set of replicates as the covariate. Typically, we would round the slope to give an interpretable transformation. Sometimes we use $(y_i^\lambda - 1)/\lambda$ as the power transformation for theoretical reasons, because it is continuous at $\lambda = 0$ and subtracting 1 and dividing by λ does not change the significance results of the tests of Equation 14.1. For details, see Box et al. (2005, Section 8.2).

N2 **Robustness of linear regression to nonnormality:** Box and Watson (1962) show "that it is 'the extent of non-normality' in the regression variables (the x's), which determines sensitivity to non-normality in the observations (the y's)." They then developed a measure of 'nonnormality' of the covariates that is a function of the sum of the square of the diagonal of the hat matrix (after excluding the intercept and standardizing all other columns of $\mathbf{X}$ to sum to 0). The measure detects nonnormality if a sort of multivariate-kurtosis is far from normality, so that if the covariates are balanced there is generally not a problem (for example, a set of covariates that represents indicators for group membership are generally not a large problem if there are close to equal numbers in each group and not too many groups, despite the fact that those covariates are obviously not normal). See Ali and Sharma (1996) and its references for details about different ways that researchers have studied robustness of the linear model to its normality assumptions.

N3 **Malaria blood assay:** One method to determine how many malaria parasites are circulating in an individual's blood is to look at a thick blood smear under a microscope and count the number of parasites (y) and the number of white blood cells (x) within the field of view. Then measure the concentration of white blood cells per microliter of blood (γ) by some other measure, so that the concentration of parasites per microliter of blood is measured as $\lambda = y\gamma/x$. If the parasites and white blood cells are randomly distributed in the blood, then the volume of blood in the field of view is approximately $t = x/\gamma$.

N4 **GEE working variance (page 265):** Let $\xi = [\alpha, \phi]$. Let $\text{Var}(Y_{ij}) = \phi v_{ij}$ where ϕ is a scalar dispersion parameter and v_{ij} is the modeled variance that depends on the mean. For example, for Y_{ij} binary, $v_{ij} = \mu_{ij}^*(1 - \mu_{ij}^*)$ where $\text{E}(Y_{ij}) = \mu_{ij}^*$. Let $\mathbf{A}_i = diag(v_{ij})$. Then

$$V_i(\beta^*, \xi) = \frac{1}{\phi} \mathbf{A}_i^{-1/2} \mathbf{R}(\alpha) \mathbf{A}_i^{-1/2},$$

where $\mathbf{R}(\alpha)$ is the working correlation matrix.

N5 **Population averaged vs subject-specific parameters:** For comparing mixed effects parameters (i.e., subject-specific) to marginal parameters (i.e., population averaged), Zeger et al. (1988, p. 1054) give conversion functions for normal random effects distribution, with a different conversion function depending on the link function. See the end of Section 14.5.3 for more explanation and notation. For the log link, we have

$$\exp(\mathbf{x}_{ij}^\top \beta^*) = \mu_{ij}^* = \exp\left(\mathbf{x}_{ij}^\top \beta + \frac{1}{2}\mathbf{c}_{ij}^\top \mathbf{G}\mathbf{c}_{ij}\right).$$

For the probit link, $g = \Phi^{-1}$, we have

$$\Phi(\mathbf{x}_{ij}^\top \beta^*) = \mu_{ij}^* = \Phi\left(\left|\mathbf{G}\mathbf{c}_{ij}\mathbf{c}_{ij}^\top + \mathbf{I}\right|^{-r/2} \mathbf{x}_{ij}^\top \beta\right),$$

where r is the dimension of $\mathbf{b}_i$. For the logit link, let $\mathrm{expit}(\eta)$ be the inverse logit function, then

$$\mathrm{expit}(\mathbf{x}_{ij}^\top \beta^*) = \mu_{ij}^* \approx \mathrm{expit}\left(\left|\left(\frac{256}{75\pi^2}\right)\mathbf{G}\mathbf{c}_{ij}\mathbf{c}_{ij}^\top + \mathbf{I}\right|^{-r/2} \mathbf{x}_{ij}^\top \beta\right).$$

So if $r = 1$, $\mathbf{G} = \sigma_b^2$, and $\mathbf{c}_{ij} = 1$ then with the logit link we have

$$\beta^* \approx \frac{\beta}{\sqrt{1 + 0.346\sigma_b^2}}.$$

See also Fitzmaurice et al. (2004, p. 363).

14.13 Problems

P1 Write an R function to estimate β from the linear model inputting $\mathbf{y}$ and $\mathbf{x}$, test each $\beta_j = 0$, and give unadjusted 95% confidence intervals for β_j. Compare the results from the lm function in base R, using the summary and confint functions.

P2 Simulate a data set with: $n = 100$ standard normal responses, 5 independent covariate vectors of binary covariates with 50 $1s$ and 50 $0s$, and 5 covariate vectors, each with $n = 100$ independent standard normal random variables. Fit a linear model with those 10 covariates and an intercept. Count how many of the 10 covariates indicate a significant effect at the 5% level, without adjusting for multiple comparisons. Repeat this 1,000 times. What proportion of the data sets have at least one significant effect?

P3 Create a data.frame in R with a variable y that is standard normal, and a variable *trt* that is a factor with four levels. The simple linear model in R with an intercept and the *trt* effect is lm(y~trt). The test of the overall *trt* effect is anova(lm(y~1),lm(y~trt)). Write an R program to test the overall *trt* effect using Equation 14.1, and show it gives the same result.

P4 Show that in logistic regression, $\exp(\beta_j)$ is the odds ratio comparing an individual with one unit added to the jth covariate to another individual with the original jth covariate, where all other covariates are held constant. Hints: $E(Y_i|\mathbf{x}_i) = \Pr(Y_i = 1|\mathbf{x}_{ij})$. Recall, for an event A, $Odds(A) = \Pr[A]/(1 - \Pr[A])$.

P5 Simulate binary response data with two binary covariates, A and B, from the logistic model: $\text{logit}(\mu_i) = \beta_1 + \beta_2 I(A = 1) + \beta_3 I(B = 1)$. Fit the model using glm in R. Now make a table of the data, and put the data into binomial form, so that for each of the four possible covariate values you count the number of events (ystar) and number of individuals at risk (m). Fit the model using glm but with the formula cbind(ystar,m-ystar)~A+B. Are the results the same?

P6 Fit a simple logistic regression to the data from Example 7.4 to test for a treatment effect (see page 257). Does the model converge? Why do you think so or not? What is an alternative analysis?

P7 Show that the equivalence between the two forms of the proportional hazards model (Equations 14.2 and 14.3). Hint: $\lambda(t) = -f(t)/S(t) = -\partial \log(S(t))/\partial t$, for a continuous distribution with pdf $f(t)$, survival $S(t)$ and hazard $\lambda(t)$.

P8 Define the vector $\mathbf{x}_{ij}$, such that

$$
\mathbf{x}_{ij}^{\mathsf{T}} \begin{bmatrix} \beta_1 \\ \beta_2 \\ \vdots \\ \beta_{m+1} \end{bmatrix} = \gamma_i + z_{ij}\beta
$$

with $\gamma_i = \beta_i$ for $i = 1, \ldots, m$ and $\beta = \beta_{m+1}$. Note, the β notation may be confusing. On the right-hand side of the equation β is a scalar, but the left-hand side uses notation like the usual linear model, and in the usual linear model we would write $\beta = [\beta_1, \ldots, \beta_{m+1}]$ as a vector with the same dimensions as $\mathbf{x}_{ij}$.

15

Causality

15.1 Overview

We adopt the view championed by Pearl, that there is a "crisp" distinction between statistics and causality (see, e.g., Pearl, 2009b, pp. 331–332). Under this usage, "statistical" inferences are things you can measure using probability distributions (such as parameters), but causal inferences require more than just estimates of parameters and distributions. We do not mean for this use of "statistics" to suggest that statisticians have not had legitimate ways to estimate causality since Fisher and Neyman were first developing many important statistical ideas. In fact, under this limited definition of statistics, ideas like randomization that were championed by the most famous statistician, Fisher, are deemed nonstatistical. Randomization is nonstatistical in the sense that if we observe two random variables X and Y, and one of them is randomized, we cannot tell which one is randomized even if we know the bivariate distribution of (X, Y). We have already seen this distinction in Section 12.3.2 on Simpson's paradox. There we showed that on the same data table, we can perform two kinds of statistical inferences about the effect of a treatment, and we use extra-statistical information to decide on which of the two statistical inferences lead to the proper causal inferences. This chapter gives a more formal way to think about causal inferences.

Because causality is so important for scientific research, there have been several ways it has been formalized, some more common in different disciplines. Here we focus on two approaches, the potential outcome approach as in Imbens and Rubin (2015), and the graphical approach using directed acyclic graphs and structural equations as in Pearl (2009b).

We begin with the potential outcomes approach when we discuss experimental studies, that is, studies where we intervene on the individual units of research. In this approach randomization is important, and many of the basic tools used have been part of clinical trials methodology for many decades. The second part of the chapter is observational studies, studies where we do not directly impose an intervention, but only observe what has happened.

The topic of causality is expansive, and we cannot give enough details to perform any but the simplest hypothesis tests. The purpose of this chapter is to outline some important ideas, to point out potential shortfalls of naïve analyses, and to point the reader to the literature for better analyses when appropriate.

15.2 Causal Estimands

In this section, z_i represents an intervention applied on the ith unit. To keep this simple, suppose that z_i only has two possible values, 0 and 1. For each unit, we consider a pair of *potential outcomes*, the outcome that unit would have gotten if $z_i = 0$, $y_i(0)$, and the outcome that unit would have gotten if $z_i = 1$, $y_i(1)$. This notation implicitly assumes that there is no interference, meaning that the potential outcome of the ith individual does not depend on interventions applied to other individuals (see Section 15.5). Typically, we consider studies where we only perform one of the two interventions on each unit, so we only observe one of the two potential outcomes, $y_i(z_i)$. In most of our examples the unit is an individual person or animal.

Example 15.1 *Consider a study on a vaccine for a disease. Let $z_i = 1$ if the ith subject gets the vaccine, and $z_i = 0$ if not. Let $y_i(z_i) = 1$ if the subject gets the disease during the study, and $y_i(z_i) = 0$ otherwise. For this study, the ith subject has two potential outcomes: $y_i(0)$ (the outcome that person would get if they did not get the vaccine) and $y_i(1)$ (their outcome if they got the vaccine).*

We want to know if the intervention has an effect, and we typically use the difference to measure that effect. Let $D_i = Y_i(1) - Y_i(0)$, where capitalization denotes random variables. The values $D_1, \ldots, D_n$ are known as unit-level causal effects, because the effect is defined on each unit. In the typical experiment we cannot observe D_i.

Example 15.1 *(continued) Since the ith subject either gets vaccine ($z_i = 1$) or not ($z_i = 0$), we cannot observe both potential outcomes. So we cannot observe $D_i = Y_i(1) - Y_i(0)$. Nevertheless, $D_i = -1$, indicates that the ith subject would not get the disease if vaccinated, but would get the disease if not vaccinated. So for the ith individual, a value of $D_i = -1$ denotes an effective vaccine, of $D_i = 1$ denotes a harmful vaccine, and of $D_i = 0$ denotes no effect of the vaccine since either the ith subject would never have gotten the disease or would always get the disease during the study, regardless of whether or not they got the vaccine.*

Although we cannot observe D_i for any unit, we can define its expected value in a population. Throughout the rest of this section we make the following set of assumptions, unless stated otherwise. Let $\mathbf{Y}_i = [Y_i(0), Y_i(1)]$ be the random vector for the pair of potential outcomes for the ith unit. Assume that we have n independent random pairs, and for all i, $\mathbf{Y}_i \sim F$. Further assume that we have a randomized experiment, where Z_i is the intervention which is independent of all potential outcomes. We make the *stable unit treatment value assumption* (SUTVA): there is no interference (i.e., the potential outcomes for the ith unit are not affected by the Z_j for any $j \neq i$), and there are not different versions of treatment levels that give different potential outcomes (see Imbens and Rubin, 2015, p. 10). As with any assumption, we should think about our specific study to see if the assumption is reasonable.

Example 15.1a *Suppose the vaccine study is testing malaria in the United States, where there are almost no malaria infections. So people will have an almost zero probability of getting infected through natural exposure (the primary reservoir for malaria is people and it is spread by mosquitoes, so a natural infection requires a mosquito biting an infected individual then subsequently biting an uninfected one). For the study, suppose subjects are* *randomized to either $z_i = 1$, denoting an experimental malaria vaccine or $z_i = 0$, denoting a tetanus vaccine used as a control (it is not uncommon for vaccine studies to incorporate a vaccine that targets a very different disease as a control, so that all participants potentially benefit and also potentially have a mild reaction at the site of the injection). Suppose the study design is that one month after vaccination (with either vaccine), subjects are challenged by injection with Plasmodium falciparum sporozoites (the stage of the parasite that infects humans). This is considered ethical because malaria by Plasmodium falciparum is treatable. In this situation, the SUTVA assumption is reasonable. The treatment and control vaccines are quite stable and created using good manufacturing practices, so within each arm they are very similar for all subjects. Further, it is unlikely that the vaccine gotten by other participants will affect the outcome of any particular subject.*

Example 15.1b *Now suppose the vaccine study is to take place in a village in Africa, where each year about 80% of the population gets exposed to mosquito bites that contain Plasmodium falciparum sporozoites (sadly, there are villages such as this). Suppose the design is that subjects are randomized to the experimental malaria vaccine or tetanus vaccine, and then followed for a year to see who naturally (i.e., by wild mosquito bite(s)) gets malaria sickness from Plasmodium falciparum. In this situation, the SUTVA assumption is not reasonable. In this situation, if the malaria vaccine works, it is quite likely that the vaccine gotten by other participants will affect the outcome of a participant. If the vaccine works, then fewer people in the village will be infected, and it is less likely that each mosquito will be infected, and hence less likely that a particular participant will be infected. In other words, there may be indirect effects of the vaccine, and more care needs to be made in defining the causal parameter. For details on these issues in vaccine studies see Halloran et al. (2010).*

A common causal estimand of interest is the average causal difference (also called the superpopulation average causal effect),

$$\Delta = E(D_i) = E(Y_i(1) - Y_i(0)) = E(Y_i(1)) - E(Y_i(0)), \tag{15.1}$$

where the expectations are taken over the bivariate potential outcome distribution, F. Importantly, the expectations for $Y_i(0)$ and $Y_i(1)$ are done separately, and we can define Δ using only the marginal distributions, F_0 and F_1, where $Y_i(z) \sim F_z$.

An alternative estimand is to define the estimand with respect to the n units in our study. The study-specific average causal difference (also called the finite sample average treatment effect) is

$$\Delta_s = \frac{1}{n} \sum_{i=1}^{n} \{Y_i(1) - Y_i(0)\}.$$

Inferences on Δ_s typically use randomization inference, where the statistical variability comes through the distribution of the Z_i. Although randomization is sometimes presented as model-free, there are still assumptions needed for inferences (Note N1).

In this section we focus on superpopulation causal effects (e.g., Δ), rather than finite sample ones (e.g., Δ_s). Superpopulation and finite sample effects are terms from survey statistics, where often the estimand of interest is a finite sample value (e.g., the number of people that will vote for a certain candidate), rather than being a function of parameters on a distribution and having the observations be a sample from that distribution. In the latter case, the population of interest is not some finite group of individuals (e.g., eligible voters), but a theoretically infinite population (i.e., the superpopulation) that would define the distribution. The previous assumption that $\mathbf{Y}_i \sim F$ is a superpopulation approach. In most real-life situations, for the ith individual we can only observe one of the potential outcomes in $\mathbf{Y}_i = [Y_i(0), Y_i(1)]$. Thus, in those situations the bivariate distribution F is not identifiable, we can only identify estimands that are functionals of the two marginal distributions F_0 and F_1, as we did in the case of the average causal difference, Δ.

Some other identifiable causal effects are expectations of unit-level causal effects. For example, suppose $Y_i(z) \in (0, \infty)$ for all i and $z = 0, 1$. Then we can define a unit-level causal effect, $D_{Li} = \log(Y_i(1)) - \log(Y_i(0))$ and $E(D_{Li}) = \log(\mu_{g1}) - \log(\mu_{g0}) = \log(\mu_{g1}/\mu_{g0})$, where μ_{gz} is the geometric mean of $Y_i(z)$. Another possible causal estimand is the difference in medians (or some other quantile).

A more involved example is related to the Mann–Whitney parameter. Recall that the Mann–Whitney parameter is closely related to the Wilcoxon–Mann–Whitney test, where the responses from two groups are pooled and ranked, and the difference in mean ranks between the groups is $n\left(\hat{\phi} - \frac{1}{2}\right)$, where $\hat{\phi}$ is the estimate of the Mann–Whitney parameter (see Chapter 9, Problem P6). In this chapter the two groups are the two sets of potential outcomes. Instead of ranking the sample, we want a rank-like function that can be used on random variables. Let

$$G(y) = \Pr[Y_{Gi} < y] + 0.5\Pr[Y_{Gi} = y], \tag{15.2}$$

where $Y_{Gi} = BY_i(1) + (1 - B)Y_i(0)$ and B is a Bernoulli random variable with parameter 0.5. Then $G(\cdot)$ is like a ranking function that gives the rank of y among the population of all potential outcomes (both the $Y_i(0)$ and the $Y_i(1)$), and then scales the ranks to be in the interval $[0, 1]$. Now consider the unit-level causal effect, $D_{Gi} = G(Y_i(1)) - G(Y_i(0))$. Because the function G may be represented as a function of F_0 and F_1, it is identifiable, and $E(D_{Gi}) = h_{MW}(F_0, F_1) - \frac{1}{2} = \phi - \frac{1}{2}$ (Problem P1).

Not all causal estimands are identifiable functions of F_0 and F_1 only. For example, consider the unit-level causal effect representing whether the ith individual would have a higher ($D_{Ii} = 1$), lower ($D_{Ii} = 0$), or equal ($D_{Ii} = 1/2$) response under $z_i = 1$ than under $z_i = 0$, where

$$D_{Ii} = I(Y_i(0) < Y_i(1)) + 0.5I(Y_i(0) = Y_i(1)).$$

When the $Y_i(z)$ values are binary, then $D_{Ii} = 0.5(Y_i(1) - Y_i(0)) + 0.5$ and $E(D_{Ii}) = 0.5\Delta + 0.5$, and $E(D_{Ii})$ is identifiable; however, for orderable values of $Y_i(z)$ with more than two possible values, then

$$E(D_{Ii}) = \Pr[Y_i(0) < Y_i(1)] + \frac{1}{2}\Pr[Y_i(0) = Y_i(1)] \equiv \eta, \tag{15.3}$$

is not identifiable. The parameter η represents the proportion of individuals that have larger potential outcomes when $z_i = 1$ than when $z_i = 0$ (plus a correction for ties). We show nonidentifiability by a straightforward example.

Example 15.2 *Consider two populations, each with only three types of individuals, and each type is equally represented in each population.*

	Population 1			Population 2	
Type	$Y(0)$	$Y(1)$	Type	$Y(0)$	$Y(1)$
A	1	2	D	1	6
B	3	4	E	3	2
C	5	6	F	5	4

Both populations have the same marginal distributions. For both populations, F_0 represents a distribution with $\Pr[Y(0) = 1] = \Pr[Y(0) = 3] = \Pr[Y(0) = 5] = 1/3$ and F_1 represents a distribution with $\Pr[Y(1) = 2] = \Pr[Y(1) = 4] = \Pr[Y(1) = 6] = 1/3$. But even though the two populations have the same marginal distributions, F_0 and F_1, they have different values of η. In Population 1, $Y_i(1) - Y_i(0) = 1$ for everyone in the population (Types A, B, and C), while in Population 2, $Y_i(1) - Y_i(0) = 5$ for Type D individuals and $Y_i(1) - Y_i(0) = -1$ for Type E and F individuals. In Population 1, $\eta = 1$ but in Population 2, $\eta = 1/3$.

Beware, it is common to mistake η for the Mann–Whitney parameter, ϕ, but they are different parameters (see Section 9.5.3).

To highlight that causal effects are not always expectations of unit-level causal effects, consider ratio effects in binary potential outcomes, which are used with vaccine efficacy as a causal effect. First, consider Example 15.1a, a randomized vaccine study where we can accept the SUTVA. In this case, we can treat the two arms of the study as if $Y_i(z) \sim F_z$ for i with $z_i = z$, where F_z is Bernoulli with parameter θ_z. Then a causal parameter of interest is $1 - \theta_1/\theta_0$, the *vaccine efficacy*.[1] For example, a vaccine efficacy of 93% means that the ratio of expected number of infections in the treated compared to the control arm is $\theta_1/\theta_0 = 7\%$. The vaccine efficacy itself is not the expected value of unit-level causal effects (the ratio of two expectations is not the expectation of unit-level ratios), but nevertheless is an estimable causal effect.

15.3 Why Causal Inference Is Difficult for Observational Studies

The US Centers for Disease Control and Prevention recommends a measles vaccine (actually a vaccine for three diseases: measles, mumps, and rubella) at ages 12 to 15 months, with a booster at 4 to 6 years of age. For simplicity, assume that the ith subject in the population of interest either follows the vaccine schedule ($z_i^{obs} = 1$) or does not get the vaccine at all ($z_i^{obs} = 0$). Here we represent the factor of interest as z_i^{obs} to emphasize that it is not randomly assigned but is just observed, and it may be related to many other factors that are also related to the potential outcome vector, $\mathbf{Y}_i = [Y_i(0), Y_i(1)]$ (where here $Y_i(1)$ represents the potential outcome when $z_i^{obs} = 1$). A simple observational study might have the binary response be the indicator of whether a child got measles between 16 months and their fourth

---

[1] Vaccine efficacy is defined as 1 minus a ratio effect (vaccinated over unvaccinated) of an infectious disease. The ratio could represent a ratio of proportions (Equation 15.4), of rates (see Example 14.5, page 258), of hazards, or some other measure of infectious disease. See Halloran et al. (2010).

birthday ($Y = 1$) or not ($Y = 0$). Then an effect of interest could be the vaccine efficacy. By simply adopting the estimate used in a randomized study, a naïve estimate of vaccine efficacy is 1 minus a ratio of proportions,

$$\widehat{VE} = 1 - \frac{\hat{p}_1}{\hat{p}_0}, \tag{15.4}$$

where

$$\hat{p}_a = \frac{\sum_{i:z_i^{obs}=a} y_i(a)}{n_a},$$

and $z_i^{obs} = 1$ if the ith subject was observed to get the measles vaccine and $z_i^{obs} = 0$ if not, and n_1 and n_0 are the numbers who got the measles vaccine or did not, respectively. So if no one who was vaccinated for measles got measles, but some who were not vaccinated for it got it, then the estimated vaccine efficacy would be 1, or 100% effective. Even if we could observe all the 4-year-olds in the population of interest, this study may have a problem in determining if the vaccine is the cause of the prevention of measles.

We consider two issues with using $\widehat{VE}$ to estimate vaccine efficacy from an observational study. First, there may be some variables [observed or unobserved] that if you could adjust for them (by, for example, estimating a weighted average of effects for each level of the variable) will give a more interpretable estimator of the treatment effect (e.g., vaccine efficacy). We call this problem confounding, and informally, we call variables *confounders* if better causal estimates result after adjusting for them (in Section 15.7.3 we give a more formal definition for confounders related to the backdoor criterion). The second issue is interference, where, for example, the ith individual's response depends not just on whether they got the vaccine, but also on whether others got the vaccine.

Consider the confounding issue. For the ith subject, let $\mathbf{x}_i$ be the set of measured covariates and let $\mathbf{u}_i$ denote the set of all unknown covariates, and both sets describe the subject at 12 months (before any measles vaccine occurred). For an observational study, it is possible that some factor (either measured or unknown) affects both the choice to get the vaccine (i.e., the value of z_i^{obs}) as well as the result (the value of $y_i(z_i)$). For example, suppose r_i represents the incidence rate of measles in the year before the ith subject was born in their county of residence. These rates could be measured (part of $\mathbf{x}_i$) or not (part of $\mathbf{u}_i$). It could be that if r_i is high, then it is more likely that both $y_i(0) = 1$ (i.e., the subject would get measles if not vaccinated) and that $z_i^{obs} = 1$ (the subject would get vaccinated, because their parents knew of neighbors who got measles). If the vaccine is not 100% efficacious, then high values of r_i would also make it more likely that $y_i(1) = 1$.

Example 15.3   *Fake measles vaccine data (and incorrect analysis): Consider the following fake data. First, we present it supposing that r_i is an unobserved variable, so we collapse over all counties, because we do not suspect it is important to adjust for county.*

	No. cases	Rate per 100,000	No. at risk (percent)
Unvaccinated	1,240	4.960	2.5×10^7 (25%)
Vaccinated	3,912	5.216	7.5×10^7 (75%)

Here 75% of the overall population is vaccinated, but the vaccinated population has a higher incidence of disease. The apparent vaccine efficacy is $1 - 5.216/4.960 = -5.16\%$, and it looks like the vaccine is causing more cases of measles than not getting vaccinated.

Now suppose r_i values are observed, and we break up the data into the low-risk counties and the high-risk counties:

Low risk

	No. cases	Rate per 100,000	No. at risk (percent)
Unvaccinated	240	1.0	2.4×10^7 (30%)
Vaccinated	112	0.2	5.6×10^7 (70%)

High risk

	No. cases	Rate per 100,000	No. at risk (percent)
Unvaccinated	1,000	100.0	0.1×10^7 (5%)
Vaccinated	3,800	20.0	1.9×10^7 (95%)

Although this is fake data, there are plausible explanations. In counties with high infection rates, doctors and public health officials are more likely to strongly advocate for vaccines and parents are more likely to listen, and hence the vaccination rates would be higher (95% rather than 70%). Also, although there are fewer subjects in the high-risk counties, they account for most of the actual cases. Within each stratum the vaccine efficacy is 80%. So this is another example of Simpson's paradox. If (!) we could assume that the rate of vaccination of the neighbors did not affect the probability of getting infected (i.e., if we can assume no interference), and if there were no unobserved confounders (e.g., other variables like r_i, but that we do not know about), then the vaccine efficacy of 80% would be a good estimate. But, alas, for vaccines the no interference assumption is not even close to being true. Finally, while we set up these fake data to illustrate some concepts, they actually are implausible since it is highly unlikely that the vaccine efficacy would be the same between high- and low-risk communities, because of the indirect effects of vaccine (e.g., you are protected because you live in an area with high vaccination rates) which can be higher than the direct effects (e.g., you are protected because you got vaccinated yourself).

We hold off on examining specific solutions to violations of the no interference assumption until Section 15.5, but now we turn to two very important solutions to unmeasured confounders: experimentation and randomization.

15.4 Causality and Experimental Studies

15.4.1 Experimentation without Randomization

One way to measure the causal effect of a "treatment" ($z_i = 1$) compared to a "control" ($z_i = 0$) is to set up an experiment. Let $\mathbf{y}_i = [y_i(0), y_i(1)]$ be the pair of potential outcomes, and let $\mathbf{x}_i$ be a vector of all the variables for the ith individual measured at baseline (i.e., before the intervention z_i [treatment or control] is applied) that affect the potential outcomes $\mathbf{y}_i$. We allow random error for the ith individual as long as it is independent of $\mathbf{x}_i$ and z_i and the error of other individuals. One experimental design is to find pairs of individuals with identical values of $\mathbf{x}$ and assign one member of the pair to get the treatment ($z = 1$)

and the other to not get it ($z = 0$). Let the indices for the mth matching pair be j_{m0} (when the individual got control) and j_{m1} (when the individual got treatment), so that $z_{j_{m0}} = 0$ and $z_{j_{m1}} = 1$ for all m. Under this design, since within each pair the $\mathbf{x}$ vectors are the same (i.e., $\mathbf{x}_{j_{m0}} = \mathbf{x}_{j_{m1}} \equiv \mathbf{x}_{j_m}$ for all m), and those vectors give all variables that could systematically affect y except treatment, then we can estimate treatment effects within each matching pair by comparing $y_{j_{m0}}$ and $y_{j_{m1}}$. For example, we could estimate $\Delta_m = E(Y_{j_{m1}}|\mathbf{x}_{j_m}) - E(Y_{j_{m0}}|\mathbf{x}_{j_m})$ for the mth pair using $y_{j_{m1}} - y_{j_{m0}}$. We can estimate some overall effect, $\Delta_M = M^{-1} \sum_{m=1}^{M} \Delta_m$, which depends on the set of covariates that defines the M matching pairs, $\{\mathbf{x}_{j_1}, \dots, \mathbf{x}_{j_M}\}$.

These kinds of experiments can work well for physical systems and in some biological cases, but for most biological, sociological, or economic systems it is very hard to know all the variables that may affect the response. In some cases, there are a few variables that have a strong effect and other variables can be modeled into the error distribution. When those kinds of models are not feasible, one strategy is a randomized experiment.

15.4.2 A Simple Randomized Experiment

To start talking about randomization, we consider a simple example of a population of individuals who have a noncontagious disease, and we want to know if a new treatment ($z = 1$) works better than the standard one ($z = 0$). As before let the set of potential outcomes for the ith individual be $\mathbf{y}_i = [y_i(0), y_i(1)]$, and we only observe $y_i = y_i(z_i)$ for each individual, the response after getting the treatment associated with z_i. Consider the binary response case, where $y_i = 1$ is a positive response (e.g., cured) and $y_i = 0$ is a negative response (e.g., not cured).

We are studying a population of individuals and we are trying to estimate $\Delta = E(Y_i(1)) - E(Y_i(0))$, the proportion of the population that would be cured under the new treatment minus the proportion that would be cured under the standard treatment. For an observational study or an experimental study designed as in Section 15.4.1, we need to know all other variables that can affect the response and the treatment variables. If those other variables are unknown, an alternative approach is to perform a randomized study. Suppose we randomly assign half of the individuals in the study the new treatment ($z_i = 1$), and the other half of the individuals get the standard treatment ($z_i = 0$), and suppose everyone complies and takes the treatment assigned to them. Then because of the random assignment, we can use the means of those assigned to each treatment to get an unbiased estimate of Δ, say $\hat{\Delta} = \bar{y}(1) - \bar{y}(0)$, where $\bar{y}(z)$ is the sample mean of those individuals with $z_i = z$. In this case, we do not need to know the values or even the existence of variables that may affect the response, because by randomization we know that all of those variables are independent of treatment assignment (z_i).

Suppose there was one variable, x, measured before treatment is applied, such that if the ith individual has a large value of x_i, it is more likely that they will have a positive response. In a randomized study, while it is possible that more individuals with large values of x_i will be in one of the treatment arms, it is equally possible (in a 1 : 1 randomization) that those values would be in the other arm. Further, as the sample size gets large, it is unlikely that a much larger proportion of the large values of x_i will be in one of the treatment arms. Since we know the probability distribution of z_i induced by the randomization, we can properly account for these kinds of unbalanced allocations using proper statistical inferences. Nevertheless, even if an imbalance in x is very unlikely, if we do see it in a study, then some may have lingering doubts about the results of the study. For example, we could find that a test of $H_0 : \Delta \leq 0$ can be rejected with a low p-value (e.g., $p \approx 0.01$), so that with 99% confidence we can state that the study shows that $\Delta > 0$ (new treatment is better). But if there is an observed imbalance in the x_i values between arms, and more large values of x_i were in the new treatment arm, then some might have some doubts about the conclusion that $\Delta > 0$, since it could have been due to an "unlucky" (i.e., unbalanced in x) randomization.

To avoid these doubt-inducing situations, especially for small studies, we sometimes perform a *stratified randomization* whereby we randomize such that the allocation is about $1 : 1$ into the two arms within each level of x (or within grouped levels created from a partition of the possible values of x). Stratified randomization will make "unlucky" randomizations (at least with respect to x) impossible. Alternatively, we can perform the unstratified randomization, and then at the end of the study estimate Δ using an estimate adjusted for x, such as a weighted average of within-stratum-level effects. For example, suppose x has four levels, then we could estimate Δ using

$$\hat{\Delta}_{x_{adj}} = \sum_{\ell=1}^{4} w_\ell \hat{\Delta}_\ell, \tag{15.5}$$

where w_ℓ and $\hat{\Delta}_\ell$ are the weight and difference in means associated with the ℓth level. Typically, w_ℓ might be proportion of individuals in the sample with x_i at level ℓ. Alternatively, if we are interested in Δ in a population, then w_ℓ would be an estimate of the proportion of the population with x_i at level ℓ.

Now consider the case where s_i represents a different variable that was measured after the treatment was applied but before the response was measured. Let $\hat{\Delta}_{s_{adj}}$ be an estimator of Δ analogous to $\hat{\Delta}_{x_{adj}}$, except adjusting for levels of s_i. Then $\hat{\Delta}_{s_{adj}}$ may be a poor estimate of Δ because s_i is measured after the treatment is applied, and it could have changed in response to treatment.

Example 15.4 *(Immunotherapy) For example, if the treatment is immunotherapy for a cancer and s_i is a binary measure of immune response to that cancer by some biomarker ($s_i = 1$ positive immune response, $s_i = 0$ no immune response), then positive s_i can lead to better responses, but the cause of those higher values may be highly related to which treatment arm the individual received. If the effect of the treatment was to increase the proportion of positive immune responses, and that positive immune response was the mediator of the effect of the treatment on the outcome, then adjusting for s_i through an estimator like $\hat{\Delta}_{s_{adj}}$ may totally remove all the effect of the treatment on the outcome.*

Thus, traditionally, we do not adjust for postrandomization variables when comparing randomized arms, in a typical analysis of covariance. However, the causal literature has introduced strategies where adjustment for postbaseline variables can be used for targeted purposes, but only if done using specialized methods that are designed to provide causal estimates as described in Sections 15.4.3 and 15.4.4.

Description	Notation	Number at risk	Number of deaths	Mortality rate (per 1,000)
Control arm	$Z_i = 0$	11,588	74	6.4
Treatment arm	$Z_i = 1$	12,094	46	3.8
Trt arm + complied	$Z_i = 1$ & $W_i(1) = 1$	9,675	12	1.2
Trt arm + did not comply	$Z_i = 1$ & $W_i(1) = 0$	2,419	34	14.1

Table 15.1 *Vitamin A supplement study data (Sommer and Zeger, 1991)*

15.4.3 Causality with Imperfect Compliance

Suppose we want to know the causal effect of a treatment compared to control and do a randomized study to explore that, but not all the individuals randomized to the treatment arm actually take the treatment. This is a common problem, and we explore it with a famous data set given by Sommer and Zeger (1991).

The data are derived from a cluster randomized trial in Indonesia on vitamin A supplements in children. In the study, some villages were randomized to receive treatment (two large doses of vitamin A), and the other villages received no special intervention (i.e., were in the control group). To simplify the example, we ignore clustering effects. The data are given in Table 15.1.

First, consider the effect of getting randomized to the treatment arm. The mortality rate of the treatment arm is lower (3.8/1,000 versus 6.4/1,000), and the difference is statistically significant (ignoring clustering and using a binomial melding, the estimated difference is $-2.6/1,000$; 95% confidence interval is $[-4.5/1,000, -0.7/1,000]$, two-sided p-value is 0.007). An effect such as this is known as an intention-to-treat (ITT) effect. But what if we want to know the effect not of intending to treat (i.e., getting randomized to the treatment arm), but the effect of taking the treatment among those who would comply? Those effects are different in this case because not everyone randomized to the treatment arm took the treatment.

One could naïvely expect that the overall mortality rate in the control group should approximately match the mortality rate among those in the treatment arm who did not take the treatment; however, the rates are very different (6.4/1,000 versus 14.1/1,000). It seems reasonable to expect that just getting randomized to the treatment arm would only have an effect on an individual's survival if that individual actually took the treatment. The problem is selection bias: among people in the treatment arm, there is something different about the individuals that choose to take treatment versus those that do not, and that something affects mortality risk. In this example, perhaps children of parents that encourage taking the vitamin A when offered have other advantages that affect mortality. Thus, a more proper analysis would compare those in the treated arm that complied with the treatment with those in the control arm that would have complied with treatment if they had been randomized to the treatment arm.

Let us introduce notation to more precisely study this. For the ith individual let y_i, z_i, and t_i be respectively, the response ($y_i = 1$ is death, $y_i = 0$ is alive), the randomization allocation ($z_i = 1$ is assigned to treatment arm, $z_i = 0$ is assigned to control arm), and t_i is treatment taken ($t_i = 1$ is treatment taken, $t_i = 0$ is no treatment taken). Instead of a pair of potential outcomes, we consider a quartet, $\mathbf{Y}_i = [Y_i(0,0), Y_i(0,1), Y_i(1,0), Y_i(1,1)]$, where $Y_i(z,t)$ is the outcome the ith individual would have if $z_i = z$ and $t_i = t$. Just as it is useful to imagine each individual having a vector of potential outcomes at baseline, we can also imagine a vector of potential treatments at baseline. Let $\mathbf{T}_i = [T_i(0), T_i(1)]$, where $T_i(z) = 1$ if the ith individual takes treatment if assigned $z_i = z$, and $T_i(z) = 0$ if they take control (i.e., do not take treatment). We assume that no individual will get treatment if assigned to control, and we would never observe $Y_i(0,1)$. In other words, $T_i(0) = 0$ for all i. Thus, if $T_i(1) = 1$ then the ith individual is a complier, and $T_i(1) = 1$ represents a characteristic of individual i that is present at baseline (they would take the treatment if assigned to it), even though it is not observable at baseline. An intent-to-treat analysis compares $Y_i(1, T_i(1))$ to $Y_i(0, T_i(0))$, for example, the average causal effect of getting randomized to the treatment arm,

$$\Delta_{ITT} = \Pr[Y_i(1, T_i(1)) = 1] - \Pr[Y_i(0, T_i(0)) = 1].$$

Without complete compliance, some individuals can have $T_i(1) = 0$, and for those individuals the effect of getting randomized to treatment is not the same as the effect of taking treatment. So, it is of key interest to estimate the effect of treatment, T, not the effect of getting randomized to treatment, Z. For example, $\Pr[Y_i(z,1) = 1] - \Pr[Y_i(z,0) = 1]$ for either $z = 0$ or $z = 1$. But we cannot force $T_i = t$, we can only encourage treatment by offering it by randomization; when $z_i = 1$ we expect that $T_i(z_i)$ will more often be 1.

We make the following assumption: random allocation affects response only through its effect on treatment. We restate that assumption (when there is no interference), $Y_i(z,t) = Y_i(z',t)$ for all z, z' and all i. This assumption (even without also assuming no interference) is called the *exclusion restriction assumption* (see, e.g., Angrist et al., 1996, Assumption 3). Under the exclusion restriction assumption, we can estimate a treatment effect on a subset of the population using for example, $\Pr[Y_i(1,1) = 1] - \Pr[Y_i(0,0) = 1]$. The problem is that in a randomized trial without perfect compliance we do not observe only $Y_i(1,1)$ or $Y_i(0,0)$, we sometimes observe $Y_i(1,0)$ or $Y_i(0,1)$.

In the subset of the population that complies, we do observe, depending on the random allocation, either $Y_i(1,1)$ or $Y_i(0,0)$. The problem is that we only observe those who are the compliers in the treatment arm, in the control arm we do not know who would have taken treatment if they had been randomized to the treatment arm. Nevertheless, we can still define an estimand based on that subset. Let the complier average causal difference estimand be

$$\Delta_c = \Pr[Y_i(1,1) = 1 | T_i(1) = 1, T_i(0) = 0] - \Pr[Y_i(0,0) = 1 | T_i(1) = 1, T_i(0) = 0].$$

Because for binary responses, $\Delta_c = \mathrm{E}(Y_i(1,1)|T_i(1) = 1, T_i(0) = 0) - \mathrm{E}(Y_i(0,0)|T_i(1) = 1, T_i(0) = 0)$, this estimand is also known as the complier average causal effect (see, e.g., Imbens and Rubin, 2015). Alternatively, we can define a complier causal ratio effect using

$$\beta_c = \frac{\Pr[Y_i(1,1) = 1 | T_i(1) = 1, T_i(0) = 0]}{\Pr[Y_i(0,0) = 1 | T_i(1) = 1, T_i(0) = 0]}.$$

For making inferences, we need estimators of

$$\theta_{11, T(1)=1} = \Pr[Y_i(1,1) = 1 | T_i(1) = 1]$$

and

$$\theta_{00,T(1)=1} = \Pr[Y_i(0,0) = 1 | T_i(1) = 1]$$

(in this study $T_i(0) = 0$ for everyone, so we do not need to condition on that). We can estimate $\theta_{11,T(1)=1}$ directly using sample proportions. For the vitamin A study this is $\hat{\theta}_{11,T(1)=1} = 12/9,675 \approx 0.00124$. The estimate of $\theta_{00,T(1)=1}$ is less straightforward. Let π be the proportion of the population that are compliers and let $\theta_{00} = \Pr[Y_i(0,0) = 1]$. Then

$$\theta_{00} = \pi \theta_{00,T(1)=1} + (1 - \pi)\theta_{00,T(1)=0}.$$

By the exclusion restriction assumption $\theta_{00,T(1)=0} = \theta_{10,T(1)=0}$, so making that substitution and solving for $\theta_{00,T(1)=1}$, we get

$$\theta_{00,T(1)=1} = \frac{\theta_{00} - (1 - \pi)\theta_{10,T(1)=0}}{\pi}. \tag{15.6}$$

Now we can estimate each parameter with sample proportions. By randomization, the proportion of compliers in the population is the same as in either arm, so we can use the proportion of compliers in the treatment arm to estimate π, giving $\hat{\pi} = 9,675/12,094$. For the other two parameters, we simply use the sample proportions giving $\hat{\theta}_{00} = 74/11,588$, and $\hat{\theta}_{10,T(1)=0} = 34/2,419$. We define $\hat{\theta}_{00,T(1)=1}$ by substituting those estimates into Equation 15.6, giving

$$\hat{\theta}_{00,T(1)=1} = \frac{\hat{\theta}_{00} - (1 - \hat{\pi})\hat{\theta}_{10,T(1)=0}}{\hat{\pi}} \approx 0.00447. \tag{15.7}$$

Then we can estimate Δ_c or β_c by using the estimates, giving

$$\hat{\Delta}_c = \hat{\theta}_{11,T(1)=1} - \hat{\theta}_{00,T(1)=1} \approx 0.00124 - 0.00447 = -0.00323$$

and

$$\hat{\beta}_c = \frac{\hat{\theta}_{11,T(1)=1}}{\hat{\theta}_{00,T(1)=1}} \approx \frac{0.00124}{0.00447} \approx 0.277.$$

For inferences Sommer and Zeger (1991) use asymptotic normality with the delta method. For inferences of Δ_c see also Imbens and Rubin (2015), who use an instrumental variable approach (see Section 15.8).

15.4.4 Principled Adjustments

In this section we present the principled adjustment estimator, $\hat{\Delta}_{pa}$, which is an estimate of the average causal difference, Δ, for a randomized experiment. The principled adjustment estimator allows use of responses from individuals in the control arm to adjust the response in the treatment arm for baseline covariates and vice versa. The key is that it uses two separate models of predicted response (one for each arm). When those models are correctly specified, $\hat{\Delta}_{pa}$ minimizes the asymptotic variance, but even if those models are misspecified, then $\hat{\Delta}_{pa}$ still consistently estimates Δ. This is a rare case where postrandomization measurements may be used to properly adjust an estimator of Δ.

For the ith subject of the randomized study, let $\mathbf{x}_i$ be a vector of baseline (i.e., just prior to treatment application) variables, and let z_i be the randomized treatment variable, assuming perfect compliance. Suppose there are n total individuals, with n_1 randomized to $z_i = 1$ and $n_0 = n - n_1$ randomized to $z_i = 0$. Let $\mathbf{y}_i = [y_i(0), y_i(1)]$ be the potential outcome vector, with observed value $y_i = y_i(z_i)$. Let bars denote means: $\bar{z} = n^{-1} \sum z_i = n_1/n$, $\bar{y}(1) = n_1^{-1} \sum z_i y_i(1)$ and $\bar{y}(0) = n_0^{-1} \sum (1 - z_i) y_i(0)$ with summations denoting $i = 1$ to n. The principled adjustment estimator is given by

$$\hat{\Delta}_{pa} = \bar{y}(1) - \bar{y}(0) - \sum_{i=1}^{n} (z_i - \bar{z}) \left(\frac{\hat{y}^{(1)}(\mathbf{x}_i)}{n_1} + \frac{\hat{y}^{(0)}(\mathbf{x}_i)}{n_0} \right), \qquad (15.8)$$

where $\hat{y}^{(z)}(\mathbf{x}_i)$ is an estimator of $y_i(z)$, and the function $\hat{y}^{(z)}(\cdot)$ is formed using baseline data and response data from only the individuals with $z_i = z$. Although the estimator function $\hat{y}^{(1)}(\mathbf{x}_i)$ is developed using only data from individuals with $z_i = 1$, the function can be applied to all individuals by inputting their baseline values, $\mathbf{x}_i$, but not their responses or any postrandomization measurements. The key to the principled adjustment estimator is that each prediction function is estimated separately; $\hat{y}^{(0)}(\cdot)$ is defined using only the control group data (those with $z_i = 0$) and $\hat{y}^{(1)}(\cdot)$ is defined using only the treatment group data (those with $z_i = 1$). If the prediction models $\hat{y}^{(z)}(\cdot)$ for $z = 0, 1$ are correctly specified, then $\hat{\Delta}_{pa}$ is optimal in the sense that it has the smallest asymptotic variance. But whether or not the prediction models are correctly specified, $\hat{\Delta}_{pa}$ is a consistent and asymptotically normal estimator of Δ. Inferences using $\hat{\Delta}_{pa}$ use the sandwich estimator of variance and asymptotic normality (see Tsiatis et al., 2008, for details).

15.5 Causality and Interference

Hudgens and Halloran (2008) develop a theory for causal estimands when there is interference (i.e., when the response of an individual depends not only on their own treatment assignment, but on the treatment assignment of their neighbors). These estimands are defined based on studies where individuals are clustered into groups, and the only interference is within-cluster interference; there is no interference between clusters. The causal estimands are defined analogously to the way the study-specific average causal difference is defined. It is defined for the specific population that partook in the study by averaging over different allocations of treatment arms to that population. There are different kinds of causal effects:

direct effects – the change in outcome averaged over individuals due to changing treatment within that individual only,

indirect effects – the averaged individual change in outcome due to different treatment assignment mechanisms to the group, holding the treatment for each individual fixed,

total effects – the sum of the individual direct and individual indirect effects, and finally,

overall effects – comparing an intervention program against no intervention.

Hudgens and Halloran (2008) develop estimators of these causal estimands using a two-stage randomization (stage 1: randomize groups to an individual randomization scheme, stage 2: randomize individuals within each group). The details are notationally cumbersome, so we refer readers to Hudgens and Halloran (2008) or Tchetgen Tchetgen and VanderWeele (2012) for details.

15.6 Causality and Observational Studies: Overview

Suppose one is interested in comparing treatment to control, but the decision to get treatment is related to factors that are also related to outcome. Should one control for those variables? Match on those variables? Unfortunately, it does not always make sense to control for all those variables, sometimes controlling for variables leads to a better estimator of the causal estimand; however, sometimes controlling for a variable can lead to a worse estimator.

In this book we will not have enough space to teach one how to make every adjustment that may be needed, but we attempt to give enough information to do simple adjustments, and to recognize when those simple adjustments will not work. We give a brief overview. In Section 15.7 we discuss methods for use when confounders are measured (including how to formally define confounders). Section 15.7.1 discusses the simple use of propensity scores as used with baseline variables measured prior to the intervention of interest. Section 15.7.2 introduces directed acyclic graphs (DAGs) that are used to describe conditional independence or dependence between variables in structural causal models. In Section 15.7.3 we use DAGs to define the backdoor criterion, which is used to formally define sets of confounding variables. Section 15.7.4 briefly mentions mediation analysis that attempts to see how much a causal effect is mediated through specific variables. Finally, in Section 15.8 we discuss instrumental variables that can be used when not all the confounders are measured.

15.7 Observational Studies: Adjustments When Confounders Are Measured

15.7.1 Propensity Scores

We consider the important case when we are interested in the effect of an intervention, but the intervention is determined by factors out of the control of the researchers. Suppose we know all the variables that affect the choice of the intervention and additionally affect the potential outcomes. Since these variables affect the intervention, they must be measured prior to the intervention.

We will only discuss the simple case where the intervention is a binary choice. For concreteness, we will describe the analysis of an observational study of treatment of a noncontagious disease where the ith individual chooses one of two intervention options, treatment ($z_i = 1$, e.g., an experimental intervention) or control ($z_i = 0$, e.g., no intervention or standard care). Let $\mathbf{x}_i$ denote a vector of variables that affect both the choice between interventions and the potential outcome vector and let $\mathbf{y}_i = [y_i(0), y_i(1)]$ be the potential outcomes of the ith individual. As usual, let $\mathbf{X}_i$, Z_i, $Y_i(0)$, and $Y_i(1)$ represent the associated random variables for the ith individual. Further, assume we take a simple random sample from the population of interest, so that the random vectors, $[\mathbf{X}_i, Z_i, Y_i(0), Y_i(1)]$ for $i = 1, \ldots, n$ are all independent and identically distributed. We require two more assumptions:

conditional exchangeability: $[Y(0), Y(1)]$ is independent of Z, given $\mathbf{X}$. This can also be written as $\Pr[Z = 1 | Y(0), Y(1), \mathbf{X}] = \Pr[Z = 1 | \mathbf{X}]$.

positivity: $0 < \Pr[Z = 1 | \mathbf{x}] < 1$ for all $\mathbf{x}$ with positive support (i.e., all $\mathbf{x}$ with $f(\mathbf{x}) > 0$, where for discrete $\mathbf{x}$, $f(\cdot)$ is the probability mass function and $f(\mathbf{x}) = \Pr[\mathbf{X} = \mathbf{x}]$, and for continuous $\mathbf{x}$, $f(\cdot)$ is the probability density function).

The independence assumption plus these two additional assumptions are called *strongly ignorable treatment assignment* given **x**.

Example 15.5 *Consider the question of whether taking hormone replacement therapy upon the onset of menopause affects the risk of coronary heart disease (CHD) within 10 years of start of menopause. In this case, $z_i = 1$ is taking hormone replacement therapy within a year of onset of menopause and continuing to take it for the next 10 years and $z_i = 0$ is not taking it at all during that time. The potential outcomes are whether any occurrence of CHD would have happened within 10 years after start of menopause after taking hormone replacement therapy ($y_i(1)$) or not ($y_i(0)$). Suppose one took a simple random sample of a population of women who had just reached menopause, then $\mathbf{x}_i$ would be a vector of variables that affect both the decision to take hormone replacement therapy and coronary heart disease. This could include a healthy lifestyle and financial resources. It is easy to see that in a real example, it is difficult to envision that we will measure all the factors that affect both the intervention and the potential outcomes. As a thought experiment, assume that $\mathbf{x} = [h,r]$ is made up of only two variables, h represents healthy lifestyle, and r represents financial resources, and both variables are defined by ordered categories with 10 levels going from lowest to highest. The positivity assumption says that at every one of the combinations of the levels of the two variables for which there are women in the population of interest, the probability of taking hormone therapy is neither 0 nor 1. For example, if there are some women in the lowest levels of both categories in the population and none of them get hormone replacement therapy, then the positivity assumption is violated.*

Although it is difficult to meet these assumptions in practice, suppose all assumptions are met. If the variables in **x** are discrete, then we can estimate $\Delta = E(Y(1)) - E(Y(0))$ by using an adjusted estimator similar to Equation 15.5. Here are the details. Suppose there are L unique combinations of the values of the variables in **x** that are present in the population of interest, and we denote them by the vectors $\mathcal{X}_\ell$ for $\ell = 1, \ldots, L$, and let n_ℓ be the number of individuals sampled with $\mathbf{x}_i = \mathcal{X}_\ell$. Let $\bar{y}_\ell(z)$ be the mean of y_i among individuals with $z_i = z$ and $\mathbf{x}_i = \mathcal{X}_\ell$. Then the adjusted estimator is

$$\hat{\Delta}_{x_{adj}} = \sum_{\ell=1}^{L} \{\bar{y}_\ell(1) - \bar{y}_\ell(0)\} \frac{n_\ell}{n} \tag{15.9}$$

which estimates,

$$\Delta_{x_{adj}} = \sum_{\ell=1}^{L} \{E(Y(1)|\mathbf{X} = \mathcal{X}_\ell) - E(Y(0)|\mathbf{X} = \mathcal{X}_\ell)\} \Pr[\mathbf{X} = \mathcal{X}_\ell] \tag{15.10}$$

and by the assumptions $\Delta_{x_{adj}} = \Delta = E(Y(1)) - E(Y(0))$, the average causal difference (Problem P2).

Suppose there are many variables in **x** and L is large. In this case there may not be enough individuals within each unique level ℓ to have both an individual with $z_i = 1$ and one with $z_i = 0$. There are many methods for creating matched groups with similar covariate values. Stuart (2010) reviews matching methods for causal inference, and one of the most useful uses the propensity score. The propensity score for individual i is $e(\mathbf{x}_i) \equiv \Pr[Z_i = 1| \mathbf{X}_i = \mathbf{x}_i]$, i.e., it is the propensity for that individual to choose $Z_i = 1$. Conditional on the propensity score, Z is independent of $\mathbf{X}$. Thus, using Bayes theorem, within each stratum defined by equal propensity scores, the distribution of $\mathbf{X}$ will be the same between the treatment and control groups (see, e.g., Rosenbaum, 2002, Proposition 29, p. 298). So instead of standardizing by the many dimensional $\mathbf{x}_i$ values, we can standardize based on the one-dimensional propensity scores, $e(\mathbf{x}_i)$, for $i = 1, \dots, n$. In practice, we typically estimate propensity scores based on logistic regression, to get $\hat{e}(\mathbf{x}_i)$, which is the predicted $\Pr[Z_i = 1|\mathbf{x}_i]$ from the logistic model. Then we often standardize based on deciles of those estimated propensity scores.

Additionally, an alternative is to use inverse probability weighting, where the treatment effect is estimated by

$$\hat{\Delta}_{ipw} = \frac{1}{n} \sum_{i=1}^{n} \frac{z_i y_i}{\hat{e}(\mathbf{x}_i)} - \frac{1}{n} \sum_{i=1}^{n} \frac{(1 - z_i) y_i}{\{1 - \hat{e}(\mathbf{x}_i)\}}.$$

An issue with this estimator is that if $\hat{e}(\mathbf{x}_i)$ is close to 0 or 1, then that individual may have an extreme influence on the estimate. Lunceford and Davidian (2004) compare many different methods that use propensity scores. An important one is the double-robust estimator which, if correctly specified, is an efficient estimator (i.e., has smaller variance than $\hat{\Delta}_{ipw}$). It is

$$\hat{\Delta}_{DR} = \frac{1}{n} \sum_{i=1}^{n} \left\{ \frac{z_i y_i - (z_i - \hat{e}(\mathbf{x}_i)) \hat{y}^{(1)}(\mathbf{x}_i)}{\hat{e}(\mathbf{x}_i)} \right\}$$
$$- \frac{1}{n} \sum_{i=1}^{n} \left\{ \frac{(1 - z_i) y_i - (z_i - \hat{e}(\mathbf{x}_i)) \hat{y}^{(0)}(\mathbf{x}_i)}{1 - \hat{e}(\mathbf{x}_i)} \right\},$$

where $\hat{y}^{(z)}(\mathbf{x}_i)$ for $z = 0, 1$ is defined as in the principled adjustment estimator (Equation 15.8). In fact, in the randomized trial when the estimated propensity score is n_1/n for all subjects, $\hat{\Delta}_{DR}$ equals the principled adjustment estimator (Problem P3). This is referred to as doubly robust because the estimator is consistent if either the $\hat{e}(\mathbf{x}_i)$ is correctly specified, or if $\hat{y}^{(z)}(\mathbf{x}_i)$ for $z = 0, 1$ are correctly specified; if only one of those two models is misspecified then $\hat{\Delta}_{DR}$ is still consistent.

Another approach is stabilized weighting, see Cole and Hernán (2008) and Hernán and Robins (2020).

15.7.2 *Structural Causal Models and Directed Acyclic Graphs*

Suppose we are interested in a causal effect of some variable, Z, on a response, Y. One way to model this effect is through a structural causal model. We consider several simple models with three variables, Z, Y, and one other. It is helpful to delay naming the other variable in these simple models because depending on the model, the third variable acts differently with respect to the causal interpretation of Z on Y.

A first model of the effect of Z on Y is a mutually dependent causal model, where there is some "fork" variable, W, that causes both Z and Y. In terms of a structural causal model, it is:

$$W = f_w(\epsilon_w)$$
$$Z = f_z(W, \epsilon_z) \tag{15.11}$$
$$Y = f_y(Z, W, \epsilon_y),$$

where the variables ϵ_w, ϵ_z, and ϵ_y represent some unmeasured errors which are independent of each other and of W, Z, and Y, and f_w, f_z, and f_y represent functions giving the relationships between the errors and the W, Z, and Y. We can represent this structural causal model using a directed acyclic graph (DAG), where the functions are not given and usually the unmeasured error variables are not depicted in the DAG. A DAG is directed (it has arrows from causes to effects) and acyclic (there are no feedback loops, meaning if you start from any variable and follow the arrows, then none lead back to that variable). In general, we restrict the causal models to acyclic ones so that if we know the error variables, then all other variables in the model can be determined. Here is the DAG associated with the structural causal model of Expression 15.11:

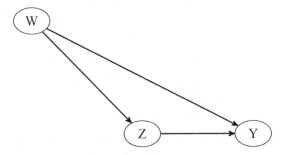

In this case, it is appropriate to control for W. The term "control for W" is vague, but loosely it means to estimate the overall causal effect by combining causal effect estimates from different levels of W (analogous to adjustment estimators like Equation 15.9). Consider the case of the average causal effect, Δ, when Z is either 0 or 1. In that case, to estimate an average causal effect of Z on Y controlling for W, we take the expectation of the conditional causal effects over the values of W. For example, suppose Z is binary taking values 0 and 1, and Y and W are discrete numerical. Write the potential outcome for $Z = z$ as $Y(z)$, then we can reformulate $\Delta = E(Y(1)) - E(Y(0))$ controlling for W as

$$\Delta_{w_{adj}} = E\{E(Y_i(1)|W)\} - E\{E(Y_i(0)|W)\}$$
$$= \sum_w \sum_y y\Pr[Y(1) = y|W = w]\Pr[W = w] \tag{15.12}$$
$$- \sum_w \sum_y y\Pr[Y(0) = y|W = w]\Pr[W = w].$$

Another structural causal model is when the effect of Z on Y passes through some *mediator* (or *chain*) M. We write the structural causal model as

$$Z = f_z(\epsilon_z)$$
$$M = f_m(Z, \epsilon_m) \tag{15.13}$$
$$Y = f_y(M, \epsilon_y),$$

where again the error variables are independent of each other and all other variables. We can represent this structural causal model using a DAG, as $Z \to M \to Y$.

In the mediator model with three variables (Equations 15.13), we should not control for the mediator when looking for the effect of Z on Y. For example, suppose you are measuring the effect of a drug Z taken orally on a response Y, and the effect of the drug taken orally has its effect by getting to the target organ through the blood stream. A mediator is the amount of drug measured in the blood after the drug is taken, and if you control for the mediator, then it may completely erase any of the effect of the drug on the outcome.

Finally, there is the collider model, where the third variable C is a collider:

$$Z = f_z(\epsilon_z)$$
$$Y = f_y(Z, \epsilon_y), \qquad\qquad\qquad (15.14)$$
$$C = f_c(Z, Y, \epsilon_w),$$

with the same independence assumptions on the errors. The associated DAG is:

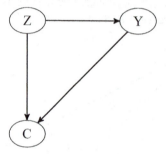

For the collider model, we do not control for C when trying to estimate the effect of Z on Y. Even if there is no causal effect between Z and Y, if we control for a collider, then it can induce an apparent effect (see Problem P4).

For real examples, we can have complicated models that contain all three types of variables, mediators, forks, and colliders (see Figure 15.1).

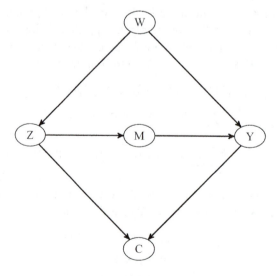

Figure 15.1 Directed acyclic graph example. With respect to the relationship between Z and Y: M is a mediator, W is a fork (or mutually dependent cause), and C is a collider (mutual causation).

15.7.3 The Backdoor Criterion

It is easy to see how things can get very complicated. What if we want to know if A has any effect in causing B? There could be variables that affect both A and B, or variables that are caused by both A and B. In a DAG, there can be all kinds of variables that can make the effect between A and B less straightforward.[2] In this section, we discuss the backdoor criterion for deciding on whether to adjust a set of variables or not when we are interested in a causal relationship between two variables in a DAG.

To describe the backdoor criterion, we first need to define some terms related to DAGs. Recall, DAG is a *directed acyclic graph*, where 'graph' here is used in its technical mathematical meaning. The variables in a DAG are nodes or vertices, and the lines connecting them are edges. In DAGs the edges are usually directed edges (ones with arrows), denoting causation. A *path* from nodes A to B is any series of nodes connected by edges, starting from A and ending at B, regardless of the direction of the edges. A *directed path* from nodes A to B is a path where all the edges are directed in a chain from A to B. For example, A, X, Y, B is a directed path in this graph: $A \rightarrow X \rightarrow Y \rightarrow B$, but is only a path in this graph: $A \rightarrow X \leftarrow Y \rightarrow B$. In $A \rightarrow B$ we say that A is the parent and B is the child. Following in this analogy, any node in a directed path from A is a descendant of A, and any node in a directed path to A is an ancestor of A.

A more abstract definition is whether a set of nodes *blocks* a path. We first describe this with simple three-node DAGs, each with paths from A to B with X in the middle. There are four possible DAGs, where in terms of the relationship between A and B, the node X acts like either: a mediator ($A \rightarrow X \rightarrow B$ or $A \leftarrow X \leftarrow B$), a fork ($A \leftarrow X \rightarrow B$), or a collider ($A \rightarrow X \leftarrow B$). In these three-node DAGs, the path from A to B blocked by X in the first three cases but not in the last case, when X is a collider. From the definition of the structural causal models, when X is a collider then A and B are independent, and in the three other cases A and B may be dependent. When you condition on X and X is not a collider, then you block the path, and A and B are independent *conditional on X*. When you condition on X and X is a collider, then you unblock the path, and the conditioning induces dependence between A and B. In summary, in a three-node DAG with the path (A, X, B), conditioning on X unblocks when X is a collider node, but blocks when X is not a collider.

Now consider a general definition of blocking for a set of nodes, S. In this definition, when S is the empty set, and we condition on it, this is the same as not conditioning on anything.

Definition 15.1 Blocked path: *Consider a set of nodes, S, which may be the empty set. A path p in a DAG is blocked by S if and only if p contains a three-node path, (A, X, B) where either:*

(1) *X is not a collider, and $X \in S$, or*
(2) *X is a collider, and $X \notin S$, and no descendant of X is in S.*

(Pearl et al., 2016, p. 46)

---

[2] Further, in real biological systems, there can be feedback loops which cannot be expressed as DAGs at all.

If S blocks every path between A and B, then we say A and B are *d-separated* conditional on S. Further, if S blocks every path between two nodes A and B, then A and B are independent conditional on S (see Pearl, 2009b, Theorem 1.2.4, p. 18).

We now define the backdoor criterion, which allows us to identify a sufficient set of variables for which we need to adjust. The setting is that we are interested in the causal effect of A on B, and we have a DAG where there is at least one directed path from A to B, but there are other paths between A and B. If we can find a set of nodes, S, that satisfy the backdoor criterion, then if we do a conditional adjustment on S, that will allow us to estimate the causal effect of A on B.

Definition 15.2   Backdoor criterion: *Suppose we have a DAG that contains a directed path between A and B, then a set of nodes, S, satisfies the backdoor criterion relative to (A, B), if no node in S is a descendant of A, and S blocks all paths between A and B that contain an arrow into A (Pearl, 2009b; Pearl et al., 2016).*

If the set of variables in S meet the backdoor criterion, then an adjustment estimand on all levels of S (analogous to the adjustment estimand for levels of W in Equation 15.12) is equivalent to the average causal effect from A to B (Pearl, 2009b, Theorem 3.3.2).

The "backdoor" name comes from blocking paths with arrows going into A, which is like entering into the A-to-B directed path through the back door. Following Hernán and Robins (2020) we define a *backdoor path* as a path from A (treatment) to B (outcome) that contains no descendants of A. Finally, we can formally define confounder. Hernán and Robins (2020) discuss the many ways that the term "confounder" has been used in the literature, and they settle on the following: a *confounder* is "a variable that can be used to block an (otherwise open) backdoor path between treatment and outcome." An interactive illustration of these concepts can be found on `www.dagitty.net`. This allows one to easily construct a DAG on the screen; any backdoor path (labeled biasing path) is highlighted, and a list of minimal sufficient adjustment sets for an unbiased estimate of the effect of an intervention or exposure on an outcome is provided. Exploration of different DAGs should facilitate understanding of the definitions in this section.

For the more complicated front-door criterion and its associated adjustment, see Pearl (2009b) or Pearl et al. (2016).

15.7.4 Mediation Analysis

Another type of analysis is understanding the partial effects of an event that may have multiple causes. Consider Example 14.13 (page 270) where we were interested in the effect of height (Z) on the incidence of Type II diabetes (Y) apart from its effect on weight (X). In Figure 15.2 we give a simple DAG for this example. If we are interested in the total effect of height on the incidence of diabetes, we would not control for weight, because some of the effect of height may be mediated through weight. But since we are interested in the effect of height on diabetes *apart from its effect on weight*, we do adjust for the weight variable. These types of analyses are beyond the scope of this book, see VanderWeele (2015).

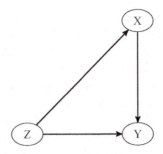

Figure 15.2 Directed acyclic graph example. With respect to the relationship between Z (height) and Y (Type II diabetes), X (weight) is partially determined by height and partially determines Type II diabetes.

15.8 Observational Studies: Analysis with Unmeasured Confounders Using Instrumental Variables

If one has an observational study but does not observe all the confounding variables, then sometimes one can use an instrumental variable analysis. An *instrumental variable* is a variable that (1) has an effect on treatment (e.g., a variable that encourages getting treatment), (2) does not affect the outcome except through its influence on the treatment, and (3) is independent of confounders (in other words, there are no variables [measured or unmeasured] that affect both the instrumental variable and the response). In Note N2, we formally write these assumptions in the case of binary instruments and treatments.

We consider the case of the binary instrumental variable and treatment where the usual instrumental variable assumptions hold (Note N2). For the ith individual, we write Z_i as the potential instrument, $T_i(z)$ as the treatment (or other intervention) when $Z_i = z$, and $Y_i(z,t)$ as the potential outcome when $Z_i = z$ and $T_i = t$.

In Section 15.4.3 we already discussed a randomized trial with imperfect adherence, an example of an instrumental variable (but did not identify it as such at the time). If it is known in advance that there will be substantial nonadherence, such a trial may be called a *randomized encouragement design*.

Example 15.6 *Angrist et al. (1996) give an example related to military service ($T_i = 1$ is served in the military for the United States in Vietnam) or not ($T_i = 0$) and its effect on civilian mortality ($Y_i = 1$ died between 1974 and 1983 [the United States withdrew troops from Vietnam in 1973]) or not ($Y_i = 0$). Let $Z_i = 1$ if the ith man had a lottery number indicating they were drafted into the military, and $Z_i = 0$ if not drafted (between 1970 and 1973, a lottery was used to select men [women were not in the lottery] for service in the military). This is a randomized encouragement study, where the instrumental variable (i.e., the lottery-based draft) represents the encouraged randomization into military service that can be used to study the effect of military service on civilian mortality (among those in the lottery who were civilians from 1974 on).*

Example 15.7 *Baiocchi et al. (2014) give an example on the effect of high-level neonatal intensive care units (NICU) ($T_i = 1$) compared to low-level ones ($T_i = 0$) on premature infant mortality ($Y_i = 1$ is died in hospital) and ($Y_i = 0$ did not die in hospital). The proposed* *instrumental variable is excess travel time (Z_i), which is how much excess time to travel from the mother's residence to the nearest high-level NICU than to the nearest low-level NICU (negative excess time means that the local NICU is high level). The instrument does not have to be binary, although it can be dichotomized with $Z_i = 1$ for lower excess travel time and $Z_i = 0$ for higher excess time so that we would expect positive correlation with T_i. Excess travel time appears to be a good instrument because the location of residence in relation to the NICU's would (1) encourage going to the high-level NICU if it was closer and the mother expected complications, (2) probably be essentially unrelated to the infant mortality aside from the type of hospital the baby was born in, and (3) probably not be strongly related to other factors that would affect the infant mortality.*

See Baiocchi et al. (2014) and Rosenbaum (2010, Section 5.3) for other common sources of instrumental variables.

Under the five instrumental variable assumptions (Note N2), then the complier average causal effect (also known as the local average treatment effect),

$$\Delta_c = E(Y_i(1,1)|T_i(1) = 1, T_i(0) = 0) - E(Y_i(0,0)|T_i(1) = 1, T_i(0) = 0),$$

is estimated by

$$\hat{\Delta}_{c,iv} = \frac{\bar{y}(1) - \bar{y}(0)}{\bar{t}(1) - \bar{t}(0)},$$

where $\bar{y}(1) = \frac{\sum z_i y_i}{\sum z_i}$, $\bar{y}(0) = \frac{\sum (1-z_i) y_i}{\sum (1-z_i)}$, $\bar{t}(1) = \frac{\sum z_i t_i}{\sum z_i}$, and $\bar{t}(0) = \frac{\sum (1-z_i) t_i}{\sum (1-z_i)}$.

If we allow T_i and Z_i to be numeric and nonbinary and we use the linear model with a constant treatment effect

$$Y_i = \mu + \beta T_i + \epsilon_i, \tag{15.15}$$

where ϵ_i is independent of Z_i. Then we can get an instrumental variable estimator of β using (see Problem P5)

$$\hat{\beta}_{iv} = \frac{\widehat{Cov}(Y_i, Z_i)}{\widehat{Cov}(T_i, Z_i)}.$$

The causal interpretation is less straightforward, see Baiocchi et al. (2014) and Morgan and Winship (2015) for details.

15.9 Summary

This chapter introduces the concept of potential outcomes for each individual: $y_i(0)$ is the outcome the individual would have had if they received treatment 0, and $y_i(1)$ is the outcome the individual would have had if they received treatment 1. A causal effect at the individual level is defined as $D_i = Y_i(1) - Y_i(0)$. However, generally, it is only possible to observe one of these two potential outcomes. Thus, the goal of casual inference is to estimate the expected causal effect in a population, either through randomized experiments or observational data. Causal inference usually requires assumptions such as the stable unit treatment value assumption (i.e., no interference between study subjects and there is a single version of the intervention). Different or additional assumptions are required for different causal inference settings.

Estimation of the causal estimate is straightforward in the context of randomized experiments, as one reasonable estimate is simply the difference in outcomes between randomized groups. Additionally, since randomization does not preclude some imbalance between the groups, causal estimates can also use stratification to better control for important prognostic covariates. One should not use this simple form of stratification, however, to control for postbaseline variables, as this would no longer necessarily provide a consistent estimator of the true causal effect. However, suppose we want to know the causal treatment effect in those who comply with the treatment. It would not be appropriate to just compare the outcome of compliers in the treatment group with the full control group, since compliers may be intrinsically different from the group as a whole, so this would not be a fair comparison. However, as introduced in Section 15.4.3, there is an approach for using causal principles along with some extra assumptions, to derive a fair comparison of compliers in the treatment group to a latent cohort of control group who would have complied to treatment if they had been randomized to treatment.

Not surprisingly, causal inference in the context of observational data is considerably more challenging because of the potential for confounders. Under the assumption that all confounders are known and measured, there are several very useful techniques from the causal literature. The method of propensity scores allows us to reduce the factors that predict treatment into a single index, which facilitates the comparison of the treatment groups in a stratified analysis, and potentially uses matching to better mimic a randomized trial. There are several other methods that utilize propensity scores such as inverse proportional weighting and the related double-robust estimator which has the advantage that the causal estimate is consistent even if one of the two models it employs is not correctly specified.

We discuss Directed Acyclic Graphs (DAGs), a powerful device for determining which variables should be adjusted for in the observational data setting. This framework also provides a formal definition for a confounder. It illustrates that it is not a good practice to adjust for any or all variables that might seem relevant or important, because adjusting variables that are not confounders can attenuate the true causal effect, or even potentially overestimate it.

Finally, we present a brief overview of instrumental variables, a device that is useful in the observational setting where we are not confident that all confounders are measured. This method identifies a variable that is strongly associated with treatment, such as a variable that encourages getting the intervention. Further assumptions include: (1) the instrumental variable should not affect the outcome except through its influence on getting treatment; and (2) there are no confounders that affect both the instrumental variable and outcome. When the instrument and treatment are binary, the corresponding causal estimate can be viewed as the mean difference on the outcome in the two groups defined by the instrumental variable, adjusted by the mean differences in treatment rates in the two instrumental variable groups. So, if the instrumental variable were a perfect classifier of treatment, then the causal estimate would simply be the difference on outcome in the two instrumental variable categories. Of course, it is often not possible to identify an instrumental variable.

15.10 Extensions and Bibliographic Notes

There are several excellent books on causal inference, and most, if not all, are by researchers who have done important fundamental work on the topic. Imbens and Rubin (2015) give a very good description of the potential outcomes approach, including randomization inferences, propensity scores, and instrumental analysis. Pearl (2009b) gives a theoretical approach using the directed acyclic graph (DAG) approach, the emphasis is on understanding what you are estimating with little emphasis on practical details of statistical analysis. Pearl et al. (2016) give an easier introduction to the DAG approach. Hernán and Robins (2020) is also quite comprehensive and includes references to computer code. Morgan and Winship (2015) is also comprehensive and focuses more on sociological examples. Rosenbaum (2010) gives practical details on designing observational studies as well as theoretical details on measuring causal effects and on measuring sensitivity to the causal assumptions.

For a review article on causality, see Pearl (2009a).

15.11 Notes

N1 **Randomization inference:** Making randomization inferences on Δ_s is sometimes presented as a model-free way of doing inferences (see, e.g., Ding et al., 2016); however, certain assumptions (in addition to assuming a proper randomization) are still needed to perform the inferences. The problem is that we cannot observe both potential outcomes for any individual, so we must rely either on very strong assumptions (e.g., $D_i = Y_i(1) - Y_i(0) = D$ for all i), or on asymptotic methods. The asymptotic methods rely on some assumptions about the distributions of the $Y_i(1)$ and $Y_i(0)$ to ensure that there is not a very small probability that an extremely large value of $Y_i(z)$ is unobserved. So although the method is model-free, there are still some restrictions on the form of the distribution for the responses. See Section 9.3 for the need for similar assumptions under more standard nonrandomization inferences.

N2 **Instrumental variable assumptions for binary instruments and interventions:** For the ith individual, let Z_i represent a possible instrumental variable and T_i represent the effect (e.g., treatment) of interest. Let $\mathbf{Z} = [Z_1, \ldots, Z_n]$ and similarly define $\mathbf{T}$. We write $T_i(\mathbf{Z})$ to note that in general T_i may depend on $\mathbf{Z}$; however, sometimes we write $T_i(Z_i)$ where Z_i is a scalar to indicate that $T_i(\mathbf{Z}) = T_i(\mathbf{Z}')$ whenever the $Z_i = Z'_i$. Write the set of potential outcome vectors as $\mathbf{Y}(\mathbf{Z}, \mathbf{T}) = [Y_1(\mathbf{Z}, \mathbf{T}), \ldots, Y_n(\mathbf{Z}, \mathbf{T})]$ where $Y_i(\mathbf{Z}, \mathbf{T})$ is the potential outcome vector of individual i.

An instrumental variable formally needs four assumptions (Angrist et al., 1996; Baiocchi et al., 2014):

stable unit treatment value assumption (SUTVA): If $Z_i = Z'_i$, then $T_i(\mathbf{Z}) = T_i(\mathbf{Z}')$, and if $Z_i = Z'_i$ and $T_i = T'_i$, then $Y_i(\mathbf{Z}, \mathbf{T}) = Y_i(\mathbf{Z}', \mathbf{T}')$).

positive correlation: Z_i and T_i are positively correlated, so that $E(T_i(1)) > E(T_i(0))$.

independence from unmeasured confounders: Z is independent of $[T(0), T(1), Y(1,1), Y(1,0), Y(0,1), Y(0,0)]$.

exclusion restriction: $\mathbf{Y}(\mathbf{Z}, \mathbf{T}) = \mathbf{Y}(\mathbf{Z}', \mathbf{T})$ for all $\mathbf{Z}, \mathbf{Z}'$ and $\mathbf{T}$.

A fifth assumption is often added but not always (Baiocchi et al., 2014, p. 2307):

monotonicity: $T_i(1) \geq T_i(0)$ for all i.

15.12 Problems

P1 Consider the case where $\mathbf{Y}_i$ are independent from F, with marginal distributions F_0 and F_1. Show that G (Equation 15.2) may be written as a function of F_0 and F_1, and that $E(D_{Gi}) = \phi - \frac{1}{2}$, where $h_{MW}(F_0, F_1) = \phi$ is the Mann–Whitney parameter. Hint: show separately for continuous and ordinal distributions (see Fay et al., 2018a, for answers).

P2 Show that $\Delta_{x_{adj}} = \Delta$ (see Equation 15.10 and after). You may additionally assume that the potential outcomes are discrete. Hint: when A is independent of B conditional on C, then $\Pr[A = a|C, B] = \Pr[A = a|C]$ for all a.

P3 Show that $\hat{\Delta}_{DR} = \hat{\Delta}_{pa}$ in a randomized trial (when the propensity score is the same for all individuals and equal to n_1/n).

P4 Consider $n = 1,000$ pairs of independent bivariate uncorrelated normal random variables, (Z_i, Y_i) with $E(Y_i) = E(Z_i) = 0$ and $\text{Var}(Y_i) = \text{Var}(Z_i) = 1$. Let $C_i = I(Z_i > 0)I(Y_i > 0)$. Simulate $n = 1,000$ observations (Z_i, Y_i, C_i). Calculate the sample correlation of Z_i and Y_i when $C_i = 1$ and when $C_i = 0$. Then calculate the correlation of Z_i and Y_i controlling for C_i. For describing the causal effect between Z and Y, what structural causal model from Section 15.7.2 most closely describes this model? Draw a DAG of this model. Should you control for C to estimate the causal effect between Z and Y?

P5 Motivate the estimator $\hat{\beta}_{iv}$ from Equation 15.15. Hint: use the definition of covariance and Equation 15.15 (Morgan and Winship, 2015).

16

Censoring

16.1 Overview: Types of Censoring

Censored responses are not observed exactly but are only partially observed. Typically, censoring occurs in the context where responses are the time until some event, and a common example is time from entering the study until death. For this reason, these types of analyses are often called *survival analyses*. For the ith individual, let X_i be the time from the start time at 0 until the event of interest. A censored observation is not observed exactly, but only known to be within some subset of the positive real line. The ith response is "an observed event time" when we know X_i, otherwise it is

right censored if we only know that $X_i \in (L_i, \infty)$,
left censored if we only know that $X_i \in (0, R_i]$, and
interval censored if we only know that $X_i \in (L_i, R_i]$

for values $0 < L_i < R_i < \infty$. Thus, for an individual with a right-censored event time we only know that the event time is to the right of some censoring time L_i, and for left-censored event times we know that the event time is to the left of some censoring time R_i. When we say "right-censored data," we mean a mixture of observed event times and right-censored event times (e.g., a time-to-death study where not everyone dies before the end of the study). Often "interval-censored data" denotes a mixture of interval-censored, right-censored, and left-censored observations, with no observed event times. For example, that type of interval-censored data would occur if the event time cannot be observed through continuous assessment (e.g., time until HIV infection), but requires discrete assessment times (e.g., blood draws for indications of infection). The term *current status data* denotes a mixture of only right- and left-censored observations; for example, in a mouse study of time until development of lung cancer where mice must be sacrificed to determine if a lung cancer had begun to develop or not. Current status data is also known as Case one interval censoring, because each individual has only one assessment done to determine whether the event has happened or not.

In this chapter we deal mostly with one- and two-sample problems, although we discuss semiparametric models as well. In the one-sample problem, we assume that X_i is distributed according to some distribution F, and we want to make inferences about the distribution at some time point t, say $F(t) = \Pr[X_i \leq t]$. Alternatively, we often make inferences about the *survival distribution*, $S(t) = 1 - F(t) = \Pr[X_i > t]$. For the two-sample problem, we want to make inferences about the distributions of two groups, F_0 and F_1 (or the associated survival distributions).

With censored data, it is important to be clear about the assumptions about why the data are censored. Those assumptions about censoring will be different for different types of censoring. Further, mathematically it will make sense to focus separately on right-censored data and interval-censored data, because right-censored data is more common and is simpler.

Truncation is related to selection of individuals based on their event time and is discussed in Section 16.6.

16.2 One Sample with Right Censoring

16.2.1 Independent Censoring Assumption

Although the time to event may be the time to any event of interest (positive or negative), for ease of exposition we call the event a failure, and survival to time t denotes that the failure has not occurred by t. For the ith subject let X_i be the failure time and C_i be the censoring time. With right-censored data, for the ith individual we observe only $T_i = \min(X_i, C_i)$ and $\delta_i = I(X_i \leq C_i)$. Most inferences assume *independent censoring*, which informally is when the instantaneous probability of failure at time t given survival until just before t is the same as the instantaneous probability of failure at time t given all the information about failures and censoring up until time t. In other words, if knowing the censoring times before t tells us no information about when a failure will occur after t, then the censoring is considered independent of survival. Note N1 gives a more formal definition and discusses other censoring assumptions.

Example 16.1 *Consider a randomized trial where for the ith individual, X_i is the time from randomization until death, and C_i is the time from randomization until the study ends; that is, no one drops out of the study prematurely. If individuals enroll in the study slowly over time, and each individual is randomized at the time of enrollment, then C_i is related to when the individual is randomized and the end of the study. If there is no calendar effect on survival time (i.e., the length of X_i does not depend on when the individual entered the study), then the independent censoring assumption is reasonable. The assumption would fail if sicker people were more likely to enroll earlier in calendar time.*

Example 16.2 *Consider the same example as Example 16.1, except that some individuals drop out of the study early before the end of the study. For those individuals, if the reason they drop out is related to X_i then the independent censoring assumption fails, otherwise it still holds. For example, if some individuals drop out of the study because they feel very sick and want to go home to die, then the independent censoring assumption will not hold.*

When the censoring is dependent on event time, we give references in Section 16.8 for some approaches of analysis.

16.2.2 Kaplan–Meier Estimator of S

Under independent censoring (or more generally, noninformative censoring, see Note N1) an important nonparametric maximum likelihood estimator of the survival distribution is the Kaplan–Meier estimator, also known as the product-limit estimator. We introduce notation to write it allowing for tied observations. Let $t_1, t_2, \ldots, t_n$ be the observed values associated with $T_1, \ldots, T_n$. If there are ties, then there are only $k < n$ unique values of $t_1, \ldots, t_n$, and we write them as $t_1^* < t_2^* < \cdots < t_k^*$. Let $d_j = \sum_{i=1}^{n} I(t_i = t_j^*)\delta_i$ be the number of events at time t_j^* and $r_j = \sum_{i=1}^{n} I(t_i \geq t_j^*)$ be the number at risk just before t_j^* (so $r_1 = n$). Then the *Kaplan–Meier estimator* of $S(t)$, is

$$\hat{S}(t) = \prod_{j:t_j^* \leq t} \left(1 - \frac{d_j}{r_j} \right). \tag{16.1}$$

The Kaplan–Meier estimator only changes when there are observed event times (i.e., at t_j^* if $d_j > 0$), and it is sometimes left undefined for $t > t_k^*$. Unlike Equation 16.1, typically, the Kaplan–Meier estimator is written by only indexing the times with observed events (i.e., t_j^* with $d_j > 0$), but this notation will be useful for describing confidence intervals for $S(t)$. For example, the beta product confidence procedure (BPCP) allows the confidence intervals to change at each t_j^* even if there are no events observed there (i.e., even if $d_j = 0$). Allowing those changes lets the BPCP (Equation 16.6) provide better coverage than standard methods for the confidence intervals at the latter parts of the survival curve as fewer individuals remain at risk.

We motivate the Kaplan–Meier estimator by imagining the time scale is partitioned into very small intervals, $[0, \tau_1), [\tau_1, \tau_2), [\tau_2, \tau_3), \ldots$, where for all $j = 1, 2, \ldots$, the value $\tau_j = j\Delta$, where $\Delta = \tau_j - \tau_{j-1}$ is constant and small.[1] For example, suppose that deaths are measured only to the nearest day, and time is measured in days, then $\Delta = 1$. Further imagine that the event times only occur at the discrete times $\tau_1, \tau_2, \ldots$. Suppose $\Pr[X \geq \tau_\ell] > 0$, then

$$\Pr[X \geq \tau_{\ell+1}] = \frac{\Pr[X \geq \tau_{\ell+1}]}{\Pr[X \geq \tau_\ell]} \frac{\Pr[X \geq \tau_\ell]}{\Pr[X \geq \tau_{\ell-1}]} \cdots \frac{\Pr[X \geq \tau_2]}{\Pr[X \geq \tau_1]}$$

$$= \prod_{h=1}^{\ell} \Pr[X \geq \tau_{h+1} | X \geq \tau_h], \tag{16.2}$$

where the last step comes from the conditional probability definition (for events A and B,

$$\Pr[A|B] = \frac{\Pr[A \text{ and } B]}{\Pr[B]},$$

when $\Pr[B] > 0$), and the fact that

$$\Pr[X \geq \tau_{h+1} \text{ and } X \geq \tau_h] = \Pr[X \geq \tau_{h+1}].$$

The Kaplan–Meier estimator estimates $\Pr[X \geq \tau_{h+1} | X \geq \tau_h]$ with the sample proportion of individuals who were at risk just prior to τ_h that survived to at least τ_{h+1}. All of these proportions are 1 except at times where $t_j^* = \tau_h$ and $d_j > 0$, and in those cases the proportion is $(r_j - d_j)/r_j$. In other words, using those sample proportions, Equation 16.2 reduces to the Kaplan–Meier estimator (Equation 16.1) evaluated when t is the time just prior to τ_{h+1},

---

[1] There are more mathematically elegant ways to derive the Kaplan–Meier estimator as a nonparametric maximum likelihood estimator using cumulative hazard functions, counting processes, and the product integral, but they are beyond the scope of this book (see, e.g., Andersen et al., 1993; Aalen et al., 2008)

which we write as $\hat{S}(\tau_{h+1}-)$. Loosely speaking, we can imagine Kaplan–Meier estimator for continuous event times using this motivation and letting $\Delta \to 0$.

Kaplan and Meier (1958) used the Greenwood formula to derive an asymptotic estimator for the variance of $\hat{S}(t)$ that may be used to get confidence intervals for $S(t)$ for large samples,

$$\widehat{var}\{\hat{S}(t)\} = \{\hat{S}(t)\}^2 \prod_{j:t_j^* \le t} \frac{d_j}{r_j(r_j - d_j)}.$$

A simple useful $100(1 - \alpha)\%$ confidence interval uses the Greenwood variance estimator and asymptotic normality using a log transformation and applying the delta method (Section 10.6):

$$\exp\left\{ \log(\hat{S}(t)) \pm \Phi^{-1}(1 - \alpha/2) \frac{\sqrt{\widehat{var}(\hat{S}(t))}}{\hat{S}(t)} \right\}. \tag{16.3}$$

Fay et al. (2013) and Fay and Brittain (2016) reviewed many confidence intervals for $S(t)$ and developed the beta product confidence procedure (BPCP) that outperformed all prior methods with respect to ensuring proper coverage. The BPCP intervals are designed to be valid for all sample sizes, and they are asymptotically equivalent to traditional confidence intervals for $S(t)$ (e.g., Equation 16.3). When there is no censoring, the BPCP intervals reduce to the exact central intervals on the binomial parameter (Problem P2). This relationship makes sense because with no censoring the Kaplan–Meier estimator of $S(t)$ is the proportion of events greater than t, so that $n\hat{S}(t) = \sum_{i=1}^{n} I(x_i > t)$ is binomial with parameters n and $S(t)$ (Problem P1).

We first introduce the BPCP for $S(t)$ using confidence distribution random variables (CDRVs), see Section 10.11, then we give the motivation. For $t \in [t_j^*, t_{j+1}^*)$ the upper and lower CDRVs are

$$W_{Uj} = \prod_{h=1}^{j} B(r_h - d_h + 1, d_h), \tag{16.4}$$

and

$$W_{Lj} = W_{Uj} B(r_{j+1}, 1), \tag{16.5}$$

where $B(a, b)$ is a beta random variable with parameters a and b, all beta random variables are independent, and $B(a, 0)$ with $a > 0$ is defined as a point mass at 1. Then the $100(1-\alpha)\%$ BPCP interval is

$$\{q\left(\alpha/2, W_{Lj}\right), q\left(1 - \alpha/2, W_{Uj}\right)\}, \tag{16.6}$$

where $q(a, X)$ is the ath quantile of a random variable X (see Note N2).

For a motivation of the BPCP, we use the probability integral transformation and order statistics. Briefly, suppose there is no censoring and we have a sample of n independent event times $X_1, \ldots, X_n$ with distribution $F(t)$. The probability integral transformation says for continuous failure times that $F(X_i)$ is uniform. Let $X_{(1)} < X_{(2)} < \cdots < X_{(n)}$ be the ordered event times, then $F(X_{(j)})$ is the jth largest of a sample of n independent uniform distributions. Using the theory of order statistics (see, e.g., Casella and Berger, 2002), $F(X_{(j)}) \sim B(j, n - j + 1)$ and $S(X_{(j)}) = 1 - F(X_{(j)}) \sim B(n - j + 1, j)$. We can use this trick repeatedly, except now on the conditional survival distributions. If we have a sample of r_j independent event times that survive until just before t_j^*, then the d_jth largest failure of that sample will have conditional survival distribution distributed beta with parameters $r_j - d_j + 1$ and d_j. We imagine that the d_jth largest failure occurs after t_j^* but before $t_j^* + \Delta$ (here again Δ is some small value that creates the small interval that induces ties by grouping the continuous data). Using these beta distributions and comparing Equation 16.2 to Equation 16.4, we get a loose motivation of the BPCP. Equation 16.5 is designed to be conservative since the next event (the one after the d_j events) could occur immediately after $t_j^* + \Delta$, so the $B(r_{j+1}, 1)$ factor in the equation gives a conservative lower bound. For a more rigorous motivation and the details on how the beta product distributions are estimated see Fay et al. (2013) and Fay and Brittain (2016). For an alternate derivation of the BPCP using generalized fiducial inference see Cui and Hannig (2019).

In Figure 16.1 we plot the Kaplan–Meier survival estimator together with 95% pointwise confidence intervals using the Greenwood (with a log transformation) or BPCP methods. The two confidence interval methods closely match for most of the curve except at the end when the sample size gets smaller. The BPCP interval decreases drastically at the end since very few individuals are left after year 16 (< 30) and the BPCP makes no assumptions about the shape of the survival curve (e.g., it allows that $S(t)$ can drop suddenly after year 16). The BPCP is designed to be a central interval, so for example, the 95% interval would have both the lower and the upper error rates no more than 2.5%. Fay et al. (2013) ran simulations with sample sizes of 30 comparing the lower and upper one-sided error rates of the BPCP to other confidence interval methods such as the Greenwood method with log transformation. The 95% BPCP had simulated lower error rates close to 0, while 95% Greenwood intervals with log transformation had some simulated lower error rates over 10%, where if the error is balanced, we expect 2.5% one-sided error rates. The simulation showed that the BPCP appears to guarantee coverage, while other methods do not, but the natural consequence of this is much wider intervals on the low side. Narrower intervals are possible if we force more structure to the curve and use a parametric model (e.g., a Weibull model) (see, e.g., Kalbfleisch and Prentice, 2002), but of course they only have good coverage if the parametric model holds.

The BPCP intervals provide pointwise confidence intervals for any $S(t)$ with fixed t. Alternatively, one might want $100(1 - \alpha)\%$ simultaneous confidence bands, $(L(\cdot; \mathbf{Y}), U(\cdot; \mathbf{Y}))$, such that

$$\Pr[L(t; \mathbf{Y}) \leq S(t) \leq U(t; \mathbf{Y}) \text{ for all } t \in (0, \tau]] \geq 1 - \alpha,$$

where $\mathbf{Y} = [Y_1, \ldots, Y_n]$ with $Y_i = [T_i, \delta_i]$ are the right-censored data and τ is a constant. Hall and Wellner (1980) derived asymptotic simultaneous confidence bands for right-censored data (see also Andersen et al., 1993, Section IV.1.3.2).

16.2.3 Median Failure Time

For right-censored data we typically do not estimate or perform tests on the mean failure time, because we censor large values and hence cannot get a reasonable estimator of the

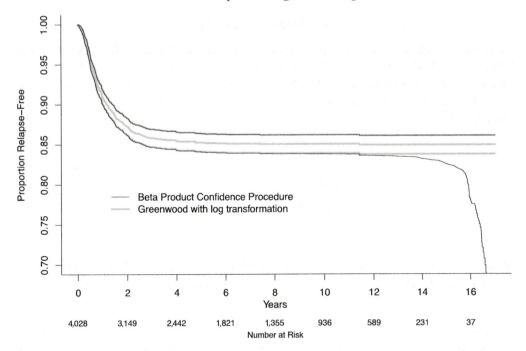

Figure 16.1 Kaplan–Meier survival estimator with 95% pointwise confidence intervals for $S(t)$ for time to relapse in the National Wilms Tumor Study Group data (Breslow and Chatterjee, 1999). Data and Greenwood confidence interval with the log transformation are from the survival R package, and BPCP is calculated with the bpcp R package (same as Figure 3 of Fay et al. (2013)).

mean; however, we can sometimes reasonably estimate the median. If the Kaplan–Meier survival estimator includes some t with $\hat{S}(t) \le 0.5$, then we can estimate the median as

$$\hat{\theta} = \inf\{t : \hat{S}(t) \le 0.50\}.$$

We create a confidence interval on θ by inverting a series of hypothesis tests: $H_0 : S(\theta_0) = 0.50$ versus $H_1 : S(\theta_0) \ne 0.50$ for different null values of θ_0, where we reject each null hypothesis at level α if the $100(1-\alpha)\%$ confidence interval for $S(\theta_0)$ does not cover 0.50. In other words, if the $100(1-\alpha)\%$ confidence interval procedure for $S(t)$ is $\{L_{S(t)}(\mathbf{y}), U_{S(t)}(\mathbf{y})\}$, then the $100(1-\alpha)\%$ confidence interval for the median θ is

$$\{t : L_{S(t)}(\mathbf{y}) \le 0.50 \le U_{S(t)}(\mathbf{y})\},$$

and if the confidence interval for $S(t_k^*)$ contains 0.50, then set the upper limit to ∞, where t_k^* is the largest t_j.

   If $\hat{S}(t_k^*) > 0.50$, then we have no good nonparametric estimator of the median. An option is to fit a parametric model and extrapolate beyond the observed data, but that is rarely appropriate. Sometimes the best option is to give a confidence interval on the median with an upper limit of infinity.

16.3 Hazard Function

As previously mentioned in survival analysis, instead of dealing with the distribution function at t, $F(t)$, we alternatively use the survival function $S(t) = 1 - F(t)$, which contains all the information in F. Another important alternative function to represent the information in the distribution is the hazard function, which is the instantaneous rate of an event at t given survival to t,

$$
\begin{aligned}
\lambda(t) &= \lim_{\Delta \to 0} \frac{\Pr[t \le X < t + \Delta | X \ge t]}{\Delta} \\
&= \frac{\lim_{\Delta \to 0} \frac{\Pr[t \le X < t + \Delta]}{\Delta}}{\Pr[X \ge t]} \\
&= \frac{f(t)}{S(t-)},
\end{aligned}
$$

where $f(t)$ is either the probability density function (for continuous X) or the probability mass function (for discrete X), and $S(t-) = \Pr[X \ge t]$. The cumulative hazard is

$$
\Lambda(t) = \int_0^t \lambda(t) dt,
$$

for continuous X, and for discrete X with the possible event times, $u_1 < u_2 < \cdots$ the cumulative hazard is

$$
\Lambda(t) = \sum_{u_j \le t} \lambda(u_j).
$$

Then $S(t)$ may be written as a product integral[2] as

$$
\begin{aligned}
S(t) &= \prod_{u \le t} \{1 - d\Lambda(u)\} \\
&= \begin{cases} \exp(-\Lambda(t)) & \text{for } X \text{ continuous} \\ \prod_{u_j \le t} \{1 - \lambda(u_j)\} & \text{for } X \text{ discrete.} \end{cases}
\end{aligned}
\tag{16.7}
$$

A common estimator of $\Lambda(t)$ is the *Nelson–Aalen estimator*,

$$
\hat{\Lambda}(t) = \sum_{j: t_j^* \le t} \frac{d_j}{r_j},
$$

which when transformed to a survival distribution by Equation 16.7 reduces to the Kaplan–Meier estimator (see, e.g., Aalen et al., 2008, p. 99). An alternative estimator of survival, the Breslow estimator, is $\tilde{S}(t) = \exp(-\hat{\Lambda}(t))$ (Breslow, 1972; Therneau and Grambsch, 2000).

---

[2] Just as the usual integral is like an infinite sum of infinitely small changes in a function, the product integral is like an infinite product of 1 plus infinitely small changes in a function. See Andersen et al. (1993) for details.

16.4 Two Samples with Right Censoring

16.4.1 How Do We Measure a Group Effect?

Suppose we are comparing the time to event in two groups, and the associated distributions are F_0 (for the control group, those with $z_i = 0$) and F_1 (for the treatment group, those with $z_i = 1$). For noncensored data, there are many ways of quantifying a comparison of two distributions: (1) summarize each distribution with a scalar parameter that requires almost no assumptions about the distribution such as the mean or median, then compare the two distributions with the difference or ratio (e.g., difference in means); (2) compare distributions based on the Mann–Whitney parameter, which is a functional of the two distributions; or (3) compare distributions by making some assumptions about the class of distributions (e.g., Poisson distributions or distributions related by proportional odds) and comparing parameters within that class. For censored data, because we cannot observe the whole range of responses (i.e., for right-censored data we may not be able to observe any events after the end of the study), there is a slight difference in emphasis of which methods are typically used.

In Section 16.2.3 we discussed how because of censoring the means of distributions are not generally used, and the medians are more often used. But if we have a study and neither of the groups has Kaplan–Meier survival estimators less than 0.50 before the end of the study, it would be difficult to compare medians. Instead of the mean or median, in censored data we might summarize a distribution with its value at a specific time, for example $F(t)$ or $S(t)$ at a fixed time t of interest (e.g., 1 year survival).

With censored time-to-event data, we often focus on the hazard function, and for comparing two groups, the hazard ratio. This is because the hazard has the nice interpretation as the instantaneous risk of an event given survival to just before the event. The Mann–Whitney parameter is less often used with censored data, because similar to the mean and median it cannot be estimated well from right-censored data without adding more structure to the problem (e.g., using either a parametric or a semiparametric model).

Consider when to use the simple method of comparing survival curves at a fixed time t, for example using $S_1(t) - S_0(t)$, where S_1 and S_0 are the survival functions for the treated and control individuals respectively. We may use $S_1(t) - S_0(t)$, even though this ignores all of the information about S_1 and S_0 at times other than t. Focusing on $S_1(t) - S_0(t)$ is especially appropriate when the survival curves may be expected to cross. For example, consider a randomized trial of a new treatment versus control (standard treatment) for a very serious cancer in which most individuals die within a year under control. Suppose a new treatment is especially harsh and leads to earlier deaths under the new treatment than under control in some individuals, but for those that survive the harsh treatment, they tend to have longer remission than when on the control. Of course, for each individual we only observe their outcome under either the new treatment or control, but we can compare the survival curves of the two arms. In designing the study, it is quite possible that, measuring time in months, we could get Kaplan–Meier curves with $\hat{S}_1(1) < \hat{S}_0(1)$ (new treatment looks worse at 1 month), but $\hat{S}_1(12) > \hat{S}_0(12)$ (new treatment looks better at 12 months). Then comparing survival at 12 months between the two groups may be an appropriate measure of the usefulness of the treatment, since many individuals would choose to take a treatment that increases the probability of dying in the short term if the treatment had a longer expected survival in the long term. We explore inferences on $S_1(t) - S_0(t)$ in Section 16.4.2.

Now we consider parametric and semiparametric models, which are useful because we can estimate parameters of interest even when the length of the study is such that only a small proportion of the individuals will have the event by the end of the study. Although certain parametric models are often used for right-censored data (see Section 16.4.3), semiparametric models are more commonly used (Sections 16.4.4 and 16.4.5). We consider two semiparametric models, the proportional odds model and the much more popular proportional hazards model (also called Cox regression). The semiparametric models are so named because there is an infinite dimensional nuisance parameter (e.g., the hazard function or the survival distribution for the control group) and a finite dimensional parameter vector of interest (e.g., the hazard ratio from the proportional hazards model). Because of the infinite dimensional nuisance parameter, semiparametric models are very flexible. For estimating a semiparametric parameter such as the hazard ratio, we can use information about the survival distributions throughout the entire length of the study. This can give good power when the semiparametric model approximately fits the data.

16.4.2 Comparing $S(t)$ between Samples

For time-to-event data with no censoring, a simple comparison between two groups is to compare the proportion that have not yet had the event by a certain time, say t. For example, let the sample proportion of individuals that survived to time t without yet having the event in the control group be $\hat{S}_0(t)$ and in the treatment group be $\hat{S}_1(t)$. With no censoring we can use the methods of Chapter 7, for example the central Fisher's exact test with associated confidence intervals (see Section 7.3.1).

With right-censored data, now $\hat{S}_0(t)$ and $\hat{S}_1(t)$ represent Kaplan–Meier estimates of $S(t)$ in the two groups. We can use the BPCP for $S(t)$ together with the melding method for combining confidence intervals in the two-group case (Section 10.11) to test for equality and calculate confidence intervals on the difference, $S_1(t) - S_0(t)$, on the ratio, $S_1(t)/S_0(t)$, or on the odds ratio, $\text{Odds}(X > t|z = 1)/\text{Odds}(X > t|z = 0)$, where $\text{Odds}(A) = \Pr[A]/(1 - \Pr[A])$ for event A.

Example 16.3 *Freireich et al. (1963) reported on a clinical trial comparing 6-mercaptopurine (6-MP) versus placebo for treatment of acute leukemia. The response time is number of weeks from treatment, started when the disease was in remission, until the remission ended. Following Gehan (1965) we analyze the data (given in the* bpcp *R package) as if it was a fixed sample trial with n = 21 in each arm. The Kaplan–Meier curves are given in Figure 16.2. Using the* bpcp2samp *function in the* bpcp *R package, we estimate the difference in survival function at 20 weeks, $S_1(20) - S_0(20)$ as 0.53 (95% CI:0.15, 0.78) with a two-sided p-value of $p = 0.003$.*

16.4.3 Parametric Models

Let $X_i \sim F(\cdot|z_i, \Theta)$, where Θ is a vector of parameters that fully defines F and z_i is a covariate vector (e.g., a treatment indicator), and let f and S denote the corresponding density and survival functions. Assume (X_i, C_i, z_i) for $i = 1, \ldots, n$ are independent and the independent censoring assumption holds (Section 16.2.1), then the likelihood is proportional to

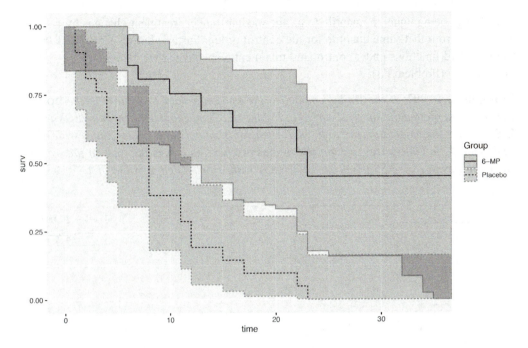

Figure 16.2 Kaplan–Meier survival estimators with 95% pointwise confidence intervals for $S(t)$ by BPCP for time (in weeks) from start to end of remission in a study comparing 6-mercaptopurine (6-MP) to placebo in acute leukemia. There where 21 matched pairs, but we ignored the pairing as in Gehan (1965).

$$L(\Theta; \mathbf{t}, \delta, \mathbf{z}) = \prod_{i=1}^{n} f(t_i; z_i, \Theta)^{\delta_i} S(t_i -; z_i, \Theta)^{1-\delta_i}.$$

For inferences on Θ, we can use the usual asymptotic likelihood methods, such as the likelihood ratio, score, or Wald methods (Section 10.7).

There are many parametric distributions that are suitable for survival analysis (see, e.g., Kalbfleisch and Prentice, 2002). We consider the Weibull distribution because it is simple, useful, and it is both an *accelerated failure time model* and a proportional hazards model. Accelerated failure time models have parameters that multiplicatively change the predicted failure times, and proportional hazards models multiplicatively change the hazard function. In this model the failure times are distributed Weibull, which is equivalent to the log failure times being modeled as

$$\log(X_i) = \beta_0 + \beta_1 z_i + \sigma \epsilon_i,$$

where ϵ_i is distributed by an extreme value distribution with $\epsilon \sim F_\epsilon$ and $dF_\epsilon(w) = f_\epsilon(w) = \exp(w - exp(w))$. This leads to the predicted survival function,

$$S(t; z) \equiv S(t; z, \beta_0, \beta_1, \sigma) = \exp\left(-\left\{t \exp(-\beta_0 - \beta_1 z)\right\}^{1/\sigma}\right).$$

We show this is an accelerated failure time model because for any $q \in (0, 1)$, we have if $S(t; 0) = q$, then $S(t \exp(\beta_1); 1) = q$. For example, in the Weibull two-group model, all other things being equal, a quantile (e.g., the median) under treatment changes by a factor of $\exp(\beta_1)$ from that same quantile for the control group. Further, letting $\lambda(t;0)$ and $\lambda(t;1)$ be the hazard functions under control and treatment respectively, we have $\lambda(t;1)/\lambda(t;0) = \exp(-\beta_1/\sigma)$ (Problem P3).

Example 16.4 *We fit the Weibull model to the leukemia data (available in the* bpcp *R package) and plotted the survival curves in Figure 16.3, and comparing them to the Kaplan–Meier curves the model appears to fit well. Using maximum likelihood estimates with Wald-type methods for inferences, we get the treatment effect increases the time to failure by a factor of* $\exp(\hat{\beta}_1) = 3.55$ *(95% CI: 1.93,6.53) with a two-sided p-value of* $p < 0.0001$.

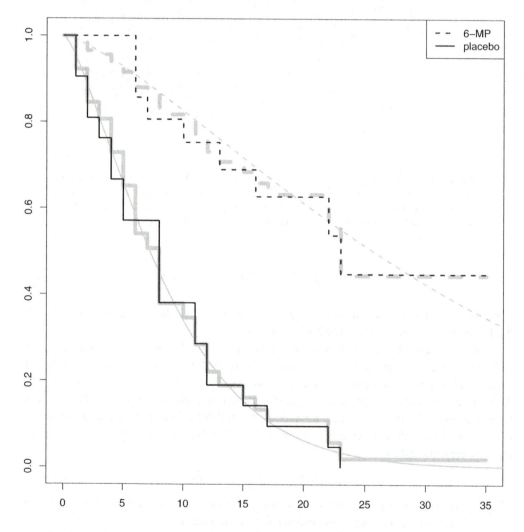

Figure 16.3 Survival estimates over time (in weeks) for remission study comparing 6-mercaptopurine (6-MP) to placebo in acute leukemia. Solid lines are placebo and dotted lines are 6-MP. Kaplan–Meier survival estimates are black, predicted survival from Weibull model are smooth thin gray lines, and proportional hazards model survival estimates (see Section 16.4.5) are thick gray lines.

For example, the median for the placebo is estimated as 7.24 weeks under control and 7.243 × 3.551 = 25.72 weeks under treatment. The hazard ratio is $\exp(-\hat{\beta}_1/\hat{\sigma}) = 0.177$ and using a delta method on the log(hazard ratio) and transforming it back to hazard we get 95% confidence interval for the hazard ratio of $(0.079, 0.398)$.

16.4.4 Proportional Hazards or Proportional Odds?

Consider two common semiparametric models applied to the two-group problem. Let S_0 and S_1 be the survival functions of the control and treatment groups, respectively. The proportional hazards model is

$$S_1(t) = S_0(t)^{\theta},$$

where θ is the hazard ratio. In terms of hazard functions and writing $\theta = \exp(\beta)$, the model is

$$\lambda(t; z_i) = \lambda_0(t) \exp(z_i \beta),$$

where $\lambda(t; 0) = \lambda_0(t)$ is the baseline hazard function, that in this case is the hazard for the control group ($z_i = 0$), and $\lambda(t; 1)$ is the hazard for the treatment group ($z_i = 1$).

The proportional odds model is

$$\frac{1 - S_1(t)}{S_1(t)} = \psi \left(\frac{1 - S_0(t)}{S_0(t)} \right)$$

or equivalently,

$$S_1(t) = \frac{S_0(t)}{S_0(t) + \psi(1 - S_0(t))}.$$

For interpretability, we can translate either of the semiparametric parameters into the Mann–Whitney parameter, $\phi = \Pr[W_0 < W_1] + 0.5\Pr[W_0 = W_1]$, where $W_a \sim F_a$ for $a = 0, 1$. For continuous responses, then from the proportional hazards model $\phi = 1/(1+\theta)$. Further, the hazard ratio is sometimes known as the loss ratio, because it is $\theta = (1 - \phi)/\phi$, and the numerator denotes $\Pr[W_0 > W_1]$ (the probability that treatment loses when the events are failures) and the denominator denotes $\Pr[W_0 < W_1]$ (the probability that treatment wins). With continuous responses the proportional odds model translates to ϕ in the following way: if $\psi = 0$, then $\phi = 0$, if $\psi = 1$, then $\phi = 0.5$, if $\psi = \infty$, then $\phi = 1$, and otherwise

$$\phi = \frac{\psi \{\psi - 1 - \log(\psi)\}}{(\psi - 1)^2} \tag{16.8}$$

(see Problem P4).

In Figure 16.4 we compare the two models. We plot both models with $F_0 = 1 - S_0$ equal to an exponential distribution with rate $= 1$, and the Mann–Whitney statistic equal to $\phi = 0.60$. Notice how in the proportional hazards model the hazard functions have the same constant ratio over time, but in the proportional odds model the hazards get closer together over time.

In Figure 16.5 we again compare the two models with $\phi = 0.60$ for both, but now S_0 has a more unusual pattern. So we see the great flexibility of these semiparametric models (because there are essentially no restrictions on S_0), but there is also a lot of structure (because once S_0 is known, then one parameter determines S_1).

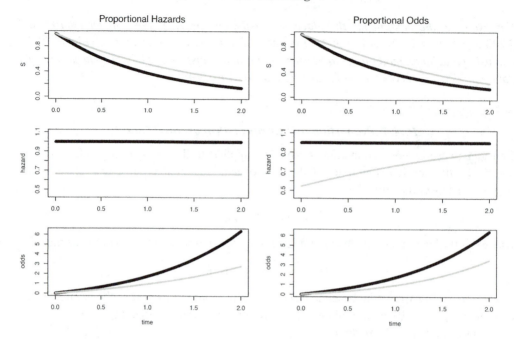

Figure 16.4 Two distributions related by either proportional hazards (left column) or proportional odds (right column). $S_0(t)$ is black and $S_1(t)$ is gray. The top row is survival $S(t)$, middle row is hazard $\lambda(t)$, and bottom row is $odds(t) = F(t)/S(t)$. For both models the baseline survival is an exponential with rate of 1, and the Mann–Whitney parameter is $\phi = 0.60$.

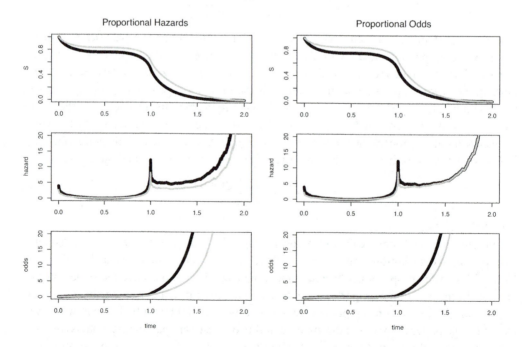

Figure 16.5 Same as Figure 16.4, except with a nonstandard baseline survival function.

16.4.5 Inferences from Semiparametric Models

Consider first the proportional hazards model, where here we allow that $\mathbf{z}_i$ may be a vector of covariates (for example, the first term could be treatment assignment at randomization and the other terms could be baseline variables). We can write the proportional hazards model as

$$\lambda(t; \mathbf{z}_i) = \lambda_0(t) \exp(\mathbf{z}_i^\top \beta).$$

To perform inferences on proportional hazards models, we use a partial likelihood, and with no tied failure times, it is

$$L_p(\beta) = \prod_{i=1}^n \left\{ \frac{\exp(\mathbf{z}_i^\top \beta)}{\sum_{j:t_j \geq t_i} \exp(\mathbf{z}_j^\top \beta)} \right\}^{\delta_i}.$$

If there are ties but the number of failures at any time point is a small proportion of the total failures, then $L_p(\beta)$ may still be used for inferences, and is called *Breslow's tie adjusted method*, but Efron developed a different approximation for ties that appears to perform better (Kalbfleisch and Prentice, 2002, p. 105–106). When there are many ties, then an "exact" method[3] can be used that is equivalent to a conditional logistic model (Therneau, 2015). The partial likelihood (without ties) can loosely be motivated as a conditional likelihood where at each observed failure time we condition on the others at risk, and the contribution to the partial likelihood represents the probability that the failure occurs on the individual who actually failed in the risk set (Cox, 1972). The likelihood is not technically a conditional likelihood, and more rigorous derivations and needed because there is dependence between the risk sets at different failure times. Nevertheless, it can be shown using either counting processes and martingales (Andersen et al., 1993) or profile likelihoods (Murphy and van der Vaart, 2000), that the usual asymptotic likelihood-based methods (Section 10.7) can be applied to the partial likelihood. Typically, confidence intervals on parameters are calculated using Wald-type methods, because those are the simplest, and these are adequate with moderate sample sizes and correctly specified models (i.e., when the proportional hazards assumption holds). Robust methods that apply asymptotically when the proportional hazards assumption fails have been developed and rely on modifying the partial likelihood so that the terms become asymptotically independent (Lin and Wei, 1989), and small sample methods have been developed for that (Fay and Graubard, 2001). Samuelsen (2003) discusses some possibilities and difficulties of exact (i.e., valid) inference for the proportional hazards model.

Inference for the proportional odds model and other semiparametric models is used less often because there is no simplifying partial likelihood. Although profile likelihood methods may be used (Murphy et al., 1997), the software is not as well developed.

In the next section, we discuss score-type tests for the two- or k-sample semiparametric models, which are logrank tests for the proportional hazards model, and a specific weighted logrank test for the proportional odds model.

Example 16.4 *(continued) We return to the leukemia example of Figure 16.3, where $\mathbf{z}_i$ is a binary treatment indicator. We estimate the hazard ratio as* $\exp(\hat{\beta}) = 0.208$. *Using the*

---

[3] Here "exact" means that the likelihood at each failure time incorporates all permutations of the covariates in the risk set, but the resulting inferences are still asymptotic and do not ensure validity (e.g., the $1 - \alpha$ confidence intervals are not guaranteed to cover the true parameter with probability at least $1 - \alpha$).

coxph *function in the* survival *R package, we get 95% confidence interval using the Wald method of* (0.093, 0.466) *or using the robust method,* (0.099, 0.435). *Because these methods require fewer assumptions than the Weibull model, the proportional hazards model is more often used.*

16.4.6 Logrank Test and Weighted Versions

A popular two- or k-sample test for right-censored data is the logrank test. Versions of the logrank test were developed using methods similar to the Mantel–Haenzel test (Section 12.3.3) on 2×2 tables of failures among those at risk at each observed failure times by Mantel (1966), and as a rank test by Peto and Peto (1972). A version of the logrank test can also be derived as a score-type test (Section 10.7) for the proportional hazards model. The test can be generalized to the proportional odds model, and this gives a right-censored data version of the Wilcoxon–Mann–Whitney test; the generalized logrank tests (including the original) are known as weighted logrank tests. The asymptotic theory for the weighted logrank tests can be developed using counting processes (Fleming and Harrington, 1991; Andersen et al., 1993). We can also use permutation methods on the rank scores (which can be defined in different ways, see Peto and Peto (1972) and Kalbfleisch and Prentice (2002)) to get valid p-values for small samples. At this point, we are not aware of compatible estimates and confidence intervals for the semiparametric parameters with the permutation weighted logrank test p-values for right-censored data, although there are compatible confidence intervals when there is no censoring (see Fay and Malinovsky (2018) for the proportional odds parameter). Thus, unfortunately, the weighted logrank p-values are typically presented without estimates and confidence intervals. Nevertheless, we still discuss weighted logrank tests because they have valid versions (e.g., the Type I error rate is bounded using permutation methods), and because the weighted logrank formulation gives some intuition on how the proportional odds and the proportional hazard models differ.

We can think of the logrank statistic for the treatment group as measuring if the treatment group fails more often than expected under the null hypothesis of no treatment effect. We measure this with a weighted average of the observed number of failures minus the expected number. Let $t_1^*, \ldots, t_k^*$ be the observed failure times in both groups (treatment and control). At t_j^* let the number of failures in both groups be d_j with d_{j0} and d_{j1} being the failures in the control and treatment groups respectively. Let $r_j, r_{j0},$ and r_{j1} be the analogous numbers at risk just before t_j^*. Let $e_j = d_j r_{j1}/r_j$ be the expected number of failures in the treatment group at t_j^* given that: d_j in both groups failed, there were r_j at risk in both groups, and r_{j1} at risk in the treatment group. Then the weighted logrank statistic for the treatment group is

$$U_1(w) = \sum_{j=1}^{k} w(t_j^*)(d_{j1} - e_j), \qquad (16.9)$$

where w_j is the weight associated with t_j^*. The usual (i.e., unweighted) logrank test uses $w(t) = 1$ for all t, and this is the score statistic associated with proportional hazards. The score statistic associated with the proportional odds model uses the weight function $w(t) = \hat{S}(t-)$, where $\hat{S}$ is the Kaplan–Meier estimate of survival. When there is no censoring, the resulting test reduces to the Wilcoxon–Mann–Whitney test. We call it the

Peto–Prentice–Wilcoxon test after Peto and Peto (1972) and Prentice (1978); although, unfortunately, Peto and Peto (1972) and Prentice (1978) used slightly different estimators of $S(t)$ for their weights, but the differences matter little in practice.[4] These two weighted logrank tests are sometimes called the G^ρ class of tests after the counting process formulation by Harrington and Fleming (1982), which in our notation is equal to $w(t) = \hat{S}(t-)^\rho$, so that $\rho = 0$ gives a logrank test and $\rho = 1$ gives a Peto–Prentice–Wilcoxon test.

The statistic $U_1(w)$ is asymptotically normal, and the variance can be estimated using counting process methods. Alternatively, $U_1(w)$ may be written as the sum,

$$U_1(w) = \sum_{i=1}^{n} c_i z_i, \tag{16.10}$$

where c_i is a rank-like score which differs depending on the model (Note N3) and p-values can be calculated using permutation methods.

Here are two ways of deciding between the two tests. In terms of the observed-minus-expected formulation (Equation 16.9), we can choose based on how we want to weigh failures; the Peto–Prentice–Wilcoxon test weighs early failures more than later failures, but the logrank test weighs them equally. Alternatively, we can choose based on which is the appropriate model: both tests are locally most powerful rank tests for either the proportional hazards model (logrank) or the proportional odds model (Peto–Prentice–Wilcoxon) (Andersen et al., 1993).

16.4.7 Causal Interpretation of the Proportional Hazards Model

Care should be made interpreting the hazard ratio from a proportional hazards model. Consider a randomized trial comparing treatment to control with the event being failure. Suppose every individual has two potential outcome distributions for failure times (one under treatment and one under control), each individual has their own baseline hazard, and all pairs of potential outcome distributions are related by the same hazard ratio. This situation does *not* imply that the population distributions of the treatment and control groups are related by proportional hazards.

Here is a demonstration of the issue using a gamma frailty model (a *frailty model* is a survival model with a random effect for each individual). Consider a causal model, and let $\mathbf{Y}_i = [Y_i(0), Y_i(1)]$ be the random variable vector of potential outcomes for the ith individual, where $Y_i(0)$ is the outcome on control and $Y_i(1)$ is the outcome on treatment. Assume that $Y_i(a) \sim F_{ai} = 1 - S_{ai}$ for $a = 0, 1$ and suppose $Y_i(a)$ are continuous. Further, suppose that within each individual the potential outcome distributions are related by proportional hazards:

$$S_{1i}(t) = \exp\left(-\int_0^t \lambda_i^*(t)\right) = \exp\left(-\int_0^t \theta^* \lambda_i^*(t)\right) = S_{0i}(t)^{\theta^*},$$

---

[4] We do not call it a generalized WMW test, since there is another generalization of the WMW test by Gehan (1965) that is no longer recommended (see Prentice and Marek, 1979).

where $\lambda_i^*(t)$ is the hazard function on control for the ith individual, and θ^* is the *individual* proportional hazards parameter. The ratio of hazard functions within each individual is $\theta^*\lambda_i(t)/\lambda_i(t) = \theta^*$ and has the interpretation as an individual causal hazard ratio.

Suppose that $\lambda_i^*(t) = G_i\lambda^*(t)$, where G_i is a frailty parameter distributed gamma with mean 1 and variance ν. Then the population hazard rate at t for the control group is

$$\lambda(t) = \frac{\lambda^*(t)}{1 + \nu\Lambda^*(t)},$$

where $\Lambda^*(t) = \int_0^t \lambda^*(u)du$ (see Aalen et al., 2008, Equation 6.8). In other words, if there is variability in the distributions between individuals, then the mean of the individual hazard functions (recall $E(G_i) = 1$) does not equal the population hazard. Further the mean of the individual causal hazard ratios, θ^*, does not equal the population hazard ratio,

$$\theta(t) = \theta^* \left(\frac{1 + \nu\Lambda^*(t)}{1 + \theta^*\nu\Lambda^*(t)} \right), \tag{16.11}$$

and the population hazard ratio changes over time and hence is not proportional hazards (see Aalen et al., 2008, Equation 6.20). We see that for $\nu > 0$ then if $\lim_{t\to\infty} \Lambda^*(t) = \infty$, then $\lim_{t\to\infty} \theta(t) = 1$.

Examining Equation 16.11, if $\nu\Lambda^*(t)$ is very small compared to 1, then $\theta(t) \approx \theta^*$. So if the variability of the frailty is small or the individual cumulative hazard function is small at t, then the proportional hazard parameter is approximately the individual causal hazard ratio. However, if $\nu\Lambda^*(t)$ grows substantially over time, then the population hazard effect, $\theta(t)$, will approach no effect ($\theta(t) = 1$) as time progresses. This means that if one observes the diminishing difference between arms over time that is not necessarily due to a weakening of the treatment effect over time but could be due to constant individual causal hazard ratios and heterogeneity of individual hazards. The intuition of this latter scenario is that if the treatment decreases the individual hazards by a constant amount over all time, then early in the study a higher percentage of frail individuals in the control group will fail and hence be removed from the risk set, so that later comparisons of those left at risk in the control and treatment groups are not comparable. See Aalen et al. (2008) and Aalen et al. (2015) for more details.

16.5 Interval Censoring

16.5.1 Censoring Assumptions

With interval-censored data, we have an unobservable event time for the ith individual, say X_i. For example, X_i could be the time from the start of the study until when an HIV virus is cleared (i.e., is no longer detectable in the blood). With interval-censored data, we do not have continuous monitoring of the event time, we just have discrete assessment times (e.g., blood draws), and at each assessment time we discover whether the event has happened or not. So one assumption for the standard analysis of interval-censored data is *event permanence*: once the event has been assessed to have occurred, we will assess that it has occurred at any subsequent assessment time. Event permanence is obvious for death (once an individual is assessed as dead, they stay dead), but for other events it may be an assumption that requires some thought. For example, if the event is HIV clearance, event permanence assumes that if the HIV virus is cleared at time t, then it will not reappear at some later time $t^* > t$, where t^* is any time before the end of the study.

Suppose the ith individual has m_i assessments scheduled at $\mathbf{A}_i = [A_{i1}, \ldots, A_{im_i}]$ and let $A_{i0} = 0$ and $A_{i,m_i+1} = \infty$. Then under the event permanence assumption, there is a last assessment done before the event has occurred, say $L_i \equiv A_{i\ell_i}$, and the first assessment done after the event has occurred, say $R_i \equiv A_{i,\ell_i+1}$, and we allow left censoring (where $A_{i\ell_i} = 0$) and right censoring (where $A_{i,\ell_i+1} = \infty$). In other words, the assessments tell us that $L_i < X_i \leq R_i$. Under the event permanence assumption, no assessments need to be done after R_i.

Suppose we assume that $X_1, \ldots, X_n$ are independent and let z_i be a treatment indicator. Then assume that $X_i | z_i = z \sim F_z$, with $S_z = 1 - F_z$. A standard assumption that simplifies analyses is that the timing and occurrence of assessments are noninformative for S_z (or F_z), which means all the information that the data tell us about S_0 and S_1 we get from the likelihood function,[5]

$$L(S_0, S_1; \mathbf{L}, \mathbf{R}, \mathbf{z}) = \prod_{i=1}^{n} \Pr[X_i \in (L_i, R_i] | z_i]$$

$$= \prod_{i=1}^{n} \left\{ S_{z_i}(L_i) - S_{z_i}(R_i) \right\}.$$

One sufficient assumption to meet the noninformative assumption is that given z_i then $\mathbf{A}_i$ is independent of X_i.[6] For more formal conditions for noninformativeness in interval censored data, see Oller et al. (2007).

Under event permanence and noninformativeness, we only need the values L_i and R_i, as the other assessment times are not informative for S_{z_i}.

16.5.2 One Sample

In this section, we have only one sample, so that $z_i = 0$ for all $i = 1, \ldots, n$, and we write $S_0(t) \equiv S(t)$. Suppose we want the nonparametric maximum likelihood estimator (NPMLE) of S, which is the most common estimator under the event permanence and noninformative assumption. This is the interval-censored analog to the Kaplan–Meier estimator for right-censored data.

The E-M algorithm is a simple method to calculate the NPMLE that also provides intuition about the estimate. There are many faster algorithms, see Section 16.8 for references.

Here is a brief description of the E-M algorithm as it applies here. The first step is to partition the event time space. Let $0 \equiv a_0 < a_1 < a_2 < \cdots < a_{k-1} < a_k \equiv \infty$ be the unique ordered set of all the values $\{0, \infty, L_1, R_1, \ldots, L_n, R_n\}$. Then the event space is partitioned into k parts, $(a_{j-1}, a_j]$ for $j = 1, \ldots, k$. The problem is reduced to finding the

---

[5] In this chapter we use L for likelihood, and lower confidence limit, and a left-censored observation. It should be clear from the context which is being used. In this case the first L is the likelihood function and $\mathbf{L}$ is the vector of left endpoints.

[6] This is not the same as $(L_i, R_i]$ being independent of X_i given z_i, which is an unrealistic and much stronger assumption. We can get dependence between $(L_i, R_i]$ and X_i even when $\mathbf{A}_i$ and X_i are independent, because L_i and R_i are picked from $\mathbf{A}_i$ (plus 0 and ∞) based on X_i.

$k - 1$ parameters $f_j = F(a_j) - F(a_{j-1}) = S(a_{j-1}) - S(a_j)$ for $j = 1, \ldots, k - 1$ and $f_k = 1 - \sum_{j=1}^{k-1} f_j$. The second step is to posit an initial guess of $\mathbf{f} = [f_1, \ldots, f_k]$, say

$$\mathbf{f}^{(1)} = [f_1^{(1)}, \ldots, f_k^{(1)}] = \left[\frac{1}{k}, \ldots \frac{1}{k} \right].$$

The third step is to do the E-M iteration. Suppose we are in the hth iteration, and let $\mathbf{f}^{(h)}$ be the estimate of $\mathbf{f}$ from the previous iteration (except we use $\mathbf{f}^{(1)}$ for the first iteration). Each iteration has two parts, the E-step and the M-step. For the E-step of the hth iteration, we define a $k \times 1$ vector for the ith individual, $\mathbf{f}_i^{(h)} = \left[f_{i1}^{(h)}, \ldots, f_{i,k}^{(h)} \right]$ that depends on $\mathbf{f}^{(h)}$, where

$$\hat{f}_{ij}^{(h)} \equiv \hat{f}_{ij}(L_i, R_i; \mathbf{f}^{(h)}) = \Pr[X_i \in (a_{j-1}, a_j] | X_i \in (L_i, R_i], \mathbf{f}^{(h)}]$$

$$= \begin{cases} 0 & a_j \notin (L_i, R_i] \\ \dfrac{f_j^{(h)}}{\sum_{g:a_g \in (L_i, R_i]} f_g^{(h)}} & a_j \in (L_i, R_i]. \end{cases}$$

In other words, we estimate the conditional probability of the ith individual at each of the partitioned intervals assuming $\mathbf{f}^{(h)}$ is the distribution. The M-step defines the $(h + 1)$th estimate, $\mathbf{f}^{(h+1)} = n^{-1} \sum_{i=1}^n \mathbf{f}_i^{(h)}$. We repeat this until $\mathbf{f}^{(h+1)} \approx \mathbf{f}^{(h)}$, based on some convergence criterion (e.g., $\max_j \left| f_j^{(h+1)} - f_j^{(h)} \right| < \epsilon$, for some small ϵ). At convergence, let $\hat{\mathbf{f}} = \mathbf{f}^{(h+1)}$ and we call this the *self-consistent estimator* of $\mathbf{f}$. The self-consistent estimator will be the maximum likelihood estimator if it meets the Kuhn–Tucker conditions (see Gentleman and Geyer, 1994), which are easily checked (the interval R package does this check). Theoretically, the maximum likelihood estimator of $\mathbf{f}$ may not be unique, which Gentleman and Vandal (2002) call *mixture non-uniqueness,* but there is an easy condition to check for uniqueness (Gentleman and Geyer, 1994).

We can speed up the algorithm by setting $f_j^{(1)} = 0$ for all intervals j that do not have an $L_i = a_{j-1}$ or an $R_{i'} = a_j$ (here i' may equal i but does not need to), and after setting those values to 0, restandardizing so that $\sum f_j^{(1)} = 1$ (Problem P6).

The vector $\mathbf{f}$ defines $F(a_j) = 1 - S(a_j) = \sum_{g=1}^j f_g$ for $j = 1, \ldots, k$ (and $F(0) = 1 - S(0) = 0$ by definition), but $\mathbf{f}$ does not define $F(t)$ for all t. For example, if $a_{j-1} < t < a_j$, then all we know about $F(t)$ from $\mathbf{f}$ is that $F(a_{j-1}) \leq F(t) \leq F(a_j)$. Thus, $\hat{\mathbf{f}}$, the NPMLE of $\mathbf{f}$, only defines a class of distributions that have values $\hat{F}(a_j)$ for $j = 0, \ldots, k$ known, but other values possibly only known within a range. Sometimes $\hat{F}(t)$ or $\hat{S}(t)$ are plotted with rectangles representing the entire class but other times they are plotted as a line connecting the points $\hat{F}(a_0), \hat{F}(a_1), \ldots, \hat{F}(a_{k-1})$.

There are many areas of development for getting confidence intervals on $S(t)$, but no clear recommended approach. Vandal et al. (2005) discuss an empirical likelihood and a bootstrap method (see also Sen and Xu (2015) on the bootstrap) and Sen and Banerjee (2007) discuss likelihood ratio, pseudolikelihood, and subsampling methods. For the special case of current status data (also called Case 1 interval censoring), when each individual is assessed only once, there are some useful and interesting methods (Banerjee and Wellner, 2005; Tang et al., 2012; Kim et al., 2021).

16.5.3 Logrank Tests and Weighted Versions

For interval-censored data, under the noninformative assumption (with event permanence), we can compare the two distributions in the two-sample case using weighted logrank-type statistics. Although the theoretical derivation is different (for review, see Fay and Shaw (2010)), we can think of the logrank test in a similar way as in the right-censored case. We can write the statistic in the form of Equation 16.9, where the observed and expected failures are replaced with estimators of them under the null hypothesis (Fay, 1999b). Alternatively, we can write the score statistic for the new treatment as $\sum c_i z_i$ (see Equation 16.10), and in the interval-censored case, a logrank version has

$$
c_i = \begin{cases} \dfrac{\hat{S}(L_i) \log \left(\hat{S}(L_i) \right) - \hat{S}(R_i) \log \left(\hat{S}(R_i) \right)}{\hat{S}(L_i) - \hat{S}(R_i)} & \text{if } R_i < \infty \\[2em] \dfrac{\hat{S}(L_i) \log \left(\hat{S}(L_i) \right)}{\hat{S}(L_i)} & \text{if } R_i = \infty \end{cases}
$$

and a Peto–Prentice–Wilcoxon version has

$$
c_i = \hat{S}(L_i) + \hat{S}(R_i) - 1,
$$

where in both cases $\hat{S}$ is the NPMLE (Fay, 1999b) (Note N3). *p*-Values for the test can be calculated using permutation methods, score methods, or a special case of within-cluster resampling (also called multiple outputation) (Huang et al., 2008). If the assessment is independent of treatment, we can get valid (i.e., exact) *p*-values by using exact implementations of the permutation test. Suppose the assessment distribution depends on the treatment only (for example, if one treatment induced adverse events that increased the likelihood of clinic visits, which in turn increased the likelihood of assessments); however, within each treatment group, the assessments are noninformative. In this case, Fay and Shih (2012) showed that the tests are approximately valid using either a permutational central limit theorem or using within-cluster resampling. Fay and Shih (2012) also reviewed and showed by simulation that the naïve method of replacing the interval-censored endpoint with its midpoint (or right endpoint), can severely inflate the Type I error rate when assessments depend on treatment.

16.6 Truncation

We have dealt with censored responses in this chapter. A related type of response is a truncated response. The most common type is left-truncated data, where the ith individual is only observed if $X_i \in (\mathcal{L}_i, \infty)$. Consider the case with both interval censoring (i.e., we know $X_i \in (L_i, R_i]$) and left truncation. Then under noninformative censoring and the case where the left truncation is independent of the event time, we write the likelihood in terms of the survival function S as

$$
L(S) = \prod_{i=1}^{n} \frac{S(L_i) - S(R_i)}{S(\mathcal{L}_i)}.
$$

Then we can estimate the NPMLE, but with left truncation the NPMLE is not guaranteed to exist (see Hudgens, 2005). This same likelihood will work under the same assumptions for right-censored data, where observed and right-censored data at t_i are represented as $(t_i - , t_i]$ and (t_i, ∞), respectively, and $S(\infty) \equiv 0$.

16.7 Summary

Censoring is defined by responses that are only partially observed. Typically, responses are the time until some event, and since a common example is time to death, analyses of time-to-event data are often call survival analyses. So, for example, if an individual is followed for one year and does not die, we only know that their survival time is greater than one year. There are three forms of censoring: right, left, and interval censoring. The chapter focuses on methods for one and two samples with right-censored data (where we only know the lower bound to the response) and/or with interval-censored data (where we only know a lower and upper bound to the response); these are, by far, the most common forms of censoring in our experience.

16.7.1 Right-Censoring Summary

Under right censoring, we observe the time which is the minimum of the individuals censored time and event time, and whether this was an event or not. Most inference assumes independent censoring, which is essentially met when censoring is not informative for events.

The Kaplan–Meier estimator is an important nonparametric maximum likelihood estimator of the survival distribution in a single sample. While this estimator has been very widely used for many decades, associated confidence intervals have been evolving since. The Greenwood formula along with a log transformation is a common asymptotic estimator of the variance. In contrast, the beta product confidence procedure (BPCP); (see Fay et al. (2013) was developed to be valid for all sample sizes and is particularly useful with heavy censoring. Based on simulations, the BPCP appears to guarantee coverage, in contrast to competing methods, and thus we recommend its use, especially in settings where guaranteed coverage is important. The Kaplan–Meier can also be used to estimate the median failure time, provided the lowest estimate of survival time is less than 0.50. We also present estimation of the hazard function, which represents the instantaneous rate of an event at time t, given survival to t.

Two primary metrics for comparing two samples are i) the hazard ratio, which compares the hazard functions in the two groups, and ii) the difference in survival at some fixed time t, where the choice of metric depends on the scientific question. For example, an aggressive treatment of a very serious disease might lead to early deaths, but in the long run is associated with much improved survival; then, the survival at five years may be the more clinically relevant endpoint, and more powerful statistically.

Parametric models such as the Weibull distribution can be used for survival analyses; while these methods can be powerful, their validity depends on assumptions that may be difficult to verify. Semiparametric models such as proportional hazards and proportional odds are much more robust, and thus more commonly used. Furthermore, the proportional hazards model (also called Cox regression) is in much greater use than the proportional odds model because of its simplified form of the partial likelihood.

When covariate adjustment is not needed, weighted logrank tests are very popular. They are based on a statistic that is a weighted average across all failure times of the observed number of failures in the treated group minus the expected number. The usual logrank test is defined such that all the observed-minus-expected differences are weighted equally, and it may be derived from the proportional hazards model. A common variant, the Peto–Prentice–Wilcoxon, weighs each observed-minus-expected difference by its associated survival estimate, so that the early events count more than later events. This variant is associated with the proportional odds model and reduces to the Wilcoxon–Mann–Whitney test with no censoring. Valid tests of weighted logrank tests are possible through permutation; but, unfortunately, at present there are no compatible estimates and confidence intervals.

The hazard ratio should be interpreted carefully. Using a causal framework, we show that even if each individual has the same hazard ratio with respect to their potential outcomes, the population hazard ratio, *even in a randomized trial*, can change over time. If we observe decreasing treatment differences over time, this might not be due to a true loss of efficacy over time but could arise from constant individual causal hazard ratios with heterogeneity of the individual hazards. That is, if a high percentage of frail individuals fail in the control group early and thus are removed from the risk set, the two arms will not be comparable at later comparisons.

16.7.2 Interval-Censoring Summary

Interval censoring occurs when we only know an event has happened sometime within an interval, such as when someones status changes between two consecutive assessments. Sometimes data sets will have a mix of right-censored and interval-censored observations, in which case we should use methods suitable for interval censoring. An analogue of the Kaplan–Meier has been developed for interval censoring, which is the nonparametric maximum likelihood estimator. There has also been work on the corresponding confidence interval, however there is little consensus on the best approach. Weighted logrank tests to compare two samples have been generalized to handle interval-censored data, where permutation tests can provide valid *p*-values. When the timing of assessments depends on the treatment, imputing interval-censored failure times as the middle of intervals and naïvely using the right-censored versions of the weighted logrank tests can inflate the Type I error rate, and it is more valid to use weighted logrank test versions developed for interval-censored data.

16.8 Extensions and Bibliographic Notes

There are many fine books on survival analysis, and we just mention a few. Kalbfleisch and Prentice (2002) give a good introduction to the theoretical and practical statistical issues for censored data. There is an extensive asymptotic theory for survival analysis that uses martingale theory and counting processes that has not been discussed in this book (see, e.g., Fleming and Harrington, 1991; Andersen et al., 1993; Aalen et al., 2008). The proportional hazards model may be extended many ways to accommodate time varying covariates, and repeated events within an individual (see, e.g., Therneau and Grambsch, 2000).

We have not dealt with dependent censoring in this chapter. See, for example, Finkelstein et al. (2002) for interval-censored data. In some cases, there may be planned assessments that are independent of the event, and unplanned assessments that may depend on the event, and in these cases, it may make sense to not use the information from the unplanned assessments in order to be able to create a valid test (Freidlin et al., 2007).

Gentleman and Vandal (2001) discuss some methods on calculating the NPMLE for interval-censored data and Groeneboom et al. (2008) developed a fast algorithm.

16.9 R Functions and Packages

The survival R package provides many methods for censored data, including the Kaplan–Meier estimator (survfit), logrank test (survdiff), proportional hazards regression (coxph), and many parametric models (survreg) that allow for interval-censored data. The bpcp R package calculates the Kaplan–Meier estimator with the BPCP intervals, as well as many of the standard asymptotic intervals. It also provides estimators and confidence intervals for quantiles, including the median, using the Kaplan–Meier and the BPCP intervals, and performs two-sample tests comparing $S(t)$.

For calculating the NPMLE for interval-censored data see the interval R package. For faster methods see the Icens and MLEcens packages. For logrank tests for interval-censored data see the interval R package. The icenReg R package does proportional hazards regression with interval-censored data (Anderson-Bergman, 2017).

16.10 Notes

N1 **Independent (right) censoring:** For right-censored data the independent censoring assumption does not require that the $C_1, \ldots, C_n$ are independent of the $X_1, \ldots, X_n$. It is more general than that. The definition of independent censoring from Kalbfleisch and Prentice (2002, p. 13) using our notation, is a kind of conditional independence and it says that at each time t

$$\lim_{\Delta \to 0} \frac{\Pr[t \leq X_i < t + \Delta | z_i, X_i \geq t]}{\Delta} = \lim_{\Delta \to 0} \frac{\Pr[t \leq X_i < t + \Delta | z_i, X_i \geq t, C_i \geq t]}{\Delta},$$

where z_i is a covariate vector. The independent censoring assumption includes several more specific types of censoring. Type I censoring is when the censoring variable, C_i, is fixed in advance for all individuals. A simple example is when all individuals start at the same time and only individuals still at risk at the end of the study are right censored. Type II censoring is when a study of n individuals continues without censoring until the mth failure, and then all surviving individuals are censored at that point. Progressive Type II censoring is when, immediately after each observed event time, a proportion of the remaining subjects are risk are selected at random for censoring (see, e.g., Kalbfleisch and Prentice, 2002). The simple random censorship model is when the $C_1, \ldots, C_n$ are from the same distribution and are independent of the $X_1, \ldots, X_n$ (Andersen, 2005). Another very general term of which independent censoring is a special case, is right censoring that is *noninformative* for β (the parameters related to F) and is defined as when the likelihood factors such that the censoring gives no information about β (see, e.g., Andersen et al., 1993, p. 151).

N2 **BPCP with ties:** We relate Equation 16.6 to the way the beta product confidence interval for $S(t)$ is given in Equation 4 of Fay and Brittain (2016). Let d_j, r_j, and t_j^* be as defined in Section 16.2. As in Fay and Brittain (2016), we partition the time line into intervals. Let the $2k + 1$ intervals be $(g_h, g_{h+1}]$ for $h = 0, \ldots, 2k + 1$ and let $g_0 = 0$, $g_{2k+1} = \infty$, and $g_{2j} = t_j^*$ and $g_{2j-1} = t_j^* - \Delta$, for $j = 1, \ldots, k$ where Δ represents the precision of the time measurements (e.g., $\Delta = 1$ measures to the nearest integer). For the $2j$th interval, $(g_{2j-1}, g_{2j}] = (t_j^* - \Delta, t_j^*]$, the value d_j is the number of events observed and r_j is the number at risk at the start of that interval. For the $(2j - 1)$th interval, $(g_{2j-2}, g_{2j-1}] = (t_{j-1}^*, t_j^* - \Delta]$, there are 0 events observed and r_j is the number at risk at the start of that interval. Following Fay and Brittain (2016) define $W^-(g_0) = W^-(g_1) = 1$ and, for $j = 1, \ldots, k$,

$$W^-(g_{2j}) = W^-(g_{2j+1}) = \prod_{h=1}^{j} B(r_h - d_h + 1, d_h).$$

Also,

$$W^+(g_{2j}) = W^+(g_{2j+1}) = W^-(g_{2j})B(r_{j+1}, 1).$$

The $100(1 - \alpha)\%$ beta product confidence procedure (BPCP) for $S(t_j^*)$ has the same confidence interval for all $t \in [g_{2j}, g_{2j+1}) \equiv [t_j^*, t_{j+1}^* - \Delta)$,

$$\left\{ q\left(\alpha/2, W^+(g_{2j+1})\right), q\left(1 - \alpha/2, W^-(g_{2j})\right)\right\}$$
$$\Rightarrow \left\{ q\left(\alpha/2, W^+(t_j^*)\right), q\left(1 - \alpha/2, W^-(t_j^*)\right)\right\},$$

where $q(a, X)$ is the ath quantile of a random variable X. For nearly continuous data, we let $\Delta \to 0$, and the intervals in the form $t \in [g_{2j-1}, g_{2j}) \equiv [t_j^* - \Delta, t_j^*)$ go to 0 and are not needed.

N3 Section 16.5.3 gives two forms for c_i for a logrank test and the Peto–Prentice–Wilcoxon tests for interval-censored data. For slightly different forms and derivations of the rank-like scores for weighted logrank tests for right-censored and/or interval-censored data, see Peto and Peto (1972), Prentice and Marek (1979) (right censoring only), or Fay and Shaw (2010).

16.11 Problems

P1 Show that when there is no censoring, the Kaplan–Meier estimator of $S(t)$ reduces to the sample proportion of events greater than t.

P2 Show that when there is no censoring that the BPCP confidence interval reduces to the exact central interval for a binomial. Hint: for constants a, b, and c, $B(a+b, c)B(a, b) = B(a, b+c)$ (Fay et al., 2013).

P3 Show that the Weibull model is a proportional hazards model, and in the two-group model $\lambda(t; 1)/\lambda(t; 0) = \exp(-\beta_1/\sigma)$.

P4 Show that the two-sample proportional odds model with continuous data may be written as a location shift in a standard logistic distribution. Then derive Equation 16.8 using integration or a computer algebra system (Fay and Malinovsky, 2018).

P5 Write R code to reproduce the analyses of Example 16.4.

P6 Show why the method for defining $\mathbf{f}^{(1)}$ by setting some values to 0 (see the end of Section 16.5.2) will lead to the same converged value of $\hat{\mathbf{f}}$. Hint: think of why the parameters change at each step of the E-M algorithm.

17

Missing Data

17.1 Overview

When analyzing real data, missingness often must be addressed. This is a brief chapter on some of the issues and terminology, providing only simple examples. In Section 17.2, we start with three important assumptions of missingness: missing completely at random (MCAR) (where we can safely ignore the missing data and analyze the individuals with complete data), missing at random (MAR) (where there are many methods to model the missing data), and missing not at random (MNAR) (where we cannot reliably predict the data that is missing and often run sensitivity analyses).

Because MNAR is the least restrictive assumption, we discuss that first in Section 17.3. We consider only a few strategies for some simple situations. For MAR data we briefly discuss inverse probability weighting and regression imputation (also called single imputation models) and give an example as a warning to how these models can break down (Section 17.4). Multiple imputation was developed using Bayesian arguments, but since it is very useful and has good frequentist properties, we cover the basic idea in Section 17.5.

Finally, although we will not address it in the body of the chapter, we emphasize here the importance of designing your study to reduce missing data. For example, in randomized clinical trials, where ethically we must allow individuals to quit the study at any point, you may unintentionally cause people to quit your study by subjecting them to many invasive tests. If feasible, individuals who withdraw from study treatment should be strongly encouraged to continue to have the primary endpoint assessed. In the design phase, researchers should anticipate problems with collecting the primary endpoint. A pretty good primary endpoint that is easy to assess in everyone is almost surely preferable to a primary endpoint that is theoretically the most relevant but may be difficult to ascertain. Similarly, researchers might be able to modify the primary endpoint to accommodate all possible outcomes. For example, for a study of multidrug resistant tuberculosis, the time to "cure" may not be a feasible outcome because some individuals may die before cure during the course of the study. In that case, the primary endpoint could be redefined as a ranking that ranks earlier cures as better than later cures, but ranks all deaths as worse than not being cured. See Shaw (2018) for ways to handle that kind of endpoint and see chapter 2 of National Research Council (2010) for more details related to designing studies to reduce missingness.

17.2 Missing Data Terminology

Let **Y** be the vector of all responses from all n individuals, whether observed or not. Partition **Y** into the observed responses, $\mathbf{Y}^{obs}$, and the missing ones, $\mathbf{Y}^{mis}$. We keep the notation general, so this could indicate a vector of single responses for each individual, or in more complicated studies, a concatenation of n individual vectors of several responses measured over time on each individual. Let **M** be the corresponding vector of missing indicators, where if an element of **Y** is missing, then the corresponding element of **M** is 1, and otherwise it is 0. Let **X** be all of the covariates needed for the primary hypothesis if there were no missing responses, for example, treatment indicators and baseline variables. Assume that none of **X** are missing. Let **W** be auxiliary variables that may be used to model $\mathbf{Y}^{mis}$. The auxiliary variables could be measured at baseline or postbaseline. They may be missing, $\mathbf{W}^{mis}$, or not, $\mathbf{W}^{obs}$, where their missingness pattern may or may not be related to the missingness pattern of the responses.

First, consider the missing completely at random (MCAR) assumption where we assume the missingness vector, **M**, is independent of all other variables, both observed and unobserved. Mathematically,

$$\text{MCAR:} \quad \Pr\left[\mathbf{M}|\mathbf{Y}^{obs}, \mathbf{Y}^{mis}, \mathbf{X}, \mathbf{W}^{obs}, \mathbf{W}^{mis}\right] = \Pr\left[\mathbf{M}\right].$$

The MCAR assumption is the most convenient assumption, because then we can just perform a complete case analysis, where we ignore individuals with any missing data and treat the observed data as if it is the full data. Thus, *if* we can assume MCAR, then we can use methods developed in other parts of this book.

A less strict assumption is missing at random (MAR), where the missingness vector only depends on observed data,

$$\text{MAR:} \quad \Pr\left[\mathbf{M}|\mathbf{Y}^{obs}, \mathbf{Y}^{mis}, \mathbf{X}, \mathbf{W}^{obs}, \mathbf{W}^{mis}\right] = \Pr\left[\mathbf{M}|\mathbf{Y}^{obs}, \mathbf{X}, \mathbf{W}^{obs}\right].$$

Because missingness only depends on observed data when MAR is the correct assumption, there are many methods available to handle that missingness to get reasonable (e.g., consistent) estimators of effects and make inferences.

Finally, if the MAR assumption is not tenable, the data are missing not at random (MNAR),

$$\text{MNAR:} \quad \Pr\left[\mathbf{M}|\mathbf{Y}^{obs}, \mathbf{Y}^{mis}, \mathbf{X}, \mathbf{W}^{obs}, \mathbf{W}^{mis}\right] \neq \Pr\left[\mathbf{M}|\mathbf{Y}^{obs}, \mathbf{X}, \mathbf{W}^{obs}\right].$$

Under MNAR we often do sensitivity analyses to see a range of possible inferences.

Example 17.1 *Consider a two-arm randomized trial comparing treatment versus control for some chronic disease. For the ith individual, let $X_i = [Z_i]$ be the treatment indicator, $Z_i = 1$ is treatment, and $Z_i = 0$ is control. Let $W_i = 1$ if after one week the individual tolerates their intervention and $W_i = 0$ if they cannot tolerate it. Let Y_i be the response after one year and $M_i = 1$ if the response is missing. Suppose all variables are observed except a few individuals are missing Y_i. Consider three scenarios. First, suppose all of the individuals missing Y_i have moved out of the area where the study is being done and suppose that the moves were not related to any aspect of the study (e.g., the health of the individual). Under this scenario, the data are MCAR. Second, suppose that all individuals missing responses* *had $W_i = 0$. Suppose that a certain proportion of people that cannot tolerate the drug will refuse to come back to the clinic to get the final outcome measured and that the probability that they will come back is unrelated to the value of Y_i. Under that scenario, the data are MAR. Finally, suppose that all individuals missing responses had $W_i = 0$, but those that had a particularly poor response were less likely to return to have the final measurement. In that case the data are MNAR, since the missingness is related to the unobserved $\mathbf{Y}^{mis}$ values. Since the data contain no information about whether the MAR or MNAR assumption is correct, we recommend performing analyses under several different sets of assumptions to evaluate the robustness of the results.*

Another way to classify missing data models is based on how we factor the probability model of the responses and missing data indicators. For simplicity, consider the case with no auxiliary variables. Up until now we have been using the following factorization,

$$\Pr[\mathbf{M}, \mathbf{Y}^{obs}, \mathbf{Y}^{mis}|\mathbf{X}] = \Pr[\mathbf{M}|\mathbf{Y}^{obs}, \mathbf{Y}^{mis}, \mathbf{X}]\Pr[\mathbf{Y}^{obs}, \mathbf{Y}^{mis}|\mathbf{X}],$$

where the right-hand-side of the equation is called the *selection model*. Alternatively, we can use a *pattern mixture model*,

$$\Pr[\mathbf{M}, \mathbf{Y}^{obs}, \mathbf{Y}^{mis}|\mathbf{X}] = \Pr[\mathbf{Y}^{mis}|\mathbf{M}, \mathbf{Y}^{obs}, \mathbf{X}]\Pr[\mathbf{Y}^{obs}|\mathbf{M}, \mathbf{X}]\Pr[\mathbf{M}|\mathbf{X}].$$

Pattern mixture models can be useful for analysis of MNAR data (National Research Council, 2010).

17.3 Sensitivity Analyses for Data Missing Not at Random

17.3.1 Worst Case and Opposite Arm Imputation

In this section, we only consider a two-sample experiment, where a small proportion of the responses may be missing. A simple MNAR adjustment is a "worst-case" analysis, where the missing values are replaced with values that are least likely to lead to significant effects.

Example 17.2 *Ranked responses: Consider a randomized clinical trial with 50 individuals randomized to the new treatment and 50 randomized to a control. Suppose two subjects in the treatment arm and one subject in the control arm are missing the response, and it is possible that the missingness is related to the response. When we only analyze the complete cases (48 in treatment arm and 49 in control arm), then we get a highly significant effect with a Wilcoxon–Mann–Whitney test, with the individuals in the treated arm tending to have larger values implying that the treatment is significantly better. A "worst-case" analysis in this situation is to impute the lowest rank responses for the two subjects missing responses in the treatment arm and impute the highest rank response for the subject missing a response in the control arm. If the Wilcoxon–Mann–Whitney test result on the data with those imputed values is still highly significant, then we can feel confident that we can reject the null hypothesis regardless of the values of the missing responses.*

Worst-case analysis is an extreme adjustment, and it would rarely be required; however, in some cases when the effect is large and the amount of missingness is small it can demonstrate robustness of results, eliminating concerns about the missing data. The worst-case analysis cannot be used for all tests. For example, we cannot perform a worst-case analysis for a two-sample t-test, because a worst case is not defined when there is no bound on the responses. In that case, even if there was an extremely large effect and extremely small p-value on the observed data which ignored the one missing response, there always exists an imputed value that can erase that apparent significant effect (Problem P1).

For the two-sample binary response case, there are conservative but less extreme imputation schemes (Proschan et al., 2001). First, consider the case with no missing data, where we use a normal approximation to perform a Wald-type test of the difference in proportions. The approximation allows us to impute fractional responses, which would not be possible with exact methods (e.g., Fisher's exact test). Let $\hat{\theta}_1$ and $\hat{\theta}_2$ be the sample proportions, and n_1 and n_2 be the sample sizes from the two groups. A test statistic for the difference in means is

$$T_Z \equiv t_Z(\hat{\theta}_1, \hat{\theta}_2, n_1, n_2) = \frac{\hat{\theta}_2 - \hat{\theta}_1}{\sqrt{\hat{\theta}(1 - \hat{\theta})\left(\frac{1}{n_1} + \frac{1}{n_2}\right)}},$$

where $\hat{\theta} = (n_1\hat{\theta}_1 + n_2\hat{\theta}_2)/(n_1 + n_2)$. We treat T_Z as asymptotically standard normal under the null hypothesis so that the two-sided p-value is $p = 2(1 - \Phi(|T_Z|))$ and the $100(1 - \alpha)\%$ confidence interval on $\theta_2 - \theta_1$ is

$$\hat{\theta}_2 - \hat{\theta}_1 \pm \Phi^{-1}(1 - \alpha/2)\sqrt{\hat{\theta}(1 - \hat{\theta})\left(\frac{1}{n_1} + \frac{1}{n_2}\right)}.$$

Suppose that there are missing responses (n_1^{mis} and n_2^{mis} missing in the two arms) and let the sample proportion of the observed responses be $\hat{\theta}_1^{obs}$ and $\hat{\theta}_2^{obs}$. The MCAR analysis would act as if the observed values were the complete data and use $t_Z(\hat{\theta}_1^{obs}, \hat{\theta}_2^{obs}, n_1 - n_1^{mis}, n_2 - n_2^{mis})$ as the test statistic. A conservative analysis is *opposite arm imputation*, where we impute each missing value with the proportion of positive responses from the opposite arm, and then calculate the Wald test statistic as if that is complete data. In other words, use $t_Z(\tilde{\theta}_1^o, \tilde{\theta}_2^o, n_1, n_2)$ as a Wald test statistic, where

$$\tilde{\theta}_1^o = \frac{(n_1 - n_1^{mis})\hat{\theta}_1^{obs} + n_1^{mis}\hat{\theta}_2^{obs}}{n_1} \text{ and } \tilde{\theta}_2^o = \frac{(n_2 - n_2^{mis})\hat{\theta}_2^{obs} + n_2^{mis}\hat{\theta}_1^{obs}}{n_2}.$$

Because the imputed missing values are pulling the test statistic toward more likely values under the null hypothesis, this is a conservative imputation. A less conservative approach is *pooled imputation*, where we use $t_Z(\tilde{\theta}_1^p, \tilde{\theta}_2^p, n_1, n_2)$ as a Wald test statistic, where

$$\tilde{\theta}_1^p = \frac{(n_1 - n_1^{mis})\hat{\theta}_1^{obs} + n_1^{mis}\tilde{\theta}^p}{n_1}, \text{ and } \tilde{\theta}_2^p = \frac{(n_2 - n_2^{mis})\hat{\theta}_2^{obs} + n_2^{mis}\tilde{\theta}^p}{n_2}, \text{ where}$$

$$\tilde{\theta}^p = \frac{(n_1 - n_1^{mis})\hat{\theta}_1^{obs} + (n_2 - n_2^{mis})\hat{\theta}_2^{obs}}{(n_1 - n_1^{mis}) + (n_2 - n_2^{mis})}.$$

Example 17.3   *Proschan et al. (2001) considered an example with $\hat{\theta}_1^{obs} = 75/193$, $\hat{\theta}_2^{obs} = 105/192$, $n_1 = 204$, $n_2 = 212$, $n_1^{mis} = 11$, and $n_2^{mis} = 20$. The p-values are: $p = 0.002$*

for the MCAR analysis, $p = 0.131$ for the worst-case analysis, $p = 0.006$ for the opposite arm imputation analysis, and $p = 0.003$ for the pooled imputation analysis. Thus, at the $\alpha = 0.05$ level we could not reject the null hypothesis for the worst case analysis, but we could for the opposite arm and pooled imputation analyses.

17.3.2 Tipping Point Analysis

Yan et al. (2009) proposed a sensitivity analysis method known as the *tipping point analysis*. The idea is to perform a series of sensitivity analyses, and the tipping point is the analysis where the study conclusion changes (e.g., where the p-value changes from significant to not significant).

First, consider the two-sample study with binary responses, where we present a slight modification of the method of Yan et al. (2009). For binary responses, we can easily enumerate all possible responses for the missing data. In Example 17.3 there are $n_1^{mis} = 11$ and $n_2^{mis} = 20$ missing values in the two groups. That means there are $(11+1)(20+1) = 252$ possible ways to impute the missing responses, and hence we can calculate a p-value for each of those ways. In Figure 17.1 we plot all those 252 p-values, each calculated using the

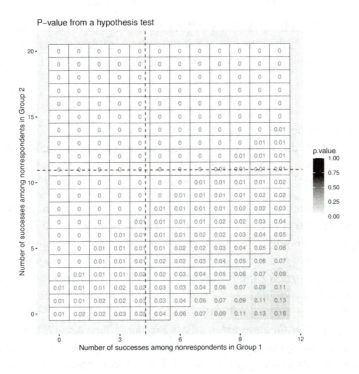

Figure 17.1 p-Values for different values of the missing responses (rounded to the nearest hundredth). Data from Example 17.3. p-Values calculated by central Fisher's exact test, dotted lines are plotted at the expected value using the sample proportions for the observed data (e.g., for Group 1, the value is $11(75/193) = 4.27$). p-Values in black boxes are ≤ 0.05. Plot created by modifying the source code for the TippingPoint R package to plot in grayscale and calculate the central Fisher's exact p-value instead of the asymptotic approximation.

central Fisher's exact test on the observed data and each possible set of missing responses. The plot shows p-values inside black boxes if they are less than 0.05, and the edge of the black boxes is a tipping point. See Liublinska and Rubin (2014) for further enhancements to that type of tipping point plot.

Now we consider a two-sample randomized experiment where we are interested in the difference in means. Let n_z, n_z^{obs}, and n_z^{mis} be the total sample sizes, number observed, and number missing in treatment arm $z = 1, 2$. For the ith individual, let Y_i, Z_i, and R_i denote the response, treatment arm, and indicator of observed response. We consider a pattern mixture model, where we assume that for treatment arm $z = 1, 2$,

$$\mu_z^{mis} = E(Y|Z = z, R = 0) = E(Y|Z = z, R = 1) + \delta_z = \mu_z^{obs} + \delta_z,$$

and δ_z is a sensitivity parameter representing how much different the means of the missing observations are from the means of the observed observations. Let $\pi_z = \Pr[R = 1|Z = z]$, then

$$\mu_z = \pi_z \mu_z^{obs} + (1 - \pi_z)\left(\mu_z^{obs} + \delta_z\right) = \mu_z^{obs} + (1 - \pi_z)\delta_z.$$

We are interested in

$$\Delta = \mu_2 - \mu_1 = \mu_2^{obs} - \mu_1^{obs} + \xi,$$

where $\xi = (1 - \pi_2)\delta_2 - (1 - \pi_1)\delta_1$. Let $(L_\Delta^{obs}, U_\Delta^{obs})$ be a $100(1 - \alpha)\%$ confidence interval on $\Delta^{obs} = \mu_2^{obs} - \mu_1^{obs}$, for example, it could be the confidence interval associated with Welch's t-test on the observed data only. Then if ξ is known, a $100(1 - \alpha)\%$ confidence interval on Δ is

$$\left(L_\Delta^{obs} + \xi, U_\Delta^{obs} + \xi\right).$$

A tipping point would be the value of ξ such that either $L_\Delta^{obs} + \xi = 0$ or $U_\Delta^{obs} + \xi = 0$. A problem is that ξ does not have a straightforward interpretation. A simplifying convenience assumption is that $1 - \pi_z = n_z^{mis}/n_z$ for $z = 1, 2$, so that we can write $\xi = \left(n_2^{mis}/n_2\right)\delta_2 - \left(n_1^{mis}/n_1\right)\delta_1$ (i.e., it is a function of δ_1 and δ_2 only).

Example 17.4 *Consider a randomized trial of treatments for anemia (low hemoglobin in the blood), where the responses are change from baseline hemoglobin to day 28 hemoglobin, so positive responses are good. The mean response within each arm measures the average increase in hemoglobin in grams per liter of blood (anemia in males is defined as $< 130g/L$). Let z_i be the treatment arm for the ith individual, with $z_i = 1$ denoting randomized to the control group (standard care) and $z_i = 2$ randomized to the treatment. Suppose $\hat{\mu}_2^{obs} = 11.1$, $\hat{\mu}_1^{obs} = 6.2$, so that $\hat{\mu}_2^{obs} - \hat{\mu}_1^{obs} = 4.9$ and with 95% confidence interval on $\mu_2^{obs} - \mu_1^{obs}$ equal to $(1.7, 8.1)$. Suppose that the proportion missing in the groups are $n_1^{mis}/n_1 = 11/232$ and $n_2^{mis}/n_2 = 16/241$. The tipping point for the significance of the difference in mean increase between the two groups is the value $\xi = -1.7$, the value where the lower limit on the difference becomes 0. Treating the sample proportions of missing as the true probabilities, and solving for δ_2, we get*

$$-1.7 = \left(\frac{16}{241}\right)\delta_2 - \left(\frac{11}{232}\right)\delta_1$$

$$\Rightarrow \quad \delta_2 = -25.6 + 0.714\delta_1.$$

Further, if we assume that in the control arm there is no difference in the means between missing and nonmissing responses (i.e., $\mu_1^{mis} - \mu_1^{obs} = \delta_1 = 0$), then we would need for the difference in means between the missing and non-missing responses in the treatment group to be $\mu_2^{mis} - \mu_2^{obs} = \delta_2 = -25.6$ or less in order to not find a significant difference.

17.4 Methods for Missing at Random Data

17.4.1 Simple MAR Methods

Consider one of the simplest MAR situations. Let $Y_1, \ldots, Y_n$ be a simple random sample from a population, and suppose we are interested in estimating $E(Y_i) = \mu$. Let W_i be an auxiliary variable that is categorical with J levels, where

$$\mu_j = E[Y_i | W_i = j],$$

and let the proportion of the population with each value of W_i be

$$p_j = \Pr[W_i = j].$$

Then we can write

$$\mu = E[Y_i] = \sum_{j=1}^{J} E[Y_i | W_i = j] \Pr[W_i = j] = \sum_{j=1}^{J} p_j \mu_j.$$

When there is no missing data, it does not matter if we adjust for the baseline variable or not when estimating μ, since the sample mean is equal to the weighted average of the sample means within each level of the baseline variable,

$$\hat{\mu} = \frac{1}{n} \sum_{i=1}^{n} Y_i = \sum_{j=1}^{J} \hat{p}_j \hat{\mu}_j, \tag{17.1}$$

where $\hat{p}_j$ is the proportion of the sample with $W_i = j$ and $\hat{\mu}_j$ is the average of observations with $W_i = j$.

Now suppose that there are missing observations. Let $R_i = 1 - M_i = 1$ if Y_i is observed and $R_i = 0$ otherwise. Suppose the data are missing at random and the probability of observing Y_i depends only on the baseline variable W_i and $\Pr[R_i = 1 | W_i = j] = \pi(W_i) = \pi(j)$. Let the average of the observed values of Y_i be

$$\hat{\mu}^{obs} = \frac{\sum_i R_i Y_i}{\sum_i R_i}.$$

Then the expected value of $\hat{\mu}^{obs}$ is

$$E[\hat{\mu}^{obs}] = \sum_j p_j \pi(j) \mu_j \neq \sum_j p_j \mu_j = \mu.$$

But under the MAR assumption, the sample means of the observed responses within the jth level of the baseline variable, say $\hat{\mu}_j^{obs}$, are unbiased, meaning

$$E[\hat{\mu}_j^{obs}] = \mu_j$$

so that an unbiased estimator of μ is

$$\hat{\mu}_{MAR} = \sum_j \hat{p}_j \hat{\mu}_j^{obs}, \qquad (17.2)$$

where the subscript MAR indicates missing at random.

For this simple example, we derive $\hat{\mu}_{MAR}$ using several different methods to give intuition on the methods; however, for more complicated MAR adjustments each method may generalize differently. First, consider *regression imputation*. We use the observed data and the baseline variables to model a predicted value for each observation, so that for the ith response our prediction depends only on W_i and the model, $m(W_i; \hat{\gamma})$, where $\hat{\gamma}$ is a set of parameter estimates from $\mathbf{Y}^{obs}$. For this example, $\hat{\gamma} = \left[\hat{\mu}_1^{obs}, \dots, \hat{\mu}_J^{obs} \right]$ and $m(j; \hat{\gamma}) = \hat{\mu}_j^{obs}$. Then the regression imputation estimator is

$$\hat{\mu}_{RI} = \frac{1}{n} \sum_i m(W_i; \hat{\gamma}), \qquad (17.3)$$

where in this case $\hat{\mu}_{MAR} = \hat{\mu}_{RI}$. In this fully parametrized model,

$$\hat{\mu}_{RI} = \frac{1}{n} \sum_i \left\{ R_i Y_i + (1 - R_i) m(W_i; \hat{\gamma}) \right\}. \qquad (17.4)$$

and we just impute the missing observations using the predicted value based on each level.

Another way to derive $\hat{\mu}_{MAR}$ is to weight the ith observed response by $1/\pi(W_i; \hat{\alpha})$, where $\pi(j; \hat{\alpha}) = \hat{p}_j$ is the proportion of observed responses among individuals with $W_i = j$ and $\hat{\alpha} = [\hat{p}_1, \dots, \hat{p}_J]$. So if only $1/3$ of the responses are observed with $W_i = j$, then each observed response when $W_i = j$ counts for three observations. This gives the *inverse probability weighted* estimator,

$$\hat{\mu}_{IPW} = \frac{1}{n} \sum_i \frac{R_i Y_i}{\hat{\pi}(W_i)}, \qquad (17.5)$$

where $\hat{\pi}(W_i) \equiv \pi(W_i; \hat{\alpha})$. This estimator can also be motivated by noting that under the MAR assumption $E(R_i Y_i | W_i) = \pi(W_i) E(Y_i)$. This formulation shows how $\hat{\mu}_{MAR}$ can be very variable if $\pi(j)$ is small and μ_j is very different from the μ_k for $k \neq j$.

17.4.2 More Complicated MAR Models

In the example in Section 17.4.1, the auxiliary variable, W_i, had only J levels, and the model for the mean, $m(W_i; \gamma)$, just estimated a parameter for each of the J levels. The model for the nonmissingness, $\pi(W_i; \alpha)$, also estimated a parameter for each level. For many practical examples, both of those models would be more complicated. For example, the auxiliary variables, $\mathbf{W}_i$, and the variables of interest, $\mathbf{X}_i$, could both be vectors. The model of the mean, $E(Y_i | \mathbf{W}_i, \mathbf{X}_i) \equiv m(\mathbf{W}_i, \mathbf{X}_i; \gamma)$, could be a generalized linear model while the model of the missingness, $\Pr[R_i = 1 | \mathbf{W}_i] = \pi(\mathbf{W}_i; \alpha)$, might be a logistic regression.

We explore some of the difficulties in this modeling by considering two scenarios that are only slightly more complicated than the example of the previous section. First, the responses are:

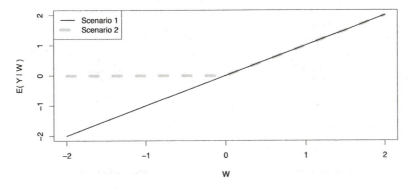

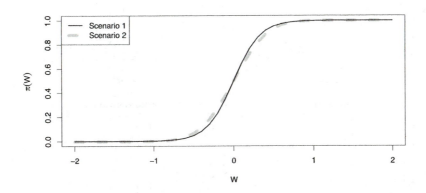

Figure 17.2 $E(Y|W)$ and $\pi(W)$ by W from the two scenarios.

$$\text{Scenario 1:} \quad Y_i = W_i + \epsilon_i$$
$$\text{and}$$
$$\text{Scenario 2:} \quad Y_i = \delta(W_i) + \epsilon_i,$$

where $\delta(W) = W$ if $W > 0$ and $\delta(W) = 0$ otherwise, and where W_i and ϵ_i are independently normally distributed with $W_i \sim N(0,4)$ and $\epsilon_i \sim N(0,16)$. For example, Y_i could represent the final response, and W_i could be a surrogate response that is correlated with Y_i. Second, the missingness models which model the probability of observing the response are:

$$\text{Scenario 1:} \quad \pi(W_i) = \text{expit}\,(6W_i)$$
$$\text{and}$$
$$\text{Scenario 2:} \quad \pi(W_i) = \Phi(3W_i),$$

where $\text{expit}(\eta) = 1/(1 + \exp(-\eta))$ and $\Phi(\cdot)$ are the cumulative distributions for a standard logistic (which is the inverse link function for logistic regression) and a standard normal distribution. We plot $E(Y|W)$ and $\pi(W)$ for the two scenarios in Figure 17.2.

We plot two simulated data sets from the two scenarios in Figure 17.3. Since the data look similar, we use the same models for both: a linear model for the expected response,

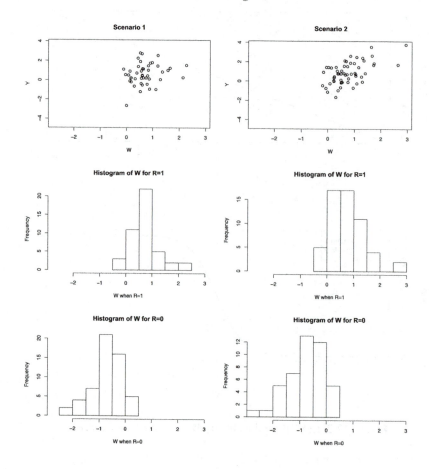

Figure 17.3 Simulated data from the two scenarios. Top panels are scatter plots of the observed responses by the W_i, middle panels are histograms of W_i for individuals with observed responses, and bottom panels are histograms of W_i for individuals with missing responses.

$$\mathrm{E}(Y_i | W_i) = \gamma_0 + \gamma_1 W_i,$$

and a logistic model for the missingness model,

$$\mathrm{E}(R_i | W_i) \equiv \pi(W_i) = \mathrm{expit}\,(\alpha_0 + \alpha_1 W_i).$$

For Scenario 1 the models are correctly specified, but for Scenario 2 the models are misspecified. From Figure 17.3 we see that there is not much overlap in the W_i between the individuals with observed responses and those with missing responses. This is a warning that we will need to extrapolate in the model of the mean outside of the range of the covariate, W_i, for the observed responses. So if we guess correctly on the model of the mean (Scenario 1), then the regression imputation should work well, but if we do not (Scenario 2) then it may not work well.

We simulate 1,000 data sets from the two scenarios, fit the models, and estimate the mean of Y by four methods: (1) $\hat{\mu}_{ALL}$ is the mean of all responses (both observed and missing),

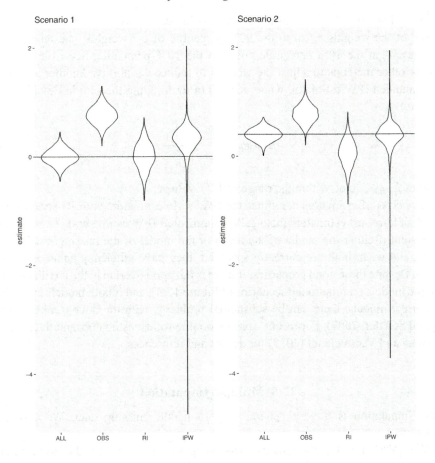

Figure 17.4 Violin plots of simulated estimates from the two scenarios, based on 1,000 simulated data sets. $ALL \equiv \hat{\mu}_{ALL}$ = mean of all responses, $OBS \equiv \hat{\mu}_{OBS}$ = mean of observed responses, $RI \equiv \hat{\mu}_{RI}$ = regression imputation method, $IPW \equiv \hat{\mu}_{IPW}$ = inverse probability weighting. Horizontal lines are the true values.

(2) $\hat{\mu}_{OBS}$ is the mean of the nonmissing responses, (3) $\hat{\mu}_{RI}$ is the regression imputation model, and (4) $\hat{\mu}_{IPW}$ is the inverse probability weighting model. In Figure 17.4 we give violin plots (like smoothed vertical symmetrized histograms) of the resulting estimates from the simulation. $\hat{\mu}_{ALL}$ uses missing data so it cannot be used in practice but gives an idea of the best we can do. $\hat{\mu}_{OBS}$ is biased because the data are MAR, with lower values of Y and W more likely to be missing. $\hat{\mu}_{RI}$ appears unbiased with larger variance than $\hat{\mu}_{ALL}$ in Scenario 1 when the mean model is correctly specified but has a bias in Scenario 2 when the mean model is misspecified. Finally, $\hat{\mu}_{IPW}$ can have some very extreme values due to large weights. In Scenario 1 it appears that the median of $\hat{\mu}_{IPW}$ is too large even though both models are correctly specified, while in Scenario 2 it happens that the median of $\hat{\mu}_{IPW}$ looks close to the true value, but this is lucky since both the model of the mean and the model of missingness are misspecified.

Because of the high variability of the inverse probability weighted (IPW) estimator, many methods have been developed to keep a few individuals from overly influencing the results due to very high weights. For example, we could set the weights greater than the 90th percentile of the weights equal to the 90% percentile of the weights, and analogously set weights less than the 10th percentile equal to the 10% percentile. See Cole and Hernán (2008) for other methods to adjust the weights to reduce variability. Another adjustment is to use a bounded IPW estimator, where instead of multiplying the sum in Equation 17.5 by $1/n$, we use

$$\hat{\mu}_{IPWb} = \frac{\sum_i \frac{R_i Y_i}{\hat{\pi}(W_i)}}{\sum_j \frac{R_j}{\hat{\pi}(W_j)}}. \tag{17.6}$$

This forces $\hat{\mu}_{IPWb}$ to be within the range of Y_i^{obs} values.

Another class of estimators combines the IPW-style estimators with RI-style estimators, to give double-robust estimators (also called augmented IPW estimators). Those estimators are consistent if either the model of the mean or the model of the missingness is correctly specified, and when both are correctly specified, they have efficiency gains over the IPW estimator. Despite these good properties, if there is not good overlap in the auxiliary variables used in the models (similar to the scenarios of Figure 17.3), and if both models are only close to the correct models, there can be substantial problems with the double-robust estimator (Kang and Schafer, 2007). For recent work on double-robust estimators and their inferences see Seaman and Vansteelandt (2018) for details and references.

17.5 Multiple Imputation

Multiple imputation is a very general way to handle missing data. We give a brief introduction.

Let $\mathbf{Y} = [\mathbf{Y}^{obs}, \mathbf{Y}^{mis}]$ be the response data. Suppose we are interested in some parameter, β, and suppose that if we had no missing data we would estimate β with the estimator $\hat{\beta}(\mathbf{Y})$ and its variance as $\hat{V}(\mathbf{Y})$. Now suppose we had missing data and some imputation model where we can impute the missing values repeatedly. Let $\mathbf{Y}^{*j} = [\mathbf{Y}^{obs}, \mathbf{Y}^{mis,j}]$ be the jth imputation out of J total. Let $\hat{\beta}_j = \hat{\beta}(\mathbf{Y}^{*j})$ and $\hat{V}_j \equiv \hat{V}(\mathbf{Y}^{*j})$. Then the multiple imputation (MI) estimator is $\bar{\beta} = \frac{1}{J} \sum_j \hat{\beta}_j$, and the MI variance estimator is

$$\hat{V}_{MI} = \bar{V} + \left(1 + \frac{1}{J}\right)\hat{B},$$

where

$$\bar{V} = \frac{1}{J}\sum_{j=1}^{J} \hat{V}_j, \text{ and}$$

$$\hat{B} = \frac{1}{J-1}\sum_{j=1}^{J}\left(\hat{\beta}_j - \bar{\beta}\right)^2.$$

For inferences we assume that (at least approximately),

$$\frac{\bar{\beta} - \beta}{\sqrt{\hat{V}_{MI}}} \sim t_d,$$

where t_d is a t-distribution with d degrees of freedom, with

$$d = (J - 1)\left\{1 + \left(\frac{J}{1+J}\right)\left(\frac{\bar{V}}{\hat{B}}\right)\right\}^2.$$

Carpenter and Kenward (2013) recommend using $J = 100$, unless a very small number of imputations gives enough precision.

Multiple imputation was originally developed for survey methods by Rubin as a Bayesian method, but we have included it because of its usefulness, wide use, as well as good frequentist properties. Multiple imputation can be expanded to vector parameters and much more complex situations. While the method is appropriate for MAR data, it can also be used in MNAR situations, but there are many ways to define the model under MNAR and applications would typically require much thought to produce those models. For some examples see Carpenter and Kenward (2013, chapter 10) or see Little et al. (2016) that does a multiple imputation for a time to event endpoint using tipping points.

17.6 Summary

We introduce the terminology of missingness assumptions: Missing Completely at Random (MCAR), Missing at Random (MAR), and Missing Not at Random (MNAR). MCAR signifies benign missingness such that an analysis of complete data will be unbiased. If that assumption is not tenable, then the data might be MAR, which means that missingness is not completely random, but the probability of missingness depends on observable data. While this limits the analysis methods that we can use to get reasonable results, there are many methods available to handle MAR data. If the previous two categories are not tenable, then we have to assume the data are MNAR. Under MNAR we need to do a range of analyses to assess the robustness of results; there is no one correct analysis. Unfortunately, there is no way to use the data to fully verify the appropriateness of the missingness assumptions. There may be settings where making assumptions such as MCAR or MAR are highly reasonable. However, often it is less clear, in which case the best approach is to plan to do a variety of analyses.

We describe several analyses to consider when data might be MNAR. We could take an extreme approach that assumes that all missing control patients were successful and that all missing treatment patients failed, for example. In some scenarios, even this extreme method will still result in a statistically significant result, providing full confidence in the outcome. Less extreme imputations can also be employed. We present tipping point analysis which, at its simplest implementation, shows exactly how different the results would need to be among the missing subjects in the two samples, for the statistically significant result to hold up. This is a very transparent method that nonstatisticians find easy to understand. Finally, pattern mixture models are useful for sensitivity analyses, see National Research Council (2010) for simple examples.

A relatively simple strategy for MAR data is to use inverse proportional weighting. The basic idea is that observed subjects are heavily overweighted if a high proportion of similar subjects have missing outcomes, in this way, the subjects with assessments are standing in for the similar missing subjects. Because the inverse proportional weighting method is susceptible to outliers having too much influence, there are strategies to address this, including a simple approach that truncates the weights at the 10th and 90th percentiles.

Multiple imputation is a popular method for MAR data. Essentially, we determine an imputation model based on the observed data. We sample from this imputation model repeatedly to create multiple complete data sets; a standard analysis is done within each data set. Then the means and variances of the per data set parameters are combined to get an overall result. Multiple imputation could also be modified to analyze data assumed to MNAR, but this is more complicated.

As a final comment not discussed in the chapter, repeated measures designs, while certainly useful, are not always the panacea for missing data that they are sometimes purported to be. That is, repeated measure designs and analyses are often recommended to address missingness at the final assessment time (i.e., the primary endpoint). It is tempting to assume the data are MAR; however, this often may not be the case, for example, in clinical trials, it is reasonable to suspect that subject drop out is often informative beyond what is observed. If MAR is assumed, the subjects who drop out might effectively be assumed to mimic those who stay in the study, which could lead to a very misleading result. This illustrates why it is important to conduct a range of sensitivity analyses.

17.7 Extensions and Bibliographic Notes

Little and Rubin (2020) is a classic book on the topic. Tsiatis (2006) gives a theoretical introduction to missing data models including double-robust estimators, and frequentist properties of multiple imputation estimators. Carpenter and Kenward (2013) discuss applications of multiple imputation with a good discussion of practical theory. National Research Council (2010) is a free, comprehensive look at the handling of missing data for clinical trials written by a panel of experts chaired by R. Little. Seaman and Vansteelandt (2018) is a good place to start for references on MAR methods such as regression imputation, IPW, and double-robust methods, and we used it heavily for Section 17.4. The website `https://rmisstastic.netlify.com/` also has some great resources, see especially the general lecture notes by Marie Davidian.

17.8 R Functions and Packages

See `https://CRAN.R-project.org/view=MissingData` for a list of many useful R packages related to missing data. The TippingPoint R package (version 1.1.0) was used to create Figure 17.1, except the source code was modified to calculate the central Fisher's exact p-value instead of using prop.test (although at that sample size and precision there was virtually no difference).

17.9 Problems

P1 Create a fake data set such that when you apply the two-sample t-test of Welch you get a p-value less than 10^{-6}. Add one additional observation that is either extremely large or extremely small and calculate the resulting p-value. Explore this further by plotting p-value versus the additional observation and using many different values for the additional observation going from extremely large to extremely small. Does the p-value become greater than 0.05 for any values of the additional observation? Explore this by plotting the t statistic, its numerator and denominator, and the degrees of freedom estimator.

P2 Show that Equation 17.1 is true.

P3 In Section 17.4.1, show that $\hat{\mu}_{MAR} = \hat{\mu}_{RI} = \hat{\mu}_{IPW}$. In other words, show that Equations 17.2, 17.3, 17.4, and 17.5 are all equivalent.

18

Group Sequential and Related Adaptive Methods

18.1 Overview

In Chapter 13 we looked at the problem of multiple testing. Here we consider a special kind of multiple testing, repeated testing of the same hypothesis from the same study as more data become available. The issue is that there is strong correlation between test statistics when we repeatedly test by successively including more data.

Consider the example of a one-sample t-test. Suppose we collected the data one observation at a time and performed a one-sample t-test of the current data in the study after each observation was added to the study. Further, suppose that we conducted each test at the two-sided α-level and stopped testing if any test was significant. Then if the total sample size is 101, and we tested up to 100 times: after 2 subjects are in the study, after 3 subjects, and so on until all $n = 101$ subjects are included. We call this sequential testing. Consider the overall Type I error rate for this type of study, the probability of a study declaring a significant effect by this sequential test method when the null hypothesis is true. Under the usual normality assumption and the null hypothesis that the mean is 0, if each test was tested at the nominal $\alpha = 0.05$ level, then the simulated overall Type I error rate is about 39% (we simulated 10,000 studies). If we use a Bonferroni correction and test each test in the sequence at the nominal $\alpha = 0.05/100 = 0.0005$, then the simulated false positive rate is about 1%. (Recall, we can do a Bonferroni correction regardless of the correlation among the tests and it bounds the overall Type I error rate.) If we repeat this simulation except going up to $n = 1,001$ with 1,000 tests, then the simulated false positive rate is about 54% for nominal $\alpha = 0.05$ or about 0.2% for a Bonferroni nominal $\alpha = 0.05/1,000 = 0.00005$. Thus, if we want the overall Type I error rate to be about 5%, then the Bonferroni correction is too strict because the tests are highly correlated, but the uncorrected overall Type I error rate is much too large.

For many studies, sequential testing after adding every individual is not feasible, but often there may be several planned interim analyses. We call this *group sequential methods*. For example, for a two-year study of a new treatment versus standard of care for a serious disease, in addition to the planned final analysis at the end of the study, there may be a planned interim analysis at the end of the first year. Suppose we test a one-sided null hypothesis at each planned analysis, and rejecting a null hypothesis suggests that the new treatment is better. In that case, the interim analysis could allow the study to stop after one year if there is enough evidence by that point to reject the null hypothesis; this has the advantage of saving research dollars and potentially getting critical treatments to patients sooner rather than later.

In Section 18.2 we discuss making k pre-planned interim analyses, in a special case when the estimator of the parameter of interest is normally distributed, with independent contributions for each part of the study. Under this special case, we show how to adjust the interim analyses in a valid way so that by the end of the planned study the Type I error rate is equal to the predefined α-level. Asymptotically, many standard test statistics approach the distribution of the special case (e.g., standardized difference in proportions, difference in means, or the logrank statistic). We show in Section 18.3 that the special case can be generalized whenever the vector of test statistics for the interim analyses has a multivariate normal distribution that can be transformed to a standard Brownian motion process with a drift (see Equation 18.4). Section 18.3 shows that with test statistics that transform to such a Brownian process, we can do an arbitrary number of interim analyses in a valid way. In Section 18.4 we discuss valid methods or better approximations when the Brownian motion approximation does not apply. Section 18.5 shows how to get estimates and confidence intervals after an interim analysis.

It is possible to design studies where the final maximum sample size is not known. Wald (1947) developed fully sequential tests where we have two hypotheses and we keep sampling observations until we can decide on one of the hypotheses, and there is no limit to the number of observations we sample; however, because of the impracticality of not having a maximum possible sample size, we do not cover that method (Note N1). Instead, in Section 18.6 we study a closely related idea of adaptive study designs. We look at adaptive methods where we run a preliminary study and use that study to design a follow-up study, and then combine the results of both studies in a valid way.

18.2 *k* Preplanned Interim Analyses

We first consider a general case where our estimator is approximately normal, and we have equally spaced interim analyses. We show that the estimates at the interim analyses have a regular covariance structure.

Consider a study where we are interested in estimating the parameter β. Suppose the planned total sample size is n and there are $k = n/m$ planned analyses, where m and k are integers. Let $\hat{\beta}_1, \ldots, \hat{\beta}_k$ be the k independent estimates of β, where $\hat{\beta}_1$ is estimated from the first m individuals, $\hat{\beta}_2$ is from the next m individuals, and so on. Assume that $\hat{\beta}_j \sim N(\beta, V_m)$. For now, we ignore the variability in estimating the variance and assume V_m is known.

Consider the cumulative estimate of β. Let $\hat{\beta}(j/k)$ be the estimate after the jth interim analysis when j/k of the study is completed, where

$$\hat{\beta}(j/k) = \frac{1}{j} \sum_{h=1}^{j} \hat{\beta}_h.$$

We can show that for $i < j$ (Problem P1),

$$\begin{bmatrix} \hat{\beta}(i/k) \\ \hat{\beta}(j/k) \end{bmatrix} \sim N \left(\begin{bmatrix} \beta \\ \beta \end{bmatrix}, V_m \begin{bmatrix} 1/i & 1/j \\ 1/j & 1/j \end{bmatrix} \right). \tag{18.1}$$

Consider a series of Wald tests of $H_0 : \beta = 0$, where at the jth interim analysis we compare

$$Z(j/k) = \frac{\hat{\beta}(j/k)}{\sqrt{\hat{V}(j/k)}}$$

to a standard normal distribution, where $\hat{V}(j/k)$ is the estimate of the variance of $\hat{\beta}(j/k)$, say $V(j/k)$, which is $j^{-1}V_m$ in this case. If $\hat{V}(j/k)$ equals $V(j/k)$, then (see Problem P1)

$$\begin{bmatrix} Z(i/k) \\ Z(j/k) \end{bmatrix} \sim N \left(\begin{bmatrix} \beta/\sqrt{V(i/k)} \\ \beta/\sqrt{V(j/k)} \end{bmatrix}, \begin{bmatrix} 1 & \sqrt{i/j} \\ \sqrt{i/j} & 1 \end{bmatrix} \right), \tag{18.2}$$

where i, j are integers with $1 \le i < j \le k$. The *Pocock procedure* compares all k Z statistics (i.e., $Z(1/k), Z(2/k), \ldots, Z(k/k)$) to the same constant $c_{k,\alpha}^P$, such that the overall probability that at least one of the k tests rejects the null is bounded by α under the assumption that $\hat{V}(j/k) = V(j/k)$. In other words, define $c_{k,\alpha}^P$ such that

$$1 - \Pr\left[|Z(j/k)| \le c_{k,\alpha}^P \text{ for } j = 1, \ldots, k \right] = \alpha \tag{18.3}$$

assuming Equation 18.2 holds under the null hypothesis that $\beta_j = 0$ for $j = 1, \ldots, k$. We can also transform the constant to represent the nominal (per test) α-level, say $\alpha_k^P = 2\{1 - \Phi(c_{k,\alpha}^P)\}$. We compare the corresponding nominal Bonferroni α-levels in Table 18.1, where we see that for the overall Type I error bound of $\alpha = 0.05$, the nominal Pocock α-levels at $k = 10$ are larger than the Bonferroni ones by a factor of about 2 (i.e., $\alpha_{10}^P/\alpha_{10}^B \approx 2$), and at $k = 20$ by a factor of about 3.

k	α_k^B	α_k^P	$c_{k,0.05}^P$
1	0.0500	0.0500	1.96
2	0.0250	0.0294	2.18
3	0.0167	0.0221	2.29
5	0.0100	0.0158	2.41
10	0.0050	0.0106	2.56
20	0.0025	0.0075	2.67
30	0.0017	0.0063	2.73

Table 18.1 *Nominal α-levels for each of k tests for Bonferroni (α_k^B) and Pocock (α_k^P) adjustments to ensure an overall Type I error rate of ≤ 0.05. Last column gives $c_{k,\alpha}^P$ for $\alpha = 0.05$, the constant that is used as a rejection threshold for the Pocock method on the absolute value of the Z statistics*

From Table 18.1, we see the cost of multiple looks at the data when using the Pocock procedure. For $k = 5$ looks, by the end of the study, if our nominal final p-value (i.e., uncorrected p-value at the end of the study) is greater than 0.0158 but less than 0.05, we would not reject by Pocock's procedure but would have rejected if we had not needed to correct for the four earlier looks. Of course, the corrections are necessary to retain the overall

Type I error rate, and the corrections are less conservative than the Bonferroni ones. An additional advantage is that if there is a substantial effect, then we have a large probability of stopping the study early and rejecting the null hypothesis.

In practice, our estimates, $\hat{\beta}_j$, are typically only asymptotically normal. Further, the variances of $\hat{\beta}_j$ are almost never known, and $Z(j/k)$ will typically be closer to a t-distribution than a normal distribution (see e.g., the derivation of the t-test in Example 10.1, page 164). For many typical Wald statistics (e.g., the standardized difference in means or proportions) and for some other statistics (e.g., the logrank test), the vector of test statistics asymptotically approaches the multivariate normality assumption for the k Z statistics of Equation 18.2.

Example 18.1 *Pocock procedure: Suppose we planned a two-sample block randomized trial to compare treatment to control. Let $\beta = \mu_2 - \mu_1$, where μ_2 and μ_1 are the mean responses in the treatment and control arms. Suppose the accrual happens over a few years, but once an individual is enrolled in the trial and randomized, their response is measured within a few days. Suppose that the total planned sample size is 2,000 and we plan $k = 5$ looks at the data after every $m = 400$ (200 in each arm). Because $m = 400$ is large, we assume the normality assumption of Equation 18.2 approximately holds. We test $H_0 : \beta = 0$. Under the null hypothesis, the approximate probability of stopping and rejecting at the five looks are (in order): 1.6%, 1.2%, 0.9%, 0.7%, and 0.6%, so that the total probability of rejecting is approximately 5%. Under an alternative with 90% power of rejecting the null under this design, our approximate probability of stopping after each of the five looks are: 20.6%, 26.0%, 20.9%, 14.0%, and 8.5%. So if there is a substantial effect, there is a high probability we will stop the study before all 2,000 individuals are enrolled.*

In practice, we usually do not use the Pocock boundaries, because the probability of stopping early under the null hypothesis is thought to be too high, and more importantly, by the end of the trial the nominal p-value is compared to the nominal α_k^P levels which can be substantially less than the overall α-level (e.g., for $k = 5$ $\alpha_5^P = 0.016$ for $\alpha = 0.05$).

A more commonly used method is the *O'Brien–Fleming procedure* which rejects based on B-values which are a function of the Z statistics. For a trial with k equally spaced looks, we say the jth look occurs at *information time* j/k, when j/k of the trial has been completed. The B-value at information time j/k is $B(j/k) = \sqrt{j/k}Z(j/k)$. The O'Brien–Fleming test rejects when any of $|B(j/k)| \geq c_{k,\alpha}^{OF}$, where analogously to Equation 18.3, we assume Equation 18.2 and solve for $c_{k,\alpha}^{OF}$ so that under the null hypothesis, the total probability of stopping at any time is α. Let the nominal α-level for the jth out of k O'Brien–Fleming boundaries be $\alpha_{k,j}^{OF}$. We give the nominal α-levels for $k = 2, 3, 4, 5$ in Table 18.2. We see that the final nominal α-level, $\alpha_{k,k}^{OF}$ is much closer to 0.05 than that of Pocock's method, so there is less of a price to pay for the early looks. However, the probability of stopping early is smaller also.

Example 18.1 *(continued) O'Brien–Fleming boundary: Under the null hypothesis, the approximate probability of stopping and rejecting at the five equally spaced looks in order are: 0.001%, 0.125%, 0.764%, 1.668%, and 2.442%, which sum to 5%. Under an alternative*

k	$c_{k,0.05}^{OF}$	$\alpha_{k,1}^{OF}$	$\alpha_{k,2}^{OF}$	$\alpha_{k,3}^{OF}$	$\alpha_{k,4}^{OF}$	$\alpha_{k,5}^{OF}$
2	1.98	0.005166	0.048			
3	2.00	0.000518	0.014	0.045		
4	2.02	0.000052	0.004	0.019	0.043	
5	2.04	0.000005	0.001	0.008	0.023	0.041

Table 18.2 *Nominal α-levels for the jth of k tests using the O'Brien–Fleming boundaries ($\alpha_{k,j}^{OF}$), made to ensure an overall Type I error rate of ≤ 0.05*

with 90% power, our approximate probability of stopping after each of five equally spaced looks is: 0.1%, 12.4%, 34.2%, 28.4%, and 14.8%.

The methods of this section are based on approximations that work well for large samples for several standard tests, such as one-sample tests of means, or two-sample tests for difference in means for numeric or binary observations. The approximations may be applied whenever the interim test statistics can be transformed so that Equation 18.2 approximately holds. Even for studies of time-to-event endpoints (see Chapter 16), we can use the Brownian motion approximation. In this case, the amount of information in the trial is not only based on the number of individuals enrolled, but also on how long each individual has been followed and the relationship of that follow-up time to the probability of an event occurring. It turns out that for right-censored data, if we plan a study based on the total number of observed events by the end of the study, then we can treat the number of observed events as proportional to the amount of information. So if the trial has 5 planned looks and 100 planned events at the end (e.g., the study is designed to continue until you observe 100 deaths), then the looks would occur after $20, 40, 60, 80$, and 100 events (see, e.g., Proschan et al., 2006, Section 2.1.3). We discuss situations where that approximation does not hold (e.g., small samples) in Section 18.4.

18.3 Arbitrary Number of Interim Analyses

Lan and DeMets (1983) generalize the earlier development to arbitrary (e.g., not equally spaced) interim analyses. Let t be the *information time*, which is the fraction of the study that is completed in terms of information. Let $\hat{\beta}(t)$ be the estimate of β at information time t, with $0 < t \leq 1$. So $\hat{\beta}(1/2)$ and $\hat{\beta}(1)$ are the estimates after the study is halfway completed and totally completed. Assume that $\hat{\beta}(t)$ is approximately normal with $\hat{\beta}(t) \sim N(\beta, V(t))$. We can relate this to Fisher information. Let the Fisher information at (information) time t be $I(t) = 1/V(t)$ (see Expression 10.5, page 171), then the information fraction is $I(t)/I(1) = t$ and $V(t) = \frac{1}{t}V(1)$. Further, $Cov\big(\hat{\beta}(t_a), \hat{\beta}(t_b)\big) = \frac{1}{\max(t_a, t_b)}V(1)$. The O'Brien–Fleming boundaries reject if the *B-value* is extreme, and the B-value in this general notation (assuming that $V(t)$ is known) is

$$B(t) = \sqrt{t}Z(t) = \frac{\sqrt{t}\hat{\beta}(t)}{\sqrt{V(t)}}.$$

Thus,

$$\begin{bmatrix} B(t_a) \\ B(t_b) \end{bmatrix} = N\left(\begin{bmatrix} t_a\beta^* \\ t_b\beta^* \end{bmatrix}, \begin{bmatrix} t_a & \min(t_a,t_b) \\ \min(t_a,t_b) & t_b \end{bmatrix} \right), \tag{18.4}$$

where $\beta^* = \beta/\sqrt{V(1)}$ is a standardized version of β. Equation 18.4 gives a property of a standard Brownian motion process with a drift parameter β^* (see Note N2 for formal definitions). That is why we call $B(t)$ a B-value (Lan and Wittes, 1988). Recall that the O'Brien–Fleming boundary for equally spaced looks solved

$$1 - \Pr\left[|B(j/k)| \le c_{k,\alpha}^{OF} \text{ for } j = 1, \ldots, k\right] = \alpha \tag{18.5}$$

for $c_{k,\alpha}^{OF}$ when $\beta = 0$. We can obtain an O'Brien–Fleming-like boundary if we find a constant, c_{tb}, such that the probability is α that a standard Brownian motion crosses either of the two boundaries $\pm c_{tb}$ before time 1, i.e.,

$$1 - \Pr\left[|B(t)| \le c_{tb} \text{ for some } t \le 1\right] = \alpha,$$

where the subscript $_{tb}$ refers to two-sided boundaries. It is more traditional and easier to calculate the one-sided boundary,

$$\Pr\left[B(t) > c_{ob} \text{ for some } t \le 1\right] = \frac{\alpha}{2}, \tag{18.6}$$

for one-sided level $\alpha/2$. We get $c_{tb} \approx c_{ob} = \Phi^{-1}(1 - \alpha/4)$ (Problem P3). For example, for two-sided boundaries with $\alpha = 0.05$ or a one-sided boundary with significance level of $\alpha/2 = 0.025$ we use $c_{tb} \approx c_{ob} = \Phi^{-1}(0.9875)$. The cumulative probability of crossing one boundary at any time before t is

$$\alpha_{ob}^*(t) = 2\left\{1 - \Phi\left(\frac{\Phi^{-1}(1 - \alpha/4)}{\sqrt{t}}\right)\right\} \tag{18.7}$$

(Problem P4). By the end of the study, the cumulative α that was spent on the one-sided test is $\alpha_{ob}^*(1) = \alpha/2$.

We can use Equation 18.7 to allow arbitrary looks at the data. We call $\alpha_{ob}^*(t)$ a *spending function* (specifically, an O'Brien–Fleming-like spending function). The overall one-sided Type I error of the study as bounded at $\alpha/2$. Each time we perform an interim analysis, we have a positive probability of stopping the study, so it is like we are *spending* some of the allowable error. With the spending function approach, we define the spending function during the design phase, then during the course of the study, we use that spending function to determine whether or not to stop at each interim analysis. A spending function can be any function that is monotonic in t with $\alpha^*(0) = 0$ and $\alpha^*(1)$ equal to the overall α-level used for the entire study. (Here we drop the $_{ob}$ subscript, because the spending function can represent either a one-sided boundary or a two-sided boundary.) For example, if we want to bound the total Type I error of the study at α, we could use $\alpha^*(t) = t\alpha$, a linear spending function. It is difficult to get a Pocock-like spending function that is a good approximation of the Pocock boundaries for large k, but for reasonably small k, one such spending function has been proposed (Proschan et al., 2006, chapter 5).

Here is how to use a spending function. First, choose the timing of the interim analyses independent of the data. Suppose we perform the first interim analysis at information time

t_1, then we allow ourselves to spend $\alpha^*(t_1)$ of the error. In other words, we stop and reject the one-sided test if the nominal one-sided p-value, say $p(t_1)$, is less than or equal to $\alpha^*(t_1)$. If we then continue on to information time t_2 to make a second interim analysis, we reject at a level such that $\alpha^*(t_2)$ is the cumulative probability of rejecting at *either* t_1 or t_2. This is calculated using the Brownian motion approximation. We continue in that manner for each additional interim analysis until the end of the study (for details see Problems P5 and P6).

The spending function approach gives great flexibility of when to choose the interim analyses, but its usual application requires that the Brownian motion approximation holds and that the decision on whether to perform an interim analysis should be independent of $B(t)$ for all times up until the proposed interim analysis. Thankfully, if a data monitoring committee decides it needs to do an interim analysis early after looking at the data, this should not inflate the Type I error rate by very much (Proschan et al., 1992).

There are many possible spending functions and sequential methods, we mention another popular one, suggested by Haybittle (1971) and Peto et al. (1976). The Haybittle–Peto sequential method is typically defined as performing a few interim analyses at a very low nominal α-level (e.g., three interim analyses each at the nominal α-level 0.001) but making no correction for multiple comparisons at the end (e.g., just use $\alpha = 0.05$ for the final level). As long as the nominal α-level is low enough and there are not too many interim analyses, then the anti-conservativeness of that method will not be too large. For example, suppose we perform three valid interim tests at level 0.001 and then a final valid test at level 0.05, then using the Bonferroni inequality the overall Type I error rate is $\leq 0.05 + 3 * (0.001) = 0.053$. Some users can accept the slight anti-conservatism in order to retain the simplicity of making no adjustment to the final test. Alternatively, we could bound the overall Type I error rate at 0.05 by using a final nominal level of 0.047 instead of 0.05. See Section 18.4 (page 349), for the application of Bonferroni's inequality to any spending function.

18.4 When the Brownian Motion Approximation Does Not Hold

Sections 18.2 and 18.3 were developed using an approximation that we have called the Brownian motion approximation. In order to improve the approximation, especially for applications with small samples, we can sometimes use the p-values appropriate for the sample size and stop and reject at each look by comparing the nominal p-value[1] to the nominal α_k^P (or $\alpha_{k,j}^{OF}$) levels. This is approximately correct whenever the vector of interim test statistics can be transformed so that Equation 18.2 approximately holds. For example, for a small sample size for the two-sample test of difference in means, we can use the t-test nominal p-values. Proschan et al. (2006, chapter 8) did simulations that gave approximately nominal Type I error rates for O'Brien–Fleming or Pocock boundaries even with five or fewer observations per arm between looks, with a slightly better approximation for O'Brien–Fleming boundaries.

If the responses are binary for the two-sample test, we suspect that if we use the nominal p-values from the central Fisher's exact test in the same manner, this will be a conservative method because those exact p-values tend to be larger than those from

---

[1] The nominal p-value is the calculated p-value from the test without making any adjustments for it being an interim analysis.

the normal approximation of using the Wald test statistics on standardized differences in proportions which asymptotically follow Equation 18.2. To check this, we simulated a two-sample study with five equally spaced looks, and m binomial observations for each sample between each look. Using one-sided Fisher's exact p-values on the cumulative data and comparing them to one-sided O'Brien–Fleming nominal α-levels with overall α-level of 0.025, we performed two sets of 100 simulations (one set with $m = 10$ and one with $m = 5$) with 10,000 replications in each simulation. For each simulation we drew one random uniform value for the mean of both binomial distributions. The simulated Type I error rate was less than the target 0.025 for all 100 simulations for each set, with the range of simulated rates equal to and (0.0096 to 0.0151) for $m = 5$ and (0.0122 to 0.0175) for $m = 10$. See Proschan et al. (2006, chapter 8) or Jennison and Turnbull (2000, chapter 12) for valid but more complicated methods for binary response studies. Proschan et al. (2006) also present valid permutation methods which can be applied to numeric or ranked responses.

Another relatively simple method is based on the Bonferroni inequality. For any spending function $\alpha^*(t)$ (i.e., a montonic function with $\alpha^*(0) = 0$ and $\alpha^*(1) = \alpha$), and any valid testing procedure that can be applied at interim times, we can use the Bonferroni inequality to get a valid sequential procedure regardless of whether the Brownian motion approximation holds. If the interim tests are performed at information times $t_1 < \cdots < t_k = 1$, then use nominal α-levels of $\alpha^*(t_1)$, $\alpha^*(t_2) - \alpha^*(t_1), \ldots, \alpha^*(t_k) - \alpha^*(t_{k-1})$. By the Bonferroni inequality the Type I error rate is bounded at $\sum_{j=1}^{k} \left\{ \alpha^*(t_j) - \alpha^*(t_{j-1}) \right\} = \alpha$, where $t_0 \equiv 0$. Even when the Brownian motion approximation holds, the loss in power may not be too large. For example, for the O'Brien–Fleming or Pocock boundaries with k equally spaced looks, one-sided 0.025 α-level, and power 90%, the Bonferroni method has power at least 80% for all $k \leq 10$, with less loss in power using O'Brien–Fleming-type (or even Haybittle–Peto-type) boundaries than for the Pocock-type ones (Proschan et al., 2006, pp. 149–150).

18.5 Estimates and Confidence Intervals after Interim Analyses

After stopping a study early after an interim analysis, it is not straightforward to present the results of the study in terms of a p-value, estimate, and confidence interval. To examine this issue, we consider the one-sample case of a binary response over time with interim analyses (see, e.g., Ignatova et al. (2012) or Kirk and Fay (2014)). The advantage of going through this simple case first is that the sample space of the boundary (the number of possible ways we can stop the study) is finite and small, so we can more easily describe the issues.

Suppose we take a series of independent Bernoulli random variables, say $Y_1, Y_2, \ldots$, each with mean θ. Suppose we are testing $H_0 : \theta = 0.5$ versus $H_1 : \theta \neq 0.5$, and we use an O'Brien–Fleming-like stopping rule with $k = 3$ equally spaced looks with a maximum sample size of $N_{max} = 30$. So the interim analyses are at $N = 10, 20$, and 30 observations, and we use capital N to denote that N is a random variable. Throughout we use the **binseqtest** R package to calculate the exact probabilities, rather than the approximate probabilities using the Brownian motion approximation. Under this stopping rule, there are 31 possible responses from the study, which we plot in Figure 18.1. After N observations, let $S = \sum_{i=1}^{N} Y_i$. In terms of the pair (S, N), let the set of 31 possible stopping places for this boundary be denoted,

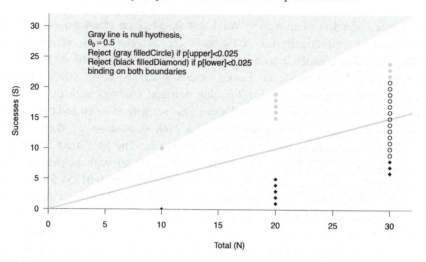

Figure 18.1 O'Brien–Fleming-type boundaries for a series of Bernoulli variables testing $H_0 : \theta = 0.5$. The outcomes that lead to rejection are presented as filled circles or diamonds for testing after 10, 20, and 30 individuals are completed, and open circles denote failing to reject after all 30 individuals are completed. Calculated using designOBF(30,k=3,theta0=0.5) in the binseqtest R package.

$$\mathcal{B} = \{(\mathcal{S}_1, \mathcal{N}_1), \dots, (\mathcal{S}_{31}, \mathcal{N}_{31})\},$$

where $\mathcal{S}_j$ is the number of successes (i.e., $\sum_{i=1}^{\mathcal{N}_j} Y_i$) out of $\mathcal{N}_j$ for the jth possible stopping place. Then

$$f_{\mathcal{B}}((s,n);\theta) = \Pr_{\mathcal{B}}[S = s, N = n|\theta] = K_{\mathcal{B}}(s,n)\theta^s(1-\theta)^{n-s},$$

where $K_{\mathcal{B}}(s,n)$ is the number of ways to get to (s,n) under the boundary, $\mathcal{B}$ (there is not a simple formula, but $K_{\mathcal{B}}(s,n)$ does not depend on θ). Let $\hat{\theta}(s,n) = s/n$ be the maximum likelihood estimate of θ when we stop at (s,n).

As usual, we use central p-values equal to two times the minimum of the one-sided p-values (bounded above by 1) (see Section 2.3.4). Consider first the one-sided p-value for testing $H_0 : \theta \geq 0.5$ versus $H_1 : \theta < 0.5$. Let $T(s,n)$ be some function of (s,n) such that smaller values of $T(s,n)$ suggest smaller values of θ. Then we can use $T(s,n)$ to order the sample space, and define the p-value as equal or more extreme than the null hypothesis as

$$p_L((s,n);\theta_0) = \Pr_{\mathcal{B}}[T(S,N) \leq T(s,n)|\theta = \theta_0], \qquad (18.8)$$

where $\theta_0 = 0.5$. The seemingly obvious choice is to use $T(s,n) = \hat{\theta}(s,n) = s/n$ (i.e., the maximum likelihood estimates [MLEs]) to order the data. There can be a practical problem with this choice. Suppose $(s,n) = (5,20)$ so that $\hat{\theta} = 5/20 = 0.25$. Then the possible stopping values with equal or smaller values of $\hat{\theta}(s,n)$ are: $0/10, 1/20, 2/20, 3/20, 4/20, 6/30, 7/30$, and $5/20$. Here is the problem: the sample space of the possible more extreme values for the study includes points with $N = 30$. What if you had continued the study beyond $N = 20$, something occurred, and the plan could not be followed and the next look happened at $N = 31$ or $N = 35$. Then the p-value would change, based on changes in the design that might happen in the future. This is not too large a problem for very strict stopping boundaries; however, if we are using a spending function, there will clearly be a problem. Another problem is that there may be multiple reasons for stopping that have nothing to do with the primary endpoint from the current study. For example, if either there are bad adverse events or if another study with definitive results was published between the start of the study and the interim analysis, then the stopping rules may not be followed. For all these reasons, we prefer the p-value definition not to depend on following the study design in the future. (For a related point, see Note N3.)

A way around this problem is to use stagewise ordering. Stagewise ordering is hierarchical: we first order by type of boundary, then by stage within boundary type, then by some other statistic (e.g., $\hat{\theta}(s, n)$) within stage/boundary type. Here are the details for the boundary of Figure 18.1. First, if there is a two-sided boundary, break it up into boundary types: the lower boundary, the end boundary, and the upper boundary. Points in different boundary types are ordered by lower < end < upper. In Figure 18.1 the lower boundary is points with $N < 30$ and $S/N < 0.5$, the end boundary is points with $N = 30$, and the upper boundary is points with $N < 30$ and $S/N > 0.5$. Second, for breaking ties within a type of boundary, break ties by stage, where the stage is given by the information time (e.g., N). For lower boundary points, earlier stages (N lower) are ordered lower, while for upper boundary points, earlier stages are ordered higher. The end boundary cannot break ties by stage, since all are at the same stage. Third, within stage/boundary-type break ties with some other method, like $\hat{\theta}(s/n)$.

We can generalize the one-sided lower p-values by for example testing $H_0 : \theta \geq \theta_0$ versus $H_1 : \theta < \theta_0$ and using Equation 18.8. Analogously, we define the one-sided upper p-values for testing $H_0 : \theta \leq \theta_0$ versus $H_1 : \theta > \theta_0$ by using

$$p_U((s,n); \theta_0) = \Pr_B [T(S,N) \geq T(s,n) | \theta = \theta_0]. \tag{18.9}$$

Then we can create a $100(1 - \alpha)\%$ central confidence interval by solving for (θ_L, θ_U), as the values that solve $p_L((s,n); \theta_L) = \alpha/2$ and $p_U((s,n); \theta_U) = \alpha/2$ (if possible, otherwise let $\theta_L = 0$ or $\theta_U = 1$). Because the $100(1 - \alpha)\%$ central confidence interval is the intersection of two $100(1 - \alpha/2)\%$ one-sided intervals, we write the limits of the $100(1 - \alpha)\%$ central interval, (θ_L, θ_U), as $\theta_L(1 - \alpha/2; s, n)$ and $\theta_U(1 - \alpha/2; s, n)$.

Finally, we consider estimation of the parameter θ. With early stopping, the maximum likelihood estimator (MLE) is biased. The problem is that especially large (or small) values of $\hat{\theta}$ will tend to stop early on the upper (or lower) boundary. So with interim analyses, the MLEs after early stopping will tend to be more extreme than the MLEs for a fixed sample size. An alternative estimator is a median unbiased estimator, defined as

$$\tilde{\theta}(s,n) = \frac{\theta_L(0.50; s, n) + \theta_U(0.50; s, n)}{2}.$$

For the general case based on approximate Brownian motion (see Sections 18.2 and 18.3), these same issues of ordering the sample space for defining p-values and bias of the MLE occur. See Proschan et al. (2006) for details.

18.6 Two-Stage Adaptive Designs

So far in this chapter, we have been discussing designing a large study and allowing for stopping early if there is a strong signal partway through the implementation of the study, but another conceptual strategy is to start with a smaller study and use the analysis of the smaller study as a basis for potentially increasing the sample size. In this section we consider two-stage adaptive designs, where we run the first stage of the study and use information from the first stage to either (1) decide to stop and reject the null hypothesis, (2) decide to stop because it is very unlikely that we will reject the null hypothesis after running a second stage, or (3) decide to continue by designing and running the second stage. We consider adaptive designs that bound the overall Type I error rate at α.

Consider the case where we use only the p-values from the two stages and both p-values test the same null hypothesis. Let p_1 and p_2 be the p-value from the first stage, and the p-value from the second stage data alone. Under the null hypothesis, both p-values are independent and uniformly distributed. There are many ways to combine such p-values (Loughin, 2004). Fisher's method is to multiply them together and reject at overall level α if

$$p_1 p_2 \leq c_\alpha, \tag{18.10}$$

where $c_\alpha = \exp\left\{-0.5 F_{\chi_4^2}^{-1}(1-\alpha)\right\}$ and $F_{\chi_4^2}^{-1}(q)$ is the qth quantile of a chi-squared distribution with four degrees of freedom. Bauer and Köhne (1994) defined a two-stage adaptive design that bounds the overall Type I error at α. First, prespecify α and a futility level γ. Then find the Stage 1 significance level α_1 that is the solution to

$$\alpha = \alpha_1 + c_\alpha \left\{\log(\gamma) - \log(\alpha_1)\right\}. \tag{18.11}$$

In other words, we find the α_1 such that the overall Type I error is α (Problem P7). Then do one of three things:

(1) if $p_1 \leq \alpha_1$, then stop and reject the null hypothesis after Stage 1; or
(2) if $p_1 \geq \gamma$, then stop for futility (fail to reject the null) at the end of Stage 1; or
(3) if $\alpha_1 < p_1 < \gamma$, then continue to Stage 2 and reject the null hypothesis when $p_1 p_2 \leq c_\alpha$, otherwise fail to reject.

The data from the two stages are not combined except through the p-values, and Bauer and Köhne (1994) show that the procedure bounds the overall Type I error rate at α even when we design the second stage based on information from the first stage. In fact, the null hypotheses may be defined differently in the two stages (see, e.g., Bauer et al., 2016).

Proschan and Hunsberger (1995) independently developed a two-stage adaptive method in a more general manner. Let the parameters be defined as previously: α is the overall significance level, α_1 and γ are the Stage 1 boundaries for stopping for efficacy (reject the null hypothesis) and futility (fail to reject and do not continue), respectively. Let $A_p(p_1)$ be the *conditional error function* in terms of p_1, where we reject when $p_2 \leq A_p(p_1)$, and assume P_1 and P_2, the random variables associated with p_1 and p_2, are independent and uniformly distributed p-values under the null hypothesis. (Proschan and Hunsberger wrote the conditional error function in terms of the Z statistic from Stage 1, but the idea is the same.) If $A_p(p_1) = 1$ (or 0), then we reject (fail to reject) regardless of p_2 and therefore do not continue onto Stage 2. We allow any conditional error function such that the overall probability of rejecting the null hypothesis under the null hypothesis using this procedure is

$$\alpha = \int_0^1 A_p(p_1)dp_1.$$

A slight adaptation of the Bauer–Köhne two-stage adaptive procedure uses,

$$A_p(p_1) = \begin{cases} 0 & \text{if } p_1 \geq \gamma \\ \frac{c}{p_1} & \text{if } \alpha_1 < p_1 < \gamma \\ 1 & \text{if } p_1 \leq \alpha_1, \end{cases} \tag{18.12}$$

where $c = (\alpha - \alpha_1)/\{\log(\gamma) - \log(\alpha_1)\}$ is solved to give overall Type I error of α given α_1 and γ (instead of fixing c and solving for α_1 as we did earlier). Proschan and Hunsberger (1995) studied one-sided p-values and used a circular conditional error function,

$$A_p(p_1) = \begin{cases} 0 & \text{if } p_1 \geq \gamma \\ 1 - \Phi\left(\sqrt{\{\Phi^{-1}(1-\alpha_1)\}^2 - \{\Phi^{-1}(1-p_1)\}^2}\right) & \text{if } \alpha_1 < p_1 < \gamma \\ 1 & \text{if } p_1 \leq \alpha_1. \end{cases}$$

Just as with the Bauer–Köhne procedure, these general procedures allow you to design your Stage 2 sample size based on Stage 1 results. The important issue is that the $A_p(\cdot)$ function must be prespecified and cannot change based on Stage 1 data. If we prespecify the conditional error function, we can use conditional power based on Stage 1 data to determine the Stage 2 sample size. For example, we could solve for the Stage 2 sample size such that under the current trend the power to reject is about 90%, but when we do the final analysis, we must use the prespecified conditional error function to decide on the rejection based on the Stage 2 p-value. This means that even if we decide to make the Stage 2 sample size 10 times larger than the Stage 1 sample size, the two stages are treated according to the prespecified $A_p(\cdot)$ that does not weight test statistics for each stage by sample sizes. Nevertheless, there is great flexibility in deciding on $A_p(\cdot)$. We can prespecify an $A_p(\cdot)$ function such that the Stage 2 data are weighted 10 times more than the Stage 1 data in the final analysis, but the weighting must be prespecified and cannot change based on the Stage 1 data. For example, still under the assumption that we are using independent uniformly distributed p-values from each stage, we can define $A_p(\cdot)$ to weight the first stage with weight w_1 and the second stage with w_2, then combine the p-values by first transforming them to standard normal and weighting them accordingly. Allowing early stopping for efficacy (using α_1) or futility (using γ), we create

$$A_p(p_1) = \begin{cases} 0 & \text{if } p_1 \geq \gamma \\ 1 - \Phi\left(\frac{c\sqrt{w_1^2+w_2^2} - w_1\Phi^{-1}(1-p_1)}{w_2}\right) & \text{if } \alpha_1 < p_1 < \gamma \\ 1 & \text{if } p_1 \leq \alpha_1, \end{cases} \tag{18.13}$$

where c is a constant determined so that the overall Type I error rate is α (Problem P9).

Proschan and Hunsberger (1995) presented their method in terms of normally distributed test statistics and setting the conditional power under the null hypothesis equal to $A_z(\cdot)$, where $A_z(\cdot)$ is the $A_p(\cdot)$ in terms of the normally distributed test statistics. That presentation allows them to show the relationship of the conditional error functions to conditional power. We present one simple approach based on conditional power described in Chen et al. (2004), which shows that if your study looks promising, you can increase the sample size and the final unadjusted p-value will be valid.

Here are the details. Suppose we plan a study such that the final estimate of the effect is $\hat{\beta}(1) \sim N(\beta, V(1))$, assuming that $V(1)$ is known. Using the notation of Section 18.3, suppose we look at the data partway through the study, at a time when the information fraction is t. For example, after we collect $1/3$ of the total planned data, then $t = 1/3$. Let $\hat{\beta}(t)$ be our estimate at information fraction t, and assume $\hat{\beta}(t) \sim N(\beta, V(t))$, where $V(t) = V(1)/t$. Let $Z(t) = \hat{\beta}/\sqrt{V(t)}$ and assume the Brownian motion assumptions of Section 18.3.] Let $CP(Z(t); \beta^*)$ be the conditional power at information time t of rejecting the one-sided null hypothesis, $H_0 : \beta \leq 0$, at level α, given we observe the standardized effect of $\beta^* = \beta/\sqrt{V(1)}$, which is

$$CP(Z(t); \beta^*) = 1 - \Phi \left(\frac{\Phi^{-1}(1 - \alpha) - \sqrt{t}Z(t) - (1-t)\beta^*}{\sqrt{1-t}} \right).$$

So the probability of finding a significant result at the end assuming the data keep following the trend seen so far, is

$$CP(Z(t); \hat{\beta}^*) = 1 - \Phi \left(\frac{\Phi^{-1}(1 - \alpha) - Z(t)/\sqrt{t}}{\sqrt{1-t}} \right),$$

where $\hat{\beta}^* = \hat{\beta}(t)/\sqrt{V(1)}$. Chen et al. (2004) show that if $CP(Z(t); \hat{\beta}^*) > 0.50$, then you can increase your sample size by any amount and then analyze the study at the end as if it was a fixed study and the resulting unadjusted final p-value is valid.

There is much debate about how to design a study with flexibility. Should we design with a large, planned sample size and allow early stopping (i.e., use group sequential methods), or design a smaller study but allow a subsequent study to add information to the first study (i.e., use a two-stage adaptive design)? See Jennison and Turnbull (2007) and Bauer et al. (2016) for illuminating discussion comparing the two approaches. While there are many statistical and nonstatistical considerations unique to each study setting, it is often useful to perform simulations to evaluate the statistical efficiency of any of several specific approaches.

18.7 Summary

If there is sufficient evidence of a treatment effect early in a study, it would be advantageous to stop the study then and report the conclusions. It is common for clinical trials to employ methods that test the results at one or a few interim time points. These methods address the multiple testing by properly and efficiently accounting for the correlation between test statistics across time. This is called group sequential testing.

The traditional Pocock and the OBrien–Fleming procedures ensure that the overall Type I error rate is bounded at the nominal α-level for equally spaced tests, when each interim test statistic is standard normal (i.e., are Z statistics), and all interim test statistics together follow a particular multivariate normal related to Brownian motion. In practice, many Z statistics can be put into B-value form (the Z statistic multiplied by the square root of the trial fraction) that asymptotically approach a standard Brownian motion (with a shift under the alternative). The Pocock method uses a constant critical value for Z statistics across test times, while the OBrien–Fleming procedure uses a constant critical value for B-values across test times. With three equally spaced tests and an overall two-sided α-level of 0.05, the required p-value at each test for the Pocock boundary is 0.022, whereas for the OBrien–Fleming, the stringency of the test lessens over time: 0.0005, 0.014, 0.045. Thus, if the study did not stop early, then the Pocock method requires a final p-value less than 0.022, whereas the OBrien–Flemings threshold is 0.045. Designers of trials generally desire methods where there is relatively little adjustment at the final test relative to the single fixed time test, thus the O'Brien–Fleming procedure is generally preferred over and used much more than the Pocock procedure.

In practice, the timing of tests will often not be equally spaced. A pre-specified spending function allows for arbitrary testing times, and this approach is in widespread use. The spending function can be any function that is monotonic in information time t (the fraction of the study that is completed) such that the function when t = 0 is 0, and when t = 1 is the overall α-level. There is such a function that is very consistent with the traditional OBrien–Fleming boundary, but a variety of spending functions are in use. Spending functions allow great flexibility in timing of the tests, although they technically require that the test timing be unrelated to previously observed critical values.

These standard methods may not be appropriate if the Brownian motion assumption does not hold, but fortunately for most conventional statistical tests it does approximately hold. Nevertheless, sometimes Brownian motion does not hold. Generally, this can occur if the test statistic is not normally distributed, the sample size is small, or successive test statistics are not based on independent increments. There are a variety of strategies to employ in such settings. We describe two conservative strategies in Section 18.4, one ad hoc method and another method known to be valid. The ad hoc strategy is to use the nominal alpha levels from the Brownian motion methods but use them with conservative p-values derived from non-Brownian motion arguments. The valid strategy is to simply use a Bonferroni adjustment which will be conservative, but often only slightly so. Another strategy not explicitly discussed is to use permutation tests which are valid and generally less conservative but are more computationally complex than those two strategies.

If a study is stopped early, the usual estimates and confidence intervals can be biased. The possible solutions may be complicated, and some are described in Section 18.5.

When there is uncertainty about the treatment effect, one approach is to start with a large sample size and possibly stop early (i.e., group sequential testing) and an alternate approach is to start with a small sample size and potentially increase the sample size based on the interim treatment effect (i.e., an adaptive design). There is considerable debate about the relative merits of these strategies. For comparing any two particular trial designs, a simulation would probably be prudent when comparing design options. We discuss flexible methods that add sample size based on the interim treatment effect, while controlling the overall Type I error rate. Using the approach of Bauer and Köhne (1994), one conducts a small study, and if that is not sufficient, one designs a second study using information from the first. If two studies are conducted, their respective p-values are appropriately combined to get an overall p-value. We may express these kinds of adaptive designs using a conditional error function introduced by Proschan and Hunsberger (1995), to show the generality of options. Another approach proposed by Chen et al. (2004) is extremely simple.

If the probability that the study with the original sample size will ultimately be statistically significant (calculated assuming that the remaining data will follow the same treatment difference observed at the interim look) exceeds 50%, then the sample size can be increased to any size beyond the original sample size, and the final unadjusted p-value will be valid.

Use of the methods described in this chapter should be considered when designing important medical studies, as they have the potential to enhance efficiency and to be more ethical. These designs are more commonly used in settings where there is urgent need to know the study result. If it is important to get a good estimate of effect, regardless of its statistical significance or if the study will provide other critical results beyond the primary test of the effect, then a fixed sample design may be the better choice.

18.8 Extensions and Bibliographic Notes

Proschan et al. (2006) is a very nice book on group sequential testing, and we used it heavily in writing this chapter. Jennison and Turnbull (2000) is also a good book and it gives more details on repeated confidence intervals and exact calculation for binary data. Demets and Lan (1994) review the alpha-spending function approach for group sequential testing discussed in Section 18.3. Posch and Bauer (1999) is helpful for comparing the Bauer and Köhne (1994) approach to the Proschan and Hunsberger (1995) approach (and others) for two-stage adaptive designs. See Bauer et al. (2016) for a review of recent work on adaptive designs, including references showing the close relationship of adaptive designs to group sequential designs.

18.9 R Functions and Packages

For the Pocock and O'Brien–Fleming boundaries, we used the gsDesign R package. For the spending function examples we used the ldbounds R package. For the single sample binary response exact sequential boundaries, see the binseqtest R package.

18.10 Notes

N1 **Sequential probability ratio test:** Wald (1947) develops the sequential probability ratio test (SPRT) for testing two-point hypotheses (e.g., $H_0 : \theta = \theta_0$ and $H_1 : \theta = \theta_1$), where there is no maximum sample size for the study. Despite this lack of a maximum sample size, the probability that the study will end is 1. Further, among tests with bounded Type I and II error rates at α and β, respectively, the SPRT minimizes the expected sample size. For details and further references see, for example, Ghosh (2006).

N2 **Brownian motion:** Formally, a Brownian motion process is a stochastic process with $\{B(t); t \geq 0\}$ with the following properties:

(a) $B(t+a) - B(t) \sim N(0, a\sigma^2)$, for $a > 0$ and $\sigma > 0$ fixed.

(b) For any k disjoint time intervals, $[t_1, t_1^*], \ldots, [t_k, t_k^*]$, with $t_1 < t_1^* \leq t_2 < t_2^* \leq \cdots \leq t_k < t_k^*$, then the k random variables $B(t_1^*) - B(t_1), \ldots, B(t_k^*) - B(t_k)$ are independent with $B(t_j^*) - B(t_j) \sim N(0, (t_j^* - t_j)\sigma^2)$.

(c) $B(0) = 0$ and $B(t)$ is continuous at $t = 0$.

It is a standard Brownian motion process when $\sigma = 1$. If $B(t)$ is a Brownian motion process, then $B(t) + \beta^* t$ is a Brownian motion process with drift parameter β^* (see, e.g., Karlin and Taylor, 1975).

N3 **Defying stopping boundaries:** Sometimes stopping boundaries are not followed, even if they are crossed. For example, in a randomized trial if a new treatment crosses a boundary for efficacy, and it significantly shows that it is better than standard of care, the data safety monitoring committee of the trial may still want the trial to continue if there are safety concerns. The extra information from continuing the trial may help in balancing whether the extra benefit from the efficacy is worth the increased incidence of the adverse events.

18.11 Problems

P1 Prove Expression 18.1. Hint: premultiply the left-hand side of Expression 18.1 by

$$\begin{bmatrix} \frac{1}{\sqrt{V(i/k)}} & 0 \\ 0 & \frac{1}{\sqrt{V(j/k)}} \end{bmatrix},$$

and find $\mathrm{Cov}(Z(i/k), Z(j/k))$ when $\hat{V}(h/k) = V(h/k)$, the variance of $\hat{\beta}(h/k)$.

P2 (Section 18.3) Use Equation 18.1 to derive Equation 18.4.

P3 Use the fact that for a standard Brownian motion, $B(t)$,

$$\Pr[B(s) > c \text{ for some } s \leq t] = 2\Pr[B(t) > c] \tag{18.14}$$

(see Proschan et al., 2006, p. 84) to show that $c_1 = \Phi^{-1}(1 - \alpha/4)$, where c_1 is defined in Equation 18.6. Proschan (1999) shows that $c \approx c_1$ is a very good approximation for typical levels of α (e.g., 0.05).

P4 Use Equation 18.14 to show Equation 18.7.

P5 We traditionally use the $Z(t)$ form of the statistics to calculate the cumulative probability of rejecting at time t_j given earlier looks at time $t_1, \ldots, t_{j-1}$ and using a spending function $\alpha_1^*(t)$. The constants for rejecting at $Z(t_1), \ldots, Z(t_j)$ are calculated iteratively. Let c_i be the boundary where we reject if $Z(t_i) > c_i$ for $i = 1, \ldots, j - 1$, and suppose we have already calculated them. Then find c_j such that

$$\Pr\left[\cup_{i=1}^{j} Z(t_i) > c_i\right] = \alpha_1^*(t_j),$$

i.e., set to $\alpha_1^*(t_j)$ the probability that we could have rejected at any of the earlier times or at t_j. Show that this is equivalent to finding c_j such that $\alpha_1^*(t_j) - \alpha_1^*(t_{j-1})$ is equal to the probability that we do not reject at any of the earlier times *and* reject at t_j (Proschan et al., 2006, p. 83).

P6 Use the **ldbounds** R package to show that the constants for one-sided boundaries in terms of the $Z(t)$ function using the O'Brien–Fleming-type spending function at information times $t_1 = 0.2$, $t_2 = 0.5$ and $t_3 = 1$ are respectively, $c_1 = 4.877$, $c_2 = 2.963$, and $c_3 = 1.969$. What are the probabilities of rejecting the null (under the null hypothesis) at each of those three time points? What are the constants when $t_1 = 0.7$ and $t_2 = 1$?

P7 Show Equation 18.11. Hint: assume $\alpha_1 < c_\alpha < \gamma$. Then write the probability of rejecting in any stage as

$$\alpha = \int_0^1 \int_0^1 I(p_1 < \gamma)I(p_1 p_2 \le c_\alpha)dp_2 dp_1.$$

Then partition the integration over p_1 into three parts: $(0, \alpha_1]$, (α, γ), and $[\gamma, 1)$.

P8 Derive c in Expression 18.12.

P9 Letting $\alpha_1 = 0.01$, $\gamma = 0.20$, and $\alpha = 0.025$, solve for c in Expression 18.13. Hint: there may not be a closed-form solution, and you may have to use a computer, for example using the uniroot function in R. How does c compare to $\Phi^{-1}(1 - \alpha)$? Also, compare the two-stage procedure defined by Expression 18.13 with rejecting based on the usual weighted combination of normals (with one-sided tests and one-sided level α), where we reject when

$$\frac{w_1 \Phi^{-1}(1 - p_1) + w_2 \Phi^{-1}(1 - p_2)}{\sqrt{w_1^2 + w_2^2}} \ge \Phi^{-1}(1 - \alpha).$$

19

Testing Fit, Equivalence, and Noninferiority

19.1 Overview

As we have shown throughout the book, the typical aim of a statistical hypothesis test is to reject the null hypothesis. The language of tests emphasizes that. We show a "significant effect" when we reject a null hypothesis of no effect. Failing to reject the null hypothesis does not necessarily mean that the null hypothesis is true, it could be that there are not enough data to show that the null hypothesis is unlikely. But what if our aim is to show that the null hypothesis is true? That is what this chapter is about.

There are two large classes of situations where we want to show that the usual null hypothesis is true. The first class is goodness-of-fit tests, where the null hypothesis is that the model fits the data well and the alternative is usually not formally specified and includes all the different ways that the model does not fit. The second class is equivalence tests, where we want to show two parameters are very close. Usually, when comparing two parameters with a two-sided test, the null hypothesis is that the two parameters are equal, and the alternative is that they are not equal. This is appropriate when we want to show that the parameters are not equal, but not when we want to show that they are equal. In order to do this using a statistical hypothesis test, we need to reframe the question by switching hypotheses, i.e., do something like making the former null hypothesis into the alternative hypothesis and the former alternative hypothesis into the null hypothesis. We also discuss noninferiority tests, which are closely related to equivalence tests.

Consider first a special type of goodness-of-fit test, testing whether a sample comes from a certain class of distributions. For example, suppose we have a model that assumes that the errors are normal, and we want to demonstrate that the errors are in fact normal. Can we frame the hypotheses such that the null hypothesis is that the error distribution is one of a set of any nonnormal distributions, and the alternative is the set of normal distributions? Unfortunately, under that framing, it is not possible to create a test with power substantially greater than the α-level (Note N1). We can, however, test the null hypothesis that the errors are normal. The problem with these kinds of tests is twofold. First, the alternative hypothesis is not what we want to show. We can demonstrate that a distribution is not normal, but not show that it is normal. Second, in some cases when the true distribution is not normal, it will be close enough to normal for our purposes (see, e.g., Example 8.3, page 127). In nearly all cases where we assume normality for convenience, the methods will still be useful if the true distribution is close to normal. But if we have enough data, we can reject the null hypothesis of normality, even when the distribution is close enough to normal. In that case, the hypothesis test of normality is not necessarily helpful. That is, if the sample size is small, it will be difficult to demonstrate deviation from normality and vice versa but concerns about nonnormality are much greater if the sample size is small. Thus, loosely speaking, it is easy to detect nonnormality when it does not matter, and hard to detect nonnormality when it does matter.

Similar issues arise for more general testing of the goodness of fit of a model. It is easy to test the null hypothesis that the model fits the data well, so that we can show the alternative of a lack of fit. Showing a lack of model fit is sometimes used to decide on whether to expand the model to a more complicated one. In that case, the alternative is the more complicated model, and rejecting the null hypothesis may be used to justify greater complexity. But analogously to the case of testing normality, with large samples we may show lack of fit of a null model even when that null model remains useful because the fit is good enough for your purposes. So Section 19.2 discusses testing distributions and fit of models, but recommends against performing those tests indiscriminately in practice. Sometimes a graphical test can be used to show that the fit is close enough.

Now consider the class of equivalence tests, where because we want to show that two parameters are equal, we put the equality of the two parameters in the alternative hypothesis space. Instead of framing this as a null and alternative hypothesis, consider the closely related problem of calculating a 95% central confidence interval on the difference in the parameters. If we can show with 95% confidence that the difference is 0, then we have shown with the same degree of confidence that the two parameters are equal. Now we see the problem. With real data and when the parameter space is continuous (as is almost always the case in practice), then getting a 95% confidence interval of $[0, 0]$ for the difference in two parameters is not possible. So to try to show with 95% confidence that the two parameters are equal is a fool's errand. The best we can do is to show that the difference is within some margin of equivalence. Let M be a prespecified margin, then if the 95% central confidence interval for the difference in the parameters is (L, U) and $-M < L < U < M$, then we are 95% confident that the true difference in means is within the equivalence margin. Because confidence intervals can be described as a series of hypothesis tests, we can interpret when the confidence interval is entirely contained within the margins as a rejection of the null hypothesis that the difference in means is outside $[-M, M]$. This is what we mean by "switching the null and alternative," because the null hypothesis usually contains the value of no effect (e.g., the difference in parameters equals 0).

To rigorously perform an equivalence test, the margin should be prespecified. In fact, if there is no prespecified margin and you want to show two parameters are equal, we recommend not performing an equivalence test at all. Instead, we suggest presenting a 95% confidence interval on the difference to show how precisely we can be confident about the difference. Even if the confidence interval includes 0, that inclusion does not give all the relevant information; a large 95% confidence interval covering 0 is very different from a very tight 95% confidence interval covering 0.

The reason why this problem of showing equivalence of two parameters (within a margin) deserves its own section, is that there are many challenges in designing and interpreting such a study. In Sections 19.3 and 19.4 we discuss several of those issues with respect to comparing a new therapy to the standard therapy for some application. In Section 19.3 we discuss showing equivalence in two therapies, for example, if a new drug is a generic version of the standard treatment drug. In Section 19.4 we consider the more common situation where, when comparing a new treatment to a standard, we typically just want to show noninferiority, i.e., that the new treatment is not worse than the standard treatment within some margin. Some of the complications with these types of studies are the following: (1) how to decide on the margin, (2) how changing the study population can lead to inflation in the Type I error rate, and (3) similarly, poor adherence to treatments might also lead to such inflation.

We present all the noninferiority (or equivalence) hypotheses in terms of the difference in two parameters, but other parameterizations are analogous and are not presented. For example, operationalizing the equivalence or noninferiority test using the ratio of two positive parameters, so that the parameters are equal when their ratio is 1, and the equivalence margins may be $1/M^*$ and M^*.

19.2 Testing Distributions and Lack of Fit

19.2.1 Testing for Normality

Suppose we have a sample of 100 observations, and we want to know whether it comes from a normal distribution. Recall that if $Y \sim N(\mu, \sigma^2)$, then we can write it as a scaled and shifted standard normal distribution, meaning that $Y = \mu + \sigma X$, where $X \sim N(0, 1)$. Thus, when we plot the histogram of a normal distribution, its shape will look the same regardless of the values of μ and σ. Those parameter values just change the axis labels by stretching them according to σ and shifting them according to μ. When we test for normality, we are not testing a particular normal distribution, but testing a class of distributions with any values of μ or σ.

In Figure 19.1 we plot histograms from simulated data from three different distributions: a normal distribution, a t-distribution with five degrees of freedom, and a log-gamma distribution with shape parameter 0.2. For determining nonnormality, we only need to look at the shape of the plot and can ignore the axis labels. Histograms are crude instruments to detect nonnormality. In Figure 19.1, the histogram from the normal sample looks like it might be slightly asymmetrical, but it is not clear if that is just what one would expect from a sample from a normal distribution. Histograms can pick up clearly nonnormal data (the one from the from the log-gamma distribution looks much too asymmetrical to be normal), but data that are close to normal but are nonnormal are hard to detect by histogram (t-distribution looks like it has perhaps slightly bigger tails than a normal but could that be due to randomness?).

An easier plot to detect nonnormality is a Q-Q plot. In the Q-Q plots of Figure 19.1 we plot the sample quantiles by the predicted quantiles given a hypothesized normal distribution. As discussed in Section 5.2, with discrete distributions quantiles are not unique. Hyndman and Fan (1996) define nine ways of calculating sample quantiles, and we discuss two of them in this section. For simplicity, we consider the case where there are no ties. Let $\hat{F}(t) = n^{-1} \sum I(y_i \leq t)$ be the empirical distribution evaluated at t from a sample $y_1, \ldots, y_n$. Let $y_{(j)}$ be the jth largest of the n values in the sample, then $\hat{F}(y_{(j)}) = j/n$, and we can define the (j/n)th quantile of the sample as $y_{(j)}$. We see a kind of asymmetry, where the largest value is the 100th percentile, but the smallest value is not the 0th percentile, but is the $100 * \left(\frac{1}{n}\right)$th percentile. To obtain a type of symmetry, we can use the normalized empirical distribution,[1] $\hat{F}^*(t) = n^{-1} \sum \{I(y_i < t) + 0.5I(y_i = t)\}$, so that $y_{(j)}$ is the $\{(j - 0.5)/n\}$th

---

[1] The term *normalized* comes from Brunner and Munzel (2000) and is not related to the normal distribution.

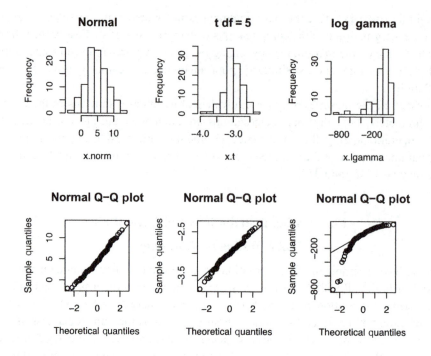

Figure 19.1 Histograms and Q-Q normal plots of simulated data ($n = 100$) from three distributions. The simulated data is the same in each column.

quantile of $\hat{F}^*$. This is the method used in Figure 19.1 where the Q-Q normal plots used the following R code from base R (where y is vector of observations): qqnorm(y); qqline(y). The qqnorm(y) plots the points $(\Phi^{-1}(\frac{j-0.5}{n}), y_{(j)})$, where $\Phi^{-1}(q)$ is the qth quantile of the standard normal distribution, and qqline(y) draws the line that connects the 25th and 75th percentile points.

In Figure 19.1 we see that as expected, the Q-Q plot from the normal sample looks like a line through almost all of the points, the Q-Q plot for the sample from the t-distribution shows that the extreme points do not match the expected normals well, and the Q-Q plot for the sample from the log-gamma distribution clearly shows lack of fit with the expected normal distribution.

There are many formal tests for normality, and a good one is the Shapiro–Wilk test. We relegate the discussion of that test to Note N2, because we want to emphasize again that although we want to show that the data are approximately normal, the test can only show significant lack of fit from normality. Further, even if the data are significantly different from normal, it may still be close enough to normal for your purposes, hence graphical tests such as the Q-Q normal plots are useful. Sometimes the data can even be very far from normal but that is close enough. For example, in Section 5.3.2, we spoke about the one-sample t-test, that is derived using normality. In that case, we showed that even when the data appear nonnormal (such as t-distribution with very low degrees of freedom, or a log-gamma distribution), the Type I error rate for some one-sided tests on the mean may still be conservative for the t-test and hence will bound the Type I error rate. The two-sample t-test also can perform quite reasonably with a large sample from nonnormal data, even though it is derived under the assumption of normality.

19.2.2 Testing Any Distribution

Consider the case where $Y_1, \ldots, Y_n$ are independently distributed from F, and we want to test $H_0 : F = F_0$, where F_0 is known. In this case an important test is the *Kolmogorov–Smirnov test*, which rejects based on the maximum distance of the difference between the empirical distribution $\hat{F}$ and F_0. Let

$$T_n = \sup_t \left| \hat{F}(t) - F_0(t) \right| \tag{19.1}$$

be the Kolmogorov–Smirnov test statistic. If F_0 is continuous, the distribution of T_n under the null hypothesis does not depend on F_0 and depends only on n, and we can calculate a constant $c_{n,\alpha}$, where $\Pr[T_n > c_{n,\alpha}] = \alpha$ when $F = F_0$. Further, if we reject when $T_n > c_{n,\alpha}$ then the Type I error rate will be $\leq \alpha$ even if F is discrete (see, e.g., Lehmann and Romano, 2005, p. 584). Although it is rare to test a known F_0, the test can be inverted to create a confidence band for F (see, e.g., Hollander et al., 2014, Section 11.5). A test for normality can be done, even though in that case we must expand the null hypothesis to include the class of all normal distributions (see Note N3).

19.2.3 Testing Model Fit

In Section 10.7 we described the likelihood ratio test, where the null hypothesis is a simpler model, and the alternative is a more complicated model that includes the simpler model as a special case. With generalized linear models such as the logistic model, the likelihood ratio test is sometimes called an *analysis of deviance* test. For these tests, if we reject the null hypothesis at level α, then we say the complicated model significantly improves the fit. Here "significantly" is used in a technical sense, meaning we reject the null hypothesis, but this does not mean that the more complicated model substantially changes the way we understand the application; it could enlighten us substantially, but it may not.

Example 19.1 *Consider a randomized trial of a new treatment versus placebo and the response is binary (improved or not). Suppose that in the design stage, there was reason to suspect that the treatment may have different effects based on sex. We design the study to have about equal numbers in each sex. We prespecify that the primary test is for a treatment effect ignoring sex, and as a secondary test, we test for a sex by treatment interaction. For the secondary endpoint, we test whether the more complicated model that allows a different treatment effect for each sex significantly improves the model fit compared to the simpler model that says the treatment works the same regardless of sex. We fit logistic regression models, and the treatment effects are in terms of odds ratios: the odds of improvement on treatment over odds of improvement on placebo. Suppose the primary test shows that the treatment is significantly better. Consider two scenarios where we reject the null hypothesis that the treatment works the same in both sexes. In the first scenario, the estimated odds ratio for females is 2.7 and for males is 1.9, suggesting that the new treatment works better than placebo for both sexes. In the second scenario, the estimated odds ratio for females is*

1.7 and for males is 0.9, suggesting that the new treatment works only for females. Although in both scenarios there is a significant difference in treatment effect between the sexes, in the first one the differential effect may not be important because the treatment appears to work in both sexes, while in the second one the differential effect is important because the treatment may not be effective in males.

The likelihood ratio test can be a useful way to test for many effects at once and to properly adjust for multiple comparisons.

Example 19.2   *Suppose we perform a follow-up study to the previous example. In this case, we suspect there is a sex difference in the treatment effect, and we wonder if the difference is working through the sex hormones. In this study all subjects of both sexes get the treatment, each subject has a binary response (improved or not), and we measure three sex hormones at baseline: estrogen, progesterone, and testosterone. There are no treatment versus placebo effects, and our focus is on whether sex hormones affect the outcome. The simple model is a logistic model with only two parameters (an intercept and a sex main effect), while the complicated model has eight parameters (one intercept, one sex main effect, three hormone main effects, and three sex-by-hormone interactions). We perform a likelihood ratio test to see whether the hormones have any effect on the treatment response. This allows us to combine into one test the effects of all three hormones and their interactions with sex.*

Consider a different type of goodness-of-fit test known as a *pure significance test* (Cox and Hinkley, 1974), where the alternative hypothesis is not defined. A famous example of a class of pure significance tests are Pearson's chi-squared goodness-of-fit tests for contingency tables. There are many ways to create such a chi-squared test for goodness of fit, and we describe the procedure in a general way to encompass many of them. Suppose we have m cells in the contingency table. For example, for a two-way table with r rows and c columns, then $m = rc$. To allow for multidimensional tables, we index the cells from 1 to m. Let n_j be the observed count in the jth cell and let N_j be the associated random variable and we assume the N_j are independent Poisson with mean μ_j. Let $N = \sum_{j=1}^{m} N_j$, and the distribution of $[N_1, \ldots, N_m]$ given $N = n$ is multinomial with parameters n and $[\mu_1/\mu., \ldots, \mu_m/\mu.]$, where $\mu. = \sum_j \mu_j$. So a Poisson model with an intercept gives equivalent inferences as a multinomial model (see, e.g., Agresti, 2013, Section 9.6.8). An application that motivates a multinomial model is a simple random sample from a population.

Consider a model with a $q \times 1$ parameter vector θ, that is estimated with $\hat{\theta}$. Estimate μ_j with $\hat{\mu}_j \equiv u_j(\hat{\theta})$. Pearson's statistic is

$$X^2 = \sum_{j=1}^{m} \frac{(n_j - \hat{\mu}_j)^2}{\hat{\mu}_j}. \tag{19.2}$$

If $\hat{\theta}$ is the maximum likelihood estimator, then under the null hypothesis that the model is correctly specified (i.e., there exists a θ such that $u_j(\theta) = \mu_j$ for all j) then the random variable associated with X^2 is asymptotically distributed chi-squared with $m - q$ degrees of freedom, as n goes to infinity. This hypothesis is a pure significance test since the alternative hypothesis is not specified. When we reject the null hypothesis, we have shown that the model does not fit the data well. This is similar to the discussion of the tests for normality; we want to show the model fits well (i.e., show that the null hypothesis is true), but we can only show when it does not fit well.

Unfortunately, Pearson's chi-squared test may have poor validity if the sample size is small compared to the number of cells, so that there are many cells with small counts (see Agresti, 2013, Section 3.2.3). For the two-way contingency table, a common model is the independence model, where the mean for each cell is the scaled product of the sum of the means for that row and the sum of the means for that column (Problem P1). For small samples we can simulate the p-value, or instead use the closely related Fisher's exact test (also called the Freeman–Halton test). The latter test is a generalization of the Fisher–Irwin test for 2×2 tables to $r \times c$ contingency tables (see Section 11.2.1, page 198). Lydersen et al. (2007) give more details, presenting a mid-p version of Fisher's exact test, and comparing by simulation several of these independence tests for $r \times c$ tables.

Hosmer and Lemeshow (1980) developed a goodness-of-fit test for logistic regression that uses a Pearson statistic. Fit the logistic regression on $y_1, \ldots, y_n$ binary responses to get corresponding predicted responses, $\hat{\pi}_1, \ldots, \hat{\pi}_n$. Because the logistic regression model may have continuous covariates or many categorical covariates, it is possible that there are n unique predicted responses, in which case we have too many categories to perform a Pearson chi-squared test. To address this issue, we partition the individuals into g groups based on their predicted response. Let J_1 the be indices of the n/g lowest (to the nearest integer) predicted responses, J_2 the indices for the next group, and so on. Create a $g \times 2$ contingency table with the ith row equal to $[n_i, n_{g+i}] \equiv [\sum_{j \in J_i} (1 - y_j), \sum_{j \in J_i} y_j]$, and the corresponding expected values equal to $[\hat{\mu}_i, \hat{\mu}_{g+i}] \equiv [\sum_{j \in J_i} (1 - \hat{\pi}_j), \sum_{j \in J_i} \hat{\pi}_j]$. Letting the number of cells equal to $m = 2g$, then the Pearson statistic is as given in Equation 19.2. The Hosmer–Lemeshow goodness-of-fit test compares that Pearson statistic to a chi-squared distribution with $g - 2$ degrees of freedom. The value of g may be adjusted based on the sample size, but usually $g = 10$ is used. This test has been generalized to multinomial and proportional odds regression models (see Fagerland and Hosmer, 2013, and its references).

19.3 Equivalence Tests

Consider two-sample tests which have the goal of demonstrating that two groups are sufficiently close with respect to some parameter. For example, we may want to demonstrate that a new generic drug's expected response (say μ_N) is both not too much larger and not too much smaller than the expected response on the associated licensed drug, which is the standard of care (μ_S). Formulated as hypotheses, these are the equivalence hypotheses,

$$H_0 : |\mu_S - \mu_N| \geq M$$
$$H_1 : |\mu_S - \mu_N| < M,$$

where M is some chosen margin.

Under traditional null hypothesis tests of equality (e.g., where $\mu_N = \mu_S$ for some class of distributions), it is easier to calculate p-values because of simplifications that occur when the parameters are equal. For example, calculating the p-value for the conditional test for the two-sample binary response test is simplified so that the noncentral hypergeometric distribution becomes the usual hypergeometric distribution under the null hypothesis that the two mean parameters are equal (see Section 7.3). That is why *p*-values (calculated under the null hypothesis) are often easier to calculate than confidence intervals. For tests of equivalence, the space of equality (e.g., distributions where $\mu_N = \mu_S$) occurs in the alternative hypothesis, so there is none of that simplification. Because there is no simplification in the null hypothesis, we present these noninferiority tests using the confidence interval on the difference in parameters, say $\Delta_{SN} = \mu_S - \mu_N$. Then we reject the null hypothesis at level α if the $100(1 - \alpha)$ two-sided confidence interval on Δ_{SN}, say (L, U), completely excludes all points in the null hypothesis. Equivalently, reject the null if $-M < L < U < M$. Thus, as long as we have a method for calculating a two-sided confidence interval on the difference in parameters, it is straightforward statistically to perform the test of equivalence given M is specified.

If we fail to show equivalence, the direction of failure may be important (i.e., whether the confidence interval covered $- M$ is practically much different from whether it covered M, or whether it covered both). Typically, the $100(1 - \alpha)$ central confidence interval is recommended so that we can bound the error at $\alpha/2$ on each side. In other words, for a valid 95% central interval on Δ_{SN}, we know that there is $\leq 2.5\%$ probability that the lower limit will be greater than Δ_{SN}, and also there is $\leq 2.5\%$ probability that the upper limit will be less than Δ_{SN}.

In the case of comparing two medical treatments, except in the special case of evaluating a generic drug, tests of equivalence are rare because often this sort of question is naturally one-sided. Say we have an existing standard treatment for ear infections. Another new treatment is developed which is much cheaper. We would want to use the new treatment in our clinical practice, as long as the proportion of patients with resolved infections is not *lower* with the new treatment than the standard. In contrast to the equivalence case above, there is obviously no problem if the success rate is higher with the new treatment. So, in this scenario, we would do a test of noninferiority to confirm there is little *loss* in efficacy. When comparing two different treatments, noninferiority testing is much more common than equivalence, and we cover that in Section 19.4. Many of the issues discussed in relation to the noninferiority tests will also apply to equivalence tests, but for ease of exposition we only discuss them in relation to noninferiority tests.

19.4 Noninferiority Tests

19.4.1 Overview

When higher values of the parameter μ imply better treatment results (e.g., proportion cured), the associated non-inferiority hypotheses are

$$H_0 : \Delta_{SN} \geq M$$
$$H_1 : \Delta_{SN} < M, \tag{19.3}$$

where $\Delta_{SN} = \mu_S - \mu_N$ is the difference between the standard therapy (μ_S) and the new therapy (μ_N). The smaller the value of the margin, M, the more stringent the test. If we chose the M to be 0 for a noninferiority test, this would be identical to a customary one-sided test of difference. That is, if we cannot tolerate any loss in efficacy whatsoever of the cheaper drug, we are actually forced to demonstrate that the new drug has superior efficacy. As with the equivalence test, once we have a confidence interval procedure for the difference in parameters, it is straightforward statistically to perform the test. For noninferiority tests we use either one-sided confidence intervals, or *central* (two-sided) confidence intervals where we know the error bounds on both sides of the interval. For example, with the Hypotheses 19.3 defined with respect to Δ_{SN}, using a 97.5% one-sided interval on Δ_{SN} is equivalent to using a 95% central interval on Δ_{SN} for the noninferiority test, and we would reject at the 2.5% level if the upper limit of either interval is less than M.

The challenge of equivalence and noninferiority tests are not with the computation of the procedure, but relate to making sure that the design, conduct, and interpretation of the tests are appropriate. First, it is especially critical that such studies be rigorously designed and conducted, and second, there needs to be consideration of what the noninferiority result implies about the comparison of the new treatment to placebo.

Consider first the consequences of poor study design and conduct: it can make it harder to show efficacy in superiority trials but can make it easier to falsely demonstrate noninferiority. At a ridiculous extreme, consider a situation where there is a subjective binary endpoint and all individuals are judged to be successes, or another situation where no individual complies with either treatment. In both cases, any hopes of showing superiority would be sunk, but in contrast, noninferiority would almost certainly be demonstrated with respect to the primary statistical analysis. In these cases, the problem would be so evident that any critical evaluation would lead to the conclusion that the results are not to be trusted. In practice, much more subtle variants of these situations could lead to apparent similarity between the arms that might disguise important differences, but these subtle problems might not be so easy to uncover. Trials that are well designed and conducted and that have ability to detect differences between trials are said to have "assay sensitivity." We discuss this issue further in Section 19.4.6.

Now consider interpretational issues with noninferiority trials and placebo comparisons. While there will always be some interest in the head-to-head comparison of two active treatments, often the primary goal of the noninferiority trial is to indirectly demonstrate that a new treatment is better than placebo. This occurs in settings where a randomized placebo-controlled trial is not considered ethical, because an approved treatment is available. Thus, the margin needs to be chosen to be small enough such that if the new treatment is demonstrated to be within M of the standard treatment, this implies that the new treatment is better than placebo. In other words, there should be existing evidence that the standard treatment is better than placebo by at least M, so that rejecting the null hypothesis in the noninferiority trial provides evidence that the new treatment is superior to placebo. In contrast to a superiority trial, the conclusions of efficacy derived from a noninferiority trial hang on external information (e.g., the size of the standard treatment effect compared to placebo), and whether this external information from the past remains relevant to the noninferiority trial. Section 19.4.3 further addresses this issue.

For the rest of this section we further address the above challenges in addition to some other common concerns with comparing an active control (i.e., the standard of care) to a new treatment with a noninferiority test as stated in Hypotheses 19.3.

19.4.2 Prespecification of Margin

First, if we do not prespecify the margin, then any declaration of noninferiority is highly suspect. Without prespecification, the naïve analyst could define M to be slightly larger than the one-sided $100(1 - \alpha)$ upper confidence limit so that regardless of the interval, the study would always reject the null hypothesis and declare noninferiority. In order to avoid the appearance that the margin was determined in such a post hoc way, the margin must be prespecified.

What about when you prespecify the test for noninferiority, but the results show a significant superiority? For example, suppose we test $H_0 : \Delta_{SN} \geq M$ where $\Delta_{SN} \equiv \mu_S - \mu_N$, and suppose U, the upper limit of the one-sided $100(1 - \alpha)$ confidence interval for Δ_{SN} is both $U < M$ (implying noninferiority at level α) but also $U < 0$ (implying superiority at level α). Is it valid to declare superiority of the new treatment at level α, given that we have not prespecified that we will do a superiority test? Yes, this is valid, because we cannot redefine the superiority margin (it is always 0!), so there can be no appearance that we have redefined the superiority margin to show significance. See Note N4 for a different justification for validly testing superiority from a noninferiority test.

19.4.3 Relating Results to Placebo Effects

As mentioned above, when selecting the margin, we need to ensure that the margin is set sufficiently small. In the case of comparing the standard treatment to a new treatment, if the margin is too large then there is a possibility of showing that the new treatment is noninferior when it is no better than placebo. To explain this consideration, define the mean response for the three groups as: μ_P for the placebo group, μ_S for the standard treatment group, and μ_N for the new treatment group, with larger means implying better average response. Let the noninferiority hypotheses be

$$H_0 : \Delta_{SN} \geq M$$
$$H_1 : \Delta_{SN} < M.$$

If we define the margin to be less than the distance between the standard treatment and placebo (i.e., $M < \Delta_{SP}$), then the alternative hypothesis implies that the new treatment is better than placebo (i.e., $\mu_N > \mu_P$). To see this, we draw two pictures:

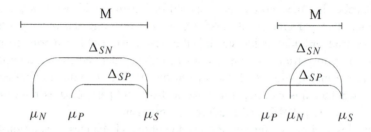

The left drawing has $M > \Delta_{SP}$ which allows the alternative hypothesis ($\Delta_{SN} < M$) to have $\mu_N < \mu_P$, while the right drawing has $M < \Delta_{SP}$ which forces $\mu_P < \mu_N$ whenever the alternative hypothesis ($\Delta_{SN} < M$) is true.

The requirement that $M < \Delta_{SP}$ gives an upper bound for M, but typically M is not selected to be close to the bound. For example, for cardiovascular outcome studies the usual practice has been to aim to use $M = \Delta_{SP}/2$ (FDA, 2016). The specific choice of M depends on the application. In practice, Δ_{SP} is not known and must be estimated from historical trials. One conservative strategy is to use the lower limit of a central 95% confidence interval for Δ_{SP}, say $L_{\Delta_{SP}}$ to define M. For example, if we want to define $M = \Delta_{SP}/2$, then use $L_{\Delta_{SP}}/2$. Here the idea is to be conservative, not to get the best estimate of Δ_{SP}. The rationale behind this conservatism is that there is always uncertainty about how well the estimate of Δ_{SP} from historical trials applies to the noninferiority trial.

19.4.4 Constancy Assumption

A major concern with noninferiority designs is that the effect of the standard treatment may change over time, due to other changes in medical care, for example. Thus, the historical standard treatment effect over placebo may be very different from that effect if it were measured today. However, one needs to be able to make the *constancy* assumption, that the treatment effect has remained constant across the time of previous studies and today. If constancy does not apply, it may not be possible to set a margin for a noninferiority study, since there is no good estimate of the standard treatment versus placebo effect.

If there are inadequate historic data to serve as a basis for a conservative estimate of Δ_{SP} that will be relevant to a noninferiority study being designed today, then the noninferiority design cannot provide indirect evidence against placebo. In such scenarios, where randomization to placebo is not ethical, trial designers should consider alternate superiority designs, including a randomized comparison between immediate and delayed treatment, or a superiority trial comparing the new treatment plus the standard versus standard alone, or, of course, a superiority trial comparing the two treatments. Novel designs, specific to a given setting, might be developed; for example, Follmann et al. (2013) proposed such a superiority approach unique to antibiotics. See Note N5 for the details.

A closely related concern is *biocreep*, where a succession of noninferiority trials could potentially lead to new treatments with weaker and weaker effects over time. This concern can be alleviated by always using the best treatment available among the previously tested treatments (D'Agostino Sr. et al., 2003).

19.4.5 Noninferiority Methods

All noninferiority tests discussed so far have been *fixed margin methods*, where the determination of the margin is fixed during the design of the noninferiority study. FDA (2016) recommends fixed margin methods in order to separate the justification of the margin using the historical data from the analysis of the noninferiority trial data. Another statistical possibility is to combine into one analysis the historical data of the standard treatment versus placebo and the noninferiority study data of the new treatment versus standard treatment. There are two methods that combine both the historical and noninferiority study data, a more traditional one called the synthesis method (see, e.g., FDA, 2016), and a more rigorous method called the full random effects (FRE) method (Brittain et al., 2012).

Because of the confusing terminology, we clarify that both the fixed margin and synthesis methods can be based on meta-analysis of historic studies assuming either fixed or random effects. Brittain et al. (2012) proposed the FRE method under random effects of historic trials that additionally accounts for variability of the true standard versus placebo effect in the *noninferiority trial* itself, where the traditional methods that assumed random effects did not. It is considerably harder to demonstrate that the new treatment is better than placebo with the FRE method than other methods, but Brittain et al. (2012) showed that conventional methods have substantial Type I error rate inflation under full random effects (i.e., random effects for both the historical trials and the noninferiority trial). We believe that if a random effects model is assumed for the historical data, then it is appropriate to also assume random effects in the noninferiority study. The low power of the FRE method makes it daunting to use in practice, but this is not a flaw of the method, but, rather, a reflection that indirectly demonstrating efficacy over placebo is very difficult when there truly are random effects across trials, especially if there are few historic trials. Further, it is difficult to rule out random effects when there are few historic trials. This is another illustration of the challenges of rigorously determining efficacy of a new drug over placebo in this framework.

19.4.6 Assay Sensitivity

The general concept of *assay sensitivity* must be considered in noninferiority trials, as it is essential. A study of a therapy has good assay sensitivity if it has "the ability to distinguish an effective therapy from an ineffective therapy" (Rothmann et al., 2012, p. 15). For example, a randomized *superiority* study comparing a new treatment to placebo has good assay sensitivity if it has adequate sample size (see Chapter 20) and is well run. Consider several situations where a superiority trial would have poor assay sensitivity. First, suppose that the population studied had many very poor compliers who usually did not take their therapy (either the new treatment or the placebo). In that case, the lack of compliance would dilute the strength of the treatment effect and diminish the assay sensitivity. Second, suppose the population for which the therapy is designed is diluted with people for which the therapy will have no effect. For example, consider a superiority trial of a new treatment against placebo for treating a lung bacterial infection that causes shortness of breath, and the response is a measure of breathing by spirometer after one day of treatment. If the study population had a large portion of people with shortness of breath due to causes other than the lung bacterial infection, then this would dilute the apparent efficacy of the new treatment in the study and decrease the assay sensitivity of the superiority trial. A third example of a case of poor assay sensitivity is when the measurement of the primary response was poor. For example, the spirometer was not functioning correctly in the previous example. For a superiority trial, all three of these situations (poor compliance, dilution of effect through population selection, and poor measurement) could decrease the power of the study when the alternative is true (i.e., could increase the Type II error rate), but it will not increase the Type I error rate. Thus, we say the superiority trials remain valid[2] in these situations, meaning that as long as all the

---

[2] Here the term "valid" means that the study is both scientifically valid (i.e., measuring what is claims to be measuring) and statistically valid (i.e., if the statistical assumptions hold then the *p*-values properly bound the Type I error rate and confidence intervals give proper coverage probability). See Section 3.3.3 for a discussion of the two meanings of valid.

statistical assumptions hold and the statistical procedures are valid under those assumptions, the rate for declaring that an ineffective therapy works will remain at or below the α-level.

For a noninferiority trial to have assay sensitivity essentially means that had there been a placebo group included, that the difference between the standard treatment and the placebo would be at least as large as the conservative estimate of Δ_{SN} determined from the historical trials (i.e., the basis for the selected noninferiority margin). The three situations discussed in the previous paragraph, which reduce power but leave trial validity intact for superiority trials, can produce validity problems for noninferiority trials. That is, any of these situations dilute the differential treatment effects between the new and standard treatments on the population of interest, thereby shrinking the estimate of the difference, Δ_{SN}, closer to 0. Since $\Delta_{SN} = 0$ is in the alternative space, any of these situations can produce invalid results, thus increasing the rate of false declarations of noninferiority.

19.4.7 Compliance

First, we review the situation with randomized superiority trials when the compliance is all-or-none (i.e., partial compliance is not possible, such as getting a transplant or not) that was previously discussed in Section 15.4.3. In that section, we gave a detailed analysis of two estimands from such a trial. The first is the average causal effect of getting randomized, say Δ_{ITT}, which is known as the intention-to-treat (ITT) effect on the outcome. The second is the average causal effect among the compliers, say Δ_c, which measures the effect of the treatment if properly taken among those who would properly take the treatment if offered. Typically, for superiority trials we prefer an analysis based on Δ_{ITT}, because it is simpler to perform a statistically valid analysis and it measures the effect of the treatment on the whole population.[3] Because the analysis based on Δ_c is more difficult and applies only to a subset of the study population, it is less often used. Further, in many situations there is partial treatment compliance over time which adds another level of complexity (see Hernán and Robins (2020) for causal analyses in those situations). In contrast, the *naïve per-protocol analysis* simply discards any subject that does not have adequate compliance and treats the remaining subjects as if they were the only ones randomized (i.e., treats the noncompliers as missing completely at random). In almost all scenarios, it would be difficult to be assured that the naïve per-protocol analysis is providing a proper analysis because the resulting two analysis groups left in each arm are not ensured to be comparable by randomization. Those two analysis groups are probably not comparable because it is unlikely that the set of subjects that fail to comply when randomized to new treatment would also have failed to comply if randomized to placebo, and vice versa.

The noninferiority trial with poor treatment compliance has all of those complexities of the superiority trial, but furthermore poor compliance dilutes the ITT standard versus new treatment effect leading to a high rate of false claims of noninferiority. Because of this, some

---

[3] Sometimes two different terms are used to measure the performance of an intervention. The "efficacy" is the performance under ideal circumstances and "effectiveness" is the performance under "real-world" conditions (Singal et al., 2014). In this situation, the estimand, Δ_{ITT}, is in the context of a clinical trial (so in some sense the intervention is not a real-world condition since the staff are often more highly trained and more time may be spent on evaluating the subject's health), but it is also not an ideal circumstance because some subjects do not comply. For this reason, we avoid either term.

have advocated using the naïve per-protocol analysis instead, suspecting that it would lead to fewer false claims of noninferiority. Besides the fact that the naïve per-protocol analysis is not protected by randomization, Brittain and Lin (2005) showed that it does not even necessarily reduce the false claims of noninferiority compared to the ITT analysis.

19.5 Summary

This chapter addresses two different cases where we want to show that the usual null hypothesis is true. In the first case, we want to demonstrate that the model fits the data well. In the second case, we want to demonstrate that two study arms are similar.

When assessing whether data are normally distributed, we encourage the use of graphical methods. While the Shapiro–Wilk test is a good formal test of normality, the test can only show significant lack of fit from normality. This is problematic because in the small sample case, where deviations from normality may be important, there will be low power to demonstrate lack of normality, and in the large sample case, small deviations from normality might be demonstrated, but such deviations are generally not important. Thus, one can make the case that testing for normality, when deciding to use a test that assumes normality, has essentially no value. As a more general tool, the Kolmogorov–Smirnov test can evaluate deviations from any distribution but suffers from the same weakness noted for the Shapiro–Wilk test.

When assessing model fit, such as whether more complicated models (with more covariates, for example) provide significantly better fit than simpler models, a likelihood ratio test is a useful way to test for many effects at once in a single test.

The Pearson's chi-squared goodness-of-fit test is very general; it is known as a pure significance test because the alternative hypothesis is not defined. Like the test for normality, it can only demonstrate that the model does not fit the data well. Further, with small sample sizes, the test may have poor validity.

The chapter then shifts gears to settings where we want to demonstrate that two arms are similar. When the goal is to show that two arms are essentially the same, not lower or higher on the endpoint, then an equivalence design would be used. This design might be used in a study of a generic drug, where the goal is to show that the generic essentially replicates the standard treatment. However, if the goal is to show that the new treatment is at least not inferior to the standard treatment, then the noninferiority design applies. The statistical demonstration of equivalence or noninferiority is straightforward, and typically employs a confidence interval that establishes that the difference between arms is smaller than a prespecified margin M. Despite this simple analysis, these designs can yield misleading conclusions. Poor trial design and conduct such as inclusion of patients who will not be affected by either treatment, poor compliance with study treatments, and noisy assessments of the primary trial endpoint would lead to lower statistical power in a superiority trial; however, in a noninferiority context, these poor designs would tend to dilute true differences between the arms, leading to an excessive likelihood of false assertions of equivalence or noninferiority.

Often noninferiority designs are employed to provide indirect evidence of superiority to placebo, when randomization to placebo is unethical. Thus, if there is evidence that the standard treatment is better than placebo on the primary endpoint by at least M and if the new treatment is statistically within M of the standard treatment, we can infer that the new treatment would very likely have been superior to the placebo arm had it been included. However, for this indirect reasoning to hold, one needs two main assumptions. First, the effect size of the standard treatment seen in historic trials is relevant to the new study. Second, the new noninferiority study has assay sensitivity, that is, it has the ability to detect differences between arms, such that if the placebo arm had been included, that the new treatment would have been superior to it. If external data are insufficient to provide a conservative estimate of the efficacy of the standard treatment, then a noninferiority trial aimed at indirect demonstration of superiority to placebo cannot be justified, and an alternative superiority design is necessary. A variety of superiority strategies when randomization to placebo is not possible are suggested. In fact, because of the many vulnerabilities of the noninferiority design in general, we recommend always considering other design options, and that the noninferiority design is the design of last resort, unless the direct head-to-head comparison of the two active agents is the true primary goal of the study.

19.6 Extensions and Bibliographic Notes

Agresti (2013) provides theoretical details for goodness-of-fit tests, and many other analyses important for categorical data. Rothmann et al. (2012) is an important resource for those designing noninferiority trials. D'Agostino Sr. et al. (2003) provide a practical review of noninferiority trials.

19.7 R Functions and Packages

See the generalhoslem R package for Hosmer–Lemeshow goodness-of-fit tests and its generalizations. See the ecdf.KS.CI function in the NSM3 R package that, despite its name, gives an approximate confidence *band* for a distribution based on the Kolmogorov–Smirnov statistic. The chisq.test and fisher.test functions in base R gives Pearson's chi-squared test for independence (including a continuity correction and simulation option) and Fisher's exact test (i.e., Freeman–Halton test) for $r \times c$ tables.

19.8 Notes

N1 Why there is no powerful test to show a distribution is normal: Suppose $Y_1, \ldots, Y_n$ are independent from the distribution F. Let $\mathcal{F}$ be the set of all continuous distributions, $\mathcal{F}_N$ be the set of all normal distributions, and $\mathcal{F}_{\bar{N}}$ be all members in set $\mathcal{F}$ that are not in $\mathcal{F}_N$. Let the null and alternative hypotheses be $H_0 : F \in \mathcal{F}_{\bar{N}}$ and $H_1 : F \in \mathcal{F}_N$. Let $R(\mathbf{Y}, \alpha)$ be a valid rejection region associated with the hypothesis test, i.e.,

$$\sup_{F \in \mathcal{F}_{\bar{N}}} \Pr_F [\mathbf{Y} \in R(\mathbf{Y}, \alpha)] \le \alpha.$$

Then, for any $F \in \mathcal{F}_N$ and any $\epsilon > 0$, let $F_{\delta(\epsilon)} = (1 - \delta(\epsilon))F + \delta(\epsilon)F^*$, where $F^* \in \mathcal{F}_{\bar{N}}$ and $0 < \delta(\epsilon) < 1$ such that $\text{Power}(\delta(\epsilon)) \equiv \Pr_{F_{\delta(\epsilon)}} [\mathbf{Y} \in R(\mathbf{Y}, \alpha)] < \alpha + \epsilon$.

To see that such an $F_{\delta(\epsilon)}$ exists, imagine that $\text{Power}(\delta(\epsilon)) > \alpha + \epsilon$, then just move $\delta(\epsilon)$ closer to 0 until $\text{Power}(\delta(\epsilon)) < \alpha + \epsilon$. So for any F in the alternative space the power cannot be $> \alpha + \epsilon$, where ϵ may be infinitesimally small.

N2 **Shapiro–Wilk test:** Shapiro and Wilk (1965) introduced a test for the null hypothesis that a sample comes from a normal distribution, and it is powerful for detecting nonnormality from many distributions. Shapiro and Wilk use a more mathematically precise derivation of the quantiles from a normal distribution than used in Figure 19.1. Suppose F denotes a continuous distribution and let $Y_{(1)} \leq Y_{(2)} \leq \cdots \leq Y_{(n)}$ be the order statistics. Then by the probability integral transformation and using order statistics methods (see, e.g., Casella and Berger, 2002, p. 230), $F(Y_{(j)}) \sim Beta(j, n - j + 1)$, which is the same as the jth order statistic from a sample of independent variables from a uniform distribution regardless of the distribution F (as long as it is continuous). Let $U_{(1)}^*, \ldots, U_{(n)}^*$ be random variables of order statistics from an independent sample of n uniform statistics (i.e., having nothing to do with our observed sample), then we can use those order statistics to create a sample of order statistics from a standard normal distribution, $Y_{(j)}^* = \Phi^{-1}(U_{(j)}^*)$. The Shapiro–Wilk test uses the means and covariances of the vector of expected normal order statistics $[Y_{(1)}^*, \ldots, Y_{(n)}^*]$ together with the observed order statistics, $[y_{(1)}, \ldots, y_{(n)}]$ and the assumption of normality to get an estimator of σ^2, say $\tilde{\sigma}^2$. Then Shapiro and Wilk create a statistic $W_n = c_n \tilde{\sigma}^2 / \hat{\sigma}^2$, where $\hat{\sigma}^2 = (n-1)^{-1} \sum (y_i - \bar{y})^2$ is the usual estimator of $\text{Var}(Y_i)$ that applies regardless of the normality assumption, and c_n is a constant that depends on n that normalizes W such that its maximum is 1. Large values of W_n are more likely when F is normal. Under the null hypothesis of normality, the distribution of W_n depends only on n. The calculation of W_n and its null distribution is approximated in practice and is complicated, but there is software (e.g., **shapiro.test** in R) that can calculate p-values for this test. If the p-value is $\leq \alpha$, then we can reject the null hypothesis that the sample comes from a normal distribution.

N3 **Lilliefors test:** Lilliefors (1967) developed an approximate test for normality using the Kolmogorov–Smirnov statistic (Equation 19.1). The test for normality has a composite null hypothesis

$$H_0 : F = F_0 \text{ for any normal distribution } F_0.$$

Nevertheless, the Lilliefors procedure uses $\hat{F}_0(t) = \Phi^{-1}((t - \bar{y})/\hat{\sigma})$ for F_0 in the definition of T_n, where $\bar{y}$ and $\hat{\sigma}$ are the sample mean and standard deviation. The distribution of $\hat{T}_n$ does not depend on the mean and variance of F, and the critical value, say $\hat{c}_{n,\alpha}$, can be found through simulation using a standard normal distribution (see also Lehmann and Romano, 2005, p. 589).

N4 **Testing for superiority after noninferiority:** Here is another justification for the validity of being able to test for superiority after testing noninferiority. Let $H_0^{(1)} : \Delta \geq \delta$ be the noninferiority null hypothesis, and $H_0^{(2)} : \Delta \geq 0$ be the superiority null hypothesis. Suppose we had set up the following sequential hypothesis testing procedure (see Section 13.7):

In other words, we first test the noninferiority null hypothesis $H_0^{(1)}$ at level α, if we fail to reject $H_0^{(1)}$ then stop, otherwise reject $H_0^{(1)}$ and pass all of the α on and the test superiority, $H_0^{(2)}$, at level α. This is a valid procedure, meaning it bounds the probability of making any false claim at level α. Since the first step of the procedure is exactly a noninferiority test, we can always implicitly use this sequential hypothesis testing procedure whenever we perform a noninferiority test and allow a "free" superiority test whenever we reject the noninferiority null hypothesis.

N5 **Superiority when justifying a noninferiority margin is difficult:** When there are insufficient external data to estimate Δ_{SP}, a noninferiority design cannot provide evidence of efficacy, so alternate designs must be used. There may be possible superiority analyses unique to specific settings, as illustrated by the following strategy. Antibiotic trials often use noninferiority designs as randomization to placebo is typically deemed unethical. However, because of the lack of historic placebo-controlled trials, justifying a margin M can sometimes be difficult. Follmann et al. (2013) developed a superiority approach to establish efficacy that compares two active treatments even where the overall success outcomes are equal for the regimens. This approach uses the bivariate individual baseline values of minimum inhibitory concentration (MIC) for both the standard and the new antibiotic to identify a subgroup where superiority of the new agent over the standard is most likely to be detected. Since this small subgroup will probably not provide sufficient power, the authors recommended a related modeling approach that uses all the data.

19.9 Problems

P1 Consider a two-way $r \times c$ contingency table with counts y_{ij} and corresponding means μ_{ij} for the ith row and jth column. Let $\sum_i \sum_j y_{ij} = n$. For the independence model, we estimate μ_{ij} with $\hat{\mu}_{ij} = n^{-1} y_{i+} y_{+j}$, where "+" denotes summation over the index. Show that the degrees of freedom of the resulting Pearson chi-squared test for independence is $(r-1)(c-1)$ (see, e.g., Agresti, 2013, p. 76).

P2 Imagine a randomized study of treatment for patients with a bacterial infection, and under one scenario, we are interested in showing the standard treatment is superior to the new treatment, whereas in a different scenario we want to show that the new treatment is noninferior to the standard, where $M = 0.10$. First simulate the probability of showing superiority as well as the probability of showing noninferiority under two different sets of true success rates: 1) $S_s = 0.80$ and $S_n = 0.70$ and 2) $S_s = 0.80$ and $S_n = 0.80$, using one-sided 97.5% confidence intervals and $n = 300$ per arm. Note that the under the first set, the superiority comparison represents the alternative, and the noninferiority comparison is under the null, and vice versa for the second set. Then, repeat the simulation but this time 20% of the enrolled patients will not have the bacterial infection and have true success rate of 0.90, regardless of treatment arm. Finally, repeat the simulation where there is still a 20% probability of lack of bacterial infection, but now in addition, 15% of failures are misclassified as successes. Assemble all the simulated estimates into a table, and this will illustrate as the trial becomes noisier, the Type I error rate is increasingly inflated with the NI analysis, but the power is improved, whereas for the superiority analysis the Type I error remains flat, but the power is diminished. (Note that the power for the noninferiority analysis when the true success rates are both 0.80, will not always improve under more noise, this will depend on the specifics.)

20

Power and Sample Size

20.1 Overview

When designing a study, we want to make sure that we will get useful information from it. Chapter 3 covered many of the design issues related to increasing the usefulness of a study. In this chapter, we tackle the more specific and technical issues of calculating power and determining sample size given we already have chosen a probability model and a hypothesis test procedure. Here, we consider studies where the final planned sample size is fixed in the design stage and there is no interim monitoring for potential early stopping of the study (see Chapter 18 for more flexible methods).

There are many possible goals in estimating the sample size for a study. The most common is to determine the minimum sample size necessary to have at least 80% or 90% power to reject the null hypothesis, given some "alternative model," a specific probability model in the alternative hypothesis. One approach to specifying the alternative model is basing it on a judgment about what minimum treatment effect would be needed to justify use of the test therapy. However, often there is no clear alternative model, so sample size calculations are done for a series of possible alternative models. In some cases, the resources available to do the study suggest the maximum sample size. In that case, we can often reverse the goal and calculate the alternative model closest to the null hypothesis space that gives the appropriate power with that sample size. In order for that not to be an exercise in justifying an underpowered study, the resulting alternative model should be examined to see if it is plausible that the alternative model could be the true data generating mechanism given preliminary information. Then the likelihood of the alternative model being correct should be balanced against the costs of doing the study. For example, even if the alternative model is a very optimistic model and unlikely to be true, if the costs of doing the study are low (and in the case of a clinical trial, if it is deemed ethical for patients to participate in such an underpowered trial), then it may not be unreasonable to spend the resources for the study. The key question is: will this study at the proposed sample size likely improve our understanding of the underlying scientific process in such a way that the information gained is worth the cost?

The following two types of sample size calculations were discussed in Section 4.4 and are not covered in this chapter: studies where we are interested in the minimum sample size necessary to obtain a certain level of precision as measured by the width of the 95% confidence interval, and studies where we want to determine the minimum sample size to determine if at least one event occurred.

The level of verification and examination needed for assumptions related to the alternative model are lower than the level needed for the assumptions of the hypothesis tests in the final model. Often sample size calculations use educated guesses of a plausible alternative model, but the guesses about the magnitude of the true treatment effect may not be based on much data. Nevertheless, the final inferences from the study are the important results, and if the guess at an effect size was poor during the planning of the study, that may affect the amount of information learned from the study, but it will not affect the validity of the study (i.e., make the Type I error rate larger than the nominal α). As we have repeatedly emphasized, failing to reject the null hypothesis is not evidence for the null hypothesis but could be due entirely to lack of evidence against it, and a low powered study will increase the Type II error rate (the rate of not rejecting the null hypothesis when the alternative is true). If the assumptions of the effect sizes during the planning are far from reality and result in an extremely underpowered study, then the study could be a big waste of resources with deleterious consequences. For example, a promising treatment may be put on hold indefinitely if an expensive underpowered study found no significant effect, and that insignificant result decreased the political or economic will to run another properly powered study.

We begin in Section 20.2 with simulation methods for power given the alternative model is known, because they are conceptually simple and can potentially handle a wide variety of hypothesis testing situations. Simulations can be programmed very quickly for simple situations; and while it could take some time to program simulations for more complicated tests, simulation may be the only, or easiest, option to determine power in those settings. Simulation for sample size is slightly more complicated and addressed in Section 20.4 after discussing normal approximations for power and sample size in Section 20.3. The approximations give simple closed-form equations for sample size and hence are very useful and fast. Further, they can approximate a variety of situations. In Section 20.5, we discuss several common complications in sample size calculations.

20.2 Power Calculations by Simulation

We use the very general notation, similar to that used in Chapter 2. Let $\mathbf{X}_n$ represent a possible data vector with sample size n. Here n could be the per group sample size, the total sample size, or could even be a vector of sample sizes for more complicated situations (e.g., cluster randomized trials). Let $P_\theta(n)$ be the parametric probability model describing the data generating mechanism, where θ may be a vector and completely describes the model. Let the null and alternative hypotheses be $H_0 : \theta \in \Theta_0$ and $H_1 : \theta \in \Theta_1$. Let $\delta(\mathbf{X}_n, \alpha)$ be the decision rule that takes on the values 1 (reject the null hypothesis) or 0 (fail to reject the null hypothesis), and α is the significance level.

To simulate the power under the model $P_{\theta_s}(n)$ for some specific model with parameter θ_s, we perform the following the following two steps:

(1) Simulate m data sets, $\mathbf{X}_n^{(1)}, \ldots, \mathbf{X}_n^{(m)}$, each one generated from $P_{\theta_s}(n)$.
(2) Estimate power with $\widehat{pow}(\theta_s, n) = \frac{1}{m} \sum_{i=1}^m \delta(\mathbf{X}_n^{(i)}, \alpha)$.

Because the $\delta(\mathbf{X}_n^{(i)}, \alpha)$ for $i = 1, \ldots, m$ are m independent binary observations with expectation $pow(\theta_s, n)$, we can use the exact central confidence intervals for the binomial parameter to evaluate the precision of $\widehat{pow}(\theta_s, n)$. For various values of m, the maximum expected width of the 95% exact central interval on $pow(\theta_s, n)$ regardless of its true value is (see Section 4.4.3)

$m = 100$	$m = 1,000$	$m = 10^4$	$m = 10^5$	$m = 10^6$
0.202	0.063	0.020	0.006	0.002

Thus, with $m = 10^5$ we can be 95% confident that our simulated power is accurate to within a percentage point of the true power.

Example 20.1 *Student's two-sample t-test compares the means of two samples and was developed for normally distributed observations equal variances. With about 10 lines of* R *code (see Note N1), we simulate the power of the two-sample standard t-test evaluated at the one-sided $\alpha = 0.025$ level when there are $n = 100$ in each group and the difference in means is one half of the standard deviation. For example, each replicate creates $\mathbf{X}_n = \{\mathbf{Y}, \mathbf{Z}\}$ with $\mathbf{Y}$ and $\mathbf{Z}$ both 200×1 vectors. Assume independent responses with $Y_i \sim N(0,1)$ when $Z_i = 0$ and $Y_i \sim N(0.5, 1)$ when $Z_i = 1$, where 100 of the Z_i values equal 1 and 100 of them equal 0. Because the t-test is invariant to location and scale changes, this simulation covers all differences in means equal to $0.5/\sigma$ for any $\sigma > 0$. With 10,000 replications the simulation takes less than two seconds. We repeated this simulation 10 times, and the simulated power ranges from 93.4% to 94.2%. In this case, there is a function in* R, **power.t.test**, *to calculate the exact power, and it is 94.04%. An advantage of the simulation is that you can very easily modify the situation. Suppose we repeat the previous simulation except if $Z_i = 1$ then $Y_i \sim N(0.5, 4)$, and we instead use Welch's t-test to accommodate the differences in variance. (The* **power.t.test** *function does not have an option for that, in* R *version 3.6.0.) The 10 simulated powers range from 59.9% to 60.9%.*

With minor modifications, the same computer code used for the power calculation under the alternative model can be used to check the Type I error rate for models under the null hypothesis.

Example 20.1 *(continued) Since Welch's t-test is based on an approximation (i.e., it and does not have the exactly correct Type I error rate), we repeat the last simulation of the example except with the difference in means equal to 0, and now with 10^5 replications. We perform one-sided tests with 2.5% level of significance, and the simulated power is 2.48%, very close to the target value.*

Although we must assume a parametric model in order to set up the simulation, power can be determined for nonparametric tests by simulation as well. For example, we can simulate the power of the Wilcoxon–Mann–Whitney test given the two samples come from two Poisson distributions. Even though the Wilcoxon–Mann–Whitney test does not require the Poisson model, it is valid under that model (since the associated null hypothesis would have equal Poisson means, and hence equal distributions). Thus, for nonparametric tests we are free to choose the alternative model from among any of the parametric models under which the nonparametric test is valid, basing the choice on plausibility of the assumption and perhaps its simplicity.

For many complicated designs, simulation is often the best way to estimate power. For example, Whidden et al. (2019) simulated power for a cluster randomized trial on an intervention to reduce mortality for children under five in rural Mali. In that case, current estimates of population were used to simulate the population size under five in the proposed study area, and those population estimates were used to check that the expected effect size of the intervention would give reasonable power. Because each study will have its own issues and complexities, we cannot cover them all, but we discuss some important complications in Section 20.5.

We can get accurate power to within one percentage point using 10^4 or 10^5 simulated data sets but estimating sample size is not as straightforward. It might seem that we need to repeat the power simulations many times with 10^4 or 10^5 replicates for each potential sample size, but we can estimate sample size combining much less accurate power estimates for many different potential sample sizes. We discuss that in Section 20.4.

20.3 Power Calculations by Normal Approximations

Because many statistics are approximately normal by the central limit theorem, we can use normal approximations in many situations to compute sample size. The advantage of these approximations is that many times they reduce to simple sample size formulas, and hence are very useful.

Let our hypotheses be,

$$H_0 : \beta \le \beta_0$$
$$H_1 : \beta > \beta_0. \tag{20.1}$$

Suppose, for a sample size of n, we have an estimator $\hat{\beta}_n \sim N\left(\beta, \frac{\tau^2}{n}\right)$, and let $\theta = [\beta, \tau]$. Let $\theta_0 = [\beta_0, \tau_0]$ be the parameters in the null hypothesis (here β_0 is at the boundary with the alternative hypothesis) and $\theta_1 = [\beta_1, \tau_1]$ be the parameters for the alternative model for which we want to calculate the power. We reject the null hypothesis when

$$\frac{\hat{\beta}_n - \beta_0}{\tau_0/\sqrt{n}} > c_\alpha,$$

where $c_a = \Phi^{-1}(1-a)$ is the $(1-a)$th quantile of a standard normal distribution. Then to get $1 - \gamma$ power assuming $\beta = \beta_1$, we solve Pr[Reject the Null] $= 1 - \gamma$ for n, and call the solution n^* (since it may not be an integer) giving

$$n^* = \frac{\left(\tau_0 c_\alpha + \tau_1 c_\gamma\right)^2}{\Delta^2}, \tag{20.2}$$

where $\Delta = \beta_1 - \beta_0$. We round up to the nearest integer to obtain the sample size calculation.

Example 20.2 *Suppose we want to know if a drug increases diastolic blood pressure (dbp). For a sample of n individuals, we will measure their change in diastolic blood pressure after taking the drug. Let y_i be that change for the ith individual, and let $E(Y_i) = \beta$ and assume $Y_i \sim N(\beta, \tau^2)$. When $\beta_0 = 0$, then rejecting the null hypothesis denotes a significant average increase in dbp. Assume the variance is the same under the null and alternative hypotheses, so $\tau_0 = \tau_1 = \tau$. Suppose that if the drug increases the dbp by at least one half of a standard deviation (i.e., $\Delta/\tau = 0.5$), then we want at least 90% power to show that the drug increases* *dbp when tested at the one-sided 2.5% level. Equation 20.2 gives a minimum sample size of* $n^* = 31.4$ *that we round up to* $n = 32$.

Equation 20.2 is an approximation when we do not know τ. In practice, we would estimate σ and use a t-test that accounts for the variability in that estimate, and the critical value would be distributed as t with degrees of freedom that is a function of the unknown sample size. Since we do not know the degrees of freedom before we compute n, an exact sample size algorithm uses a root searching algorithm, and the exact value is approximately n^*+2, where n^* is from Equation 20.2 (Problem P2).

Consider Equation 20.2 for a two-sample tests where $\tau_0 = \tau_1 = \tau$ and τ does not depend on β. Suppose we are interested in the difference in means from two groups, where the means are distributed $\hat{\mu}_0 \sim N(\mu_0, \sigma^2/n_0)$ and $\hat{\mu}_1 \sim N(\mu_1, \sigma^2/n_1)$. Let $n = n_0$ and $r = n_1/n_0$ so that $n_1 = rn$. Then letting $\hat{\beta}_n = \hat{\mu}_1 - \hat{\mu}_0$, and

$$\frac{\tau^2}{n} = \frac{\sigma^2}{n} + \frac{\sigma^2}{rn} = \frac{\sigma^2(r+1)}{rn},$$

Equation 20.2 becomes

$$n^* = \frac{(r+1)\sigma^2 \left(c_\alpha + c_\gamma\right)^2}{r\Delta^2}, \tag{20.3}$$

and with rounding up, $n_0 = \lceil n^* \rceil$ and $n_1 = \lceil rn^* \rceil$.

Equation 20.3 can be used to compare the efficiency of different randomization allocations. For example, we can compare two designs with the same power using the ratio of the total sample size of one design over the total sample size of the other. This total sample size ratio comparing the 2 : 1 allocation ($r = 2$) total sample size over the 1 : 1 allocation ($r = 1$) total gives $9/8 = 1.125$. Thus, doing a 2 : 1 allocation will require 1.125 times as many individuals as the 1 : 1 allocation to get the same power. In general, the total sample size ratio of an r : 1 allocation over a 1 : 1 allocation is $(2 + r + 1/r)/4$, which does not depend on any of the other parameters α, γ, σ, or Δ. This formula shows us that the most efficient allocation is an equal allocation.

Schouten (1999) derives an even better approximation that allows for heterogeneity of the variances and is closer to the sample size that is associated with Welch's t-test. Letting $\xi = \sigma_1^2/\sigma_0^2$ and $\sigma^2 = \sigma_0^2$, Schouten's formula is

$$n^* = \frac{(r+\xi)\sigma^2 \left(c_\alpha + c_\gamma\right)^2}{r\Delta^2} + \frac{\left(\xi^2 + r^3\right) c_\alpha}{2r \left(\xi + r\right)^2}, \tag{20.4}$$

with $n_0 = \lceil n^* \rceil$ and $n_1 = \lceil rn^* \rceil$. For example, when $\xi = 4$, $\sigma = 1$, $r = 1$, $\Delta = 0.5$, and a power of 60%, then we get $n_0 = n_1 = 99$. Recall, we simulated this in Example 20.1 and simulated a power of about 60% for sample sizes of $n_0 = n_1 = 100$.

Now consider the two-sample binomial problem, where $Y_0 \sim Binom(n_0, \pi_0)$ and $Y_1 \sim Binom(n_1, \pi_1)$ are the responses in the two arms. There are two common ways that we can use Equation 20.2 for this situation, either (1) define $\beta = \pi_1 - \pi_0$, or (2) use a variance stabilizing transformation and let $\beta = 2 \sin^{-1}(\sqrt{p_1}) - 2 \sin^{-1}(\sqrt{p_0})$. The former way is derived in Note N2. The latter method relies on the large sample approximation,

$$2 \sin^{-1}(\sqrt{\hat{p}_a}) \dot{\sim} N\left(2 \sin^{-1}(\sqrt{p_a}), \frac{1}{n}\right),$$

where $\hat{p}_a$ are sample proportions within Group a for $a = 0, 1$, and $\dot{\sim}$ denotes approximately distributed. Thus, we can use Equation 20.3 with $\Delta = 2 \sin^{-1}\left(\sqrt{\hat{p}_1}\right) - 2 \sin^{-1}\left(\sqrt{\hat{p}_0}\right)$ and $\sigma = 1$.

We can compare these to the exact minimum sample sizes for Fisher's exact test (FET in the table below) without and with a mid-p adjustment (using the exact2x2 R package). We consider three alternative models, where we calculate sample sizes for a power of 80% with a one-sided $\alpha = 0.025$ level testing the alternative that $p_1 > p_0$. The two normal approximation methods are Diff Approx (for $\beta = p_1 - p_0$) and Asin Approx (for $\beta = 2 \sin^{-1}(\sqrt{p_1}) - 2 \sin^{-1}(\sqrt{p_0})$). Sample sizes are listed as per arm sample sizes.

Method	$p_0 = 0.3$ $p_1 = 0.6$	$p_0 = 0.1$ $p_1 = 0.4$	$p_0 = 0.01$ $p_1 = 0.15$
Diff approx	42	32	58
Asin approx	42	30	45
FET	48	36	63
FET mid-p	44	32	54

It appears that the difference approximation model gives sample sizes closer to the Fisher's exact methods.

A simple approximation for a two-sample study with time to event endpoints using the logrank test is by Schoenfeld (1981) (see also Wittes, 2002). Assume proportional hazards and no censoring. Let π_0 and π_1 be the probability of an individual experiencing an event during the trial in each of the two arms for equal sample sizes in both arms. Then, in this no censoring case, the sample size per arm is

$$n^* = \frac{4}{\pi_0 + \pi_1} \frac{(c_\alpha + c_\gamma)^2}{(\log(\theta))^2}, \tag{20.5}$$

where $\theta = \log(1 - \pi_1)/\log(1 - \pi_0)$ is the hazard ratio (Note N3). The same formula holds under independent censoring when the censoring distributions are the same in the two arms, with θ still representing the hazard ratio, but now π_0 and π_1 represent the probability in each of the arms of an individual experiencing an event *before being censored* during the trial (see Schoenfeld, 1981, for details).

There are many other useful sample size formulas, with separate derivations for different tests, and many derivations using normal approximations (see, e.g., Chow et al., 2018).

20.4 Simulating the Sample Size

The simulation of sample size is statistically similar to finding the ED_P, the effective dose that will give a positive binary response for P percentage of the population, assuming that dose is a monotonic function of the proportion with a positive response. For dose finding problems, often groups of individuals are given the same specific dose, and the process is sequential, where the dose for the next group is determined using the previously collected data. In the analogy with simulating sample size, dose is the sample size, P is the target power, and a positive response is rejecting the null hypothesis. One issue is that power functions are not guaranteed to increase the power with increasing sample size (see, e.g., Figure 4.6). Nevertheless, we can use some ideas from dose-response methods to give an approximate algorithm for finding by simulation the sample size that approximately gives a specified power.

We present an algorithm that we used in the **simulateSS** function in the **asht** R package. For the algorithm, we set the one-sided α-level, the target power, and two functions: one to generate the data given the sample size, and another to calculate the decision (1 = reject the null hypothesis, or 0 = fail to reject). We additionally set some parameters to determine how many batches of data sets to simulate and the number of data sets per batch. In Note N4 we give the algorithm, which uses tools from (1) the normal approximation, (2) isotonic regression (this is the same tools that are used for estimating the nonparametric maximum likelihood for the cumulative event time distribution from current status data), and (3) the up-and-down method. This algorithm makes no claims to be optimal, but it gives the flavor that many batches of simulations at different sample sizes can be combined to estimate the sample size that meets the target power.

Let us return to the two-sample t-test of Example 20.1 (page 379). There we showed that we could get one accurate simulated power with about 10^4 to 10^5 replications. With the algorithm of Note N4 using 100 batches of 100 replicates each (i.e., 10^4 total replicates), we get relatively stable estimates of sample size. With $\alpha = 0.025$, target power of 80%, and $\Delta/\tau = 0.5$ the exact sample size is 64. With the algorithm applied 10 times, we get estimates of 63 (one time), 64 (five times), 65 (three times), and 66 (one time). When we do 100 batches of 1,000 replicates (10^5 total), we get 64.

Of course, when power programs run very fast, in practice, one might run the power program for a few sample size values, and hone in on the sample size that provides the desired power very quickly with an ad hoc trial and error approach, or by tabulating through all possible sample sizes, with no need for such an algorithm.

20.5 Nonadherence and Other Issues

Consider a randomized study comparing a new treatment to a control therapy. Suppose we expect that the new treatment will work well on about 70% of individuals randomized to that arm, but about 30% of individuals will not be able to tolerate the new treatment and will switch to the control therapy. We know from Chapter 15 that we cannot just delete the individuals that cannot tolerate the new treatment, because we have not deleted the unknown set of individuals from the control arm who would not have been able to tolerate the new treatment if they had been randomized to the new treatment arm. That is why we perform the intention-to-treat analysis, where the responses of all individuals are included in the final analysis. But how do we calculate the sample size for this situation?

One simple method is as follows. Let ρ_1 be the proportion of individuals who if randomized to the new treatment group will have responses that act as if they were in the control group, either because they cannot tolerate the new treatment and switch to the control group, or because the new treatment does not work for them, and the effect is the same as in the control group. For example, Fay et al. (2007) considered a malaria vaccine study, where they wanted the sample size to allow for the possibility that the new vaccine did not work in a proportion of the subjects, and in that case the control vaccine was for a different disease and was not expected to protect from malaria disease. Let ρ_0 be the analogous proportion of individuals who if randomized to the control group will have responses that act as if they were in the new treatment group. Consider the problem that motivated Equation 20.3, except now allow $\rho_1 > 0$ and $\rho_0 > 0$. Let "switchers" denote individuals that when given one treatment switch and respond as if they were given the other. If the observed proportion of switchers in the study match the proportion in the population, then the difference in means for the intention-to-treat analysis becomes,

$$\Delta = \{(1 - \rho_1)\mu_1 + \rho_1\mu_0\} - \{(1 - \rho_0)\mu_0 + \rho_0\mu_1\}$$
$$= (1 - \rho_1 - \rho_0)(\mu_1 - \mu_0).$$

So the original difference is multiplied by a factor of $(1 - \rho_1 - \rho_0)$ and the original sample size is multiplied by $(1 - \rho_1 - \rho_0)^2$. Note N5 gives a more complicated sample size function, when we only assume that the population values of ρ_1 and ρ_0 are fixed, but do not assume that the study sample proportions of ρ_1 and ρ_0 exactly equal the population proportions.

The intention-to-treat analysis in not the only option when there is nonadherence, we can also estimate the average complier causal effect (see Section 15.4.3), but usually the intention-to-treat analysis is the primary analysis upon which the sample size calculations are based.

If the responses are expected to be missing in a proportion of the individuals, then the easier situation is when the responses can be assumed to be missing completely at random (MCAR). Suppose that 20% of the individuals in both arms are expected to be missing the primary endpoint due to practical difficulties in collecting that data that are not related to the values of those endpoints or the variable of interest (e.g., intervention or exposure). Then we simply multiply the final endpoint by $1/(1 - 0.20) = 1.25$ so that the expected final sample size after missing 20% will equal the original calculated sample size.

Unfortunately, although MCAR is the most convenient missingness assumption, it is often not the most plausible. Chapter 17 discusses many of the other assumptions for missing data and some possible analyses. The sample size calculation should match the final planned analysis as much as is feasible. One issue is that many times there are no easy fixes for the analysis when there is a considerable amount of data that are missing not at random (the least restrictive assumption). In those cases, sensitivity analyses must be done if there is substantial missingness, and the results of the study are less clear. In that case, it may be better to spend more resources to get a smaller sample size study with less missing data, than to plan for a substantial portion of missing data and then increase the sample size by a simple factor increase.

Another issue is that typically many studies based on point estimates of parameters from preliminary trials do not achieve the target power (we suspect this is because preliminary trials that provide overestimates of treatment effects are more likely to lead to subsequent trials than those trials that produce correct estimates or underestimates). Lan and Wittes (2012) discussed this concern and recommended accounting for the uncertainty in the preliminary study in the sample size calculation. A natural way to account for uncertainty is to use a Bayesian posterior distribution in the sample size calculation. For example, consider the simulation method. For each replication, first simulate the parameter vector from the Bayesian posterior based on the preliminary data, then using the parameter vector simulate the data set. This accounts for more variability than just choosing the best point estimate from the preliminary data for the alternative model (see also Ciarleglio et al., 2016, for more details on this approach). Fay et al. (2007) discussed a frequentist approach for accounting for variability in the prior study estimates. By using either of these approaches, the sample size will typically be larger, but it is an appropriate and necessary sample size increase since it more rigorously accounts for the variability (or wide range of plausible values) of the preliminary estimates.

One simple and practical solution to address uncertainty about the true treatment effect is to plan the largest feasible study, as long as it will provide reasonable power for a plausible treatment effect estimate. While we stated at the outset that this chapter was limited to fixed sample size settings, planning the largest feasible study when there is great uncertainty is an especially useful strategy when combined with group sequential methods (see Chapter 18).

20.6 Summary

An important element of study design is to determine a sample size that will provide sufficient power to draw clear conclusions and provide actionable information. The calculation first requires specification of a treatment effect we wish the study to have good power to detect, which can be based on a minimal treatment effect of interest, or, more often, on some estimate of the true treatment effect.

We recommend simulation as a highly flexible approach for understanding power, especially as it easy to adapt these programs for important variations. It has the additional virtue of serving as a check on your understanding of the setting, and as a way of checking the Type I error rate of the planned procedure. Simulations can be very quick to set up for simple tests, and while they can be more difficult to set up for more complicated tests (such as what might be done for a cluster randomized trial), sometimes simulation is the only tractable method to compute power. Simple simulations provide power for a specified sample size, as opposed to the other way around. We provide an algorithm for going in the other direction (i.e., estimate sample size for a prespecified power) without an inordinate number of replications. That said, in some scenarios, it will be easy to compute power for the range of possible sample sizes, without using such an algorithm. There are many standard formulas and packages to determine power and sample size, and often this will be the quickest solution, especially for very standard settings.

If nonadherence to study treatments is anticipated, the sample size calculation should take this into account, and we propose several methods for this (although, if expected nonadherence is substantial, one might need to rethink the study design).

We note that studies are often underpowered because preliminary trials may tend to provide overly optimistic treatment effects; in part because the preliminary trials that produce overly optimistic treatment effects are more likely to lead to future trials. We discuss several approaches for counterbalancing this optimism. While this chapter focused exclusively on sample size and power for fixed sample design, the types of designs discussed in Chapter 18 are especially useful when there is considerable uncertainty about the effect size. Other forms of adaptive designs that reestimate sample size as a function of interim nuisance parameters such as the variance may also be helpful (see, e.g., Wittes and Brittain, 1990).

20.7 Extensions and Bibliographic Notes

Chow et al. (2018) is a comprehensive book on sample size estimation with many formulas for different kinds of study designs. Wittes (2002) reviews many of the main practical issues with sample size calculation, and we used that heavily for this chapter. Lachin (1981) is a nice review paper of normal approximation-based sample size methods.

20.8 R Functions and Packages

The R package TrialSize contains many sample size functions described in Chow et al. (2018). The algorithm of Note N4 is given in the simulateSS function in the asht R package. The R package ssanv does the sample size calculations of Fay et al. (2007).

20.9 Notes

N1 **Simulating power for two-sample t-test** Here is the R code to do the first simulation (of the first set of 10 simulations) in Example 20.1:

```
simttest<-function(nsim,n,delta,seed,varEqual=FALSE,sd2=1){
  set.seed(seed)
  # core function....7 lines
  p<-rep(NA,nsim)
  for (i in 1:nsim){
    y1<-rnorm(n)
    y2<-rnorm(n,mean=delta,sd=sd2)
    p[i]<-t.test(y1,y2,alternative="less",var.equal = varEqual)$p.value
  }
  length(p[p<=0.025])/nsim
}
simttest(1e4,100,.5,1,varEqual=TRUE)
```

N2 **Difference in proportions sample size:** Let $\beta = \pi_1 - \pi_0$ and $\beta_0 = 0$. Let $\hat{\beta}_n = \hat{\pi}_1 - \hat{\pi}_0$, where $\hat{\pi}_a = y_a/n_a$ for $a = 0, 1$. Again let $n_0 = n$ and $n_1 = rn$. Let $\theta_1 = [p_0, p_1]$ be the parameters under the alternative hypothesis model, where the subscript for θ represents the alternative hypothesis, and the subscripts for p represent the control (0) and treatment (1) arms. The boundary between the Hypotheses 20.1 is $\beta = 0$ which is $\pi_0 = \pi_1 = \pi$. We define π based on the expected value of the null hypothesis estimate under the alternative model, so that $\theta_0 = [\bar{p}, \bar{p}]$, where $\bar{p} = (1+r)^{-1}(p_0 + rp_1)$. Then $\tau_0^2 = (1 + 1/r)\bar{p}(1 - \bar{p})$ and

$$\tau_1^2 = p_0(1 - p_0) + \frac{p_1(1 - p_1)}{r}.$$

Equation 20.2 becomes

$$n^* = \frac{\left(c_\alpha \sqrt{(1 + \frac{1}{r})\bar{p}(1 - \bar{p})} + c_\gamma \sqrt{p_0(1 - p_0) + \frac{p_1(1-p_1)}{r}}\right)^2}{\Delta^2}, \qquad (20.6)$$

where $c_a = \Phi^{-1}(1 - a)$, $n_0 = \lceil n^* \rceil$, and $n_1 = \lceil rn^* \rceil$.

N3 Hazard ratio using proportion observed failures: We give a brief incomplete motivation for why we can write the hazard ratio as $\theta = \log(1-\pi_1)/\log(1-\pi_0)$, where π_0 and π_1 are the probabilities of an individual experiencing an event during the trial in each of the two arms. Note that the proportional hazards assumption is the same as there existing a monotonic transformation that takes the two distributions to exponentials (see, e.g., Fay and Malinovsky, 2018, Supplement). Since the logrank test is a rank test, its results do not change with monotonic transformations; therefore, we can base sample sizes on the exponential distribution. Let t_{end} be the end of the study, so with no censoring the probability of observing an event during the study is $\pi_a = F_a(t_{end}) = 1 - \exp(-\lambda_a t_{end})$ for arm $a = 0, 1$. So λ_1/λ_0 is $\log(1 - \pi_1)/\log(1 - \pi_0)$.

N4 Simulation for sample size algorithm: For Step 1, we pick a starting sample size, say N_1, and the number of replications within a batch, m, and the total number of batches, b_{tot}. We simulate m data sets with sample size N_1, and get the proportion of rejections, say P_1. Then we assume Equation 20.2 approximately holds with $\tau_0 = \tau_1 = \tau$, replace n^* with N_1, $1 - \gamma$ with P_1, and solve for Δ/τ. We use that Δ/τ estimate in Equation 20.2, except with the target power replacing $1 - \gamma$, to get a estimate of the target sample size, say N_{norm}. In Step 2, we replicate m data sets with sample size $N_2 \equiv N_{norm}$ to get the associated proportion of rejections, say P_2. We repeat two more batches with $N_3 = N_{norm}/2$ and $N_4 = 2N_{norm}$, to get proportions P_3, and P_4. Then in Step 3, we use isotonic regression (which forces monotonicity, power to be nondecreasing with sample size) on the four observation pairs $((N_1, P_1), \ldots, (N_4, P_4))$, and linear interpolation to get our best estimate of the sample size at the target power, N_{target}. We use that estimate of N_{target} for our sample size for the next batch of simulations. This idea is of using the best estimate of the target for the next iteration is studied in Wu (1985, see Section3). Step 4 is iterative, for the ith batch we repeat the isotonic regression, except now with N_i estimated from the first $(i - 1)$ observation pairs. We repeat Step 4 until either the number of batches is b_{tot}, or the current sample size estimate is the same as the last two estimates, in which case we switch to an up-and-down-like method. For each iteration of the up-and-down-like method, if the current proportion of rejections from the last batch of m replicates is greater than the target power, then subtract 1 from the current sample size estimate, otherwise add 1. Continue with that up-and-down-like method until we reach the number of batches equal to b_{tot}. The up-and-down-like method was added because sometimes the algorithm would get stuck in too large of a sample size estimate.

N5 Nonadherence adjustment: Fay et al. (2007) studied the sample size for the two-arm study with normal responses. Let $\bar{Y}_1$ and $\bar{Y}_0$ be the means of observations randomized to arms 1 and 0, respectively. For the ith individual, let Y_i be the response, Z_i be the arm they were randomized to, and C_{0i} and C_{1i} be the adherence variables. Suppose $Y_i|Z_i = 1, C_{1i} = 1 \sim N(\mu_1, \sigma_1^2)$ and $Y_i|Z_i = 0, C_{0i} = 1 \sim N(\mu_0, \sigma_0^2)$ for the "adherers," and $Y_i|Z_i = 1, C_{1i} = 0 \sim N(\mu_0, \sigma_0^2)$ and $Y_i|Z_i = 0, C_{0i} = 0 \sim N(\mu_1, \sigma_1^2)$ for the "nonadherers." Let the proportion of adherers to arm 1 be ρ_1 and to arm 0 be ρ_0. Assume $\bar{Y}_a \sim N((1 - \rho_a)\mu_a + \rho_a\mu_{1-a}, V_a/n_a)$ with $V_a = (1 - \rho_a)\sigma_a^2 + \rho_a\sigma_{1-a}^2 + (1 - \rho_a)\rho_a(\mu_1 - \mu_0)^2$ for $a = 0, 1$. Let $\beta = E(\bar{Y}_1) - E(\bar{Y}_0) = (1 - \rho_1 - \rho_0)(\mu_1 - \mu_0)$.

Let $\beta_0 = 0$ and test $H_0 : \beta \leq 0$ versus $H_1 : \beta > 0$. Let $n_0 = n$ and $n_1 = rn$. The sample size for the Z test (assuming the normal assumption holds) for the intention-to-treat analysis, is

$$n^* = \frac{(V_0 + V_1/r)\left(\Phi^{-1}(1-\alpha) + \Phi^{-1}(1-\gamma)\right)^2}{(1 - \rho_1 - \rho_0)^2(\mu_1 - \mu_0)^2}$$

at one-sided level α and power $1 - \gamma$.

20.10 Problems

P1 Derive Equation 20.2.

P2 Use Equation 20.2 to calculate the sample size needed for a one-sided one-sample Z test with $\alpha = 0.025$ and power of either 80% or 90% for several different values of $\Delta^* > 0$. Compare the sample size to the associated sample size using power.t.test. Show by calculating a grid of Δ^* values that if you do not round up the sample sizes to the nearest integer, the differences in calculated sample sizes are between 1.9 and 2.1 for all $\Delta^* \in (0.001, 1)$.

P3 **Two group Poisson:** Suppose that Y_i are independent and identical distributed $Pois(\lambda_a)$ for Group a, $a = 1, 2$. Using the usual normal approximation method, show that the approximate sample size needed to show a significant difference between the two groups with power $1 - \beta$ is

$$n \geq \frac{\left(\Phi^{-1}(1-\alpha/2) + \Phi^{-1}(1-\beta)\right)^2 (\lambda_1 + \lambda_2)}{(\lambda_1 - \lambda_2)^2},$$

where n is the sample size of each group. See Equation 10 of Gail (1974).

21

Bayesian Hypothesis Testing

21.1 Introduction

There are two main schools of thought in statistics, the Bayesian school and the frequentist one. Statistical hypothesis testing can be carried out within either school, but this book describes the frequentist approach. We briefly discuss some Bayesian approaches to hypothesis testing here for several reasons. First, we believe an applied statistician should be open to any statistical tool that meets a scientific need, and sometimes a Bayesian approach may feel more appropriate. Collaborators may prefer a Bayesian approach and it is good to have a general idea of how that approach works. Second, some statisticians are strongly Bayesian and may argue forcefully that frequentist hypothesis testing is misguided and only Bayesian methods make logical sense. An applied statistician should know enough Bayesian statistics to be able to speak to that person as well. We believe that these arguments are largely unfruitful, similarly to how it would be unilluminating to argue that one language is better for explaining science than another. A scientist's native language (or perhaps the language in which they first learned about a topic) is easier to understand than another language, but of course other people might be much more fluent in another language and so it is easier for them to understand that language. The analogy is not perfect, and there are very real differences in the two statistical approaches, but despite this, many problems can be usefully analyzed within either school. Nevertheless, since part of the applied statistician's job is explaining and defending their analysis choices, it is very useful to be versed in both statistical languages so that they can understand alternative approaches.

Section 21.2 gives a very brief overview on some basics of Bayesian statistics. Then we give two sections on some important ways that Bayesians have approached hypothesis testing. Section 21.3 reviews Bayesian hypothesis testing using Bayes factors. Section 21.4 reviews Bayesian hypothesis testing using decision theory. Finally, in Section 21.5 we show that for some common hypotheses (but not all!), if we pick a well-calibrated prior distribution with a preference for the null hypothesis, the p-values from the accepted standard test are equivalent to the posterior probability that the null hypothesis is true.

21.2 Brief Overview of Bayesian Statistics

In Bayesian statistics, the parameters are random variables. Consider the simple case where θ is the parameter of interest and it is the only parameter needed to describe the entire probability distribution. Prior to collecting the data from a study, the Bayesian statistician and subject matter collaborators decide on a *prior probability distribution* for θ, say $\pi(\theta)$. Let $f(x|\theta)$ be the distribution for the data x, given the parameter θ. The *posterior distribution* is

$$g(\theta|x) \propto f(x|\theta)\pi(\theta). \tag{21.1}$$

For many simple problems, there is a family of prior distributions that lead to posterior distributions from the same family. Any distribution from that family is called a *conjugate prior* for the distribution $f(x|\theta)$. Conjugate priors simplify some calculations.

Example 21.1   *Consider the one sample binomial problem, $X|\theta \sim Binomial(n,\theta)$.[1] For this problem, the conjugate prior is the beta distribution. The binomial distribution is*

$$f(x|\theta) = \binom{n}{x} \theta^x (1-\theta)^{n-x}$$

and the beta distribution with parameters $a > 0$ and $b > 0$, say $Beta(a,b)$, is

$$\pi(\theta|a,b) = \frac{\Gamma(a+b)}{\Gamma(a)\Gamma(b)} \theta^{a-1}(1-\theta)^{b-1}.$$

So the posterior is

$$g(\theta|x) \propto \theta^{x+a-1}(1-\theta)^{n-x+b-1}.$$

The right-hand side of the expression is multiplied by a constant so that $\int_0^1 g(\theta|x)d\theta = 1$ for any x, and this means the posterior is $Beta(x+a, n-x+b)$. The mean of the posterior can be written as the weighted average of the mean of the prior, say $\hat{\theta}_{prior}$, and the sample proportion, x/n, giving

$$\hat{\theta}_{post} = \left(\frac{a+b}{n+a+b}\right)\hat{\theta}_{prior} + \left(\frac{n}{n+a+b}\right)\frac{x}{n}. \tag{21.2}$$

The prior distribution could come from subject matter experts. Inferences are made using the posterior distribution. For example, a Bayesian estimate could be the mode (the most likely value), median or mean of the posterior. A Bayesian confidence interval is called the *credible interval*. For example, the central 95% credible interval is the middle 95% of the posterior distribution.

Example 21.2   *Suppose we are planning a malaria vaccine trial in a village and we want to estimate the number of new cases in the next year. Doctors in the village say that typically about 5% to 10% of the villagers get malaria every year. A short survey of 63 randomly selected villagers found that 12 out of the 63 had malaria in the previous year. A Bayesian analysis might take the information from the doctors and use a beta distribution with a mean of about 7.5%. The mean of $Beta(a,b)$ is $a/(a+b)$, so there are many beta distributions with a mean of 7.5%. By plotting several different beta distributions with that mean, a reasonable prior is chosen as $Beta(1.62, 20)$ (see Figure 21.1, top panel). This prior represents our prior distributional guess about the parameter. From Equation 21.2, we can treat the prior as if it weights the amount of prior belief as if we had seen 21.62 individuals (see how*

---

[1] Since this is a Bayesian formulation so that θ follows a distribution, we write $X|\theta$ to explicitly condition on θ.

Prior: Beta(1.62, 20)

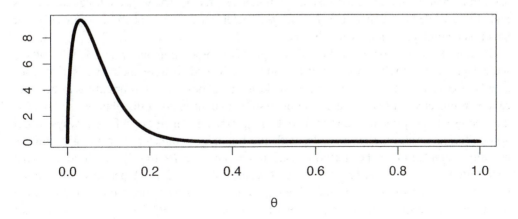

θ

Posterior: Beta(13.62, 71)

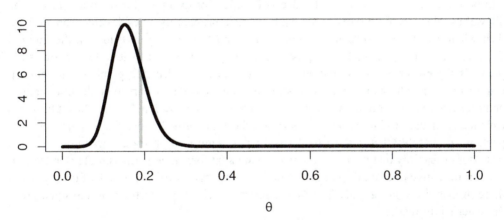

θ

Figure 21.1 Prior distribution of $Beta(1.62, 20)$ (top) and posterior distribution of $Beta(13.62, 71)$ after observing $x/n = 12/63$. The gray vertical line on the bottom panel is $12/63 = 19.0\%$

$a + b$ and n act similarly in the posterior). So although, we want to incorporate the prior information, we want the data (with $n = 63$) to have more influence. The survey alone might estimate the prevalence of the previous year as $12/63 = 19.0\%$, but incorporating the prior information gives a posterior of $Beta(13.62, 71)$ (see Figure 21.1, bottom panel). So a good posterior estimate is the mode of the posterior (16.1%) or the mean of the posterior (15.3%). (See Problem P2 for the formula for the mode.) The central 95% credible interval for θ is 9.1% to 24.6%. Notice that the posterior distribution combines information from the prior distribution and the data.

A nice feature of Bayesian analysis is that we begin with the prior, a distribution representing our prior beliefs about a parameter, and using Bayesian methods we can logically update our beliefs based on data to get a posterior distribution. The credible interval has a more natural interpretation than the confidence interval. The 95% credible interval has probability of 95% of containing the true parameter, where the probability represents our belief using both prior beliefs and the data.

Of course, where some see the advantage of incorporating our prior beliefs, others see a lack of objectivity, since each researcher will tend to have their own set of prior beliefs. Because of this, there is often a desire to choose a noninformative prior (also called an objective prior),[2] a "distribution" (sometimes it is not a distribution, see below) that represents no prior information about the parameter. There are different ways to get a noninformative prior. For example, with the binomial parameter, a seemingly obvious noninformative prior is the uniform distribution (i.e., the $Beta(1,1)$ distribution). That distribution gives an equal probability to all values of $\theta \in (0,1)$. A perhaps unsatisfying property of that uniform prior is that if we reparametrized using a one-to-one transformation on the parameter space (for example using the odds, $\beta_{odds} = \theta/(1-\theta)$), a uniform prior on the transformed parameter space would lead to different inferences on the parameter. We call a prior a *uniform prior* if it is constant for all values of the parameter. This may be an *improper prior*, meaning that the integration over the parameter space does not equal 1. This will be the case if the transformed space is infinite. For example, for the odds, β_{odds}, the allowable values are from 0 to ∞. If we put an equal probability on each value from 0 to ∞, then when we integrate over the sample space for β_{odds} (i.e., from 0 to ∞), the integral diverges (is infinite). Despite this problem, an improper prior can lead to a permissible prior[3] if the posterior (see Equation 21.1) integrates to 1. But the point for the current discussion is that although a uniform distribution on the prior appears to be the only natural noninformative prior, it is highly dependent on the parametrization. To state this a bit more technically, if the original parameter is θ and the transformed parameter is $t(\theta)$, then the posterior probability that $\theta \in (a,b)$ derived using a uniform prior on θ will not equal the posterior probability that $t(\theta) \in (t(a), t(b))$ using a uniform (and permissible) prior on $t(\theta)$. A noninformative prior on the scalar θ called Jeffreys' prior avoids this lack of invariance to those probabilities. For a scalar parameter, Jeffreys' prior is proportional to the square root of Fisher's information.

Example 21.2 *(continued) For the binomial parameter, Jeffreys' prior is $Beta(1/2, 1/2)$ which is symmetric and U-shaped, putting much of the distributional weight close to 0 and 1 (Problems P3 and P4). The 95% credible interval for θ associated with the binomial response of $x = 12$ out of $n = 63$ using Jeffreys' prior is $(10.9\%, 30.0\%)$, which is very close to the exact central binomial 95% confidence interval of $(10.2\%, 30.1\%)$.*

An important point for these Bayesian analyses is that usually as the sample size of the data gets very large, the data dominate the analysis and the prior distribution matters less and less. For example, suppose we have 100 times more data in the binomial problem with $x = 1,200$ and $n = 6,300$, then the 95% credible interval is $(18.05\%, 19.98\%)$ using the

---

[2] The class of noninformative priors are also sometimes called "reference" priors (Gelman et al., 2013). Following some literature, we reserve the term "reference prior" for a specific kind of noninformative prior first suggested by Bernardo (1979), see Section 21.4.

[3] Berger et al. (2009) define a permissible prior as a prior such that the posterior is proper (integrates to 1), together with a regularity condition (see that paper for details).

uniform prior (see Example 21.2), and is (18.09%, 20.03%) using Jeffreys' prior, and the 95% exact central confidence interval is (18.08%, 20.04%).

Now we return to the issue of improper priors. For example, suppose you have a single normal random variable with mean μ and variance 1. A noninformative prior on μ would like to treat each possible μ as equally likely, but since the possible values for μ are from $-\infty$ to $+\infty$, the integral of a constant (no matter how small) over that range does not equal 1. However, although in this case a constant is an improper prior, it is a permissible prior.

Bayesian analysis is very useful for complicated problems with many parameters. It is useful for applied statisticians because it is relatively easy using modern Monte Carlo methods to simulate from the posterior. All of this is beyond the scope of this book. We will focus on different ways to do Bayesian hypothesis testing.

21.3 Bayesian Hypothesis Testing Using Bayes Factors

Suppose we wish to choose among two different probability models of the data. The two models could represent a null and an alternative hypothesis; however, the two models need not represent the usual way we formulate the null and the alternative hypotheses. For example, the two models could have different numbers of parameters. Let M_1 and M_2 be the two models, and Θ_1 and Θ_2 be the associated sets of parameters. Let $\mathbf{x}$ be the data and $\mathbf{X}$ represent its associated random variable under either model. Let $\Pr[M_1]$ and $\Pr[M_2] = 1 - \Pr[M_1]$ be the prior probabilities for the two models. Then we can use Bayes theorem (a general form of Equation 21.3) to switch probabilities conditional on the model to probabilities conditional on the data, so for $i = 1, 2$,

$$\Pr[M_i|\mathbf{X} = \mathbf{x}] = \frac{\Pr[\mathbf{X} = \mathbf{x}|M_i]\Pr[M_i]}{\Pr[\mathbf{X} = \mathbf{x}]}, \tag{21.3}$$

where (assuming we choose one of the two models), we let

$$\Pr[\mathbf{X} = \mathbf{x}] = \Pr[\mathbf{X} = \mathbf{x}|M_1]\Pr[M_1] + \Pr[\mathbf{X} = \mathbf{x}|M_2]\Pr[M_2].$$

Then the posterior odds comparing the two models is the prior odds times the *Bayes factor*, or

$$\frac{\Pr[M_1|\mathbf{X} = \mathbf{x}]}{\Pr[M_2|\mathbf{X} = \mathbf{x}]} = \frac{\Pr[M_1]}{\Pr[M_2]} \frac{\Pr[\mathbf{X} = \mathbf{x}|M_1]}{\Pr[\mathbf{X} = \mathbf{x}|M_2]}.$$

The Bayes factor, $B_{12} = \frac{\Pr[\mathbf{X}=\mathbf{x}|M_1]}{\Pr[\mathbf{X}=\mathbf{x}|M_2]}$, is a measure of how much the prior odds is changed by the data. Then $B_{12} > 1$ implies that the data favor M_1 over M_2. Kass and Raftery (1995) suggest that $B_{12} \in (1, 3)$ is "not worth more than a bare mention" of evidence of preferring M_1 to M_2, while $B_{12} \in (3, 20)$ is positive evidence, $B_{12} \in (20, 150)$ is strong evidence, and $B_{12} > 150$ is very strong evidence. Unless the parameter sets Θ_1 and Θ_2 each have only one value, we need to put a prior distribution on the parameters (say $\pi(\theta_i|M_i)$ as a prior for θ_i), so that

$$\Pr[\mathbf{X} = \mathbf{x}|M_i] = \int \Pr[\mathbf{X} = \mathbf{x}|\theta_i, M_i]\pi(\theta_i|M_i)d\theta_i. \tag{21.4}$$

Some things to notice about this Bayes factor model comparison method. First, $B_{12} = 1/B_{21}$ so both models are treated logically the same. For frequentist hypothesis testing (i.e., most of this book), the two models are the null hypothesis and the alternative hypothesis, and both hypotheses are *not* treated equally. For example, the *p*-value is calculated with respect to the null hypothesis always, so that two hypothesis tests with the same null hypothesis but different alternative hypotheses have the same *p*-values. Second, in the Bayes approach there is no need for the models to use the same sets of parameters, or for one model to be nested within another model. This is much harder to do within the frequentist framework. Third, unlike the use of priors to get posteriors in Section 21.2 where the prior distribution is inconsequential as the sample size gets large, in the Bayes factor model comparisons the prior distributions can have a substantial effect even with large sample sizes. The issue is that the $\Pr[\mathbf{X} = \mathbf{x}|M_i]$ expressions in the numerator and denominator of the Bayes factor (see Equation 21.4), are not posterior distributions.

Example 21.3 *Jefferys (1990)[4] analyzed data with a random number generator that was supposed to give Bernoulli results with $\theta = 0.5$ and a subject attempts to influence the results using only his mind. The data are $x = 52,263,471$ and $n = 104,490,000$. The two-sided exact p-value for testing the null that $\theta = 0.5$ is $p = 0.0003$ with 95% exact central confidence intervals about θ as $(0.50008, 0.50027)$. Now consider the analysis using Bayes factors. The null hypothesis (M_1) is that $X|theta \sim Binomial(n,\theta)$ with $\theta = 0.5$ and the alternative (M_2) is the same except that $\theta \neq 0.5$. When the prior is uniform for M_2 then the Bayes factor $B_{12} = 11.9$, favoring the null hypothesis. If the prior for M_2 is Jeffreys' prior on the binomial, Beta(0.5,0.5), then $B_{12} = 5.95$. (See Problem P5). So even with an extremely large sample size, the choice between two different "noninformative" priors can have substantial impact on the Bayes factor. Jefferys (1990) tests a class of priors that puts more weight on the θ values close to 0.5. Within that class of priors, the Bayes factor B_{12} can go from less than 0.10 to greater than 10. So despite having an extreme amount of data, the choice of prior dominates the analysis.*

Although the Bayes factors can be useful (Kass and Raftery, 1995), as the example above shows, Bayes factors may be highly dependent on the prior distribution; therefore, they should be used cautiously when objectivity is important. Thus, for Bayesian hypothesis testing, another method, Bayesian decision theory, may be preferred.

21.4 Bayesian Hypothesis Testing Using Decision Theory

In the decision theoretic approach, we specify a loss function that determines the cost of making the wrong decision. The loss function could be a function of the parameter. For example,

$$\ell(t,\theta) = (t - \theta)^2$$

could be the cost of estimating θ with t. If $t = \theta$ there is no cost, and as t gets further from θ the cost becomes greater. The posterior expected loss from using t to estimate θ is

$$\bar{\ell}(t,\mathbf{x}) = \int \ell(t,\theta)g(\theta|\mathbf{x})d\theta, \tag{21.5}$$

---

[4] This Jefferys is not the Jeffreys that developed the Jeffreys' method for choosing priors.

where $g(\theta|\mathbf{x})$ is the posterior distribution for θ. The Bayes estimator is the value t that minimizes the posterior expected loss. So using the quadratic loss function, the Bayes estimator is the same as the expected value of the posterior (Problem P6).

If you transform the parameter, then the transformation of the Bayes estimator using quadratic loss function on the untransformed parameter, is not the same as the Bayes estimator using the quadratic loss function on the transformed parameter. For example, if $\tilde{\sigma}$ is the Bayes estimator of the standard deviation, σ, using Equation 21.5 with the squared error loss, then its square, $\tilde{\sigma}^2$, is not the Bayes estimator using Equation 21.5 with the squared error loss on σ^2. Other loss functions, however, are invariant to one-to-one transformations. For example, the intrinsic discrepancy loss function (see Bernardo, 2011, Definition 4) has such invariance. That loss function uses the Kullback–Leibler directed logarithmic divergence, i.e.,

$$\kappa(f_2|f_1) = \int f_1(x)\log\left(\frac{f_1(x)}{f_2(x)}\right)dx, \tag{21.6}$$

where f_1 and f_2 are two different distributions that we want to compare, and if the supports of the distributions are discrete the integral becomes a sum. Because $\kappa(f_2|f_1)$ is not equal to $\kappa(f_1|f_2)$, in order to get a symmetric loss function, the *intrinsic discrepancy* is defined as $\Delta(f_1, f_f) = \min\{\kappa(f_2|f_1), \kappa(f_1|f_2)\}$. Because this loss function compares distributions, it is invariant to reparameterizations.

Bernardo (2011) describes how a loss function such as either the squared error loss or Δ can be used to test simple (or point) null hypotheses. For now, assume no nuisance parameters and that the distribution is completely described by a scalar parameter θ. Let $H_0 : \theta = \theta_0$, then the expected loss under the posterior distribution from assuming $\theta = \theta_0$ is

$$\bar{\ell}(\theta_0, x) = \int_{\theta \in \Theta} \ell(\theta_0, \theta)g(\theta|x)d\theta,$$

where Θ is the complete parameter space (the union of the null and alternative parameter space). Then a Bayesian way of testing the null hypothesis is to reject if $\bar{\ell}(\theta_0, x)$ is greater than some constant ℓ_o that depends on the loss function and the scientific context. Bernardo (2011) suggests conventional levels for ℓ_0 when using the intrinsic discrepancy loss function to do Bayesian hypothesis testing are $\log(10)$, $\log(100)$, and $\log(1,000)$, analogous to the way 0.05, 0.01 and 0.001 are conventional significance levels for frequentist hypothesis testing.

Another way to define loss is with respect to the decision of whether to reject or fail to reject the null. Some define one loss value for Type I errors (rejecting the null hypothesis when it is true) and another loss value for Type II errors (failing to reject the null when the alternative is true). Here we formulate the loss in a similar but slightly different way following Freedman (2008). That paper discusses testing a scalar θ in a clinical trial where θ represents the difference in efficacy between two treatments, say between a new treatment and a control treatment. Although for this case sometimes researchers test the null that there is no difference between the treatments, $H_0 : \theta = 0$, it is important in this context to be able to say something about the direction of the effect (this is similar to three decision rules, see Section 2.3.4). Suppose that $\theta > 0$ indicates that the new treatment is better than the control, and $\theta < 0$ indicates the opposite. Consider two hypotheses: $H_0 : \theta \leq 0$ and $H_1 : \theta > 0$.

Instead of treating the two hypotheses as a decision between two choices, we consider three decisions:

(1) Reject H_0; this shows that the new treatment is better.
(2) Act as if H_1 is the null hypothesis and reject H_1; this shows that the control treatment is equal or better.
(3) Fail to reject both H_0 and H_1; this says that there is not enough information in the data to show which treatment is better.

Then we define three losses associated with when the three decisions are incorrect: (1) reject H_0 when H_0 is true, (2) reject H_1 when H_1 is true, and (3) fail to reject both (this last decision is always incorrect because one of the hypotheses must be true under the model). Let the losses for the three decisions be ℓ_1, ℓ_2, and ℓ_3. Treating both treatment and control similarly, we let $\ell_1 = \ell_2 \equiv \ell_0$ (representing rejecting hypotheses that are treated as null hypotheses). To perform a Bayesian decision, we choose the decision that minimizes the loss using the posterior distribution for θ. Let $g(\theta|x)$ be the posterior and let Θ_0 and Θ_1 be the sets of parameters under the null and alternative, respectively. Then the expected loss for the decision to reject H_0 is

$$\ell_0 \int_{\Theta_0} g(\theta|x)d\theta + 0 * \int_{\Theta_1} g(\theta|x)d\theta = \ell_0 \Pr[H_0|x],$$

and similarly, the expected loss for the decision to reject H_1 is $\ell_0 \Pr[H_1|x]$. The expected loss for failing to reject either hypothesis is $\ell_3 \int_{\Theta_0 \cup \Theta_1} g(\theta|x)d\theta = \ell_3$. We choose ℓ_0 (the loss from the Type I error for H_0, or the Type I error from treating H_1 as a null hypothesis as well) to be much larger than ℓ_3. If $\ell_3/\ell_0 < 0.5$ then we can write the Bayes decision as: reject H_0 if $\Pr[H_0|x] < \ell_3/\ell_0$, reject H_1 if $\Pr[H_1|x] < \ell_3/\ell_0$ or equivalently if $\Pr[H_0|x] > 1 - \ell_3/\ell_0$, and fail to reject either otherwise. Thus, $\Pr[H_0|x]$ is acting like a p-value and ℓ_3/ℓ_0 is acting like a significance level for a one-sided test. Freedman (2008) recommends using $\ell_3/\ell_0 = 0.025$ for clinical trials, where the two-sided significance level of 0.05 for frequentist hypothesis tests has become standard.

   In Section 21.5, we explore when $\Pr[H_0|x]$ is exactly equal to a p-value from a standard test.

21.5 When Can *p*-Values Be Interpreted in a Bayesian Manner?

21.5.1 One-Sample Binary Responses

To derive a Bayesian way of interpreting p-values, we need to restrict ourselves to certain types of hypotheses. So in this section, we start with the simple example of n binary responses from a single sample, and show which hypotheses are allowed a Bayesian interpretation. We also formulate the testing situation differently than the methods listed above (although it can be thought of as a particular way of selecting priors and then using the approach of Freedman (2008)). The basic idea of this formulation is to only use one-sided hypotheses and to use a *well-calibrated null preference* prior distribution. By *well calibrated*, we mean that the posterior claims about the truth of the alternative hypothesis are calibrated so that the claims are not overconfident. By *null preference*, we mean that the prior belief is nearly 100% in the null hypothesis, so that before we collect any data, we fail to reject the null hypothesis nearly 100% of the time.[5] This is different from the Bayes factor framing of the problem (see Section 21.3), where our prior belief is usually that the null and alternative are equally likely.

How would this work for the binary problem? Let x be the sum of n binary responses with mean θ, so that the associated random variable is $X|\theta \sim Binomial(n,\theta)$. Let the prior be $Beta(a,b)$, which we write as $\pi(\theta) = f_{beta}(\theta|a,b)$. Then the posterior distribution $g(\theta|x,a,b)$ is $Beta(x+a, n-x+b)$. Let $\hat{\theta} = x/n$, $\hat{\theta}_{prior} = a/(a+b)$, and $\hat{\theta}_{post} = (x+a)/(n+a+b)$ be the estimate of θ from the data, the prior, and the posterior, respectively. We can write the posterior estimate as the weighted average of $\hat{\theta}_{prior}$ and $\hat{\theta}$, with the prior getting a weight of $(a+b)/(a+b+n)$ and the data getting a weight of $n/(a+b+n)$ (see Equation 21.2). We want the weight of the prior to not overwhelm the data, so $a+b$ should be much less than n when n is large. It turns out that the well-calibrated choice uses $a+b=1$, since smaller values of $a+b$ give too much weight to the data and can lead to overconfident statements about the alternative, and larger values give too much weight to the prior (see Fay et al. (2021) for details).

Consider first the one-sided null hypotheses $H_0 : \theta \geq \theta_0$ versus $H_1 : \theta < \theta_0$. Then the prior probability that H_0 is true is

$$\Pr[\theta \geq \theta_0|\pi] = \int_{\theta_0}^{1} f_{beta}(\theta|a,b)d\theta. \tag{21.7}$$

If we let $b \to 0$, then $\Pr[\theta \geq \theta_0|\pi] \to 1$. Let $b = \epsilon$ and $a = 1 - \epsilon$, where $\epsilon > 0$ and ϵ is infinitesimal. Then the posterior distribution is $Beta(x+1-\epsilon, n-x+\epsilon)$. As $\epsilon \to 0$ the posterior probability that the null is true approaches

$$\Pr[\theta \geq \theta_0|x] = \int_{\theta_0}^{1} f_{beta}(\theta|x+1, n-x)d\theta. \tag{21.8}$$

The exact one-sided p-value is the supremum of the probability that X is equal or more extreme than x under the null, or

$$p(x,\theta_0) = \Pr[X \leq x|n,\theta_0] = F_{binom}(x|n,\theta_0). \tag{21.9}$$

It can be shown mathematically that Equation 21.8 equals Equation 21.9 (see, e.g., Casella and Berger, 2002, Exercise 2.40). In other words, by choosing this prior distribution (actually by choosing the limit of a series of prior distributions), then the one-sided exact p-value is equivalent to the posterior probability that the null hypothesis is true. This equivalence is another indication that the prior is well calibrated (Fay et al., 2021). As a shorthand for describing the prior distribution as a limit of a series of proper prior distributions, we denote the prior distribution as $Beta(1-,0+)$ and call it a null preference prior, because in the limit we approach 100% prior preference for the null. None of this creates any real problems because the limit of the posterior distributions is a proper distribution. Using this null preference prior, the associated $100(1-\alpha)\%$ one-sided credible interval is $\{0, F_{beta}(1-\alpha; x+1, n-x)\}$, where $F_{beta}(q; a,b)$ is the qth quantile of a beta distribution with parameters a and b. The upper limit of that credible interval is equivalent to the exact $100(1-\alpha)\%$ one-sided upper confidence limit.

---

[5] The "nearly" is needed because if our prior believe was 0% in the alternative hypothesis, then the posterior would continue to have 0% in the alternative hypothesis.

If we consider the other one-sided hypotheses $H_0 : \theta \leq \theta_0$ versus $H_1 : \theta > \theta_0$, the results are analogous. In this case, the prior sets $b = 1 - \epsilon$ and $a = \epsilon$ and lets $\epsilon \to 0$. The prior belief approaches 100% for $\theta = 0$, the deepest point in the null parameter space. In the limit, the posterior distribution approaches $Beta(x, n - x + 1)$. In the notation introduced above, the null preference prior is $Beta(0+, 1-)$. The p-value now becomes $p(x, \theta_0) = \Pr[X \geq x | n, \theta_0]$ and the limits of integration in Equation 21.8 become from 0 to θ_0, and the two interpretations are equivalent again: the p-value is the posterior probability that the null hypothesis is true. The lower limit of the $100(1 - \alpha)\%$ one-sided credible interval is $F_{beta}(\alpha; x, n - x + 1)$ and is equal to the associated lower confidence limit.

An important point for this formulation is that there are two different posterior distributions, $Beta(x + 1, n - x)$ for the upper limits, and $Beta(x, n - x + 1)$ for the lower limits. The reason for this is that we use null preference priors that change as we change the null hypothesis. This will only be an issue for discrete data. For continuous data, despite having different priors for the two different types of one-sided hypotheses, because they give nearly zero weight to the prior, the resulting posteriors may be the same for any specific data (see Section 21.5.3).

This formulation that makes the posterior belief in the null hypothesis equal to a p-value does not work with a two-sided point null hypothesis, for example, $H_0 : \theta = \theta_0$ versus $H_1 : \theta \neq \theta_0$. In the binomial case, if we take a limit of priors that puts all the prior belief on $\theta = \theta_0$, then the weight of $\hat{\theta}_{prior}$ in the weighted average of $\hat{\theta}$ and $\hat{\theta}_{prior}$ (Equation 21.2) goes to 1. Further, if the posterior is a beta distribution, then $\Pr[\theta = \theta_0 | x] = 0$ regardless of x.

Just as we can use the union of two one-sided 97.5% confidence intervals to get a 95% central confidence interval, we can use the union of two one-sided 97.5% *credible* intervals to get a 95% credible interval. This is perhaps not in the spirit of Bayesian statistics, because we are taking the union of two posteriors formulated with different priors. Nevertheless, it will give a useful credible interval with guaranteed frequentist coverage properties.

21.5.2 Inferences on a Single Parameter with No Nuisance Parameters

We generalize the results of the previous section. Here are some sufficient conditions to interpret the p-value as the posterior probability that the null hypothesis is true. Let $X|\theta \sim P_\theta$ and suppose that θ is a scalar. Suppose $\theta \in \Theta$, where Θ will be partitioned into two disjoint sets representing the null (Θ_0) or alternative (Θ_1) hypothesis. Let $T(x)$ be a statistic whose distribution depends only on θ with $T(x) \in [\tau_{min}, \tau_{max}]$, where τ_{min} and τ_{max} may be infinite. Let $F(t|\theta) = \Pr[T(X) \leq t|\theta]$ and $\bar{F}(t|\theta) = \Pr[T(X) \geq t|\theta] \equiv 1 - F(t - |\theta)$. Suppose that for each $t \in (\tau_{min}, \tau_{max})$, $F(t|\theta)$ is strictly monotonically decreasing in θ.

We want to test one of the following:

Alternative is less	Alternative is greater
$H_{U0} : \theta \geq \theta_0$	$H_{L0} : \theta \leq \theta_0$
$H_{U1} : \theta < \theta_0$	$H_{L1} : \theta > \theta_0.$

Then $F(t|\theta_0)$ is a valid p-value for testing H_{U0}, and $\bar{F}(t|\theta_0)$ is a valid p-value for testing H_{L0} (Problem P7).

Now we define well-calibrated null preference priors under this scenario (really it is a series of priors). Let π_ϵ be a proper prior distribution indexed by $\epsilon > 0$. We say π_ϵ has a null preference, if

$$\lim_{\epsilon \to 0} \int_{\theta \in \Theta_0} \pi_\epsilon(\theta) d\theta = 1.$$

For this scenario, π_ϵ is well calibrated to the hypotheses H_{U0} and H_{U1} if the limit of the posterior belief in the null hypothesis is equal to $F(t|\theta)$, i.e.,

$$F(t|\theta) = \lim_{\epsilon \to 0} b_0(\mathbf{x}, \pi_\epsilon), \qquad (21.10)$$

where Θ_0 is defined in H_{U0}, and

$$b_0(\mathbf{x}, \pi) = \int_{\theta \in \Theta_0} g(\theta|\mathbf{x}, \pi) d\theta$$

is the posterior belief in the null hypothesis for a prior π. Similarly, π_ϵ is well calibrated to the hypotheses H_{L0} and H_{L1} if the limit of the posterior belief in the null hypothesis H_{L0} is equal to $\bar{F}(t|\theta)$, i.e.,

$$\bar{F}(t|\theta) = \lim_{\epsilon \to 0} b_0(\mathbf{x}, \pi_\epsilon), \qquad (21.11)$$

where Θ_0 is defined in H_{L0}. Note that for discrete distributions, we need to use different priors to solve Equations 21.10 and 21.11. Additionally, if Equations 21.10 and 21.11 hold the associated one-sided confidence intervals will equal their associated one-sided credible intervals. We can more generally write that π_ϵ is well calibrated if

$$\lim_{\epsilon \to 0} \sup_{\theta \in \Theta_0} \Pr[b_0(\mathbf{X}, \pi_\epsilon) \le b_0(\mathbf{x}, \pi_\epsilon)|\theta] = \lim_{\epsilon \to 0} b_0(\mathbf{x}, \pi_\epsilon). \qquad (21.12)$$

Comparing Expression 21.12 to the definition of a valid *p*-value (Equation 2.3), it is not surprising that $\lim_{\epsilon \to 0} b_0(\mathbf{x}, \pi_\epsilon)$ leads to a valid *p*-value. Fay et al. (2021) give details, including well-calibrated null preference priors for binomial, negative binomial, and Poisson distributions. Veronese and Melilli (2015) discuss the closely related idea of "fiducial priors."

Example 21.4 Negative binomial: *Suppose we sample binary observations, each with success probability θ, and we keep sampling until the rth failure, such that the sum, X, conditional on θ, is a negative binomial distribution with $f(x|r,\theta) = \binom{x+r-1}{x} \theta^x (1-\theta)^r$ and mean $r\theta/(1-\theta)$. If we use a $Beta(a,b)$ prior for θ then the posterior is $Beta(x + a, x + r + b)$. If we used the same priors as used with the binomial case (see Section 21.5.1), then the posteriors would be the same (except with n replaced by $x + r$). But we use a different prior for the lower limit. For the upper limit we use $\lim_{b \to 0} Beta(1,b)$ for the prior, giving a posterior of $Beta(x + 1, r)$, which is the same as the binomial case. In contrast, for the lower limit we use $\lim_{a,b \to 0} Beta(a,b)$ with $a/(a + b) \to 0$ for the prior, giving a posterior of $Beta(x,r)$. Equation 21.11 motivates a different prior because $\bar{F}_{binom}(x|n,\theta) \ne \bar{F}_{negBinom}(x|r, 1-\theta)$ (see Problem P8).*

21.5.3 Other Cases

In some cases, θ may have dimension greater than 1, but we may only be interested in one element of θ, and the rest of the elements are nuisance parameters. For example, suppose we have n exchangeable normal random variables with mean μ and variance σ^2, so that $\theta = [\mu, \sigma]$. Suppose we are interested only in μ, and σ is an unknown nuisance parameter. In Chapter 9, we discussed how

$$T(X) = \frac{\bar{X} - \mu}{S/\sqrt{n}}$$

is distributed t with $n - 1$ degrees of freedom, where $\bar{X}$ and S are the random variables associated with the mean and s, the square root of the unbiased variance estimate. Gelman et al. (2013, pp. 64–66) show how using uniform priors on μ and $\log(\sigma)$, the resulting posterior on $T(X)$ is a t-distribution with $n - 1$ degrees of freedom. Because of this, the usual 95% confidence interval on μ is equivalent to the 95% credible interval on μ associated with that posterior. Fay et al. (2021) show how to form well-calibrated null preference priors in this situation, which lead to credible intervals that are equivalent to both of those intervals.

Now consider the general two-sample case, where θ_a is the parameter of interest from the ath sample, $a = 1, 2$. For example, a randomized clinical trial where subjects are randomized to two treatment groups, and θ_a is the mean of the response from the ath treatment group. Suppose we are interested in $\beta = \theta_2 - \theta_1$. For Group a, let T_{La} and T_{Ua} be random variables from the lower and upper posterior distributions associated with the lower and upper credible intervals developed throughout this section. Recall that for discrete random variables, the posterior associated with the lower and upper intervals are different. Define the one-sided hypotheses:

Alternative is less	Alternative is greater
$H_{U0} : \beta \geq \beta_0$	$H_{L0} : \beta \leq \beta_0$
$H_{U1} : \beta < \beta_0$	$H_{L1} : \beta > \beta_0.$

In terms of $\theta = [\theta_1, \theta_2]$, the null hypothesis space for H_{U0} is

$$\Theta_0(\beta_0) = \{(\theta_1, \theta_2) : \theta_2 - \theta_1 \geq \beta_0\},$$

and the alternative hypothesis space for H_{U1} is

$$\Theta_1(\beta_0) = \{(\theta_1, \theta_2) : \theta_2 - \theta_1 < \beta_0\},$$

where $\Theta_0(\beta_0)$ and $\Theta_1(\beta_0)$ are written as functions of β_0. Then to test H_{U0} versus H_{U1} using this Bayesian method, we get a posterior distribution by multiplying the distributions for T_{U2} and T_{L1} since the two samples are independent and the null hypothesis has large values of θ_2 and small values of θ_1. The posterior belief in the null hypothesis H_{U0} with $\beta_0 = 0$ uses that posterior distribution to calculate $\Pr[\theta \in \Theta_0(0)|\mathbf{x}]$, which is equivalent to the p-value associated with the melding method of Fay et al. (2015). For a $100(1 - \alpha)\%$ one-sided credible interval, the upper limit is B_U which solves $\Pr[\theta \in \Theta_1(B_U)|\mathbf{x}] = 1 - \alpha$, using the posterior distribution for θ. The methods for the alternative-is-greater hypotheses are analogous.

We consider two important examples.

Example 21.5 Two-sample binomial: *Suppose $X_a|\theta_a \sim Binomial(n_a, \theta_a)$. Then $T_{U2} \sim Beta(x_2 + 1, n_2 - x_2)$ and $T_{L1} \sim Beta(x_1, n_1 - x_1 + 1)$. The p-value associated with testing $\beta \geq 0$ is equal to the one-sided Fisher's exact test p-value, and the one-sided confidence intervals are the melded intervals.*

Example 21.6 Two-sample normal: *Suppose $X_{11}, \ldots, X_{1n_1}$ and $X_{21}, \ldots, X_{2n_2}$ are exchangeable with $X_{ai}|\mu_a, \sigma_a^2 \sim N(\mu_a, \sigma_a^2)$, let n_a, $\bar{x}_a$ and s_a^2 be the sample size, the mean, and the unbiased sample variance estimate for $a = 1, 2$. Then $T_{Ua} \sim \bar{x}_a + \frac{s_a}{\sqrt{n_a}}T_{n_a-1}$ and $T_{La} \sim \bar{x}_a - \frac{s_a}{\sqrt{n_a}}T_{n_a-1}$, where T_{n_a-1} for $a = 1, 2$ are random variables each independently distributed as a t-distribution with $n_a - 1$ degrees of freedom. In this case, because T_{n_a-1} is symmetric about 0, T_{Ua} and T_{La} have the same distribution, say T_a. The posterior distribution of $T_2 - T_1$ gives credible interval equivalent to the Behrens–Fisher intervals for $\mu_2 - \mu_1$. These intervals have been extensively studied by simulation and are thought to cover $\beta = \mu_2 - \mu_1$ with at least the nominal level for all values of σ_a, $a = 1, 2$.*

For more details and other examples, see Fay et al. (2015, 2021).

21.6 Bibliographic Notes

A very good book on Bayesian analysis is Gelman et al. (2013). For example, unlike some other books on Bayesian analysis, it addresses modeling accounting for data collection, which addresses the important point for applications on how one decides when to stop collecting data and analyze it. Many otherwise good Bayesian articles and texts cite the likelihood principle as implying that Bayesian methods may and should be applied without concern for how the data were collected. Gelman et al. (2013, chapter 8) offer a clear argument against that idea. Kass and Raftery (1995) is a nice review article about Bayes factors and includes several applications that highlight situations where the frequentist analysis may not be as straightforward as the Bayesian one.

There are many ways to get noninformative priors (Ghosh, 2011). One of the most general and useful methods is the reference method introduced by Bernardo (1979). That method uses the κ function. Although the details are mathematically difficult (Berger et al., 2009), we give a flavor of the idea. Let $\pi(\theta)$ be a prior distribution and $g(\theta|x)$ be its corresponding posterior. Then $\kappa(g(\cdot|x)|\pi)$ is the divergence between the posterior with respect to the prior (note we write $g(\cdot|x)$ because κ is a function of the density function, and the parameter θ is just an index of integration). Define the expected information as the expectation of $\kappa(g(\cdot|X)|\pi)$ over the random variable X with distribution $f(x) = \int f(x|\theta)\pi(\theta)d\theta$. The expected information is a measure of how far (although not really a distance measure) the posterior is from its prior. The idea is to imagine collecting an infinite amount of data until the posterior approaches the true distribution of the parameter. Then in that limit, the expected information represents the amount of missing information about θ with respect to the prior, $\pi(\theta)$. A noninformative "reference" prior is the limit of priors that maximize the missing expected information. This reference method for finding priors gives inferences that are invariant to reparameterizations, as does Jeffreys' method. In fact, for a scalar parameter the reference prior is Jeffreys' prior. For details see Bernardo (2011) and its references.

Two classic papers presented together with discussion on a Bayesian interpretation of frequentist hypotheses are Berger and Sellke (1987), who deal with point null hypotheses, and Casella and Berger (1987), who deal with one-sided hypotheses. Both papers predominately consider equal prior beliefs in the two hypotheses. Since Berger and Sellke (1987) are dealing with the point null hypothesis, they consider prior distributions on the continuous parameters as having point mass on the null hypothesis point, say θ_0. Using this methodology, they make statements such as "data that yield a P value of .05, when testing a normal mean, result in a posterior probability of the null of *at least* .30 for *any* objective prior distribution [italics in the original]." They use this to argue that p-values are misleading for describing evidence against the null hypothesis. Alternatively, Casella and Berger (1987) deal with one-sided hypotheses, and using equal prior belief in the null and alternative, come up with a different conclusion. They compare the p-value to the infimum of the posterior probability of H_0 and show that often the p-value is equal to or less than that infimum.

There is a long literature of determining when frequentist and Bayesian inferences match. Two important early references are Welch and Peers (1963) and Rubin (1984). Also, Meng (1994) who discusses posterior predictive p-values, which give a Bayesian interpretation of some classical p-values. Fay et al. (2021) describes in detail the well-calibrated null preference priors of Section 21.5. Veronese and Melilli (2015) studies the closely related problem of equivalence between fiducial distributions and confidence distributions for the natural exponential family with quadratic variance functions.

21.7 Problems

P1 Using R and the dbeta function, plot different beta density functions such as Jeffreys' prior for a binomial, or $Beta(0.001, 1 - 0.001)$.

P2 **Mode of a beta distribution:** Show that the mode (the most likely value) of a beta distribution with parameters $a > 1$ and $b > 1$ is $(a - 1)/(a + b - 2)$. Hint: for $\theta \sim Beta(a, b)$, set the first derivative of the density to 0 and solve for θ. Then show that there is only one peak in the density by examining the slope as a function of θ.

P3 **Jeffreys' prior and transformations:** For any distribution $f(x|\theta)$, with θ a scalar, Jeffreys' prior is proportional to $\left\{ -E_{x|\theta} \left[\frac{\partial^2}{\partial \theta^2} \log f(x|\theta) \right] \right\}^{1/2}$ (see e.g., Carlin and Louis, 2009, p. 39). Show that Jeffreys' prior leads to posteriors that are invariant to one-to-one transformation functions on the parameter for a scalar continuous parameter. Hint: write $\Pr[\theta \in (a, b)]$ using an integral and the posterior distribution, then change the integral using substitution with the transformation.

P4 Show that Jeffreys' prior for a binomial parameter is $Beta(1/2, 1/2)$.

P5 Consider the binomial example with $x = 52,263,471$ and $n = 104,490,000$. Calculate $\Pr[X = x|M_2]$, where M_2 is the alternative with $\theta \neq 0.5$ and a uniform prior. Recalculate with a prior of $Beta(1/2, 1/2)$. Hint: use the fact that a beta distribution integrates to 1. Also, in R use the lgamma and lchoose functions to avoid computational overflow.

P6 Show that the Bayes estimator using the quadratic loss function is the expected value of the posterior (see Bernardo, 2011, Section 2.1 for statement of this result).

P7 Show that $F(t|\theta_0)$ is a valid p-value for testing H_{U0}, and $\bar{F}(t|\theta_0)$ is a valid p-value for testing H_{L0}. Hint: show that if T is distributed $F(\cdot|\theta)$ then $\Pr[F(T|\theta) \leq \alpha] \leq \alpha$ for all $\alpha \in [0, 1]$ (see, e.g., Casella and Berger, 2002).

P8 Using R, pick any $t \in (0,1)$ and integer $0 \le x \le n$ with $r = n - x$ and show that pbinom(x,n,t) = pnbinom(x,r,1-t) = 1-pbeta(t,x+1,r).

P9 Prior elicitation exercise. Meet with a friend and get them to pick a prior distribution on the proportion of babies born in the United States that are girls. Suppose you were going to use that prior together with the year 2000 US census data to predict next year's proportion of live births that were girls. Before you meet think about the following: what tools would you use to help them pick their prior? What questions would you ask? After you meet, look up the data, calculate the posterior, and plot its distribution.

References

Numbers in square brackets are pages where the item is cited.

Aalen, O., Borgan, O., and Gjessing, H. (2008), *Survival and Event History Analysis*, New York: Springer. [47, 304, 308, 318, 323]

Aalen, O. O., Cook, R. J., and Røysland, K. (2015), "Does Cox analysis of a randomized survival study yield a causal treatment effect?" *Lifetime Data Analysis*, 21, 579–593. [318]

Agresti, A. (2013), *Categorical Data Analysis*, 3rd ed., Hoboken, NJ: John Wiley & Sons. [101, 121, 199, 214, 222, 273, 364, 365, 373, 375]

Agresti, A. and Caffo, B. (2000), "Simple and effective confidence intervals for proportions and differences of proportions result from adding two successes and two failures," *The American Statistician*, 54, 280–288. [66]

Agresti, A. and Min, Y. (2001), "On small-sample confidence intervals for parameters in discrete distributions," *Biometrics*, 57, 963–971. [114, 121]

—. (2002), "Unconditional small-sample confidence intervals for the odds ratio," *Biostatistics*, 3, 379–386. [114, 121]

Ali, M. M. and Sharma, S. C. (1996), "Robustness to nonnormality of regression F-tests," *Journal of Econometrics*, 71, 175–205. [274]

Andersen, P., Borgan, O., Gill, R., and Keiding, N. (1993), *Statistical Models Based on Counting Processes*, New York: Springer. [304, 306, 308, 315, 316, 317, 323, 325]

Andersen, P. K. (2005), "Censored data," *Encyclopedia of Biostatistics, 2nd ed.*, 1, 722–727. [324]

Anderson, J. M., Samake, S., Jaramillo-Gutierrez, et al. (2011), "Seasonality and prevalence of Leishmania major infection in Phlebotomus duboscqi Neveu-Lemaire from two neighboring villages in central Mali," *PLoS Neglected Tropical Diseases*, 5, e1139. [64]

Anderson-Bergman, C. (2017), "icenReg: regression models for interval censored data in R," *Journal of Statistical Software*, 81, 1–23. [324]

Angrist, J. D., Imbens, G. W., and Rubin, D. B. (1996), "Identification of causal effects using instrumental variables," *Journal of the American Statistical Association*, 91, 444–455. [287, 297, 300]

Baggerly, K. A., Morris, J. S., and Coombes, K. R. (2004), "Reproducibility of SELDI-TOF protein patterns in serum: comparing datasets from different experiments," *Bioinformatics*, 20, 777–785. [39]

Baiocchi, M., Cheng, J., and Small, D. S. (2014), "Instrumental variable methods for causal inference," *Statistics in Medicine*, 33, 2297–2340. [297, 298, 300]

Baker, S. G. (1994), "The multinomial-Poisson transformation," *The Statistician*, 43, 495–504. [198]

Banerjee, M. and Wellner, J. A. (2005), "Confidence intervals for current status data," *Scandinavian Journal of Statistics*, 32, 405–424. [320]

Barber, R. F. and Candès, E. J. (2015), "Controlling the false discovery rate via knockoffs," *The Annals of Statistics*, 43, 2055–2085. [273]

Barnhart, H. X., Haber, M. J., and Lin, L. I. (2007), "An overview on assessing agreement with continuous measurements," *Journal of Biopharmaceutical Statistics*, 17, 529–569. [101]

Basu, D. (1980), "Randomization analysis of experimental data, the Fisher randomization test (with discussion)," *Journal of the American Statistical Association*, 75, 575–595. [47]

Bauer, P., Bretz, F., Dragalin, V., König, F., and Wassmer, G. (2016), "Twenty-five years of confirmatory adaptive designs: opportunities and pitfalls," *Statistics in Medicine*, 35, 325–347. [352, 354, 356]

Bauer, P. and Köhne, K. (1994), "Evaluation of experiments with adaptive interim analyses," *Biometrics*, 1029–1041. [352, 355, 356]

Begg, C. B. (1990), "On inferences from Wei's biased coin design for clinical trials," *Biometrika*, 77, 467–473. [34]

Benjamini, Y. (2010), "Simultaneous and selective inference: current successes and future challenges," *Biometrical Journal*, 52, 708–721. [239]

—. (2016), "It's not the P-values' fault," *The American Statistician*, 70, 1–2. [xi]

Benjamini, Y. and Hochberg, Y. (1995), "Controlling the false discovery rate: a practical and powerful approach to multiple testing," *Journal of the Royal Statistical Society: Series B (Methodological)*, 57, 289–300. [241]

Benjamini, Y. and Yekutieli, D. (2001), "The control of the false discovery rate in multiple testing under dependency," *Annals of Statistics*, 1165–1188. [241, 250]

Beran, R. (1997), "Diagnosing bootstrap success," *Annals of the Institute of Statistical Mathematics*, 49, 1–24. [192]

Berger, J. O., Bernardo, J. M., and Sun, D. (2009), "The formal definition of reference priors," *The Annals of Statistics*, 905–938. [392, 401]

Berger, J. O. and Sellke, T. (1987), "Testing a point null hypothesis: the irreconcilability of P values and evidence," *Journal of the American Statistical Association*, 82, 112–122. [402]

Berger, R. L. and Boos, D. D. (1994), "P values maximized over a confidence set for the nuisance parameter," *Journal of the American Statistical Association*, 89, 1012–1016. [114]

Bernardo, J. M. (1979), "Reference posterior distributions for Bayesian inference," *Journal of the Royal Statistical Society: Series B (Methodological)*, 41, 113–147. [392, 401]

—. (2011), "Integrated objective Bayesian estimation and hypothesis testing," *Bayesian Statistics*, 9, 1–68. [395, 401, 402]

Bickel, P. and Freedman, D. (1981), "Some asymptotic theory for the bootstrap," *Annals of Statistics*, 9, 1196–1217. [192]

Bishop, Y., Fienberg, S., and Holland, P. (1975), *Discrete Multivariate Analysis: Theory and Practice*, Cambridge, MA: MIT Press. [211]

Blaker, H. (2000), "Confidence curves and improved exact confidence intervals for discrete distributions," *Canadian Journal of Statistics*, 28, 783–798. [55, 56, 63, 75, 111]

Bland, J. M. and Altman, D. G. (1999), "Measuring agreement in method comparison studies," *Statistical Methods in Medical Research*, 8, 135–160. [101]

Blyth, C. and Still, H. (1983), "Binomial confidence intervals," *Journal of the American Statistical Association*, 78, 108–116. [63]

Boos, D. D. and Stefanski, L. (2013), *Essential Statistical Inference*, New York: Springer. [170, 171, 173, 175, 183, 190, 191, 192, 193]

Boschloo, R. (1970), "Raised conditional level of significance for the 2×2-table when testing the equality of two probabilities," *Statistica Neerlandica*, 24, 1–9. [112]

Box, G. E. and Cox, D. R. (1964), "An analysis of transformations," *Journal of the Royal Statistical Society: Series B (Methodological)*, 26, 211–252. [274]

Box, G. E., Hunter, J. S., and Hunter, W. G. (2005), *Statistics for Experimenters: Design, Innovation, and Discovery*, vol. 2, New York: Wiley-Interscience. [274]

Box, G. E. and Watson, G. S. (1962), "Robustness to non-normality of regression tests," *Biometrika*, 49, 93–106. [274]

Brazzale, A. R., Davison, A. C., and Reid, N. (2007), *Applied Asymptotics: Case Studies in Small-Sample Statistics*, vol. 23, Cambridge: Cambridge University Press. [256]

Breslow, N. (1972), "Contribution to the discussion of Cox (1972)," *Journal of the Royal Statistical Society: Series B (Statistical Methodology)*, 34, 216–217. [308]

Breslow, N. and Chatterjee, N. (1999), "Design and analysis of two-phase studies with binary outcome applied to Wilms tumour prognosis," *Journal of the Royal Statistical Society: Series C (Applied Statistics)*, 48, 457–468. [307]

Breslow, N. E. (1996), "Statistics in epidemiology: the case-control study," *Journal of the American Statistical Association*, 91, 14–28. [122]

Bretz, F., Hothorn, T., and Westfall, P. (2011), *Multiple Comparisons using R*, Boca Raton, FL: CRC Press. [208, 209, 210, 211, 212, 244, 250, 251]

Bretz, F., Maurer, W., Brannath, W., and Posch, M. (2009), "A graphical approach to sequentially rejective multiple test procedures," *Statistics in Medicine*, 28, 586–604. [246, 247]

Brillinger, D. R. (1986), "The natural variability of vital rates and associated statistics (with discussion)," *Biometrics*, 42, 693–734. [229]

Brittain, E. and Lin, D. (2005), "A comparison of intent-to-treat and per-protocol results in antibiotic non-inferiority trials," *Statistics in Medicine*, 24, 1–10. [372]

Brittain, E. H., Fay, M. P., and Follmann, D. A. (2012), "A valid formulation of the analysis of noninferiority trials under random effects meta-analysis," *Biostatistics*, 13, 637–649. [227, 369, 370]

Brown, B. M. and Hettmansperger, T. P. (2002), "Kruskal–Wallis, multiple comparisons and Efron dice," *Australian & New Zealand Journal of Statistics*, 44, 427–438. [157, 159]

Brown, L. D., Cai, T. T., and DasGupta, A. (2001), "Interval estimation for a binomial proportion (with discussion)," *Statistical Science*, 16, 101–133. [60]

Brown, M. B. and Forsythe, A. B. (1974a), "372: the ANOVA and multiple comparisons for data with heterogeneous variances," *Biometrics*, 30, 719–724. [200, 211]

—. (1974b), "The small sample behavior of some statistics which test the equality of several means," *Technometrics*, 16, 129–132. [200]

Brunner, E., Konietschke, F., Pauly, M., and Puri, M. L. (2017), "Rank-based procedures in factorial designs: hypotheses about non-parametric treatment effects," *Journal of the Royal Statistical Society: Series B (Statistical Methodology)*, 79, 1463–1485. [221]

Brunner, E. and Munzel, U. (2000), "The nonparametric Behrens-Fisher problem: asymptotic theory and a small-sample approximation," *Biometrical Journal*, 42, 17–25. [95, 147, 157, 361]

Bühlmann, P., Kalisch, M., and Meier, L. (2014), "High-dimensional statistics with a view toward applications in biology," *Annual Review of Statistics and its Application*, 1, 255–278. [271]

Burnham, K. P. and Anderson, D. R. (2002), *Model Selection and Multimodel Inference: A Practical Information-Theoretic Approach, 2nd ed.*, New York: Springer. [273]

Candès, E., Fan, Y., Janson, L., and Lv, J. (2018), "Panning for gold: 'model-X' knockoffs for high dimensional controlled variable selection," *Journal of the Royal Statistical Society: Series B (Statistical Methodology)*, 80, 551–577. [273]

Carlin, B. P. and Louis, T. A. (2009), *Bayesian Methods for Data Analysis, 3rd ed.*, Boca Raton, FL: CRC Press. [40, 402]

Carpenter, J. and Kenward, M. (2013), *Multiple Imputation and its Application*, Chichester: John Wiley & Sons. [339, 340]

Caruso, J. C. and Cliff, N. (1997), "Empirical size, coverage, and power of confidence intervals for Spearman's Rho," *Educational and Psychological Measurement*, 57, 637–654. [97]

Casella, G. (1986), "Refining binomial confidence intervals," *The Canadian Journal of Statistics/La Revue Canadienne de Statistique*, 14, 113–129. [63]

—. (1989), "Refining Poisson confidence intervals," *The Canadian Journal of Statistics/La Revue Canadienne de Statistique*, 17, 45–57. [75]

Casella, G. and Berger, R. L. (1987), "Reconciling Bayesian and frequentist evidence in the one-sided testing problem," *Journal of the American Statistical Association*, 82, 106–111. [402]

—. (2002), *Statistical Inference, 2nd ed.*, Pacific Grove, CA: Duxbury Press. [xi, 102, 123, 306, 374, 397, 402]

Chen, Y. J., DeMets, D. L., and Lan, K. G. (2004), "Increasing the sample size when the unblinded interim result is promising," *Statistics in Medicine*, 23, 1023–1038. [354, 355]

Cheng, S., Wei, L., and Ying, Z. (1995), "Analysis of transformation models with censored data," *Biometrika*, 82, 835–845. [261]

Chow, S.-C., Shao, J., Wang, H., and Lokhnygina, Y. (2018), *Sample Size Calculations in Clinical Research, 3rd ed.*, Boca Raton, FL: Chapman and Hall/CRC. [382, 386]

Chung, E. and Romano, J. P. (2016), "Asymptotically valid and exact permutation tests based on two-sample U-statistics," *Journal of Statistical Planning and Inference*, 168, 97–105. [157, 160]

Ciarleglio, M. M., Arendt, C. D., and Peduzzi, P. N. (2016), "Selection of the effect size for sample size determination for a continuous response in a superiority clinical trial using a hybrid classical and Bayesian procedure," *Clinical Trials*, 13, 275–285. [385]

Cole, S. R. and Hernán, M. A. (2008), "Constructing inverse probability weights for marginal structural models," *American Journal of Epidemiology*, 168, 656–664. [292, 338]

Coulibaly, Y. I., Dembele, B., Diallo, A. A., et al. (2009), "A randomized trial of doxycycline for Mansonella perstans infection," *New England Journal of Medicine*, 361, 1448–1458. [43, 104]

Cox, D. (1972), "Regression models and life-tables (with discussion)," *Journal of the Royal Statistical Society: Series B (Statistical Methodology)*, 34, 187–220. [315]

Cox, D. and Hinkley, D. (1974), *Theoretical Statistics*, London: Chapman and Hall. [364]

Cox, D. R. (1975), "Partial likelihood," *Biometrika*, 62, 269–276. [45]

Crow, E. (1956), "Confidence intervals for a proportion," *Biometrika*, 43, 423–435. [63]

Cui, Y. and Hannig, J. (2019), "Nonparametric generalized fiducial inference for survival functions under censoring," *Biometrika*, 106, 501–518. [306]

D'Agostino Sr., R. B., Massaro, J. M., and Sullivan, L. M. (2003), "Non-inferiority trials: design concepts and issues–the encounters of academic consultants in statistics," *Statistics in Medicine*, 22, 169–186. [369, 373]

Dagum, C. (2006), "Income inequality measures," *Encyclopedia of Statistical Sciences* DOI:10.1002/0471667196.ess6030.pub2. [80]

Davidson, A. and Hinkley, D. (1997), *Bootstrap Methods and Their Application*, New York: Cambridge University Press. [181, 184, 190]

De Neve, J., Thas, O., and Gerds, T. A. (2019), "Semiparametric linear transformation models: Effect measures, estimators, and applications," *Statistics in Medicine*, 38, 1484–1501. [274]

De Veaux, R. D. and Hand, D. J. (2005), "How to lie with bad data," *Statistical Science*, 20, 231–238. [47]

Demets, D. L. and Lan, K. G. (1994), "Interim analysis: the alpha spending function approach," *Statistics in Medicine*, 13, 1341–1352. [356]

DerSimonian, R. and Kacker, R. (2007), "Random-effects model for meta-analysis of clinical trials: an update," *Contemporary Clinical Trials*, 28, 105–114. [227]

DerSimonian, R. and Laird, N. (1986), "Meta-analysis in clinical trials," *Controlled Clinical Trials*, 7, 177–188. [227]

Diaconis, P. and Efron, B. (1985), "Testing for independence in a two-way table: new interpretations of the chi-square statistic," *The Annals of Statistics*, 13, 845–874. [198]

DiCiccio, T. J. and Efron, B. (1996), "Bootstrap confidence intervals," *Statistical Science*, 11, 189–212. [185]

Ding, P., Feller, A., and Miratrix, L. (2016), "Randomization inference for treatment effect variation," *Journal of the Royal Statistical Society: Series B (Statistical Methodology)*, 78, 655–671. [300]

Druesne-Pecollo, N., Latino-Martel, P., Norat, T., et al. (2010), "Beta-carotene supplementation and cancer risk: a systematic review and metaanalysis of randomized controlled trials," *International Journal of Cancer*, 127, 172–184. [28, 47]

Dudewicz, E. and Mishra, S. (1988), *Modern Mathematical Statistics*, New York: Wiley. [82, 139]

Dudoit, S. and Van Der Laan, M. (2008), *Multiple Testing Procedures with Applications to Genomics*, New York: Springer. [245, 250]

Dunnett, C. W. (1955), "A multiple comparison procedure for comparing several treatments with a control," *Journal of the American Statistical Association*, 50, 1096–1121. [208]

Edgington, E. S. (1987), *Randomization Tests, 2nd ed.*, New York: Marcel Dekker. [34]

Efron, B. and Hinkley, D. V. (1978), "Assessing the accuracy of the maximum likelihood estimator: observed versus expected Fisher information," *Biometrika*, 65, 457–483. [171]

Efron, B. and Tibshirani, R. (1993), *An Introduction to the Bootstrap*, vol. 57, Boca Raton, FL: CRC press. [183, 184, 190]

Ellenberg, J. (2014), *How Not to Be Wrong: The Power of Mathematical Thinking*, New York: Penguin Press. [37]

Fagerland, M. W. and Hosmer, D. W. (2013), "A goodness-of-fit test for the proportional odds regression model," *Statistics in Medicine*, 32, 2235–2249. [365]

Farrington, C. P. and Manning, G. (1990), "Test statistics and sample size formulae for comparative binomial trials with null hypothesis of non-zero risk difference or non-unity relative risk," *Statistics in Medicine*, 9, 1447–1454. [116, 121]

Fay, M. P. (2010a), "Confidence intervals that match Fisher's exact or Blaker's exact tests," *Biostatistics*, 11, 373–374. [22, 122]

Fay, M. P. (1999a), "Approximate confidence intervals for rate ratios from directly standardized rates with sparse data," *Communications in Statistics-Theory and Methods*, 28, 2141–2160. [230, 234]

—. (1999b), "Comparing several score tests for interval censored data (Corr: 1999V18 p2681)," *Statistics in Medicine*, 18, 273–285. [321]

—. (2005), "Random marginal agreement coefficients: rethinking the adjustment for chance when measuring agreement," *Biostatistics*, 6, 171–180. [99, 100, 103]

—. (2010b), "Two-sided exact tests and matching confidence intervals for discrete data," *R Journal*, 2, 53–58. [22, 75, 80]

Fay, M. P. and Brittain, E. H. (2016), "Finite sample pointwise confidence intervals for a survival distribution with right-censored data," *Statistics in Medicine*, 35, 2726–2740. [115, 305, 306, 325]

Fay, M. P., Brittain, E. H., and Proschan, M. A. (2013), "Pointwise confidence intervals for a survival distribution for right censored data with small samples or heavy censoring," *Biostatistics*, 14, 723–736. [305, 306, 307, 322, 325]

Fay, M. P., Brittain, E. H., Shih, J. H., Follmann, D. A., and Gabriel, E. E. (2018a), "Causal estimands and confidence intervals associated with Wilcoxon-Mann-Whitney tests in randomized experiments," *Statistics in Medicine*, 37, 2923–2937. [146, 157, 159, 160, 301]

Fay, M. P. and Feuer, E. J. (1997), "Confidence intervals for directly standardized rates: a method based on the gamma distribution," *Statistics in Medicine*, 16, 791–801. [230]

Fay, M. P. and Follmann, D. A. (2002), "Designing Monte Carlo implementations of permutation or bootstrap hypothesis tests," *The American Statistician*, 56, 63–70. [181]

Fay, M. P., Follmann, D. A., Lynn, F., et al. (2012), "Anthrax vaccine–induced antibodies provide cross-species prediction of survival to aerosol challenge," *Science Translational Medicine*, 4, 151ra126. [47]

Fay, M. P., Freedman, L. S., Clifford, C. K., and Midthune, D. N. (1997), "Effect of different types and amounts of fat on the development of mammary tumors in rodents: a review," *Cancer Research*, 57, 3979–3988. [175]

Fay, M. P. and Graubard, B. I. (2001), "Small-sample adjustments for Wald-type tests using sandwich estimators," *Biometrics*, 57, 1198–1206. [175, 315]

Fay, M. P., Graubard, B. I., Freedman, L. S., and Midthune, D. N. (1998), "Conditional logistic regression with sandwich estimators: application to a meta-analysis," *Biometrics*, 54, 195–208. [273]

Fay, M. P., Halloran, M. E., and Follmann, D. A. (2007), "Accounting for variability in sample size estimation with applications to nonadherence and estimation of variance and effect size," *Biometrics*, 63, 465–474. [384, 385, 386, 387]

Fay, M. P. and Hunsberger, S. A. (2021), "Practical valid inferences for the two-sample binomial problem," *Statistics Surveys*, 15, 72–110. [16, 22, 121, 122]

Fay, M. P. and Kim, S. (2017), "Confidence intervals for directly standardized rates using mid-p gamma intervals," *Biometrical Journal*, 59, 377–387. [230, 232]

Fay, M. P. and Lumbard, K. (2021), "Confidence intervals for difference in proportions for matched pairs compatible with exact McNemar's or sign tests," *Statistics in Medicine*, 40, 1147–1159. [102]

Fay, M. P. and Malinovsky, Y. (2018), "Confidence intervals of the Mann-Whitney parameter that are compatible with the Wilcoxon-Mann-Whitney test," *Statistics in Medicine*, 37, 3991–4006. [5, 128, 147, 149, 157, 160, 180, 234, 259, 316, 326, 387]

Fay, M. P. and Proschan, M. A. (2010), "Wilcoxon-Mann-Whitney or t-test? On assumptions for hypothesis tests and multiple interpretations of decision rules," *Statistics Surveys*, 4, 1–39. [82, 128, 157, 158, 160]

Fay, M. P., Proschan, M. A., and Brittain, E. (2015), "Combining one-sample confidence procedures for inference in the two-sample case," *Biometrics*, 146–156. [81, 143, 190, 400, 401]

Fay, M. P., Proschan, M. A., Brittain, E. H., and Tiwari, R. (2021), "Interpreting p-values and confidence intervals using well-calibrated null preference priors," *Statistical Science*. https://imstat.org/journals-and-publications/statistical-science/statistical-science-future-papers/ [397, 399, 400, 401, 402]

Fay, M. P., Sachs, M. C., and Miura, K. (2018b), "Measuring precision in bioassays: rethinking assay validation," *Statistics in Medicine*, 37, 519–529. [79]

Fay, M. P. and Shaw, P. A. (2010), "Exact and asymptotic weighted logrank tests for interval censored data: the interval R package," *Journal of Statistical Software*, 36, 1–34. [321, 325]

Fay, M. P. and Shih, J. H. (1998), "Permutation tests using estimated distribution functions," *Journal of the American Statistical Association*, 93, 387–396. [214, 221]

—. (2012), "Weighted logrank tests for interval censored data when assessment times depend on treatment," *Statistics in Medicine*, 31, 3760–3772. [321]

Fay, M. P., Tiwari, R. C., Feuer, E. J., and Zou, Z. (2006), "Estimating average annual percent change for disease rates without assuming constant change," *Biometrics*, 62, 847–854. [234]

FDA. (2016), "Guidance for industry: non-inferiority clinical trials to establish effectiveness," *US Department of Health and Human Services and US Food and Drug Administration, Washington, DC*. [369]

Finkelstein, D. M., Goggins, W. B., and Schoenfeld, D. A. (2002), "Analysis of failure time data with dependent interval censoring," *Biometrics*, 58, 298–304. [323]

Finniss, D. G., Kaptchuk, T. J., Miller, F., and Benedetti, F. (2010), "Biological, clinical, and ethical advances of placebo effects," *The Lancet*, 375, 686–695. [47]

Firth, D. (1993), "Bias reduction of maximum likelihood estimates," *Biometrika*, 80, 27–38. [258, 273]

Fitzmaurice, G. M., Laird, N. M., and Ware, J. H. (2004), *Applied Longitudinal Analysis*, Hoboken, NJ: John Wiley & Sons. [275]

Fleiss, J. L., Levin, B., and Paik, M. C. (2003), *Statistical Methods for Rates and Proportions, 3rd ed.*, Hoboken, NJ: John Wiley & Sons. [76]

Fleming, T. and Harrington, D. (1991), *Counting Processes and Survival Analysis*, New York: Wiley. [316, 323]

Follmann, D., Brittain, E., and Powers, J. H. (2013), "Discordant minimum inhibitory concentration analysis: a new path to licensure for anti-infective drugs," *Clinical Trials*, 10, 876–885. [369, 375]

Follmann, D. and Fay, M. (2010), "Exact inference for complex clustered data using within-cluster resampling," *Journal of Biopharmaceutical Statistics*, 20, 850–869. [188]

Follmann, D., Proschan, M., and Leifer, E. (2003), "Multiple outputation: inference for complex clustered data by averaging analyses from independent data," *Biometrics*, 59, 420–429. [188]

Freedman, L. S. (2008), "An analysis of the controversy over classical one-sided tests," *Clinical Trials*, 5, 635–640. [395, 396]

Freeman, G. and Halton, J. H. (1951), "Note on an exact treatment of contingency, goodness of fit and other problems of significance," *Biometrika*, 38, 141–149. [197]

Freidlin, B., Korn, E. L., Hunsberger, S., et al. (2007), "Proposal for the use of progression-free survival in unblinded randomized trials," *Journal of Clinical Oncology*, 25, 2122–2126. [324]

Freireich, E. J., Gehan, E., Frei, E., et al. (1963), "The effect of 6-mercaptopurine on the duration of steroid-induced remissions in acute leukemia: a model for evaluation of other potentially useful therapy," *Blood*, 21, 699–716. [310]

Friedman, J., Hastie, T., and Tibshirani, R. (2010), "Regularization paths for generalized linear models via coordinate descent," *Journal of Statistical Software*, 33, 1. [273]

Gail, M. H. (1974). "Power computations for designing comparative Poisson trials," *Biometrics*, 30, 2, 231–237. [388]

Gail, M. H., Lubin, J. H., and Rubinstein, L. V. (1981), "Likelihood calculations for matched case-control studies and survival studies with tied death times," *Biometrika*, 68, 703–707. [273]

Galton, F. (1886), "Regression towards mediocrity in hereditary stature." *Journal of the Anthropological Institute of Great Britain and Ireland*, 15, 246–263. [33]

Gautret, P., Lagier, J.-C., Parola, P., et al. (2020), "Hydroxychloroquine and azithromycin as a treatment of COVID-19: results of an open-label non-randomized clinical trial," *International Journal of Antimicrobial Agents*, 56, 105949. [1, 2]

Gehan, E. A. (1965), "A generalized Wilcoxon test for comparing arbitrarily singly-censored samples," *Biometrika*, 52, 203–224. [310, 311, 317]

Gelman, A., Carlin, J., Stern, H., et al. (2013), *Bayesian Data Analysis, 3rd ed.*, New York: CRC Press. [40, 392, 400, 401]

Gelman, A. (2005), "Analysis of variancewhy it is more important than ever (with discussion)," *The Annals of Statistics*, 33, 1–53. [212]

Gentleman, R. and Geyer, C. (1994), "Maximum likelihood for interval censored data: consistency and computation," *Biometrika*, 81, 618–623. [320]

Gentleman, R. and Vandal, A. (2001), "Computational algorithms for censored-data problems using intersection graphs," *Journal of Compuational and Graphical Statistics*, 10, 403–421. [324]

—. (2002), "Nonparametric estimation of the bivariate CDF for arbitrarily censored data," *Canadian Journal of Statistics*, 30, 557–571. [320]

Ghosh, B. (2006), "Sequential analysis", in *Encyclopedia of Statistical Sciences*, eds. Kotz, S., Read, C. B., Balakrishnan, N., et al., Wiley Online Library. DOI:10.1002/0471667196.ess2398.pub2 [356]

Ghosh, M. (2011), "Objective priors: an introduction for frequentists," *Statistical Science*, 26, 187–202. [401]

Gibbons, J. (1971), *Nonparametric Statistical Inference*, New York: McGraw-Hill Book Company. [95, 96, 97]

Goeman, J. J. and Solari, A. (2014), "Multiple hypothesis testing in genomics," *Statistics in Medicine*, 33, 1946–1978. [245, 246]

Gould, S. and Norris, S. L. (2021), "Contested effects and chaotic policies: the 2020 story of (hydroxy) chloroquine for treating COVID-19," *Cochrane Database of Systematic Reviews*, 2021, 3 (ED000151), 1–5. [2]

Graybill, F. A. (1976), *Theory and Application of the Linear Model*, Pacific Grove, CA: Wadsworth Publishing Company. [190, 254]

Greenland, S. (2017), "Invited commentary: the need for cognitive science in methodology," *American Journal of Epidemiology*, 186, 639–645. [6, 21]

—. (2019), "Valid p-values behave exactly as they should: some misleading criticisms of p-values and their resolution with s-values," *American Statistician*, 73, 106–114. [9]

Grimes, D. A. and Schulz, K. F. (2002), "Uses and abuses of screening tests," *The Lancet*, 359, 881–884. [32]

Groeneboom, P., Jongbloed, G., and Wellner, J. (2008), "The support reduction algorithm for computing nonparametric function estimates in mixture models," *Scandinavian Journal of Statistics*, 35, 385–399. [324]

Guo, X., Pan, W., Connett, J. E., Hannan, P. J., and French, S. A. (2005), "Small-sample performance of the robust score test and its modifications in generalized estimating equations," *Statistics in Medicine*, 24, 3479–3495. [175]

Hall, W. J. and Wellner, J. A. (1980), "Confidence bands for a survival curve from censored data," *Biometrika*, 67, 133–143. [306]

Halloran, M. E., Longini, I. M., and Struchiner, C. J. (2010), *Design and Analysis of Vaccine Studies*, New York: Springer. [41, 279, 281]

Hampel, F. R., Ronchetti, E. M., and Rousseeuw, P. J. (1986), *Robust Statistics: The Approach Based on Influence Functions*, New York: John Wiley & Sons. [153]

Hand, D. J. (1992), "On comparing two treatments," *The American Statistician*, 46, 190–192. [160]

—. (1994), "Deconstructing statistical questions," *Journal of the Royal Statistical Society: Series A (Statistics in Society)*, 157, 317–338. [47, 128]

Hanley, J. A. and McNeil, B. J. (1982), "The meaning and use of the area under a receiver operating characteristic (ROC) curve," *Radiology*, 143, 29–36. [130]

Harrell, F., Lee, K. L., and Mark, D. B. (1996), "Tutorial in biostatistics multivariable prognostic models: issues in developing models, evaluating assumptions and adequacy, and measuring and reducing errors," *Statistics in Medicine*, 15, 361–387. [130]

Harrington, D. P. and Fleming, T. R. (1982), "A class of rank test procedures for censored survival data," *Biometrika*, 69, 553–566. [317]

Hastie, T., Tibshirani, R., and Friedman, J. (2009), *The Elements of Statistical Learning: Data Mining, Inference, and Prediction, 2nd ed.*, New York: Springer. [268, 273]

Haybittle, J. (1971), "Repeated assessment of results in clinical trials of cancer treatment," *The British Journal of Radiology*, 44, 793–797. [348]

Hayes, R. J. and Moulton, L. H. (2009), *Cluster Randomised Trials*, Boca Raton, FL: Chapman and Hall/CRC. [216, 232]

Heinze, G. (2006), "A comparative investigation of methods for logistic regression with separated or nearly separated data," *Statistics in Medicine*, 25, 4216–4226. [258]

Hennekens, C. H., Buring, J. E., Manson, J. E., et al. (1996), "Lack of effect of long-term supplementation with beta carotene on the incidence of malignant neoplasms and cardiovascular disease," *New England Journal of Medicine*, 334, 1145–1149. [47]

Hepworth, G. (1996), "Exact confidence intervals for proportions estimated by group testing," *Biometrics*, 1134–1146. [64, 65]

—. (2005), "Confidence intervals for proportions estimated by group testing with groups of unequal size," *Journal of Agricultural, Biological, and Environmental Statistics*, 10, 478–497. [64, 65]

Hernán, M. A., Alonso, A., Logan, R., et al. (2008), "Observational studies analyzed like randomized experiments: an application to postmenopausal hormone therapy and coronary heart disease," *Epidemiology*, 19, 766–779. [47]

Hernán, M. A. and Hernández-Díaz, S. (2012), "Beyond the intention-to-treat in comparative effectiveness research," *Clinical Trials*, 9, 48–55. [36]

Hernán, M. A. and Robins, J. M. (2020), *Causal Inference: What If*, Boca Raton, FL: Chapman and HallCRC. [292, 296, 300, 371]

Hirji, K. (2006), *Exact Analysis of Discrete Data*, New York: Chapman and Hall/CRC. [121]

Hochberg, Y. and Tamhane, A. C. (1987), *Multiple Comparison Procedures*, New York: Wiley. [201, 206, 207, 211, 250]

Hoffman, E. B., Sen, P. K., and Weinberg, C. R. (2001), "Within-cluster resampling," *Biometrika*, 88, 420–429. [187]

Hollander, M., Wolfe, D. A., and Chicken, E. (2014), *Nonparametric Statistical Methods, 3rd ed.*, Hoboken, NJ: John Wiley & Sons. [91, 97, 101, 102, 204, 363]

Hommel, G. (1988), "A stagewise rejective multiple test procedure based on a modified Bonferroni test," *Biometrika*, 75, 383–386. [240]

Hosmer, D. and Lemeshow, S. (1980), "Goodness of fit statistics tests for the multiple regression model," *Communications in Statistics A*, 9, 1043–1069. [365]

Hu, X., Jung, A., and Qin, G. (2020), "Interval estimation for the correlation coefficient," *The American Statistician*, 74, 29–36. [97, 101]

Huang, J., Lee, C., and Yu, Q. (2008), "A generalized log-rank test for interval-censored failure time data via multiple imputation," *Statistics in Medicine*, 27, 3217–3226. [321]

Hudgens, M. G. (2005), "On nonparametric maximum likelihood estimation with interval censoring and left truncation," *Journal of the Royal Statistical Society: Series B (Statistical Methodology)*, 67, 573–587. [321]

Hudgens, M. G. and Halloran, M. E. (2008), "Toward causal inference with interference," *Journal of the American Statistical Association*, 103, 832–842. [289]

Hyndman, R. J. and Fan, Y. (1996), "Sample quantiles in statistical packages," *The American Statistician*, 50, 361–365. [361]

Ignatova, I., Deutsch, R. C., and Edwards, D. (2012), "Closed sequential and multistage inference on binary responses with or without replacement," *The American Statistician*, 66, 163–172. [349]

Imbens, G. W. and Rubin, D. B. (2015), *Causal Inference in Statistics, Social, and Biomedical Sciences*, New York: Cambridge University Press. [277, 278, 287, 288, 300]

Ioannidis, J. P. (2005), "Why most published research findings are false," *PLoS Medicine*, 2, e124. [7]

Irwin, J. (1935), "Tests of significance for differences between percentages based on small numbers," *Metron*, 12, 84–94. [110]

Jefferys, W. H. (1990), "Bayesian analysis of random event generator data," *Journal of Scientific Exploration*, 4, 153–169. [394]

Jennison, C. and Turnbull, B. W. (2000), *Group Sequential Methods with Applications to Clinical Trials*, Boca Raton, FL: Chapman and Hall/CRC. [349, 356]

—. (2007), "Adaptive seamless designs: selection and prospective testing of hypotheses," *Journal of Biopharmaceutical Statistics*, 17, 1135–1161. [354]

Johnson, N. L., Kemp, A. W., and Kotz, S. (2005), *Univariate Discrete Distributions, 3rd edition*, New York: John Wiley & Sons. [123]

Johnson, N. L., Kotz, S., and Balakrishnan, N. (1995), *Continuous Univariate Distributions*, vol. 2, New York: John Wiley & Sons. [97]

Kahneman, D. (2011), *Thinking, Fast and Slow*, New York: Farrar, Straus, and Giroux. [32]

Kalbfleisch, J. and Prentice, R. (2002), *The Statistical Analysis of Failure Time Data, 2nd ed.*, New York: Wiley. [260, 306, 311, 315, 316, 323, 324]

Kang, J. D. and Schafer, J. L. (2007), "Demystifying double robustness: A comparison of alternative strategies for estimating a population mean from incomplete data (with discussion)," *Statistical Science*, 22, 523–539 (discussion: 540–580). [338]

Kaplan, E. and Meier, P. (1958), "Nonparametric estimation from incomplete observations," *Journal of the American Statistical Association*, 53, 457–481. [305]

Karlin, S. and Taylor, H. (1975), *A First Course in Stochastic Processes, 2nd ed.*, New York: Academic Press. [82, 357]

Kass, R. E. and Raftery, A. E. (1995), "Bayes factors," *Journal of the American Statistical Association*, 90, 773–795. [393, 394, 401]

Kauermann, G. and Carroll, R. J. (2001), "A note on the efficiency of sandwich covariance matrix estimation," *Journal of the American Statistical Association*, 96, 1387–1396. [175]

Kawaguchi, Atsushi, and Koch, Gary, G. (2015), "sanon: an R package for stratified analysis with nonparametric covariable adjustment," *Journal of Statistical Software*, 67, 9. [226]

Kim, S., Fay, M. P., and Proschan, M. A. (2021), "Valid and approximately valid confidence intervals for current status data," *Journal of the Royal Statistical Society: Series B*, DOI:10.1111/rssb.12422 [320]

Kirk, J. L. and Fay, M. P. (2014), "An introduction to practical sequential inferences via single-arm binary response studies using the binseqtest R package," *The American Statistician*, 68, 230–242. [349]

Konietschke, F., Hothorn, L. A., Brunner, E., et al. (2012), "Rank-based multiple test procedures and simultaneous confidence intervals," *Electronic Journal of Statistics*, 6, 738–759. [210, 212, 244]

Konietschke, F., Placzek, M., Schaarschmidt, F., and Hothorn, L. A. (2015), "nparcomp: an R software package for nonparametric multiple comparisons and simultaneous confidence intervals," *Journal of Statistical Software*, 64, 9, 1–17. [212]

Koopmans, L. H., Owen, D. B., and Rosenblatt, J. (1964), "Confidence intervals for the coefficient of variation for the normal and log normal distributions," *Biometrika*, 51, 25–32. [80]

Korn, E. L. and Graubard, B. I. (1999), *Analysis of Health Surveys*, vol. 323, New York: John Wiley & Sons. [47, 64]

Koziol, J. A. and Jia, Z. (2009), "The concordance index C and the Mann–Whitney parameter $\Pr(X_i, Y)$ with randomly censored data," *Biometrical Journal*, 51, 467–474. [130]

Kuznetsova, A., Brockhoff, P. B., and Christensen, R. H. B. (2017), "lmerTest package: tests in linear mixed effects models," *Journal of Statistical Software*, 82, 13, 1–26. [219]

Lachin, J. M. (1981), "Introduction to sample size determination and power analysis for clinical trials," *Controlled Clinical Trials*, 2, 93–113. [386]

Lan, K. G. and DeMets, D. L. (1983), "Discrete sequential boundaries for clinical trials," *Biometrika*, 70, 659–663. [346]

Lan, K. G. and Wittes, J. (1988), "The B-value: a tool for monitoring data," *Biometrics*, 579–585. [347]

Lan, K. G. and Wittes, J. T. (2012), "Some thoughts on sample size: a Bayesian-frequentist hybrid approach," *Clinical Trials*, 9, 561–569. [384]

Lang, Z. and Reiczigel, J. (2014), "Confidence limits for prevalence of disease adjusted for estimated sensitivity and specificity," *Preventive Veterinary Medicine*, 113, 13–22. [64]

Lehmann, E. (1975), *Nonparametrics: Statistical Methods Based on Ranks*, Oakland, CA: Holden-Day. [203, 226]

—. (1999), *Elements of Large Sample Theory*, New York: Springer. [83, 101, 153, 166, 172, 190]

Lehmann, E. and Romano, J. (2005), *Testing Statistical Hypotheses, 3rd ed.*, New York: Springer. [xii, xiii, 9, 20, 21, 71, 77, 81, 82, 151, 158, 190, 191, 363, 374]

Lilliefors, H. W. (1967), "On the Kolmogorov-Smirnov test for normality with mean and variance unknown," *Journal of the American Statistical Association*, 62, 399–402. [374]

Lin, D. Y. and Wei, L.-J. (1989), "The robust inference for the Cox proportional hazards model," *Journal of the American Statistical Association*, 84, 1074–1078. [315]

Lin, L. I.-K. (1989), "A concordance correlation coefficient to evaluate reproducibility," *Biometrics*, 45, 255–268. [99]

Lindley, D. V. and Phillips, L. (1976), "Inference for a Bernoulli process (a Bayesian view)," *The American Statistician*, 30, 112–119. [47]

Little, R. J. and Rubin, D. B. (2020), *Statistical Analysis with Missing Data, 3rd ed.*, New York: John Wiley & Sons. [340]

Little, R. J., Wang, J., Sun, X., et al. (2016), "The treatment of missing data in a large cardiovascular clinical outcomes study," *Clinical Trials*, 13, 344–351. [339]

Liublinska, V. and Rubin, D. B. (2014), "Sensitivity analysis for a partially missing binary outcome in a two-arm randomized clinical trial," *Statistics in Medicine*, 33, 4170–4185. [332]

Lloyd, C. J. (2008), "Exact p-values for discrete models obtained by estimation and maximization," *Australian & New Zealand Journal of Statistics*, 50, 329–345. [114]

Loughin, T. M. (2004), "A systematic comparison of methods for combining p-values from independent tests," *Computational Statistics & Data Analysis*, 47, 467–485. [352]

Lunceford, J. K. and Davidian, M. (2004), "Stratification and weighting via the propensity score in estimation of causal treatment effects: a comparative study," *Statistics in Medicine*, 23, 2937–2960. [39, 292]

Lydersen, S., Pradhan, V., Senchaudhuri, P., and Laake, P. (2007), "Choice of test for association in small sample unordered r × c tables," *Statistics in Medicine*, 26, 4328–4343. [365]

Mann, H. B. and Whitney, D. R. (1947), "On a test of whether one of two random variables is stochastically larger than the other," *The Annals of Mathematical Statistics*, 18, 50–60. [143]

Mantel, N. (1966), "Evaluation of survival data and two new rank order statistics arising in its consideration," *Cancer Chemotherapy Reports*, 50, 163–170. [316]

Marcus, R., Eric, P., and Gabriel, K. R. (1976), "On closed testing procedures with special reference to ordered analysis of variance," *Biometrika*, 63, 655–660. [248]

Martín Andrés, A., Sánchez Quevedo, M., and Silva Mato, A. (1998), "Fisher's mid-p-value arrangement in 2 × 2 comparative trials," *Computational Statistics & Data Analysis*, 29, 107–115. [112]

Mayo, D. G. (1996), *Error and the Growth of Experimental Knowledge*, Chicago, IL: University of Chicago Press. [46]

McCullagh, P. (1980), "Regression models for ordinal data," *Journal of the Royal Statistical Society: Series B (Methodological)*, 42, 109–142. [151]

McCullagh, P. and Nelder, J. A. (1989), *Generalized Linear Models, 2nd ed.*, London: Chapman and Hall. [133, 212, 217, 257, 259, 273]

Mee, R. W. (1990), "Confidence intervals for probabilities and tolerance regions based on a generalization of the Mann-Whitney statistic," *Journal of the American Statistical Association*, 85, 793–800. [130]

Mehta, C. R. and Patel, N. R. (1995), "Exact logistic regression: theory and examples," *Statistics in Medicine*, 14, 2143–2160. [258, 273]

Mehta, C. R., Patel, N. R., and Gray, R. (1985), "Computing an exact confidence interval for the common odds ratio in several 2 × 2 contingency tables," *Journal of the American Statistical Association*, 80, 969–973. [123]

Mehta, J. and Srinivasan, R. (1970), "On the BehrensFisher problem," *Biometrika*, 57, 649–655. [139]

Meng, X.-L. (1994), "Posterior predictive p-values," *The Annals of Statistics*, 22, 1142–1160. [402]

Michael, H., Thornton, S., Xie, M., and Tian, L. (2019), "Exact inference on the random-effects model for meta-analyses with few studies," *Biometrics*, 75, 485–493. [227, 232, 233]

Miettinen, O. and Nurminen, M. (1985), "Comparative analysis of two rates," *Statistics in Medicine*, 4, 213–226. [121]

Morgan, S. L. and Winship, C. (2015), *Counterfactuals and Causal Inference 2nd ed.*, New York: Cambridge University Press. [298, 300, 301]

Moser, B. K., Stevens, G. R., and Watts, C. L. (1989), "The two-sample t test versus Satterthwaite's approximate F test," *Communications in Statistics – Theory and Methods*, 18, 3963–3975. [139]

Mullen, G. E., Ellis, R. D., Miura, K., et al. (2008), "Phase 1 trial of AMA1-C1/Alhydrogel plus CPG 7909: an asexual blood-stage vaccine for Plasmodium falciparum malaria," *PLoS One*, 3, e2940. [131]

Murphy, S., Rossini, A., and van der Vaart, A. W. (1997), "Maximum likelihood estimation in the proportional odds model," *Journal of the American Statistical Association*, 92, 968–976. [261, 315]

Murphy, S. A. and van der Vaart, A. W. (2000), "On profile likelihood," *Journal of the American Statistical Association*, 95, 449–465. [45, 188, 315]

National Research Council. (2010), *The Prevention and Treatment of Missing Data in Clinical Trials*, Washington, DC: National Academies Press. [327, 329, 339, 340]

Nel, D. d., van der Merwe, C. A., and Moser, B. (1990), "The exact distributions of the univariate and multivariate Behrens-Fisher statistics with a comparison of several solutions in the univariate case," *Communications in Statistics – Theory and Methods*, 19, 279–298. [139]

Neubert, K. and Brunner, E. (2007), "A studentized permutation test for the non-parametric Behrens–Fisher problem," *Computational Statistics & Data Analysis*, 51, 5192–5204. [157]

Newcombe, R. G. (2006), "Confidence intervals for an effect size measure based on the Mann–Whitney statistic. Part 2: asymptotic methods and evaluation," *Statistics in Medicine*, 25, 559–573. [157]

Neyman, J. and Scott, E. L. (1948), "Consistent estimates based on partially consistent observations," *Econometrica*, 16, 1–32. [263]

Ng, H. K. T., Filardo, G., and Zheng, G. (2008), "Confidence interval estimating procedures for standardized incidence rates," *Computational Statistics & Data Analysis*, 52, 3501–3516. [230]

Oakes, D. (2016), "On the win-ratio statistic in clinical trials with multiple types of event," *Biometrika*, 103, 742–745. [260]

O'Brien, P. C. and Fleming, T. R. (1987), "A paired Prentice-Wilcoxon test for censored paired data," *Biometrics*, 43, 169–180. [103]

Oller, R., Gómez, G., Calle, M. L. (2007), "Interval censoring: identifiability and the constant-sum property," *Biometrika*, 94, 61–70. [319]

Owen, A. B. (2001), *Empirical Likelihood*, Boca Raton, FL: Chapman and Hall/CRC. [188]

Park, M. Y. and Hastie, T. (2007), "L1-regularization path algorithm for generalized linear models," *Journal of the Royal Statistical Society: Series B (Statistical Methodology)*, 69, 659–677. [273]

Paule, R. C. and Mandel, J. (1982), "Consensus values and weighting factors," *Journal of Research of the National Bureau of Standards*, 87, 377–385. [227]

Pauly, M., Asendorf, T., and Konietschke, F. (2016), "Permutation-based inference for the AUC: a unified approach for continuous and discontinuous data," *Biometrical Journal*, 58, 1319–1337. [157, 160]

Pearl, J. (2009a), "Causal inference in statistics: an overview," *Statistics Surveys*, 3, 96–146. [300]

—. (2009b), *Causality: Models, Reasoning, and Inference, 2nd ed.*, New York: Cambridge University Press. [24, 46, 222, 232, 277, 296, 300]

Pearl, J., Glymour, M., and Jewell, N. P. (2016), *Causal Inference in Statistics: A Primer*, Chichester: John Wiley & Sons. [270, 295, 296, 300]

Peikes, D. N., Moreno, L., and Orzol, S. M. (2008), "Propensity score matching: a note of caution for evaluators of social programs," *The American Statistician*, 62, 222–231. [26]

Perlman, M. and Wu, L. (1999), "The emperor's new tests (with discussion)," *Statistical Science*, 14, 355–381. [21]

Peto, R. and Peto, J. (1972), "Asymptotically efficient rank invariant test procedures," *Journal of the Royal Statistical Society A*, 135, 185–207. [316, 317, 325]

Peto, R., Pike, M., Armitage, P., et al. (1976), "Design and analysis of randomized clinical trials requiring prolonged observation of each patient. I. Introduction and design," *British Journal of Cancer*, 34, 585. [348]

Plesser, H. E. (2018), "Reproducibility vs. replicability: a brief history of a confused terminology," *Frontiers in Neuroinformatics*, 11, 76. [xi]

Popper, K. (1963), *Conjectures and Refutations: The Growth of Scientific Knowledge*, London: Routledge. [28]

Posch, M. and Bauer, P. (1999), "Adaptive two stage designs and the conditional error function," *Biometrical Journal: Journal of Mathematical Methods in Biosciences*, 41, 689–696. [356]

Pratt, J. W. (1959), "Remarks on zeros and ties in the Wilcoxon signed rank procedures," *Journal of the American Statistical Association*, 54, 655–667. [87, 89, 90]

—. (1964), "Robustness of some procedures for the two-sample location problem," *Journal of the American Statistical Association*, 59, 665–680. [157]

Prentice, R. L. (1978), "Linear rank tests with right censored data," *Biometrika*, 65, 167–179. [317]

Prentice, R. L., Langer, R., Stefanick, M. L., et al. (2005), "Combined postmenopausal hormone therapy and cardiovascular disease: toward resolving the discrepancy between observational studies and the Women's Health Initiative clinical trial," *American Journal of Epidemiology*, 162, 404–414. [25, 47]

Prentice, R. L. and Marek, P. (1979), "A qualitative discrepancy between censored data rank tests," *Biometrics*, 35, 861–867. [317, 325]

PREVAIL II Writing Group. (2016), "A randomized, controlled trial of ZMapp for Ebola virus infection," *The New England Journal of Medicine*, 375, 1448. [111]

Proschan, M. and Brittain, E. (2020), "A primer on strong versus weak control of familywise error rate," *Statistics in Medicine*, 39, 1407–1413. [213]

Proschan, M., Brittain, E., and Kammerman, L. (2011), "Minimize the use of minimization with unequal allocation," *Biometrics*, 67, 1135–1141. [34]

Proschan, M. and Follmann, D. (2008), "Cluster without fluster: the effect of correlated outcomes on inference in randomized clinical trials," *Statistics in Medicine*, 27, 795–809. [41]

Proschan, M. A. (1999), "Miscellanea. Properties of spending function boundaries," *Biometrika*, 86, 466–473. [357]

Proschan, M. A., Follmann, D. A., and Waclawiw, M. A. (1992), "Effects of assumption violations on type I error rate in group sequential monitoring," *Biometrics*, 1131–1143. [348]

Proschan, M. A. and Hunsberger, S. A. (1995), "Designed extension of studies based on conditional power," *Biometrics*, 51, 1315–1324. [352, 353, 355, 356]

Proschan, M. A., Lan, K. G., and Wittes, J. T. (2006), *Statistical Monitoring of Clinical Trials: A Unified Approach*, New York: Springer. [346, 347, 348, 349, 351, 356, 357]

Proschan, M. A., McMahon, R. P., Shih, J. H., et al. (2001), "Sensitivity analysis using an imputation method for missing binary data in clinical trials," *Journal of Statistical Planning and Inference*, 96, 155–165. [330]

Reiczigel, J., Földi, J., and Ózsvári, L. (2010), "Exact confidence limits for prevalence of a disease with an imperfect diagnostic test," *Epidemiology and Infection*, 138, 1674–1678. [64, 65]

Robins, J., Breslow, N., and Greenland, S. (1986), "Estimators of the Mantel-Haenszel variance consistent in both sparse data and large-strata limiting models," *Biometrics*, 42, 311–323. [225, 233]

Röhmel, J. (2005), "Problems with existing procedures to calculate exact unconditional p-values for non-inferiority/superiority and confidence intervals for two binomials and how to resolve them," *Biometrical Journal*, 47, 37–47. [16]

Röhmel, J. and Mansmann, U. (1999), "Unconditional non-asymptotic one-sided tests for independent binomial proportions when the interest lies in showing non-inferiority and/or superiority," *Biometrical Journal*, 41, 149–170. [113]

Rosenbaum, P. R. (2002), *Observational Studies, 2nd ed.*, New York: Springer. [47, 292]

—. (2010), *Design of Observational Studies*, New York: Springer. [47, 298, 300]

Rosendaal, F. R. (2020), "Review of: 'hydroxychloroquine and azithromycin as a treatment of COVID-19: results of an open-label non-randomized clinical trial Gautret et al 2010'," *International Journal of Antimicrobial Agents*, 56, 106063. [2]

Rothmann, M. D., Wiens, B. L., and Chan, I. S. (2012), *Design and Analysis of Non-Inferiority Trials*, Boca Raton, FL: Chapman and Hall/CRC. [370, 373]

Rubin, D. (2006), *Matched Sampling for Casual Effects*, New York: Cambridge University Press. [47]

Rubin, D. B. (1997), "Estimating causal effects from large data sets using propensity scores," *Annals of Internal Medicine*, 127, 757–763. [38]

Rubin, D. B. (1984), "Bayesianly justifiable and relevant frequency calculations for the applied statistician," *The Annals of Statistics*, 12, 1151–1172. [402]

Sadoff, J., Gray, G., Vandebosch, A., et al. (2021), "Safety and efficacy of single-dose Ad26.COV2.S vaccine against Covid-19," *New England Journal of Medicine*, 384, 2187–2201. [120]

Sagara, I., Ellis, R. D., Dicko, A., et al. (2009), "A randomized and controlled Phase 1 study of the safety and immunogenicity of the AMA1-C1/Alhydrogel® + CPG 7909 vaccine for Plasmodium falciparum malaria in semi-immune Malian adults," *Vaccine*, 27, 7292–7298. [91, 92]

Samara, B. and Randles, R. H. (1988), "A test for correlation based on kendallfs tau," *Communications in Statistics – Theory and Methods*, 17, 3191–3205. [97]

Samuelsen, S. O. (2003), "Exact inference in the proportional hazard model: possibilities and limitations," *Lifetime Data Analysis*, 9, 239–260. [315]

Sarkar, S. K. and Chang, C.-K. (1997), "The Simes method for multiple hypothesis testing with positively dependent test statistics," *Journal of the American Statistical Association*, 92, 1601–1608. [250]

Schenker, N. and Gentleman, J. F. (2001), "On judging the significance of differences by examining the overlap between confidence intervals," *The American Statistician*, 55, 182–186. [194]

Schilling, M. and Doi, J. (2014), "A coverage probability approach to finding an optimal binomial confidence procedure," *American Statistician*, 68, 133–145. [63]

Schoenfeld, D. (1981), "The asymptotic properties of nonparametric tests for comparing survival distributions," *Biometrika*, 68, 316–319. [382]

Schouten, H. J. (1999), "Sample size formula with a continuous outcome for unequal group sizes and unequal variances," *Statistics in Medicine*, 18, 87–91. [381]

Schweder, T. and Hjort, N. L. (2016), *Confidence, Likelihood, Probability: Statistical Inference with Confidence Distributions*, New York: Cambridge University Press. [190]

Seaman, S. R. and Vansteelandt, S. (2018), "Introduction to double robust methods for incomplete data," *Statistical Science*, 33, 184–197. [338, 340]

Seber, G. A. (1984), *Multivariate Observations*, Hoboken, NJ: John Wiley & Sons. [192]

Self, S. G. and Liang, K.-Y. (1987), "Asymptotic properties of maximum likelihood estimators and likelihood ratio tests under nonstandard conditions," *Journal of the American Statistical Association*, 82, 605–610. [170, 273]

Sen, B. and Banerjee, M. (2007), "A pseudolikelihood method for analyzing interval censored data," *Biometrika*, 94, 71–86. [320]

Sen, B. and Xu, G. (2015), "Model based bootstrap methods for interval censored data," *Computational Statistics & Data Analysis*, 81, 121–129. [320]

Sen, P. (1985), "Permutational Central Limit Theorems," in *Encyclopedia of Statistics*, eds. Kotz, S. and Johnson, N. L., Hoboken, NJ: Wiley, vol. 6, pp. 683–687. [192]

Serfling, R. and Mazumder, S. (2009), "Exponential probability inequality and convergence results for the median absolute deviation and its modifications," *Statistics & Probability Letters*, 79, 1767–1773. [80]

Shao, J. and Tu, D. (1995), *The Jackknife and Bootstrap*, New York: Springer. [184]

Shapiro, S. S. and Wilk, M. B. (1965), "An analysis of variance test for normality (complete samples)," *Biometrika*, 52, 591–611. [374]

Shaw, P. A. (2018), "Use of composite outcomes to assess risk–benefit in clinical trials," *Clinical Trials*, 15, 352–358. [327]

Simmons, J. P., Nelson, L. D., and Simonsohn, U. (2011), "False-positive psychology: undisclosed flexibility in data collection and analysis allows presenting anything as significant," *Psychological Science*, 22, 1359–1366. [238]

Singal, A. G., Higgins, P. D., and Waljee, A. K. (2014), "A primer on effectiveness and efficacy trials," *Clinical and Translational Gastroenterology*, 5, e45. [371]

Singh, B., Ryan, H., Kredo, T., Chaplin, M., and Fletcher, T. (2021), "Chloroquine or hydroxychloroquine for prevention and treatment of COVID-19," *The Cochrane Database of Systematic Reviews*, 2, CD013587. [3]

Skou, S. T., Roos, E. M., Laursen, M. B., et al. (2015), "A randomized, controlled trial of total knee replacement," *New England Journal of Medicine*, 373, 1597–1606. [237]

Snee, R. D. (1974), "Graphical display of two-way contingency tables," *The American Statistician*, 28, 9–12. [198]

Sommer, A. and Zeger, S. L. (1991), "On estimating efficacy from clinical trials," *Statistics in Medicine*, 10, 45–52. [286, 288]

Steering Committee for PHS. (1989), "Final report on the aspirin component of the ongoing Physicians' Health Study," *New England Journal of Medicine*, 321, 129–135. [30]

Sterne, T. E. (1954), "Some remarks on confidence or fiducial limits," *Biometrika*, 41, 1–2, 275–278. [55, 63, 64]

Strassburger, K. and Bretz, F. (2008), "Compatible simultaneous lower confidence bounds for the Holm procedure and other Bonferroni-based closed tests," *Statistics in Medicine*, 27, 4914–4927. [240]

Stuart, E. A. (2010), "Matching methods for causal inference: A review and a look forward," *Statistical Science: A Review Journal of the Institute of Mathematical Statistics*, 25, 1. [291]

Tamhane, A. C. and Gou, J. (2017), "Advances in p-value based multiple test procedures," *Journal of Biopharmaceutical Statistics*, 1–18. [250]

Tan, W. (1982), "Sampling distributions and robustness of t, F and variance-ratio in two samples and ANOVA models with respect to departure from normality," *Communications in Statistics – Theory and Methods*, 11, 2485–2511. [201]

Tang, R., Banerjee, M., Kosorok, M. R., et al. (2012), "Likelihood based inference for current status data on a grid: A boundary phenomenon and an adaptive inference procedure," *The Annals of Statistics*, 40, 45–72. [320]

Tarone, R. E. and Gart, J. J. (1980), "On the robustness of combined tests for trends in proportions," *Journal of the American Statistical Association*, 75, 110–116. [203]

Tchetgen Tchetgen, E. J. and VanderWeele, T. J. (2012), "On causal inference in the presence of interference," *Statistical Methods in Medical Research*, 21, 55–75. [289]

Thangavelu, K. and Brunner, E. (2007), "Wilcoxon–Mann–Whitney test for stratified samples and Efron's paradox dice," *Journal of Statistical Planning and Inference*, 137, 720–737. [210]

Thas, O., Neve, J. D., Clement, L., and Ottoy, J.-P. (2012), "Probabilistic index models," *Journal of the Royal Statistical Society: Series B (Statistical Methodology)*, 74, 623–671. [130, 261]

The Open Science Collaboration. (2015), "Estimating the reproducibility of psychological science," *Science*, 349, aac4716. [3]

Therneau, T. (2015), *A Package for Survival Analysis in S*, r package version 2.38. https://CRAN.R-project.org/package=survival [315]

Therneau, T. M. and Grambsch, P. M. (2000), *Modeling Survival Data: Extending the Cox Model*, New York: Springer. [308, 323]

Tibshirani, R. (1996), "Regression shrinkage and selection via the lasso," *Journal of the Royal Statistical Society: Series B (Methodological)*, 58, 267–288. [267]

—. (2011), "Regression shrinkage and selection via the lasso: a retrospective," *Journal of the Royal Statistical Society: Series B (Statistical Methodology)*, 73, 273–282. [273]

Tsiatis, A. (2006), *Semiparametric Theory and Missing Data*, New York: Springer. [340]

Tsiatis, A. A., Davidian, M., Zhang, M., and Lu, X. (2008), "Covariate adjustment for two-sample treatment comparisons in randomized clinical trials: a principled yet flexible approach," *Statistics in Medicine*, 27, 4658–4677. [289]

Væth, M. (1985), "On the use of Wald's test in exponential families," *International Statistical Review*, 53, 199–214. [172]

Vandal, A. C., Gentleman, R., and Liu, X. (2005), "Constrained estimation and likelihood intervals for censored data," *Canadian Journal of Statistics*, 33, 71–83. [320]

VanderWeele, T. (2015), *Explanation in Causal Inference: Methods for Mediation and Interaction*, New York: Oxford University Press. [296]

Veronese, P. and Melilli, E. (2015), "Fiducial and confidence distributions for real exponential families," *Scandinavian Journal of Statistics*, 42, 471–484. [399, 402]

Vonesh, E. and Chinchilli, V. M. (1997), *Linear and Nonlinear Models for the Analysis of Repeated Measurements*, New York: Marcel Dekker. [190]

Vos, P. and Hudson, S. (2008), "Problems with binomial two-sided tests and the associated confidence intervals," *Australian and New Zealand Journal of Statistics*, 50, 81–89. [56]

Wacholder, S., McLaughlin, J. K., Silverman, D. T., and Mandel, J. S. (1992), "Selection of controls in case-control studies: I. Principles," *American Journal of Epidemiology*, 135, 1019–1028. [122]

Wald, A. (1947), *Sequential Analysis*, New York: Dover. [343, 356]

Wang, R., Lagakos, S., and Gray, R. (2010), "Testing and interval estimation for two-sample survival comparisons with small sample sizes and unequal censoring," *Biostatistics*, 11, 676–692. [122]

Wang, W. (2010), "On construction of the smallest one-sided confidence interval for the difference of two proportions," *The Annals of Statistics*, 38, 1227–1243. [114]

Wang, W. and Shan, G. (2015), "Exact confidence intervals for the relative risk and the odds ratio," *Biometrics*, 71, 985–995. [114]

Wasserstein, R. L. and Lazar, N. A. (2016), "The ASA's statement on p-values: context, process, and purpose," *The American Statistician*, 70, 129–133. [xi]

Wasserstein, R. L., Schirm, A. L., and Lazar, N. A. (2019), "Moving to a World Beyond '$p < 0.05$'," *The American Statistician*, 73, 1–19. [xi, 6]

Webster, W., Walsh, D., McEwen, S. E., and Lipson, A. (1983), "Some teratogenic properties of ethanol and acetaldehyde in C57BL/6J mice: implications for the study of the fetal alcohol syndrome," *Teratology*, 27, 231–243. [215]

Welch, B. and Peers, H. (1963), "On formulae for confidence points based on integrals of weighted likelihoods," *Journal of the Royal Statistical Society: Series B (Methodological)*, 25, 318–329. [402]

Westfall, P. H. (1997), "Multiple testing of general contrasts using logical constraints and correlations," *Journal of the American Statistical Association*, 92, 299–306. [248]

Westfall, P. H., Tobias, R. D., Rom, D., Wolfinger, R. D., and Hochberg, Y. (1999), *Multiple Comparisons and Multiple Tests Using the SAS System*, Cary, NC: SAS Institute. [206]

Westfall, P. H. and Troendle, J. F. (2008), "Multiple testing with minimal assumptions," *Biometrical Journal*, 50, 745–755. [245]

Westfall, P. H. and Young, S. S. (1993), *Resampling-Based Multiple Testing: Examples and Methods for p-Value Adjustment*, New York: John Wiley & Sons. [245, 250]

Whidden, C., Treleaven, E., Liu, J., et al. (2019), "Proactive community case management and child survival: protocol for a cluster randomised controlled trial," *BMJ Open*, 9, e027487. [379]

Wilcoxon, F. (1945), "Individual comparisons by ranking methods," *Biometrics Bulletin*, 1, 80–83. [143]

Wittes, J. (2002), "Sample size calculations for randomized controlled trials," *Epidemiologic Reviews*, 24, 39–53. [382, 386]

Wittes, J., Barrett-Connor, E., Braunwald, E., et al. (2007), "Monitoring the randomized trials of the Women's Health Initiative: the experience of the Data and Safety Monitoring Board," *Clinical Trials*, 4, 218–234. [25]

Wittes, J. and Brittain, E. (1990), "The role of internal pilot studies in increasing the efficiency of clinical trials," *Statistics in Medicine*, 9, 65–72. [386]

Wu, C. J. (1985), "Efficient sequential designs with binary data," *Journal of the American Statistical Association*, 80, 974–984. [387]

Xie, M.-g. and Singh, K. (2013), "Confidence distribution, the frequentist distribution estimator of a parameter: a review (with discussion)," *International Statistical Review*, 81, 3–77. [190]

Yan, X., Lee, S., and Li, N. (2009), "Missing data handling methods in medical device clinical trials," *Journal of Biopharmaceutical Statistics*, 19, 1085–1098. [331]

Yates, F. (1984), "Tests of significance for 2×2 contingency tables," *Journal of the Royal Statistical Society: Series A (General)*, 147, 426–463. [110, 111, 121]

Zeger, S. L., Liang, K.-Y., and Albert, P. S. (1988), "Models for longitudinal data: a generalized estimating equation approach," *Biometrics*, 44, 1049–1060. [266, 275]

Zeileis, A. (2004), "Econometric computing with HC and HAC covariance matrix estimators," *Journal of Statistical Software, Articles*, 11. [255, 273]

Zeileis, A., Kleiber, C., and Jackman, S. (2008), "Regression models for count data in R," *Journal of Statistical Software*, 27, 1–25. [259]

Zhang, H., Lu, N., Feng, C., et al. (2011), "On fitting generalized linear mixed-effects models for binary responses using different statistical packages," *Statistics in Medicine*, 30, 2562–2572. [218]

Notation Index

This index lists notation used repeatedly through the text and omits most notation whose use is confined to a subsection or so. Page numbers indicate the notation's first appearance. For a parameter, say β, estimators such as $\hat{\beta}$ and $\tilde{\beta}$ are not listed unless there are several different estimators.

Concept Index